中国科协三峡科技出版资助计划

英汉天文学名词

An English – Chinese Dictionary of Astronomy

李竞 余恒 崔辰州 编

中国天文学会天文学名词审定委员会
全国科学技术名词审定委员会天文学名词审定委员会 审定

中国科学技术出版社
·北京·

图书在版编目（CIP）数据

英汉天文学名词 / 李竞，余恒，崔辰州编. —北京：
中国科学技术出版社，2015.9
（中国科协三峡科技出版资助计划）
ISBN 978 – 7 – 5046 – 6977 – 3

Ⅰ. ①英… Ⅱ. ①李… ②余… ③崔 Ⅲ. ①天文学—
名词—英、汉 Ⅳ. ① P1 – 61

中国版本图书馆 CIP 数据核字（2015）第 212532 号

总 策 划	沈爱民　林初学　刘兴平　孙志禹	责任编辑	赵　晖　夏凤金
项目策划	杨书宣　赵崇海	责任校对	刘洪岩
出 版 人	秦德继	印刷监制	李春利
编辑组组长	吕建华　赵　晖	责任印制	张建农

出　　版	中国科学技术出版社
发　　行	科学普及出版社发行部
地　　址	北京市海淀区中关村南大街 16 号
邮　　编	100081
发行电话	010 – 62103130
传　　真	010 – 62103166
网　　址	http：//www.cspbooks.com.cn

开　　本	787mm × 1092mm　1/16
字　　数	415 千字
印　　张	24.75
版　　次	2015 年 9 月第 1 版
印　　次	2015 年 9 月第 1 次印刷
印　　刷	北京盛通印刷股份有限公司

书　　号	ISBN 978 – 7 – 5046 – 6977 – 3/P · 184
定　　价	98.00 元

（凡购买本社图书，如有缺页、倒页、脱页者，本社发行部负责调换）

总　序

　　科技是人类智慧的伟大结晶，创新是文明进步的不竭动力。当今世界，科技日益深入影响经济社会发展和人们的日常生活，科技创新发展水平深刻反映着一个国家的综合国力和核心竞争力。面对新形势、新要求，我们必须牢牢把握新的科技革命和产业变革机遇，大力实施科教兴国战略和人才强国战略，全面提高自主创新能力。

　　科技著作是科研成果和自主创新能力的重要体现形式。纵观世界科技发展历史，高水平学术论著的出版常常成为科技进步和科技创新的重要里程碑。1543年，哥白尼的《天体运行论》在他逝世前夕出版，标志着人类在宇宙认识论上的一次革命，新的科学思想得以传遍欧洲，科学革命的序幕由此拉开。1687年，牛顿的代表作《自然哲学的数学原理》问世，在物理学、数学、天文学和哲学等领域产生巨大影响，标志着牛顿力学三大定律和万有引力定律的诞生。1789年，拉瓦锡出版了他的划时代名著《化学纲要》，为使化学确立为一门真正独立的学科奠定了基础，标志着化学新纪元的开端。1873年，麦克斯韦出版的《论电和磁》标志着电磁场理论的创立，该理论将电学、磁学、光学统一起来，成为19世纪物理学发展的最光辉成果。

　　这些伟大的学术论著凝聚着科学巨匠们的伟大科学思想，标志着不同时代科学技术的革命性进展，成为支撑相应学科发展宽厚、坚实的奠基石。放眼全球，科技论著的出版数量和质量，集中体现了各国科技工作者的原始创新能力，一个国家但凡拥有强大的自主创新能力，无一例外也反映到其出版的科技论著数量、质量和影响力上。出版高水平、高质量的学术著作，成为科技工作者的奋斗目标和出版工作者的不懈追求。

中国科学技术协会是中国科技工作者的群众组织，是党和政府联系科技工作者的桥梁和纽带，在组织开展学术交流、科学普及、人才举荐、决策咨询等方面，具有独特的学科智力优势和组织网络优势。中国长江三峡集团公司是中国特大型国有独资企业，是推动我国经济发展、社会进步、民生改善、科技创新和国家安全的重要力量。2011年12月，中国科学技术协会和中国长江三峡集团公司签订战略合作协议，联合设立"中国科协三峡科技出版资助计划"，资助全国从事基础研究、应用基础研究或技术开发、改造和产品研发的科技工作者出版高水平的科技学术著作，并向45岁以下青年科技工作者、中国青年科技奖获得者和全国百篇优秀博士论文获得者倾斜，重点资助科技人员出版首部学术专著。

由衷地希望，"中国科协三峡科技出版资助计划"的实施，对更好地聚集原创科研成果，推动国家科技创新和学科发展，促进科技工作者学术成长，繁荣科技出版，打造中国科学技术出版社学术出版品牌，产生积极的、重要的作用。

是为序。

序 言

本《英汉天文学名词》（以下简称《名词》）是经中国天文学会天文学名词审定委员会、全国科学技术名词审定委员会天文学名词审定委员会审定发布的英汉对照天文学名词译名汇编，供广大天文研究人员、天体物理相关专业学生、科技译者、辞书著者、天文爱好者和其他社会人士参考使用。

中国天文学会自1922年成立以来一直非常重视天文学名词的审定、统一、规范和普及。下属的译名委员会早在1934年就完成了第一版的《天文学名词》，编译英德法日中五语种对照的天文学名词1324条，由当时的民国政府教育部公布，为20世纪中国天文学一创举。新中国成立以后，天文学名词编译委员会在戴文赛委员主持下，在第一时间对英汉对照《天文学名词》进行修订和增补，于1952年出版。后来为满足形势需要，由天文学会名词编译委员会委员李竞和沈良照在1959年编辑出版俄英中三语对照版本。南京大学天文系受中国天文学会委托，在戴文赛教授的组织下于1974年完成新的修订，词条总数扩充到6000条，名称改为《英汉天文学词汇》。1983年天文学名词审定委员会成立，名词收集和审定工作从此常态化。1986年，科学出版社出版了南京大学许邦信教授主持编写的《英汉天文学词汇（第二版）》，词汇扩充到16000余条。2000年，天文学名词审定委员会主任、北京天文台李竞研究员与南京大学许邦信教授共同完成的新版《英汉天文学名词》由上海科技教育出版社出版，收录词条23000余条。2000年版《英汉天文学名词》即为本《名词》的前身。

在网络时代到来之后，国家天文台崔辰州博士从2002年开始提供词条在线检索服务，并逐步增加模糊检索、分类浏览、RSS 订阅更新、用户意见收集、新词提交、名词审定等众多功能。这极大地方便了中英文天文学名词的查询和检索，对天文学术语译名的普及和推广工作起到巨大的推动作用。天文学名词网站的网址是"http：//astrodict. china – vo. org"，已经成为天文学名词工作的重要工具和窗口。北京师范大学天文系余恒博士自 2007 年开始承担天文学名词数据库的维护更新工作。数字化技术的应用，大大加快了名词收集的速度，缩短了译名审定发布的周期，使得天文学名词审定委员会的工作成果能够在第一时间向社会公开。

此次推出的新版，是对前一版词典的系统修订和更新，也是天文学名词审定委员会近 30 年工作的全面总结。此次编撰出版引入自动化排版系统，直接将数据库转化为排版文件，完全避免了人工录入时可能出现的疏漏差错，保证了词典的质量和时效性。而且在体例上，首次加入同义词和等价词的部分，增强了词条间的内部联系，方便读者比对取舍。本书共包含英文天文术语约 26700 条（包含部分拉丁语、德语、法语、西班牙语等非英语词汇）。与 2000 年版相比，本书新增词条近 6000 个、补充注释 3400 多条，修正排版讹误、扫描错误、不规范名称等共 2100 余处。

本词典在编纂过程中所参考的图书除上文提到的各个版本之外，还包括李竞主编的《汉英天文学词汇》（上海科技教育出版社，1991），全国科学技术名词审定委员会出版的《天文学名词（定义版）》（科学出版社，2001），卞毓麟主持修订全国科学技术名词审定委员会出版的《海峡两岸天文学名词》（科学出版社，2013），英国 Ian Ripath 主编的《牛津天文学词典》（上海外语教育出版社，2002），李竞等自 1985 – 2009 年发表于期刊《天文学进展》的前后 16 批天文学名词推荐译名，天文学名词审定委员会各分组提交的学科新词，以及天文学名词网站用户在线推荐的最新词汇等。

附录中的星座名列表来自国际天文学联合会（IAU）于 1922 年公布的全天星座名称；黄道十二宫和节气译名来自 2001 年版的《天文学名词（定义版）》；三垣四象和二十八宿的译名取自 2013 年版《海峡两岸天文学名词》；月面地名列表包含国际天文学联合会公布的全部月面地貌名称（除环

形山）；天然卫星列表按天体系统给出已获正式命名的太阳系天然卫星的译名；小行星列表给出编号在 3800 以内且有中文译名的小行星；陨星坑列表收录已被确认的直径在 20 千米以上的地球陨星坑。全书的所有信息和数据截至 2015 年 6 月。

本版《名词》的审定历经从 1999 到 2015 年的第六至第九届天文学名词审定委员会。参加审定的委员有：卞毓麟、陈力、陈学雷、崔辰州、邓劲松、方成、杭恒荣、何妙福、何香涛、黄天衣、李鉴、李竞、李启斌、林元章、刘炎、刘麟仲、卢炬甫、卢仙文、陆埮、潘君骅、彭云楼、全和钧、沈良照、孙小淳、王传晋、王玉民、萧耐园、谢懿、许邦信、杨大卫、杨世杰、叶式辉、余恒、赵刚、赵君亮、赵永恒、周又元、朱慈墭等。参加本书定稿的专家有：邓劲松、卞毓麟、刘炎、谢懿、杨大卫、邹振隆等。

《名词》编辑和审定过程中，全国科学技术名词审定委员会、中国天文学会、国家天文台、紫金山天文台、上海天文台、南京大学天文与空间科学学院、北京师范大学天文系、中国科学技术大学天文学系、北京大学天文学系、科学出版社、中国科学技术出版社等单位提供了多方面的支持和协助，特别是得到了中国科协三峡科技出版资助计划的支持，在此一并表示衷心的感谢。

天文学名词审定委员会欢迎各界读者在使用《名词》的过程中提出改进意见和修订建议，并通过天文学名词网站反馈给我们。

李竞　余恒　崔辰州
2015 年 6 月于北京

目 录

定名规则

凡例

正文

    拉丁字母 ·· 1

    数字 ··· 364

    希腊字母 ·· 366

附录

    表 1 星座 ·· 368

    表 2 黄道十二宫 ·· 370

    表 3 三垣四象七曜 ·· 370

    表 4 二十八宿 ·· 370

    表 5 节气 ·· 371

    表 6 天然卫星 ·· 371

    表 7 月面地名 ·· 373

    表 8 小行星 ··· 375

    表 9 陨星坑 ··· 378

    表 10 流星群 ··· 379

    表 11 希腊字母 ··· 379

    表 12 罗马数字 ··· 379

定名规则

当今科技发展日新月异，名词审定、辞书出版的速度远远跟不上新闻媒体的报道需求。现给出天文学新词定名的一般规则，方便读者参考已有译名和规范对未收录的名词进行合理的翻译。

1. 对于名词中涉及的人名、地名：
(a) 根据人名、地名所属语种确定正确发音并翻译，一般以新华社译名为准。人名翻译参考新华通讯社译名室编《世界人名翻译大辞典》修订版（中国对外翻译出版公司，2007）；地名翻译参考周定国主编《世界地名翻译大辞典》（中国对外翻译出版公司，2008）。
(b) 尊重各学科的译名传统和规范，尽量保持连续性和一致性。如法国天文学家 Charles Messier 译为梅西叶；德国物理学家 Karl Schwarzschild 译为施瓦西；夏威夷的 Mauna Kea 山译为莫纳克亚。
2. 对于有缩写的专有名词：
(a) 通常给出全称的完整翻译。若全称中有明确限定词，不会被误解为通用名词，则直译。如 Australian Astronomical Observatory（AAO）译为"澳大利亚天文台"；若直译名称与通用名词相近，则保留字母缩写作为限定词，如美国宇航局在 20 世纪 70 年代发射了一系列 Small Scientific Satellite（SSS）卫星。若直译为"小科学卫星"会被误认为泛称，因此保留缩写译为"SSS 小科学卫星"。又如夏威夷莫纳克亚山上的射电望远镜 Sub – Millimeter Array（SMA）翻译为"SMA 亚毫米波阵"。
(b) 若缩写本身有含义，可采用缩写的翻译，并附加表明其属性的中心词作为译名。如球载望远镜 Boomerang telescope，全称为 Balloon Observations of Millimetric

Extragalactic Radiation and Geophysics telescope，直译过于冗长，因此译名定为"飞镖球载望远镜"。又如欧洲空间局的 Hipparcos 卫星，全称为 High Precision Parallax Collecting Satellite，翻译为"依巴谷天文卫星"。

（c）对于既有全称译名，又有缩写译名的名词，择一使用即可。实际使用中，可在首次出现时给出中文或英文全称，此后直接使用缩写形式。如日本天文卫星 Advanced Satellite for Cosmology and Astrophysics 缩写形式为 ASCA（在日语中意为飞鸟）。其全称翻译为"宇宙学和天体物理学高新卫星"，缩写译为"飞鸟号"。使用时结合上下文采用"飞鸟号卫星"或者"ASCA 卫星"均可。

3. 名词中包含的希腊字母、罗马数字一般不作翻译。如需要查询对应的英文或中文，可参考本书附录。

4. 天体编号通常为星表缩写加编号的形式，如梅西叶星云星团表（M），星系星云新总表（NGC）等。实际使用中只作注解不作翻译。本书只收录著名星表及其缩写，未收录具体天体。

凡　例

本词典所有条目按照如下顺序进行编排：缩写或全称、语种（仅注明非英语条目）、大陆译名、繁体台湾名（有且与大陆名的繁体形式不同）、备注说明、同义词或变体。具体规则如下：

1. 英文部分按字母序排列（大写字母在前），空格、引号、连字符等非字母符号未参与排序。含特殊拉丁字符的词条均按照对应英文拼写排序。数字、希腊字符排在英文之后。

2. 对于英语中的变体，如 disc/disk，以及美语和英语中拼写不同的单词，如：centre/center，colour/color 等，本书一般只保留常见形式（上述三个单词均保留后一种拼法）。

3. 缩写或全称放在圆括号中，用等号（＝）引出。

4. 对于部分复数变化不规则的名词，其复数形式放在圆括号中置于英文部分，不再单独列出。

　　如：**facula**（复数：**faculae**）　光斑

5. 为方便理解英语中吸收的部分非英语单词或缩写，非英语名词在全角方括号中注明语源。包括：【拉】– 拉丁语，【德】– 德语，【法】– 法语，【意】–意大利语，【西】– 西班牙语，【梵】– 梵语等。

　　如：**AG**（＝ **Astronomische Gesellschaft**）　【德】德国天文学会

6. 图书期刊等出版物名称加注书名号《》。

7. 如台湾译名暂缺或完全对应大陆译名的繁体形式，则省略不写；如台湾译名不完全对应，则以竖杠（｜）间隔，并用六角方括号注明（〔台〕）。

　　如：**Abbe comparator**　阿贝比长仪｜阿貝比對器〔台〕

8. 词条的等同译法以逗号为间隔列出，不同义项用数字圈码表示。

　　如：**appulse**　①犯，最小角距　②合

9. 中文译名中可省略的部分用半角方括号标出。

　　如：**daughter isotope**　　子［体］同位素，表示"子体同位素"或"子同位素"两种译名均可

10. 对部分字面含义不明显且不常见的名词采用备注的方式说明学科、归属、来源或者性质，放在译名之后的圆括号中。

　　如：**Abbe crater**　　阿贝环形山（月球）

11. 词条的同义词以"即"引出，变体及其他等价形式用"又见"引出。

　　如：**Bielid meteors**　　比拉流星群｜比拉流星雨〔台〕即：*Andromedids*；又见：*Bielids*

A

Å　（＝angstrom）埃（1^{-10} 米）

A1　（＝Atomic Time 1）A1 原子时

A&A　（＝Astronomy and Astrophysics）《天文学和天体物理学》|《天文學和天文物理學》〔台〕

A.A.　（＝Air Almanac）《美国航空历书》|《[美國] 航空曆書》〔台〕

AAA　（＝Astronomy and Astrophysics Abstracts）《天文学和天体物理学文摘》|《天文學和天文物理學文摘》〔台〕

AAO　① （＝Anglo-Australian Observatory）英澳天文台（Australian Astronomical Observatory 的曾用名）② （＝Australian Astronomical Observatory）澳大利亚天文台（Anglo-Australian Observatory 的现用名）

AAp　（＝Astronomy and Astrophysics）《天文学和天体物理学》|《天文學和天文物理學》〔台〕

AAPS　（＝Automated Astronomical Positioning System）自动天文定位系统

AAS　（＝American Astronomical Society）美国天文学会

AAT　① （＝Anglo-Australian Telescope）英澳望远镜（Australian Astronomical Telescope 的曾用名）② （＝Australian Astronomical Telescope）澳大利亚望远镜（Anglo-Australian Telescope 的现用名）

AAVSO　（＝American Association of Variable Star Observers）美国变星观测者协会

A band　A 谱带（夫琅和费线）

abaxial aberration　轴外像差　又见：off-axis aberration

abaxial astigmatism　轴外像散

Abbe comparator　阿贝比长仪 | 阿貝比對器〔台〕

Abbe crater　阿贝环形山（月球）

Abel crater　阿贝尔环形山（月球）

Abelian symmetry　阿贝尔对称性

Abel integral　阿贝尔积分

Abell Catalogue　艾贝尔星系团表 | 艾伯耳星系團表〔台〕

Abell cluster　艾贝尔星系团 | 艾伯耳星系團〔台〕

Abell radius　艾贝尔半径

Abell richness class　艾贝尔富度 | 艾伯耳豐級〔台〕

Abel's integral equation　阿贝尔积分方程

aberration　① 像差 ② 光行差　又见：aberration of light

aberration angle　光行差角

aberration constant　光行差常数　又见：constant of aberration

aberration day number　光行差日数

aberration ellipse　光行差椭圆

aberration image　像差像

aberration of light　光行差　又见：aberration

aberration shift　光行差位移

ablation　烧蚀（流星）

ablation age　烧蚀年龄

abnormal galaxy　① 不规则星系　又见：irregular galaxy ② 反常星系

abnormal redshift　反常红移

abnormal refraction　反常折射　又见：anomalous refraction

abnormal spectrum　反常光谱

abridged armilla　简仪　又见：Equatorial Torquetum

abridged nautical almanac　简明航海历

ABRIXAS X-ray Satellite　（＝A Broadband Imaging X-ray All-sky Survey X-ray Satellite）ABRIXAS 宽带成像 X 射线卫星

A Broadband Imaging X-ray All-sky Survey X-ray Satellite　（＝ABRIXAS X-ray Satellite）ABRIXAS 宽带成像 X 射线卫星

absolute acceleration　绝对加速度

absolute age　绝对年龄

absolute altitude　绝对高度　又见：absolute height

absolute black body　绝对黑体

absolute blue magnitude　绝对蓝星等

absolute bolometric luminosity　绝对热光度

absolute bolometric magnitude　绝对热星等　又见：bolometric absolute magnitude

absolute brightness　绝对亮度

absolute calibration 绝对定标
absolute catalogue 绝对星表　又见：*absolute star catalogue*
absolute coordinate 绝对坐标
absolute degree 绝对温度　又见：*absolute temperature*
absolute determination 绝对测定
absolute energy distribution 绝对能量分布
absolute error 绝对误差
absolute frequency 绝对频率
absolute gradient 绝对梯度
absolute height 绝对高度　又见：*absolute altitude*
absolute instability 绝对不稳定性
absolute intensity 绝对强度
absolute luminosity 绝对光度
absolute magnitude 绝对星等
absolute magnitude effect 绝对星等效应
absolute magnitude scale 绝对星等标
absolute measurement 绝对测量
absolute orbit 绝对轨道
absolute parallax 绝对视差
absolute perturbation 绝对摄动
absolute perturbation method 绝对摄动法
absolute photoelectric magnitude 绝对光电星等
absolute photographic magnitude 绝对照相星等
absolute photometry 绝对测光 | 絕對光度學，絕對測光〔台〕
absolute photovisual magnitude 绝对仿视星等
absolute position 绝对位置
absolute proper motion 绝对自行
absolute radio magnitude 绝对射电星等
absolute radiometric magnitude 绝对辐射星等
absolute red magnitude 绝对红星等
absolute spectrophotometric gradient 绝对分光光度梯度
absolute stability 绝对稳定性
absolute standard 绝对标准
absolute star catalogue 绝对星表　又见：*absolute catalogue*
absolute temperature 绝对温度　又见：*absolute degree*
absolute temperature scale 绝对温标
absolute time 绝对时间
absolute unit 绝对单位
absolute value 绝对值
absolute velocity 绝对速度
absolute visual magnitude 绝对目视星等
absolute zero 绝对零度 | 絕對零點，絕對零度〔台〕
absorbability 可吸收性　又见：*absorptivity*
absorbed energy 吸收能
absorbed light 吸收光
absorber 吸收体，吸收器
absorbing cloud 吸收云
absorbing dust mass 吸光尘埃质量
absorbing layer 吸收层
absorbing medium 吸收介质
absorbing particle 吸收粒子
absorbing power 吸收本领　又见：*absorption capacity; absorption power*
absorption 吸收
absorption band 吸收带
absorption capacity 吸收本领　又见：*absorbing power; absorption power*
absorption cell 吸收泡
absorption coefficient 吸收系数　又见：*coefficient of absorption*
absorption condition 吸收条件
absorption cross section 吸收截面
absorption depth 吸收深度
absorption edge 吸收限　又见：*absorption limit*
absorption feature 吸收特征，吸收表象
absorption flocculus 吸收谱斑
absorption-free space 无吸收空间
absorption frequency 吸收频率
absorption limit 吸收限　又见：*absorption edge*
absorption line 吸收线 | 吸收[譜]線〔台〕
absorption line profile 吸收线轮廓
absorption line system 吸收线系统
absorption loss 吸收耗损
absorption nebula 吸光星云
absorption-poor space 低吸收空间
absorption power 吸收本领　又见：*absorbing power; absorption capacity*
absorption probability 吸收概率
absorption profile 吸收线廓
absorption resonance 吸收共振
absorption spectrum 吸收谱 | 吸收[光]譜〔台〕
absorption trough 吸收槽
absorptivity 可吸收性　又见：*absorbability*
abstract index notation 抽象指标记号
abstraction reaction 提取反应
abstraction sequence 提取序列
Abulfeda crater 艾布·菲达环形山（月球）

abundance　丰度 | 豐[盛]度〔台〕
abundance anomaly　丰度异常
abundance broadening　丰度致宽
abundance classification　丰度分类
abundance effect　丰度效应
abundance gradient　丰度梯度
abundance of element　元素丰度　又见：*element abundance; elemental abundance*
abundance ratio　丰度比
abundance standard star　丰度标准星
ab variable　ab 型变星
AC　①（＝Astrographic Catalogue）照相天图星表 | AC 星表〔台〕②（＝Arctic Circle）北极圈
Ac　（＝actinium）锕（89 号元素）
Acamar　天园六（波江座 θ）
ACBAR　（＝Arcminute Cosmology Bolometer Array Receiver）角分宇宙学辐射热计阵
accelerated frame　加速参考架
accelerated motion　加速运动
Acceleration, Reconnection, Turbulence and Electrodynamics of the Moon's Interaction with the Sun　（＝ARTEMIS）阿尔忒弥斯号月球探测器（美国）
acceleration mechanism　加速机制
acceleration of following　牵连加速度
acceleration of gravity　重力加速度
accelerometer　加速度计
accepter　接收器　又见：*acceptor*
acceptor　接收器　又见：*accepter*
accessibility　可接近性
accidental accuracy　随机精度
accidental coincidence　偶然符合
accidental connection　偶然关联
accidental error　偶然误差 | 偶[然誤]差〔台〕
accreting binary　吸积双星
accretion　吸积
accretion by black hole　黑洞吸积　又见：*black hole accretion*
accretion by compact companion　致密伴星吸积
accretion by compact object　致密天体吸积
accretion by neutron star　中子星吸积
accretion by white dwarf　白矮星吸积
accretion column　吸积柱
accretion disc　吸积盘　又见：*accretion disk*
accretion disk　吸积盘　又见：*accretion disc*
accretion efficiency　吸积效率
accretion flow　吸积流　又见：*accretion stream*
accretion model　吸积模型
accretion mound　吸积堆
accretion radius　吸积半径
accretion ring　吸积环
accretion shock　吸积激波
accretion stream　吸积流　又见：*accretion flow*
accretion theory　吸积理论
accretion wake　吸积尾流
accumulated time　累积时间
accumulation　①累积 ②聚点　又见：*condensation point*
accuracy　准[确]度 | 準確度〔台〕
accuracy user　准确度用户
ACE　（＝Advanced Composition Explorer）高新化学组成探测器（美国太阳风探测飞船）
acetaldehyde　乙醛（CH_3CHO）
acetonitrile　乙腈（CH_3CN）　又见：*methyl cyanide*
acetylene　乙炔（C_2H_2）
Achernar　水委一（波江座 α）
Acheron Fossae　阿克戎堑沟（火星）
achievement factor　成就因子
Achilles　阿基里斯（小行星 588 号）
Achilles group　阿基里斯群
Achird　王良三（仙后座 η）
achondrite　无球粒陨石 | 無[球]粒陨石〔台〕
achromat　消色差透镜 | 消色[差]透鏡〔台〕　又见：*achromatic lens*
achromatic eyepiece　消色差目镜 | 消色[差]目鏡〔台〕
achromatic image　消色差像
achromatic interference coronagraph　消色散干涉星冕仪
achromatic lens　消色差透镜 | 消色[差]透鏡〔台〕　又见：*achromat*
achromatic objective　消色差物镜 | 消色[差]物鏡〔台〕
achromatic prism　消色差棱镜
achromatic telescope　消色差望远镜 | 消色[差]望遠鏡〔台〕
achromatism　消色差
Acidalia Planum　阿西达里亚平原（火星）
ACN　（＝automatic celestial navigation）自动天文导航
acnode　孤点
A coefficient　A 系数
acoustic horizon　声学视界
acoustic loss　声学损失
acoustic mode　声模

acoustic oscillation 声学振荡
acoustic peak 声学峰
acoustic velocity 声速　又见: *sound velocity; speed of sound*
acoustic wave 声波　又见: *sound wave*
acoustic wave heating 声波致热
acousto-optic image processor 声光像处理机
acousto-optic light modulator 声光调制器
acousto-optic radiospectrometer （=AORS）射电声光频谱计
acousto-optic spectrograph （=AOS）声光频谱仪
acquisition camera 导星相机
Acrab 房宿四（天蝎座 β）　又见: *Graffias*
Acraman crater 阿克拉曼陨星坑（地球）
acronical rising 偕日升, 晨出　又见: *heliacal rising; acronychal rising*
acronical setting 偕日落, 夕没 | 偕日降, 夕沒〔台〕　又见: *heliacal setting; acronychal setting*
acronychal rising 偕日升, 晨出　又见: *heliacal rising; acronical rising*
acronychal setting 偕日落, 夕没 | 偕日降, 夕沒〔台〕　又见: *heliacal setting; acronical setting*
Acrux 十字架二（南十字座 α）
ACT （=Atacama Cosmology Telescope）阿塔卡马宇宙学望远镜
Acta Astronomica 【拉】《天文学报》（波兰期刊）
actinicity 光化性
actinium （=Ac）锕（89号元素）
actinogram 测光图
actinograph 辐射仪
actinometer ①太阳辐射计 | 日射儀〔台〕②感光计 | 露光計〔台〕　又见: *sensitometer*
actinometry 辐射测量　又见: *radiometry*
actinoscope 光能测定仪
action-angle variable 作用角变量
action at a distance 超距作用
action principle 作用量原理
action space 作用量空间
action through the medium 媒递作用
action variable 作用变量
activation 激活 | 激化, 激活〔台〕
activation coefficient 激活系数 | 激化係數, 激活係數〔台〕
activation energy 激活能
active binary 活动双星 | 活躍雙星〔台〕
active center 活动中心　又见: *center of activity*
active chromosphere 活动色球 | 活躍色球〔台〕
active chromosphere binary 活动色球双星
active chromosphere region 色球活动区
active chromosphere star 活动色球星 | 活躍色球星〔台〕
active comet 活跃彗星
active complex 活动复合体 | 活躍復合體〔台〕
　即: *active nest*
active corona ①活动日冕 | 活躍日冕〔台〕②活动星冕 | 活躍星冕〔台〕
active day 活动日
active device 主动装置
active elliptical galaxy 椭圆活动星系
active filament system （=AFS）活动暗条系统 | 活躍暗條系統〔台〕
active galactic nucleus （=AGN）活动星系核 | 活躍星系核〔台〕
active galaxy 活动星系 | 活躍星系〔台〕
active gravitational mass 主动引力质量
active longitude 活动经度 | 活躍經度〔台〕
Active Magnetospheric Particle Tracer Explorer （=AMPTE）活动磁层粒子示踪探测器（美德英联合卫星任务）
active maser oscillator 有源微波激射振荡器
active mechanism 作用机制
active nest 活动穴, 活动复合体 | 活躍穴〔台〕
active nucleus 活动核
active optics 主动光学
active photospheric region 光球活动区
active pixel detector （=APS）主动像元传感器
active prominence 活动日珥 | 活躍日珥, 活動日珥〔台〕
active prominence region （=APR）活动日珥区
active region 活动区 | 活躍區, 活動區〔台〕
active-region filament 活动区暗条
active-region heating 活动区加热
active-region loop 活动区环
active satellite 有源卫星
active shielding 主动屏蔽
active star 活动星 | 活躍星〔台〕
active sun 活动太阳 | 活躍太陽〔台〕
active sunspot 活动太阳黑子 | 活躍太陽黑子〔台〕
active-sunspot prominence 活动黑子日珥
activity 活性
activity index 活动指数
activity sphere 活动范围
actual aperture ①实际口径②实际孔径

actual frequency 实际频率 | 真實頻率，實際頻率〔台〕

actual position 实际位置

actuator 触动器

Acubens 柳宿增三（巨蟹座 α）

A.D. （＝Anno Domini）公元

ADAF （＝advection-dominated accretion flow）径移占优吸积流

Adams crater 亚当斯环形山（月球）

Adams ring 亚当斯环（海王星）

Adams-Russell phenomenon 亚当斯－罗素现象（恒星大气）

adapter 适配器，转接器

adapter module 过渡舱，适配舱

adaptive optics 自适应光学 | 自調光學〔台〕

Adara 弧矢七（大犬座 ε） 又见：*Adhara*

Addams crater 亚当斯环形山（金星）

adding interferometer 相加干涉仪 又见：*simple interferometer*

additional perturbation 附加摄动

additive group 加法群

additive noise 可加噪声

Adhafera 轩辕十一（狮子座 ζ）

Adhara 弧矢七（大犬座 ε） 又见：*Adara*

adhesion model 黏连模型

Adhil 天大将军增一，天大将军增二（仙女座 ξ）

adiabat 绝热线 又见：*adiabatic line*

adiabatic baryon model 绝热重子模型

adiabatic CDM model 绝热冷暗物质模型

adiabatic change 绝热变化

adiabatic compression 绝热压缩

adiabatic condition 绝热条件

adiabatic contraction 绝热收缩

adiabatic curve 绝热曲线

adiabatic drift wave 绝热漂移波

adiabatic equation 绝热方程

adiabatic equilibrium 绝热平衡

adiabatic expansion 绝热膨胀

adiabatic exponent 绝热指数 又见：*adiabatic index*

adiabatic fluctuation 绝热涨落

adiabatic fluid 绝热流体

adiabatic HDM model （＝adiabatic hot dark matter model）绝热热暗物质模型

adiabatic hot dark matter model （＝adiabatic HDM model）绝热热暗物质模型

adiabatic index 绝热指数 又见：*adiabatic exponent*

adiabatic inflow-outflow solution （＝ADIOS）绝热内外流解

adiabatic invariance 绝热不变性

adiabatic invariant 绝热不变量

adiabatic line 绝热线 又见：*adiabat*

adiabatic perturbation 绝热扰动

adiabatic plasmon 绝热等离子体激元

adiabatic process 绝热过程

adiabatic pulsation 绝热脉动

adiabatic sound speed 绝热声速

adiabatic temperature gradient 绝热温度梯度

adiabatic theory 绝热理论

Adib 右枢，紫微右垣一（天龙座 α）
又见：*Thuban*

ADIOS （＝adiabatic inflow-outflow solution）绝热内外流解

adjacency 邻接

adjacency effect 邻接效应（底片）

adjacent galaxy 邻近星系

adjustable eyepiece 可调目镜

adjusting screw 校正螺旋 | 調整螺絲〔台〕

adjustment ① 平差 ② 校准，调整 | 校正，調整〔台〕

adjustment parameter 平差参数

adjustment residual 平差残差

ADM formalism （＝Arnowitt-Deser-Misner formalism）ADM 形式

admissible error 容许误差 又见：*permissible error; allowable error; tolerance*

admixture instability 混合不稳定性

Adonis 阿多尼斯（小行星 2101 号）

adopted latitude 纬度采用值

adopted longitude 经度采用值

ADP （＝automatic data processing）自动数据处理

Adrastea 木卫十五

ADS ① （＝Aitken's Double Star Catalogue）艾特肯双星表，ADS 双星表 ② （＝Astrophysical Data System）天体物理数据系统 | 天文物理資料系統〔台〕

AdS （＝anti-de Sitter space）反德西特空间

AdS/CFT correspondence （＝Anti-de Sitter/Conformal Field Theory correspondence）反德西特／共形场论对应性

adsorption 吸附作用

adsorption equilibrium 吸附平衡

adsorptive power 吸附本领

Advanced Composition Explorer (=ACE) 高新化学组成探测器（美国太阳风探测飞船）

Advanced Laser Interferometer Gravitational-wave Observatory (=aLIGO) 高新激光干涉仪引力波天文台

Advanced Liquid-mirror Probe for Astrophysics, Cosmology and Asteroids (=ALPACA) 阿尔帕卡望远镜

Advanced Radio Interferometer between Space and Earth (=ARISE) 高新空地射电干涉仪

Advanced Satellite for Cosmology and Astrophysics (=ASCA) 宇宙学和天体物理学高新卫星，飞鸟号（日本 X 射线天文卫星）又见: Astro-D

Advanced Technology Solar Telescope (=ATST) 高新技术太阳望远镜 | 先進技術太陽望遠鏡〔台〕

advanced T/F system 先进时频系统

Advanced Thin Ionization Calorimeter experiment (=ATIC) 高新薄离子量能器实验

Advanced X-ray Astrophysics Facility (=AXAF) 高新 X 射线天体物理台 | 先進 X 光天文物理觀測台〔台〕（Chandra X-ray Observatory 曾用名）

advance of apsidal line 拱线运动，近星点运动 | 拱線運動〔台〕 又见: apsidal motion; apsidal rotation

advance of Mercury's perihelion 水星近日点进动 又见: perihelion precession of Mercury

advance of the periastron 近星点进动 | 近星點前移〔台〕 又见: periastron advance; periastron precession

advance of the perihelion 近日点进动 | 近日點前移〔台〕 又见: perihelion precession; perihelion motion; precession of perihelion

Advances In Space Research 《空间研究进展》（英国期刊）

advancing shock front 前进激波前

advancing wave 前进波 又见: progressive wave

advection 径移

advection-dominated accretion flow (=ADAF) 径移占优吸积流

A dwarf star A 型矮星

A.E. (=astronomical ephemeris) 天文年历

Aegaeon 土卫五十三

Aegina 河神星（小行星 91 号）

Aegir 土卫三十六

Aegle 辉神星（小行星 96 号）

Aeneas crater 埃涅阿斯环形山（土卫四）

aeolosphere 各向异性球

aeolotropy 各向异性 又见: anisotropy; anisotropism

aerial camera 航空照相机

aerial telescope 航空望远镜

aerobraking 大气制动

aerodynamic coefficient 气体动力系数

aerodynamics 气体动力学 又见: gas dynamics

aerogel 气凝胶

aerolite 石陨星，石陨石 | 石質隕石〔台〕 又见: asiderite; aerolith; stony meteorite

aerolith 石陨星，石陨石 | 石質隕石〔台〕 又见: asiderite; stony meteorite; aerolite

aerolithology 陨星学，陨石学 又见: aerolitics

aerolitics 陨星学，陨石学 又见: aerolithology

aeronomy 高层大气[物理]学

aeronomy satellite 高层大气科学卫星

aerosiderite 铁陨星，铁陨石，陨铁 | 隕鐵，鐵[質]隕石〔台〕 又见: iron meteorite; siderite; meteoric iron

aerosiderolite 铁石陨星，铁石陨石 | 石陨鐵，石[質]鐵隕石〔台〕

aerosol 气溶胶

aerospace ① 宇航空间 ② 航空航天

aerosphere 大气层 又见: atmospheric layer

aerostatics 气体静力学

Ae star Ae 星，A 型发射线星

aether 以太 又见: ether

affine connection 仿射联络

affine distance 仿射距离

affine equation 仿射方程

affine geometry 仿射几何

affine parameter 仿射参量，仿射参数

affine plate constant 仿射底片常数

affine transformation 仿射变换

affinity 亲合性

Affleck-Dine mechanism 阿弗莱克－戴恩机制

afocal system 无焦系统

AFS ①（=active filament system）活动暗条系统 | 活躍暗條系統〔台〕②（=arch filament system）拱状暗条系统

after-burner effect 爆后效应

after effect 后效应

afterglow 余辉

after image 余像

AG (=Astronomische Gesellschaft)【德】德国天文学会

Ag (=silver) 银（47 号元素）

Agamemnon 阿伽梅农（小行星 911 号）

AGASA (=Akeno Giant Air Shower Array) 明野巨型空气簇射阵（日本）

AGB (=asymptotic giant branch) 渐近巨星支

AGB star (=asymptotic branch giant) 渐近支巨星，AGB 星 | 漸近支巨星〔台〕

AGB variable AGB 变星
AG catalogue AGK 星表
age characteristics 年龄特征
age dating 年龄测定 又见：*age determination*
age determination 年龄测定 又见：*age dating*
ageing 老化 又见：*aging*
ageing process 老化过程
ageing star 老化星
age-metallicity relationship （＝AMR）年龄－金属丰度关系
Agena 马腹一（半人马座β） 又见：*Hadar*
age of the earth 地球年龄
age of the moon 月龄 又见：*moon's age*
age of the universe 宇宙年龄 又见：*cosmic age*
age problem 年龄问题
agglomeration 附聚
aggregate 凝聚 又见：*aggregation*
aggregation ①凝聚 又见：*aggregate* ②团块
A giant star A 型巨星
AGILE （＝Gamma-ray Light Detector）敏捷号γ射线天文卫星
aging 老化 又见：*ageing*
agitation 扰动
AGK （＝Astronomische Gesellschaft Katalog）【德】AGK 星表 | AG 星表〔台〕
Aglaja 仁神星（小行星 47 号）
Aglaonice crater 艾格洛尼克环形山（金星）
AGN （＝active galactic nucleus）活动星系核 | 活躍星系核〔台〕
AGN beaming 活动星系核集束效应 又见：*beaming in AGN*
AGN classification 活动星系核分类
AGN emission line 活动星系核发射线
AGN evolution 活动星系核演化
AGN host galaxy 活动星系核宿主星系
A.H. （＝Anno Hegirae）伊斯兰纪元
ahargana 积日（印度历法）
Ahfa al Farkadain 勾陈四（小熊座ξ）
Ahnighito meteorite 阿尼吉托陨星
AIC （＝Akaike information criterion）赤池信息判据
Ain 毕宿一（金牛座ε） 又见：*Oculus Boreus*
Ain al Rami 人马座ν（人马座v）
AIPS （＝astronomical image-processing system）天文图像处理系统 | 天文影像處理系統〔台〕
Air Almanac （＝A.A.）《美国航空历书》|《[美國] 航空曆書》〔台〕
airborne observation 机载观测

airborne platform 机载平台
airborne telescope 机载望远镜 又见：*airplane borne telescope; airplane-based telescope*
air cap 空气帽 | [空] 氣冠〔台〕（流星）
aircraft flyover synchronization 飞机飞越同步（对钟）
air drag 大气阻力 又见：*atmospheric drag*
air fluorescence cosmic ray detector 大气荧光式宇宙线探测器
airglow 大气辉光，气辉 | 氣輝〔台〕
airlock 密封舱 又见：*airlock module; capsule*
airlock module 密封舱 又见：*airlock; capsule*
air mass 大气质量
air path 大气路径
airplane-based observatory 机载天文台
airplane-based telescope 机载望远镜 又见：*airborne telescope; airplane borne telescope*
airplane borne instrument 机载仪器
airplane borne telescope 机载望远镜 又见：*airborne telescope; airplane-based telescope*
air pollution 空气污染
air pressure 气压 又见：*barometric pressure*
air shower 大气簇射 | 空氣射叢〔台〕（宇宙线）
air wavelength 大气波长
Airy crater 艾里环形山（火星）
Airy diffraction pattern 艾里衍射图样
Airy disk 艾里斑 | 艾瑞盤〔台〕
Airy function 艾里函数
Airy transit circle 艾里子午环
AIS （＝automatic interplanetary station）自动行星际站
Aitken crater 艾特肯环形山 | 艾肯環形山〔台〕（月球）
Aitken's Double Star Catalogue （＝ADS）艾特肯双星表，ADS 双星表
Aitne 木卫三十一
AI Velorum star （＝AI Vel star）船帆 AI 型星
AI Vel star （＝AI Velorum star）船帆 AI 型星
AJ （＝Astronomical Journal）《天文学报》（美国期刊）
AJB （＝Astronomischer Jahresbericht）【德】《天文年报》
Akaike information criterion （＝AIC）赤池信息判据
Akari 光亮号（日本红外天文卫星）又见：*ASTRO-F*
Akatsuki 黎明号（日本金星探测器）

Akeno Giant Air Shower Array （=AGASA）明野巨型空气簇射阵（日本）
Akna Montes 阿克娜山脉（金星）
A.L. （=Anno Lucis）露西纪元
Al （=aluminium）铝（13号元素）
Aladfar 辇道，辇道二（天琴座 η）
Alaraph 右执法，太微右垣一（室女座 β） 又见：*Zavijava*
Alasco 勾陈增九（小熊座 η）
Alathfar 织女增三（天琴座 μ）
alautun 阿劳顿（玛雅历，2.304×10^{10} 天）
Albaldah 建三（人马座 π）
Albali 女宿一（宝瓶座 ε） 又见：*Albulaan*
Albategnius crater 巴塔尼环形山（月球）
albedo 反照率
albedometer 反照率计
albedo theory 反照率理论
Albert 阿尔伯特（小行星719号）
Albiorix 土卫二十六
Albireo 辇道增七（天鹅座 β）
Al-Biruni crater 比鲁尼环形山（月球）
Albulaan 女宿一（宝瓶座 ε） 又见：*Albali*
Alchiba 右辖（乌鸦座 α）
Alcock-Paczynski test （=AP test）阿尔科克-帕金斯基检验，AP检验
Alcor 辅，开阳增一（大熊座80）
Alcott crater 奥尔科特环形山（金星）
Alcyone 昴宿六（金牛座 η）
Aldebaran 毕宿五（金牛座 α） 又见：*Cor Tauri*
Alden crater 奥尔登环形山（月球）
Alderamin 天钩五（仙王座 α）
Alder crater 阿尔德环形山（月球）
Al Dhanab 败臼一（天鹤座 γ）
Al Dhibain 女史增一（天龙座 ψ） 又见：*Dsiban*
Aldhibain 少宰，紫微左垣三（天龙座 η）
Aldrin crater 奥尔德林环形山（月球）
Alekhin crater 阿廖欣环形山（月球）
Alexander crater 亚历山大环形山（月球）
Alexandra 哲女星（小行星54号）
Alexandra family 哲女星族
Alexandrian calendar 亚历山大历（古希腊）
ALEXIS （=Array of Low-Energy X-ray Imaging Sensors）亚历克西斯号，低能X射线成像传感器阵列（美国极紫外天文卫星）

ALFALFA （=Arecibo Legacy Fast ALFA Survey）阿雷西沃遗珍快速ALFA巡天，苜蓿巡天
Alfard 星宿一（长蛇座 α） 又见：*Alphard; Cor Hydrae*
Alfecca Meridiana 鳖六（南冕座 α）
Alfirk 上卫增一（仙王座 β）
Alfvén drift wave 阿尔文漂移波
Alfvén frequency 阿尔文频率
Alfvén number 阿尔文数
Alfvén speed 阿尔文速率 | 阿耳芬速率〔台〕
Alfvén surface 阿尔文表面
Alfvén turbulence 阿尔文湍动
Alfvén velocity 阿尔文速度
Alfvén wave 阿尔文波 | 阿耳芬波〔台〕
Alfvén wave instability 阿尔文波不稳定性
Alga 徐，天市左垣七（巨蛇座 θ） 又见：*Alya*
Algedi 摩羯座 α 又见：*Al Giedi*
Algedi Prima 牛宿增六（摩羯座 α） 又见：*Prima Giedi*
Algedi Secunda 牛宿二（摩羯座 α^2） 又见：*Secunda Giedi; Algiedi Secunda*
Algenib 壁宿一（飞马座 γ）
Algenubi 轩辕九（狮子座 ε） 又见：*Asad Australis; Ras Elased Australis*
Al Gieba 轩辕十二（狮子座 γ） 又见：*Algieba*
Algieba 轩辕十二（狮子座 γ） 又见：*Al Gieba*
Al Giedi 摩羯座 α 又见：*Algedi*
Algiedi Secunda 牛宿二（摩羯座 α^2） 又见：*Secunda Giedi; Algedi Secunda*
Algjebba 参宿增三（猎户座 η）
Algol 大陵五（英仙座 β）
Algol binary 大陵型双星 又见：*Algol-type binary; Algol system*
Algol paradox 大陵佯谬
Algols 大陵型星 又见：*Algol star*
Algol star 大陵型星 又见：*Algols*
Algol system 大陵型双星 又见：*Algol-type binary; Algol binary*
Algol-type binary 大陵型双星 又见：*Algol binary; Algol system*
Algol-type eclipsing variable 大陵型食变星
Algol-type variable 大陵型变星 又见：*Algol variable*
Algol variable 大陵型变星 又见：*Algol-type variable*
Algomeyla 南河二（小犬座 β） 又见：*Gomeisa*
Algonquin Radio Observatory （=ARO）阿尔冈金射电天文台

Algorab 轸宿三（乌鸦座 δ）
Alhena 井宿三（双子座 γ）
aliased grating ring 混杂栅环（方向图）
aliased sidelobe 混杂旁瓣
aliasing 混淆现象
aliasing frequency 混杂频率 | 假频，假讯〔台〕
alibile 钠长石
alidade ① 游标盘 | 游標盤〔台〕② 照准仪
alien creation 外星人
alignment chart 列线图 又见：*nomogram; nomograph*
alignment star 校准星
aLIGO （＝Advanced Laser Interferometer Gravitational-wave Observatory）高新激光干涉仪引力波天文台
Alioth 玉衡，北斗五（大熊座 ε）
Al Jabhah 轩辕十三（狮子座 η）
Alkaid 摇光，北斗七（大熊座 η） 又见：*Benetnash*
Al Kalb al Rai 少卫增七（仙王座 ρ）
Alkalurops 七公六（牧夫座 μ）
Alkaphrah 太阳守（大熊座 χ）
Al Kaprah 上台二，二台二（大熊座 κ） 又见：*Talitha Australis; Alphikra Australis*
Alkes 翼宿一（巨爵座 α）
Alkmene 怨女星（小行星 82 号）
Al Kurud 孙二（天鸽座 θ）
Allan variance 阿伦方差
Allegheny Observatory 阿勒格尼天文台（美国）
Allende crater 阿连德陨星坑（地球）
Allende meteorite 阿连德陨星 | 阿顏德陨石〔台〕
Allen Telescope Array （＝ATA）艾伦望远镜阵
allocation of frequency 频率分配 又见：*frequency allocation*
allowable error 容许误差 又见：*admissible error; permissible error; tolerance*
allowable transition 容许跃迁 又见：*permitted transition; allowed transition*
allowed orbit 容许轨道
allowed spectrum 容许光谱
allowed transition 容许跃迁 又见：*permitted transition; allowable transition*
All-Sky Automated Survey （＝ASAS）自动化全天巡视
all-sky camera 全天照相机
all-sky patrol 巡天 又见：*patrol survey; survey*
all-sky photography 全天照相观测

ALMA （＝Atacama Large Millimeter/sub-millimeter Array）阿塔卡马大型毫米 [/亚毫米] 波阵 | 阿塔卡瑪大型毫米 [/次毫米] 波陣〔台〕
Al Maaz 柱一（御夫座 ε）
Almach 天大将军一（仙女座 γ）
Almagest 《天文学大成》（古希腊托勒玫著）
almanac 年历，历书 | [天文] 年曆，曆書〔台〕
al-manazil 马纳吉尔（月站，阿拉伯二十七宿）
Al Minliar al Asad 轩辕七（狮子座 κ）
Al Minliar al Shuja 柳宿二（长蛇座 σ）
almost degenerate amplifier 近简并放大器
almost periodic function 殆周期函数
almucantar ① 高度－方位仪 又见：*almucantur* ② 地平纬圈，平行圈 | 地平緯圈，等高圈〔台〕 又见：*altitude circle; circle of altitude; parallel of altitude*
almucantur 高度－方位仪 又见：*almucantar*
Almuredin 东次将，太微左垣四（室女座 ε） 又见：*Vindemiatrix*
Alnair 鹤一（天鹤座 α）
Al Nasl 箕宿一（人马座 γ） 又见：*Nushaba*
Alnath 五车五（金牛座 β） 又见：*Elnath*
Alnilam 参宿二（猎户座 ε）
Alnitak 参宿一（猎户座 ζ）
Al Niyat ① 心宿一（天蝎座 σ） 又见：*Alniyat* ② 心宿三（天蝎座 τ） 又见：*Alniyat*
Alniyat ① 心宿一（天蝎座 σ） 又见：*Al Niyat* ② 心宿三（天蝎座 τ） 又见：*Al Niyat*
along-track component 沿迹分量
along-track error 沿迹误差
ALPACA （＝Advanced Liquid-mirror Probe for Astrophysics, Cosmology and Asteroids）阿尔帕卡望远镜
Alpha Cygnids 天鹅座 α 流星群
alpha element 阿尔法元素
Alpha Magnetic Spectrometer （＝AMS-02）阿尔法磁谱仪
Alphard 星宿一（长蛇座 α） 又见：*Alfard; Cor Hydrae*
Alpha Regio 阿尔法区（金星）
Alpha Scorpiids 天蝎座 α 流星群
Alphecca 贯索四（北冕座 α） 又见：*Gemma*
Alpheratz 壁宿二（仙女座 α） 又见：*Sirrah*
Alphikra Australis 上台二，二台二（大熊座 κ） 又见：*Talitha Australis; Al Kaprah*
Alphonsus crater 阿方索环形山（月球）
Alrakis 天棓增九（天龙座 μ）
Alrami 天渊三（人马座 α） 又见：*Rukbat*

Alrescha 外屏七（双鱼座 α） 又见：*Kaitain*; *Okda*
Al Rukbah al Dajajah 天津增三十四（天鹅座 ω）
Alsafi 天厨二（天龙座 σ）
Alsahm 左旗一（天箭座 α） 又见：*Sham*
Alschain 河鼓一（天鹰座 β）
Alsciaukat 上台增四（天猫座 31）
　　又见：*Mabsuthat*
ALSEP （＝Apollo Lunar Surface Experiments Package）阿波罗月面实验装置
Alshain 虚宿一（宝瓶座 β） 又见：*Fortuna Fortunarum*
Alshat 牛宿增七（摩羯座 ν）
Alsuhail 天记（船帆座 λ） 又见：*Al Suhail al Wazn*
Al Suhail al Muhlif 天社一（船帆座 γ）
　　又见：*Regor*
Al Suhail al Wazn 天记（船帆座 λ） 又见：*Alsuhail*
Altain 天厨一（天龙座 δ） 又见：*Nodus II*; *Nodus Secundus*
Altair 河鼓二，牛郎星，牵牛星（天鹰座 α）
alt-alt mounting （＝altitude-altitude mounting）地平－地平式装置
Al Tarf 柳宿增十（巨蟹座 β）
Altawabi 五车一（御夫座 ι）
altazimuth 地平经纬仪 | 經緯儀〔台〕
altazimuth mounting 地平装置，经纬仪式装置 | 地平[式]裝置〔台〕 又见：*azimuth mounting*
altazimuth reflector 地平式反射镜
altazimuth telescope 地平式望远镜
Alterf 轩辕八（狮子座 λ）
alternating satellite 交错卫星
alternative cosmology 非标准宇宙学，备择宇宙学 又见：*non-standard cosmology*
alternative gravity model 备择引力理论
　　又见：*alternative theory of gravity*
alternative theory of gravity 备择引力理论
　　又见：*alternative gravity model*
Althalimain 天弁七（天鹰座 λ）
altimeter 测高仪
altimetry 测高
altitude 地平纬度，高度
altitude-altitude mounting （＝alt-alt mounting）地平－地平式装置
altitude axis 高度轴，水平轴
altitude circle 地平纬圈，平行圈 | 地平緯圈，等高圈〔台〕 又见：*circle of altitude*; *parallel of altitude*; *almucantar*
altitude effect 高度效应（宇宙线）
Altjira 寰神星（小行星 14878 号）

Aludra 弧矢二（大犬座 η）
Alula Australis 下台二，三台六（大熊座 ξ）
Alula Borealis 下台一，三台五（大熊座 ν）
aluminium （＝Al）铝（13 号元素）
aluminium mirror 铝镜
aluminizing 镀铝 又见：*aluminum coating*
aluminum coating 镀铝 又见：*aluminizing*
aluminum-rich star 富铝星
Alwaid 天棓三（天龙座 β） 又见：*Rastaban*
Alya 徐，天市左垣七（巨蛇座 θ） 又见：*Alga*
Al-Zij-Ilkhani 《伊尔汗历表》（波斯历书）
　　又见：*Zīj-i Īlkhānī*
Alzirr 井宿四（双子座 ξ）
AM （＝amplitude modulation）调幅 | 振幅調度〔台〕
Am （＝americium）镅 | 鋂〔台〕（95 号元素）
a.m. （＝ante meridiem）【拉】上午
Amalthea ① 木卫五 ② 羊神星（小行星 113 号）
AMANDA （＝Antarctic Muon And Neutrino Detector Array）南极 μ 介子和中微子探测器阵
amateur astronomer 天文爱好者 | 業餘天文學家，天文愛好者〔台〕 又见：*astrophile*; *stargazer*
Amati relation 阿马蒂关系（伽马暴）
ambient background radiation 环境背景辐射
ambient light illumination 环境光照度
ambient medium number density 环境介质数密度
ambient radiant power 环境辐射功率
ambient temperature 环境温度
ambiguity diagram 歧义图
ambiplasma instability 双等离子体不稳定性
ambipolar diffusion 双极扩散 | 雙極性擴散〔台〕
ambipolar field 双极场
ambipolarity 双极性
AMC （＝Automatic Meridian Circle）自动子午环
AM Canum Venaticorum binary （＝AM CVn binary）猎犬 AM 型双星 | 獵犬 [座]AM 型雙星〔台〕 即：*AM CVn star*
AM Canum Venaticorum star （＝AM CVn star）猎犬 AM 型星 | 獵犬 [座]AM 型星〔台〕
AM Canum Venaticorum variable （＝AM CVn variable）猎犬 AM 型变星 即：*AM CVn star*
AM CVn binary （＝AM Canum Venaticorum binary）猎犬 AM 型双星 | 獵犬 [座]AM 型雙星〔台〕 即：*AM CVn star*
AM CVn star （＝AM Canum Venaticorum star）猎犬 AM 型星 | 獵犬 [座]AM 型星〔台〕
AM CVn variable （＝AM Canum Venaticorum variable）猎犬 AM 型变星 即：*AM CVn star*

Amelia Creek crater 阿米利亚溪陨星坑（地球）
American Association of Variable Star Observers （＝AAVSO）美国变星观测者协会
American Astronomical Society （＝AAS）美国天文学会
americium （＝Am）镅｜錼〔台〕（95号元素）
Ames Research Center 艾姆斯研究中心（美国）
AM Her binary （＝AM Herculis binary）武仙AM型双星｜武仙AM型[雙]星〔台〕
AM Herculis binary （＝AM Her binary）武仙AM型双星｜武仙AM型[雙]星〔台〕
AMiBA （＝Yuan Tseh Lee Array for Microwave Background Anisotropy）李远哲宇宙背景辐射阵｜李遠哲宇宙背景輻射陣列〔台〕
Amici crater 阿米奇环形山（月球）
Amici prism 阿米奇棱镜
amino acid 氨基酸
Ammavaru Volcano 阿玛瓦鲁火山（金星）
ammonia 氨（NH_3）
ammonia beam maser ①氨束微波激射，氨束脉泽 ②氨束微波激射器
ammonia clock 氨钟
ammonia inversion transition 氨反演跃迁
ammonia maser 氨微波激射器，氨脉泽｜氨邁射〔台〕
ammonium hydrosulphide 硫氢化铵（NH_4SH）
Amor 阿莫尔（小行星1221号）
Amor asteroid 阿莫尔型小行星 又见：*Amors*
Amor group 阿莫尔群
amorphous galaxy 无定形星系
Amors 阿莫尔型小行星 又见：*Amor asteroid*
amphiscian region 日影双向区
Amphitrite 海后星（小行星29号）
amphoteric chondrite 两性球粒陨石
amphoterite 无粒古铜橄榄陨石｜超低鐵球粒隕石〔台〕
amplification 放大率
amplification bias 放大偏差
amplification factor 放大因子 又见：*magnification factor*
amplification ratio 放大比
amplifier 放大器
amplitude ①天体出没方位角 ②幅度，振幅｜振幅，變幅〔台〕
amplitude discriminator 幅度鉴别器
amplitude interferometer 振幅干涉仪
amplitude interferometry 振幅干涉测量

amplitude modulation （＝AM）调幅｜振幅調度〔台〕
amplitude spectrum 幅谱
AMPTE （＝Active Magnetospheric Particle Tracer Explorer）活动磁层粒子示踪探测器（美德英联合卫星任务）
AMR （＝age-metallicity relationship）年龄－金属丰度关系
AMS-02 （＝Alpha Magnetic Spectrometer）阿尔法磁谱仪
Am star Am星，A型金属线星
Amundsen crater 阿蒙森环形山（月球）
AN （＝Astronomische Nachrichten）【德】《天文通报》
anaberrational reflector 消像差反射望远镜
anaberrational refractor 消像差折射望远镜
anaberrational telescope 消像差望远镜
anabibazon 白道升交点
anaemic galaxy 弱臂星系
analemma ①地球正投影仪 ②日行迹
analog signal 模拟信号
analogue 模拟装置
analyser 检偏器，分析器 又见：*analyzer*
analyzer 检偏器，分析器 又见：*analyser*
anamorphic magnification 畸变放大
Ananke 木卫十二｜木衛十二，安那喀〔台〕
anastigmat 消像散透镜
anastigmatic system 消像散系统
anastigmatism 消像散｜消像散性〔台〕
Anaximander crater 阿那克西曼德环形山（月球）
Anaximenes crater 阿那克西米尼环形山（月球）
Ancha 泣二（宝瓶座 θ）
Anchentenar 天苑九（波江座 τ） 又见：*Angetenar*
Anchises 安希塞斯（小行星1173号）
ancient astronomy 古代天文学
And （＝Andromeda）仙女座
Anderson crater 安德森环形山（月球）
Andoyer variable 安多耶变量
andrite 斜辉陨石
Andromeda （＝And）仙女座
Andromeda galaxy 仙女星系（M31, NGC 224）
Andromeda nebula 仙女星云 即：*Andromeda galaxy*
Andromeda subgroup 仙女次星系群
Andromedids 仙女流星群｜仙女[座]流星雨〔台〕
Aneas 阿涅阿斯（小行星1172号）
Angelina 安杰利纳（小行星64号）

Angetenar 天苑九（波江座 τ） 又见：*Anchentenar*
Anglee 李安（小行星 64291 号）
angle of commutation 太阳－行星角距
angle of declination 偏角，磁偏角
angle of deflection 偏转角 又见：*deflection angle*
angle of departure 出射角 又见：*angle of emergence*
angle of depression 俯角
angle of deviation 偏向角，磁偏角 | 偏角〔台〕
angle of diffraction 衍射角 又见：*diffraction angle*
angle of eccentricity 偏心角 又见：*eccentric angle*
angle of emergence 出射角 又见：*angle of departure*
angle of incidence 入射角 又见：*incidence angle; incident angle*
angle of inclination 倾角，交角，水平差 | 交角，倾角〔台〕 又见：*inclination*
angle of lag 滞后角
angle of minimum resolution 最小分辨角
angle of polarization 偏振角 又见：*polarization angle*
angle of projection 投射角，发射角
angle of reflection 反射角 又见：*reflection angle*
angle of refraction 折射角 又见：*refracting angle*
angle of rotation 转动角，自转角
angle of sight 视线角 | 视[線]角〔台〕
angle of the vertical 垂线角
angle of tilt 倾角 又见：*tilt angle; obliquity*
angle of torsion 扭转角 又见：*angle of twist*
angle of total reflection 全反射角
angle of twist 扭转角 又见：*angle of torsion*
angle variable 角变量 又见：*angular variable*
Anglo-Australian Observatory （＝AAO）英澳天文台（Australian Astronomical Observatory 的曾用名）
Anglo-Australian Planet Search 英澳行星搜寻计划
Anglo-Australian Telescope （＝AAT）英澳望远镜（Australian Astronomical Telescope 的曾用名）
angrite 无球粒钛辉陨石 | 钛輝無粒隕石〔台〕
angstrom （＝Å）埃（1^{-10} 米）
angular acceleration 角加速度
angular accuracy 角精度
angular aperture 角孔径
angular breadth 角宽度
angular coordinate 角坐标
angular correlation function 角相关函数
angular diameter 角直径 | 角徑〔台〕
angular diameter distance 角直径距离
angular diameter-redshift relation 角径－红移关系
angular diameter-redshift test 角径－红移检验
angular dimension 角大小 又见：*angular size*
angular dispersion 角色散
angular displacement 角位移
angular distance 角距离 | 角距 [離]〔台〕
angular distribution 角分布
angular element 角要素
angular elongation 距角 又见：*elongation*
angular extent 角范围 | 角範圍，角展〔台〕
angular frequency 角频率 又见：*pulsatance*
angular magnification 角放大率
angular measure 角量度
angular momentum 角动量
angular-momentum catastrophe 角动量灾难
angular momentum density 角动量密度
angular motion 角向运动
angular path length 轨迹角长度 | 角程長度〔台〕
angular power spectrum 角功率谱
angular resolution 角分辨率
angular resolving power 角分辨本领
angular scale 角尺度
angular semi-major axis 角半长径
angular semi-minor axis 角半短径
angular separation 角间距 | 角距 [離]〔台〕
angular size 角大小 又见：*angular dimension*
angular spectrum 角谱
angular variable 角变量 又见：*angle variable*
angular velocity 角速度
anharmonicity 非简谐性
anharmonic pulsation 非谐脉动
Anhui 安徽（小行星 2162 号）
animal cycle 十二生肖
anion 负离子，阴离子 又见：*negative ion*
anisotropic conductivity 各向异性传导率
anisotropic correlation function 各向异性相关函数
anisotropic cosmological model 各向异性宇宙学模型
anisotropic cosmological space-time 各向异性宇宙学时空
anisotropic cosmology 各向异性宇宙论
anisotropic inertia 各向异性惯量
anisotropic magnetohydrodynamics 各向异性磁流力学
anisotropic medium 各向异性介质

anisotropic orbit　各向异性轨道
anisotropic parameter　各向异性参数
　　又见：*anisotropy parameter*
anisotropic power spectrum　各向异性功率谱
anisotropic pressure　各向异性压力
anisotropic propagation　各向异性传播
anisotropic scattering　各向异性散射
anisotropic universe　各向异性宇宙
anisotropic velocity dispersion　各向异性速度弥散
anisotropism　各向异性　又见：*anisotropy; aeolotropy*
anisotropy　各向异性　又见：*anisotropism; aeolotropy*
anisotropy instability　各向异性不稳定性
anisotropy parameter　各向异性参数
　　又见：*anisotropic parameter*
Ankaa　火鸟六（凤凰座 α）　又见：*Nair al Zaurak*
Annales Geophysicae　【德】《地球物理学年鉴》（德国期刊）
Annefrank　安妮（5535 号小行星）
annihilation　湮灭 | 毀滅 [作用]〔台〕
annihilation of electron pair　电子对湮灭
annihilation photon　湮灭光子
annihilation process　湮灭过程
annihilation radiation　湮灭辐射
Anno Domini　（＝A.D.）公元
Anno Hegirae　（＝A.H.）伊斯兰纪元
Anno Lucis　（＝A.L.）露西纪元
annual aberration　周年光行差
annual catalogue　年星表
annual epact　年闰余
annual equation　周年差　又见：*annual inequality*
annual inequality　周年差　又见：*annual equation*
annual motion　周年运动
annual parallax　周年视差
annual period　年周期
annual precession　周年岁差 | 歲差，年進動〔台〕
annual proper motion　周年自行 | 年自行〔台〕
annual register　年鉴
Annual Review of Astronomy and Astrophysics　（＝ARAA）《天文学和天体物理学年评》|《天文學和天文物理學年評》〔台〕（美国期刊）
Annual Review of Earth and Planetary Sciences　《地球与行星科学年评》（美国期刊）
annual variation　周年变化 | [周] 年變 [化]〔台〕
annular aperture　环状孔径　又见：*ring aperture*
annular eclipse　环食 | [日] 環食〔台〕
annular gap　环状隙
annularity　环食态

annular nebula　环状星云　又见：*ring nebula*
annular phase　环食相
annular solar eclipse　日环食
annular-total eclipse　全环食　又见：*total-annular eclipse*
annulus　环带
annum　【拉】年
anomalistic mean motion　近点平运动
anomalistic month　近点月
anomalistic period　近点周期
anomalistic revolution　近点周
anomalistic year　近点年
anomalous absorption　反常吸收
anomalous cosmic ray　反常宇宙线
anomalous coupling　反常耦合
anomalous dispersion　反常色散
anomalous refraction　反常折射　又见：*abnormal refraction*
anomalous scattering　反常散射
anomalous tail　反常彗尾
anomalous transport　反常输运
anomalous X-ray pulsar　（＝AXP）反常 X 射线脉冲星
anomalous Zeeman effect　反常塞曼效应 | 異常則曼效應〔台〕
anomaly　近点角
anonymous galaxy　未名星系 | 無名星系〔台〕
anorthite　钙长石
anorthosite　斜长岩
ANS　（＝Astronomical Netherlands Satellite）荷兰天文卫星
ansa　环脊，尖脊
Anser　齐增五（狐狸座 α）
Ansgarius crater　安斯加尔环形山（月球）
Anshan　鞍山（小行星 3136 号）
Ant　（＝Antlia）唧筒座
antalgol　逆大陵变星　即：*RR Lyrae variable; RR Lyrae star*
antapex　背点
Antarctic astronomy　南极天文学
antarctic circle　南极圈
Antarctic Muon And Neutrino Detector Array　（＝AMANDA）南极 μ 介子和中微子探测器阵
antarctic observatory　南极天文台
antarctic pole　南极（地球）
antarctic zone　南极区

ANTARES （=Astronomy with a Neutrino Telescope and Abyss environmental RESearch project）天蝎座 α 中微子望远镜

Antares 心宿二，大火 | 大火〔台〕（天蝎座 α） 又见：*Cor Scorpii*

antecedence 逆行 又见：*retrograde motion; retrogression; backward movement*

ante meridiem （=a.m.）【拉】上午

antenna aperture 天线孔径

antenna array 天线阵

antenna baseline 天线基线

antenna calibration 天线校准，天线定标

antenna diagram 天线方向图 又见：*antenna pattern*

antenna efficiency 天线效率

Antennae Galaxies 触须星系（NGC 4038）

antenna element 天线单元，天线辐射元 | 天線 [單] 元〔台〕

antenna gain 天线增益

antenna illumination 天线照明

antenna impedance 天线阻抗

antenna matching 天线匹配

antenna pattern 天线方向图 又见：*antenna diagram*

antenna pattern solid angle 天线方向图立体角

antenna power pattern 天线功率图 | 天線功率方向圖〔台〕

antenna spacing 天线间距

antenna temperature 天线温度

Anthe 土卫四十九

anthelic arc 反日弧

anthelion 反假日 | 幻日〔台〕

anthropic principle 人择原理

anticenter 反银心，反银心方向 | 反銀心〔台〕 又见：*Galactic anticenter*

anticenter region 反银心区

anticoincidence 反符合

anticommutation relation 反对易关系

anticommutator 反对易子

anticommute 反对易

anticrepuscular ray 反曙暮辉

Anti-de Sitter/Conformal Field Theory correspondence （=AdS/CFT correspondence）反德西特/共形场论对应性

anti-de Sitter space （=AdS）反德西特空间

anti-dwarf nova 逆反矮新星

antielectron 正电子 又见：*positive electron; positron*

antigalaxy 反物质星系

antiglow 反日照

antigravity 反引力

antihalo 消晕

anti-inversion 逆反演

anti-jovian point 对木点

Antilochus 安蒂洛库斯（小行星 1583 号）

antimatter 反物质

antimatter cosmology 反物质宇宙论

antimatter problem 反物质问题

antimeridian 下子午圈

antimony （=Sb）锑（51 号元素）

antimony-doped germanium 锗掺锑

antineutrino 反中微子

antinode 背交点

antinucleon 反核子

Antiope 休神星（小行星 90 号）

anti-parallel 逆平行

antiparticle 反粒子

antiphase 反相 又见：*phase reversal*

antiproton 反质子

antiquark 反夸克

antireflection coating 减反射镀膜

antireflection film 减反射膜

antiresonance 反共振

antiselena 反假月 | 幻月〔台〕

antisolar point 对日点

anti-spiral theorem 反旋涡定理

antisymmetrical state 反对称态

antitail 逆向彗尾

anti-twilight 反辉

antivertex 奔离点

Antlia （=Ant）唧筒座

Antlia Dwarf Galaxy 唧筒座矮星系

Ant Nebula 蚂蚁星云（Menzel 3）

Antoniadi crater 安东尼亚迪环形山（月球）

Antoniadi scale 安东尼亚迪视宁标度

Antonov-Lebovitz theorem 安东诺夫－列波维茨定理

Anuchin crater 阿努钦环形山（月球）

AO （=Arecibo Observatory）阿雷西沃天文台

Aoede 木卫四十一

Aokeda 澳科大（小行星 200003 号）

AORS （=acousto-optic radiospectrometer）射电声光频谱计

AOS （=acousto-optic spectrograph）声光频谱仪

Aouelloul crater 奥埃洛陨星坑（地球）

Aoyunzhiyuanzhe 奥运志愿者（小行星 18639 号）

AP （=Arecibo Pulsar）阿雷西沃天文台脉冲星

Apache Point Observatory （=APO）阿帕奇天文台

apareon 远火点 | 遠火[星]點〔台〕
又见：*apomartian*

apastron 远星点 又见：*apoastron*

APC （=automatic phase corrector）自动相位改正器

Apennines crater 亚平宁环形山（月球）

aperiodic comet 非周期彗星 | 非週期[性]彗星〔台〕 又见：*non-periodic comet*

aperiodicity 非周期性

aperiodic motion 非周期运动

aperture ① 口径 ② 孔径

aperture angle 孔径角

aperture array 孔径阵

aperture blocking 孔径遮挡

aperture diaphragm 孔径光阑

aperture efficiency 孔径效率

aperture function 孔径函数

aperture illumination function 孔径照明函数

Aperture Masking Interferometry 孔径遮挡干涉测量

aperture photometry 孔径测光

aperture ratio 口径比 | 口徑比，孔徑比〔台〕

aperture synthesis 综合孔径 | 孔徑合成〔台〕
又见：*synthetic aperture; synthesis aperture; synthesized aperture*

aperture-synthesis array 综合孔径阵

aperture synthesis radiotelescope 综合孔径射电望远镜 | 孔徑合成電波望遠鏡〔台〕 又见：*synthesis radio telescope*

APEX （=Atacama Pathfinder Experiment telescope）阿塔卡马探路者实验望远镜

apex 向点

apex of the earth's motion 地球向点

apex star 向点星

APFS （=Apparent Places of Fundamental Stars）基本星视位置表

aphelic conjunction 远日点合

aphelic distance 远日[点]距 | 遠日點距〔台〕

aphelic opposition 远日点冲

aphelion 远日点

aphesperian 远金点 | 遠金[星]點〔台〕

Aphrodite Terra 阿佛洛狄忒台地（金星）

Ap index Ap 指数（地磁）

Apis 蜜蜂座（苍蝇座曾用名）

ApJ （=Astrophysical Journal）《天体物理学报》|《天文物理期刊》〔台〕（美国期刊）

ApJL （=Astrophysical Journal Letters）《天体物理学报通信》|《天文物理期刊通訊》〔台〕（美国期刊）

ApJS （=Astrophysical Journal Supplement Series）《天体物理学报增刊》（美国期刊）

aplanat 消球差透镜，齐明透镜 | 齊明透鏡〔台〕
又见：*aplanatic lens*

aplanatic lens 消球差透镜，齐明透镜 | 齊明透鏡〔台〕 又见：*aplanat*

aplanatic system 消球差系统，齐明系统 | 齊明系統〔台〕

aplanatic telescope 消球差望远镜，齐明望远镜 | 齊明望遠鏡〔台〕

aplanatism 消球差，齐明 | 齊明〔台〕

APM ① （=Automated Plate-Measuring System）底片自动测量系统 ② （=automatic plate-measuring machine）底片自动测量仪

APM galaxy catalogue （=Automatic Plate Measuring galaxy catalogue）底片自动测量星系表，APM 星系表

APM Survey （=Automated Plate Measuring Survey）底片自动测量巡天，APM 巡天

APO （=Apache Point Observatory）阿帕奇天台

apoapse 远质心点

apoapse distance 远质心距

apoapsis 远点，远拱点 | 遠拱點〔台〕

apoastron 远星点 又见：*apastron*

apocenter 远心点

apochromat 复消色差透镜 又见：*apochromatic lens*

apochromatic lens 复消色差透镜
又见：*apochromat*

apochromatic objective 复消色差物镜

apochromatic refractor 复消色差折射望远镜

apochromatism 复消色差

apocronus 远土点 | 遠土[星]點〔台〕
又见：*aposaturnian*

apocynthion 远月点 又见：*aposelene; apolune*

apodisation 切趾法 又见：*apodization*

apodization 切趾法 又见：*apodisation*

apodized-pupil Lyot coronagraph 切趾瞳李奥星冕仪

apodizing phase plate 切趾相位板

apofocus 远主焦点

apogalacteum 远银心点 又见：*apogalacticon; apogalacticum*

apogalacticon 远银心点 又见：*apogalacteum; apogalacticum*
apogalacticum 远银心点 又见：*apogalacticon; apogalacteum*
apogean tide 远地点潮 又见：*apogeic tide*
apogee 远地点
apogee distance 远地点距
apogeic tide 远地点潮 又见：*apogean tide*
apojove 远木点 | 遠木 [星] 點〔台〕
Apollinaris Patera 阿波里那托边火山（火星）
Apollo ①阿波罗（小行星1862号）②阿波罗飞船 ③北河二（双子座α） 又见：*Castor*
Apollo-Amor object 阿波罗－阿莫尔型天体
Apollo asteroid 阿波罗型小行星 又见：*Apollos*
Apollo crater 阿波罗环形山（月球）
Apollo group 阿波罗群
Apollo Lunar Surface Experiments Package (=ALSEP) 阿波罗月面实验装置
Apollo mission 阿波罗任务
Apollonius crater 阿波罗尼奥斯环形山（月球）
Apollo project 阿波罗计划
Apollos 阿波罗型小行星 又见：*Apollo asteroid*
Apollo-Soyuz Test Project (=ASTP) 阿波罗－联盟号实验计划
Apollo Telescope Mount (=ATM) 阿波罗望远镜装置
apolune 远月点 又见：*aposelene; apocynthion*
apomartian 远火点 | 遠火 [星] 點〔台〕
 又见：*apareon*
apomercurian 远水点 | 遠水 [星] 點〔台〕
Apophis 毁神星（小行星99942号）
apoplutonian 远冥 [王] 点 | 遠冥 [王星] 點〔台〕
apoposeidon 远海 [王] 点 | 遠海 [王星] 點〔台〕
aposaturnian 远土点 | 遠土 [星] 點〔台〕
 又见：*apocronus*
aposelene 远月点 又见：*apocynthion; apolune*
apouranian 远天 [王] 点 | 遠天 [王星] 點〔台〕
 又见：*apouranium*
apouranium 远天 [王] 点 | 遠天 [王星] 點〔台〕
 又见：*apouranian*
apparatus function 仪器函数
apparent absorption 视吸收
apparent altitude 视地平纬度，视高度
apparent angle 视角 又见：*viewing angle; visual angle*
apparent anomaly 视近点角
apparent antivertex 视奔离点
apparent area 视面积
apparent association 表观成协

apparent binary 视双星，光学双星 又见：*optical double*
apparent bolometric magnitude 视热星等
apparent brightness 视亮度
apparent coordinate 视坐标
apparent declination 视赤纬
apparent depression of the horizon 视地平俯角
apparent diameter 视直径 | 視徑〔台〕
 又见：*visual diameter*
apparent direction 视方向
apparent distance 视距离 | 視距 [離]〔台〕
apparent diurnal motion 周日视动
apparent diurnal path 视周日轨迹
apparent equatorial coordinate 视赤道坐标
apparent exterior contact 视外切
apparent field 可见视场
apparent flattening 视扁率
apparent flux 视流量
apparent horizon 视地平
apparent intensity 视强度
apparent interior contact 视内切
apparent latitude 视黄纬
apparent libration 视天平动
apparent longitude 视黄经
apparent luminosity function 视光度函数
apparent magnitude 视星等
apparent magnitude-color index diagram 视星等－色指数图
apparent mass 表观质量
apparent mean place 视平位置
apparent modulus 视距离模数
apparent motion 视运动 | 視 [運] 動〔台〕
apparent noon 视正午
apparent orbit 视轨道
apparent path 视轨迹 | 視路徑，視軌跡〔台〕
apparent photographic magnitude 视照相星等
apparent photovisual magnitude 视仿视星等
apparent place 视位置 | 視位 [置]〔台〕
 又见：*apparent position*
Apparent Places of Fundamental Stars (=APFS) 基本星视位置表
apparent position 视位置 | 視位 [置]〔台〕
 又见：*apparent place*
apparent radiant 视辐射点
apparent radiometric magnitude 视辐射星等
apparent revolution 视公转
apparent right ascension 视赤经

apparent rotation 视自转
apparent rotational potential 视自转势
apparent semi-diameter 视半径
apparent sidereal time 真恒星时｜视恆星時〔台〕
　　又见：*true sidereal time*
apparent sky brightness 视天空亮度
apparent solar day 视太阳日
apparent solar time 真太阳时｜视[太阳]时〔台〕
　　又见：*true solar time*
apparent sun 视太阳
apparent superluminal motion 视超光速运动
apparent superluminal velocity 视超光速
apparent temperature 视温度
apparent velocity 视速度
apparent vertex 视奔赴点
apparent visual magnitude 视目视星等
apparent zenith distance 视天顶距
apparition 出现，可见期｜顯現期〔台〕
appearance height 出现点高度　又见：*beginning height*
appearance in the day 昼见
appearance point 出现点　又见：*beginning point*
Appleton crater 阿普顿环形山（月球）
Appleton layer 阿普顿层｜阿卜吞層〔台〕（电离层 F 区）
applied astronomy 应用天文学
applied force 外加力
approximate solution 近似解
approximate value 近似值
approximation 近似法
appulse ①犯，最小角距 ②半影月食
　　又见：*penumbral lunar eclipse; lunar appulse*
APR （＝active prominence region）活动日珥区
April Lyrid meteor shower 四月天琴流星雨
APS （＝active pixel detector）主动像元传感器
Aps （＝Apus）天燕座
apse 拱点　即：*apsis*
apse line 拱线　又见：*apsidal line; line of apsides*
apse-node terms 拱交项
apsidal advance 拱线进动，近星点进动
　　又见：*apsidal precession*
apsidal constant 近星点常数
apsidal line 拱线　又见：*apse line; line of apsides*
apsidal motion 拱线运动，近星点运动｜拱線運動〔台〕　又见：*advance of apsidal line; apsidal rotation*
apsidal period 拱线转动周期，近星点运动周期｜拱線[轉動]週期〔台〕

apsidal precession 拱线进动，近星点进动
　　又见：*apsidal advance*
apsidal resonance 拱线共振，近星点共振
apsidal rotation 拱线运动，近星点运动｜拱線運動〔台〕　又见：*apsidal motion; advance of apsidal line*
apsis （复数：**apsides**）　拱点，遠近[焦]點，拱點〔台〕
ApSS （＝Astrophysics and Space Science）《天体物理和空间科学》｜《天文物理和太空科學》〔台〕
Ap star （＝peculiar A star）Ap 星，A 型特殊星
APT ①（＝Australian Patrol Telescope）澳洲巡天望远镜 ②（＝automatic photoelectric telescope）自动光电测光望远镜
AP test （＝Alcock-Paczynski test）阿尔科克－帕金斯基检验，AP 检验
Apus （＝Aps）天燕座
Aql （＝Aquila）天鹰座
Aqr ①（＝Aquarius）宝瓶座 ②（＝Aquarius）宝瓶宫，玄枵，子宫
Aquarids 宝瓶流星群｜寶瓶[座]流星雨〔台〕
Aquarius ①（＝Aqr）宝瓶座 ②（＝Aqr）宝瓶宫，玄枵，子宫
aqueous alteration 水蚀变
aqueous meteor 水流星
Aquila （＝Aql）天鹰座
Aquila Rift 天鹰座暗隙
Ar （＝argon）氩（18 号元素）
Ara ①天坛座 ②阿拉（小行星 849 号）
ARAA （＝Annual Review of Astronomy and Astrophysics）《天文学和天体物理学年评》｜《天文學和天文物理學年評》〔台〕（美国期刊）
arachnoid 蛛网地形（金星）
Arago distance 阿拉戈距离｜阿拉哥角距〔台〕
Arago point 阿拉戈点｜阿拉哥點〔台〕
Arago ring 阿拉戈环（海王星）
Araguainha crater 阿拉瓜伊尼亚陨星坑（地球）
ARC （＝Australian Research Council）澳大利亚研究委员会
arc crater 弧坑
arc degree 度　又见：*degree; degree of arc*
archaeoastronomy 考古天文学
　　又见：*archeoastronomy*
Arche 木卫四十三
archeoastronomy 考古天文学
　　又见：*archaeoastronomy*
Archeops 祖翼鸟（法国宇宙微波背景测量实验）
arch filament 拱状暗条
arch filament system （＝AFS）拱状暗条系统

Archimedes crater 阿基米德环形山（月球）
Archimedes spiral 阿基米德螺线
Archimedes spiral field 阿基米德旋涡场
arc length 弧长
arc line 弧光谱线
arcmin （＝minute of arc）角分
arc minute 角分 又见：minute of arc
Arcminute Cosmology Bolometer Array Receiver （＝ACBAR）角分宇宙学辐射热计阵
arcsec （＝second of arc）角秒
arc second 角秒 又见：second of arc
arc spectrum 弧光谱
Arctic Circle （＝AC）北极圈
arctic circle 极圈 又见：polar circle
Arctic pole 北极（地球）
Arctic zone 北极区
Arcturus 大角（牧夫座 α）
Arcturus group 大角星群
area image sensor 面成像传感器
areal velocity 面积速度，掠面速度
area-mass ratio 面质比
area of audibility 能听范围 | 能聞範圍〔台〕
area of coverage 覆盖区，观测区
area of visibility 能见范围
area photometer 面积光度计
area photometry 面源测光 又见：surface photometry
area-preserving map 保积映射 又见：equiareal mapping
area spectroscopy 面源分光
Arecibo Legacy Fast ALFA Survey （＝ALFALFA）阿雷西沃遗珍快速 ALFA 巡天，首蓿巡天
Arecibo Observatory （＝AO）阿雷西沃天文台
Arecibo Pulsar （＝AP）阿雷西沃天文台脉冲星
Arecibo radio telescope 阿雷西沃射电望远镜 | 阿雷西波電波望遠鏡〔台〕
Arend-Roland comet 阿连德－罗兰彗星（C/1956 R1）
areocentric coordinate 火心坐标 | 火 [星中] 心坐標〔台〕
areodesy 火星测地学
areographic chart 火面图 | 火 [星表] 面圖〔台〕 又见：areographic map
areographic coordinate 火面坐标 | 火 [星表] 面坐標〔台〕
areographic latitude 火面纬度 | 火 [星] 緯度〔台〕
areographic longitude 火面经度 | 火 [星] 經度〔台〕

areographic map 火面图 | 火 [星表] 面圖〔台〕 又见：areographic chart
areographic pole 火面极 | 火面極〔台〕
areography 火面学 | 火 [星表] 面學〔台〕
areology 火星学
areophysics 火星物理学 | 火星物理 [學]〔台〕
Arete 阿雷特（小行星 197 号）
Arethusa 源神星（小行星 95 号）
Argelander method 阿格兰德法
Argo （＝Argo Navis）【拉】天舟座，南船座（曾是最大的星座，后被拆分成四个单独的星座：船帆座、船底座、船尾座和罗盘座。）
argon （＝Ar）氩（18 号元素）
Argo Navis （＝Argo）【拉】天舟座，南船座（曾是最大的星座，后被拆分成四个单独的星座：船帆座、船底座、船尾座和罗盘座。）
argon-potassium method 氩钾纪年法
ARGO-YBJ （＝Astrophysical Radiation with Ground-based Observatory at YangBaJing）羊八井 ARGO 实验（中国）
argument 幅角，辐角，角距
argument of latitude 升交角距
argument of periapsis 近点幅角 | 近點輻角〔台〕
argument of pericenter 近心点幅角 | 近心點輻角〔台〕
argument of perigee 近地点幅角 | 近地點輻角〔台〕
argument of perihelion 近日点幅角 | 近日點角距〔台〕
Argyre Planitia 阿尔及尔平原（火星）
Ari ①（＝Aries）白羊座 ②（＝Aries）白羊宫，降娄，戌宫
Ariadne 爱女星（小行星 43 号）
Ariane 阿丽亚娜运载火箭
Ariel ①天卫一 | 天 [王] 衛一〔台〕②羚羊号（英美系列天文卫星）
Aries ①（＝Ari）白羊座 ②（＝Ari）白羊宫，降娄，戌宫
Arietids 白羊流星群
ARISE （＝Advanced Radio Interferometer between Space and Earth）高新空地射电干涉仪
Aristarchus crater 阿利斯塔克环形山（月球）
Aristoteles crater 亚里士多德环形山（月球）
Arizona crater 亚利桑那陨星坑 | 亞利桑那隕石坑〔台〕（地球）
Arkab Posterior 天渊一（人马座 β^2）
Arkab Prior 天渊二（人马座 β^1）
Arkangelsky crater 阿尔汉格尔斯基环形山（火星）

AR Lacertae star （＝AR Lac star）蝎虎 AR 型星
AR Lac star （＝AR Lacertae star）蝎虎 AR 型星
ARMA （＝autoregressive-moving average）自回归移动平均
arm class 臂类型
Armenian calendar 亚美尼亚历
Armenian Virtual Observatory （＝ArVO）亚美尼亚虚拟天文台
armillary sphere 浑仪
arm population 臂族，极端星族 I｜旋臂星族，极端星族 I〔台〕 又见: extreme population I
Armstrong crater 阿姆斯特朗环形山（月球）
Arneb 厕一（天兔座 α）
Arnold crater 阿诺尔德环形山（月球）
Arnold diffusion 阿诺德扩散
Arnold web 阿诺德网
Arnowitt-Deser-Misner formalism （＝ADM formalism）ADM 形式
ARO （＝Algonquin Radio Observatory）阿尔冈金射电天文台
array antennae 阵列天线
array configuration 阵列构型
array feed 阵列馈源
Array of Low-Energy X-ray Imaging Sensors （＝ALEXIS）亚历克西斯号，低能 X 射线成像传感器阵列（美国极紫外天文卫星）
array processor 阵列处理器
array telescope 阵列望远镜
arrival time 到达时间
arsenic （＝As）砷（33 号元素）
Arsia Mons 阿尔西亚山（火星）
Artamonov crater 阿尔塔莫诺夫环形山（月球）
ARTEMIS （＝Acceleration, Reconnection, Turbulence and Electrodynamics of the Moon's Interaction with the Sun）阿尔忒弥斯号月球探测器（美国）
Artemis Chasma 阿尔忒弥斯深谷（金星）
Artemis Corona 阿尔忒弥斯火山冕（金星）
Artermis 群女星（小行星 105 号）
artificial asteroid 人造小行星 又见: artificial planetoid
artificial comet 人造彗星
artificial crater 人造环形山
artificial guide star 人造引导星
artificial horizon 假地平
artificial meteor 人造流星
artificial object 人造天体
artificial planet 人造行星
artificial planetoid 人造小行星 又见: artificial asteroid
artificial radiation belt 人造辐射带
artificial satellite 人造卫星
artificial star 假星 又见: fictitious star
ArVO （＝Armenian Virtual Observatory）亚美尼亚虚拟天文台
Arzachel crater 阿尔扎赫尔环形山（月球）
As （＝arsenic）砷（33 号元素）
ASA （＝Astronomical Society of Australia）澳大利亚天文学会
Asad Australis 轩辕九（狮子座 ε） 又见: Algenubi; Ras Elased Australis
ASAS （＝All-Sky Automated Survey）自动化全天巡视
ASCA （＝Advanced Satellite for Cosmology and Astrophysics）宇宙学和天体物理学高新卫星，飞鸟号（日本 X 射线天文卫星）
Ascella 斗宿六（人马座 ζ） 又见: Axilla
ascend 上升 又见: ascension; ascent
ascendant 星位（占星）
ascending branch 上升支｜上升部份〔台〕
ascending node 升交点 又见: northbound node
ascension 上升 又见: ascend; ascent
ascent 上升 又见: ascend; ascension
Aschere 天狼[星]（大犬座 α） 又见: Sirius; Canicula; Dog Star
Ascraeus Mons 艾斯克雷尔斯山（火星）
Asellus Australis 鬼宿四（巨蟹座 δ）
Asellus Borealis 鬼宿三（巨蟹座 γ）
Asellus Primus 天枪三（牧夫座 θ）
Asellus Secundus 天枪二（牧夫座 ι）
Asellus Tertius 天枪一（牧夫座 κ）
Ashbrook crater 阿什布鲁克环形山（月球）
ashen light 灰光
Asia 亚女星，亚细亚（小行星 67 号）
asiderite 石陨星，石陨石｜石質隕石〔台〕 又见: aerolith; stony meteorite; aerolite
ASJ （＝Astronomical Society of Japan）日本天文学会
ASKAP （＝Australian Square Kilometre Array Pathfinder）澳大利亚平方千米阵探路者
Asmidiske 弧矢增十七（船尾座 ξ） 又见: Azmidiske
ASP （＝Astronomical Society of the Pacific）太平洋天文学会
aspect 视方位

aspect angle　视界角
aspect astrology　星象　又见：*star portent*
aspect of the moon　月相　又见：*phase of the moon; lunar aspect; lunar phase; moon's phase*
aspect ratio　纵横比
aspherical lens　非球面透镜 | 消球差透鏡〔台〕
aspherical mirror　非球面镜
asphericity　非球面度
aspheric surface　非球面
Aspidiske　海石二（船底座ι）　又见：*Scutulum; Turyeish; Turais*
ASSA　（＝Astronomical Society of Southern Africa）南非天文学会
assemblage　系综　又见：*ensemble*
assembly　系集 | 聚〔台〕
assembly bias　组装偏袒，组装偏差
assembly of five planets　五星连珠
assembly time　组装时间
associated Legendre function　缔合勒让德函数
association in time　时间关联
association of galaxies　星系协
Association of Universities for Research in Astronomy　（＝AURA）大学天文研究联合组织（美国）
AST　（＝Atlantic Standard Time）大西洋标准时间
A star　A 型星　又见：*A-type star*
astatine　（＝At）砈 | 砹〔台〕（85 号元素）
aster　天体　又见：*celestial body; celestial object; orb*
Asterion　常陈四（猎犬座β）　又见：*Chara*
asterism　①星官，星宿 | [星]官〔台〕②星组 ③星芒
asteroid　小行星　又见：*minor planet*
asteroidal belt　小行星带　又见：*asteroid belt; asteroid zone*
asteroidal dynamics　小行星动力学
asteroidal meteor　小行星流星
asteroidal resonance　小行星共振
asteroid belt　小行星带　又见：*asteroid zone; asteroidal belt*
asteroid family　小行星族 | 小行星[家]族〔台〕　又见：*family of asteroids; minor-planet family*
asteroid group　小行星群　又见：*group of asteroids*
asteroid-like comet　类小行星彗星
asteroid-like object　类小行星天体
asteroid-mass object　小行星质量天体
asteroid ring　小行星环

asteroid zone　小行星带　又见：*asteroid belt; asteroidal belt*
Asterope　昴宿三（金牛座21）　又见：*Sterope*
asteroseismology　星震学　又见：*stellar seismology; astroseismology*
asthenosphere　软流层
astigmatism　像散
ASTP　（＝Apollo-Soyuz Test Project）阿波罗—联盟号实验计划
Astraea　义神星（小行星 5 号）
astration　循环创造
astrionics　天文电子学
Astro-1　天星一号（美国航天飞机搭载天文台）
Astro-2　天星二号（美国航天飞机搭载天文台）
Astro-A　火鸟号（日本太阳探测器）　又见：*Hinotori*
astroarchaeology　天文考古学
Astro-B　天马号（日本 X 射线天文卫星）　又见：*Tenma*
Astrobiology　《天体生物学》（美国期刊）
astrobiology　天体生物学 | 天文生物學〔台〕
astrobleme　陨星撞迹 | 隕石撞跡〔台〕
astrobotany　天体植物学
Astro-C　银河号（日本 X 射线天文卫星）　又见：*Ginga*
astrocamera　天体照相机　又见：*astrographic camera; astrophotocamera*
astrochemistry　天体化学 | 天文化學〔台〕
astrochronology　天体年代学 | 天文年代學〔台〕
astroclimate　天文气候　又见：*astronomical climate*
astroclimatology　天文气候学
Astro-D　宇宙学和天体物理学高新卫星，飞鸟号（日本 X 射线天文卫星）　又见：*Advanced Satellite for Cosmology and Astrophysics*
astrodome　天文圆顶 | [觀測] 圓頂〔台〕
astrodynamics　天文动力学 | 太空動力學〔台〕
astroecology　宇宙生态学
Astro-EII　朱雀号（日本 X 射线天文卫星）　又见：*Suzaku*
ASTRO-F　光亮号（日本红外天文卫星）　又见：*Akari*
astrofundamental physics　天体基本物理学
astrogeodesy　天文测地学　又见：*astronomical geodesy*
astrogeodetic deflection　天文大地测量偏差
astrogeodynamics　天文地球动力学
astrogeography　天体地理学 | 天文地理學〔台〕
astrogeology　天体地质学 | 天文地質學〔台〕　又见：*astronomical geology*

astrogeophysics 天文地球物理学
astrognosy 恒星学
astrogony 恒星演化学 又见：*stellar cosmogony*
astrogram 天文电报 又见：*astronomical telegram*
astrograph 天体照相仪 | 天文照相儀〔台〕
　　又见：*astrophotograph*
astrographic atlas 照相星图 又见：*astrographic chart; astrographic map; photographic chart*
astrographic camera 天体照相机
　　又见：*astrophotocamera; astrocamera*
Astrographic Catalogue （＝AC）照相天图星表 | AC 星表〔台〕
astrographic catalogue 照相星表
　　又见：*photographic star catalogue; photographic catalogue*
astrographic chart 照相星图 又见：*astrographic map; astrographic atlas; photographic chart*
astrographic doublet 天体照相双合透镜 | 天文照相雙合透鏡〔台〕
astrographic map 照相星图 又见：*astrographic chart; astrographic atlas; photographic chart*
astrographic objective 天体照相物镜
astrographic plate 天体照相底片
astrographic position 天体照相位置
astrographic refractor 折射天体照相仪
　　又见：*photographic refractor*
astrography 天体照相学 | 天文照相學，天體照相〔台〕 又见：*astrophotography; astronomical photography*
AstroGrid （＝Virtual Observatory United Kingdom）英国虚拟天文台
Astroinformatics 天文信息学
astrolabe ①星盘 | 星[象]盤〔台〕②等高仪
astrolithology 陨星学，陨石学 | 隕石學〔台〕
　　又见：*meteoritics*
astrologer 占星术士
astrology 占星术
astromagnetism 天体磁学
astromechanics 天体力学 又见：*celestial mechanics*
astrometeorology 天体气象学 | 天文氣象學〔台〕
astrometer 天体测量仪
astrometric aspect 天体测量方位
astrometric baseline 天体测量基线
astrometric binary 天测双星，天体测量双星 | 天體測量雙星〔台〕
astrometric declination 天体测量赤纬
astrometric distance 天测距离，天体测量距离 | 天體測量距離〔台〕
astrometric error 天测误差，天体测量误差
astrometric expectation 天测期望，天体测量期望

astrometric instrument 天测仪器，天体测量仪器 | 天體測量儀器〔台〕
astrometric orbit 天测轨道，天体测量轨道 | 天體測量軌道〔台〕
astrometric photograph 天测照相，天体测量照相，天测照片
astrometric place 天体测量位置 又见：*astrometric position*
astrometric position 天体测量位置
　　又见：*astrometric place*
astrometric precision 天测精度
astrometric right ascension 天体测量赤经
astrometrist 天体测量学家
astrometry 天体测量学 | 天體測量學，測天學〔台〕
astrometry satellite 天体测量卫星
ASTRON （＝automatic astronomical station）天体号（苏联天文卫星）
astronaut 宇航员 | 太空人〔台〕 又见：*cosmonaut*
astronautics 宇航学 | 宇[宙]航[行]學，太空航行學〔台〕 又见：*cosmonautics*
astronautic ZHR 航天天顶每时出现率
astronavigation 天文导航 | 太空導航〔台〕
　　又见：*celestial guidance; celestial navigation*
astronegative 天文底片 又见：*astronomical plate*
astronette 女宇航员
astronomer 天文学家
Astronomer Royal ①皇家天文学家（英格兰）②太史令（秦汉官职） 又见：*imperial astronomer*
Astronomer Royal for Scotland 苏格兰皇家天文学家
astronomical aberration 天文光行差
astronomical almanac 天文年历
　　又见：*astronomical ephemeris; astronomical year book*
Astronomical Bureau 钦天监，司天监 | 欽天監〔台〕 又见：*Bureau of Astronomy*
Astronomical Chapter 天文志
astronomical chronicle 天文历书
astronomical climate 天文气候 又见：*astroclimate*
astronomical clock 天文钟
astronomical clock-tower 水运仪象台
astronomical compass 天文罗盘
astronomical constant 天文常数
astronomical constant system 天文常数系统
　　又见：*system of astronomical constants*
astronomical coordinate 天文坐标
astronomical coordinate system 天文坐标系统

astronomical data center 天文数据中心 | 天文資料中心〔台〕
astronomical date 天文日期
astronomical day 天文日
astronomical distance 天文距离
astronomical ephemeris （＝A.E.）天文年历
又见：*astronomical almanac; astronomical year book*
astronomical geodesy 天文测地学
又见：*astrogeodesy*
astronomical geology 天体地质学 | 天文地質學〔台〕 又见：*astrogeology*
astronomical image-processing system （＝AIPS）天文图像处理系统 | 天文影像處理系統〔台〕
astronomical image reconstruction 天文图像复原 | 天文影像重建〔台〕
astronomical instrument 天文仪器
Astronomical Journal （＝AJ）《天文学报》（美国期刊）
astronomical latitude 天文纬度
astronomical levelling 天文水准测量
astronomical limitation 天文极限
astronomical longitude 天文经度
astronomical measurement 天文测量
astronomical meridian 天文子午圈，天文子午线
astronomical method 天文方法
Astronomical Netherlands Satellite （＝ANS）荷兰天文卫星
astronomical nomenclature 天文命名
astronomical observation 天文观测
astronomical observatory 天文台
又见：*observatory*
astronomical optics 天文光学
astronomical orientation 天文定向
astronomical photography 天体照相学 | 天文照相學，天體照相〔台〕 又见：*astrophotography; astrography*
astronomical photometry 天体测光 | 天文光度學，天體測光〔台〕 又见：*astrophotometry*
astronomical plate 天文底片 又见：*astronegative*
astronomical point 天文点
astronomical polarimetry 天体偏振测量 | 天體偏振測量〔法〕〔台〕 又见：*astropolarimetry*
astronomical pyrometer 天体高温计 | 天體測溫計〔台〕
astronomical reference system 天文参考系
astronomical refraction 大气折射
又见：*atmospheric refraction*
astronomical research 天文研究

astronomical satellite 天文卫星
astronomical seeing 天文视宁度，星像视宁度
astronomical sign 天文符号
Astronomical Society of Australia （＝ASA）澳大利亚天文学会
Astronomical Society of Japan （＝ASJ）日本天文学会
Astronomical Society of Southern Africa （＝ASSA）南非天文学会
Astronomical Society of the Pacific （＝ASP）太平洋天文学会
astronomical spectrograph 天体摄谱仪
又见：*astrospectrograph*
astronomical spectrophotometer 天体分光光度计
astronomical spectrophotometry 天体分光光度测量
astronomical spectroscope 天体分光镜
又见：*astrospectroscope*
astronomical spectroscopy 天体光谱学
又见：*astrospectroscopy*
astronomical technique 天文技术
astronomical telegram 天文电报 又见：*astrogram*
astronomical telescope 天文望远镜
astronomical test 天文检验
astronomical theodolite 天文经纬仪
astronomical time 天文时
astronomical time scale 天文时标
astronomical triangle 天文三角形 | 定位三角形，天文三角形〔台〕
astronomical twilight 天文晨昏蒙影 | 天文曙暮光，天文晨昏蒙影〔台〕
astronomical unit （＝au; AU）天文单位
astronomical vertical 天文垂线
astronomical year book 天文年历
又见：*astronomical ephemeris; astronomical almanac*
astronomical zenith 天文天顶
Astronomische Gesellschaft （＝AG）【德】德国天文学会
Astronomische Gesellschaft Katalog （＝AGK）【德】AGK 星表 | AG 星表〔台〕
Astronomische Nachrichten （＝AN）【德】《天文通报》
Astronomischer Jahresbericht （＝AJB）【德】《天文年报》
astronomy 天文学
Astronomy & Geophysics 《天文学与地球物理学》（英国期刊）

Astronomy and Astrophysics （＝A&A; AAp）《天文学和天体物理学》|《天文學和天文物理學》〔台〕

Astronomy and Astrophysics Abstracts （＝AAA）《天文学和天体物理学文摘》|《天文學和天文物理學文摘》〔台〕

Astronomy and Astrophysics Review 《天文学与天体物理学评论》（美国期刊）

Astronomy Letters 《天文学通信》（俄罗斯）

Astronomy Reports 《天文学报告》（俄罗斯）

Astronomy Satellite D2B D2B 号天文卫星

Astronomy with α Neutrino Telescope and Abyss environmental RESearch project （＝ANTARES）天蝎座 α 中微子望远镜

Astroparticle Physics 《天体粒子物理学》（荷兰期刊）

astroparticle physics 天文粒子物理学，天体粒子物理学

astrophile 天文爱好者 | 業餘天文學家，天文愛好者〔台〕 又见：amateur astronomer; stargazer

astrophotocamera 天体照相机 又见：astrographic camera; astrocamera

astrophotogram 天文照片

astrophotograph 天体照相仪 | 天文照相儀〔台〕 又见：astrograph

astrophotography 天体照相学 | 天文照相學，天體照相〔台〕 又见：astronomical photography; astrography

astrophotometer 天体光度计

astrophotometry 天体测光 | 天文光度學，天體測光〔台〕 又见：astronomical photometry

astrophotonics 天文光子学

astrophysical astrometry 天体物理测量

Astrophysical Data System （＝ADS）天体物理数据系统 | 天文物理資料系統〔台〕

astrophysical jet 天体物理喷流

Astrophysical Journal （＝ApJ）《天体物理学报》|《天文物理期刊》〔台〕（美国期刊）

Astrophysical Journal Letters （＝ApJL）《天体物理学报通信》|《天文物理期刊通訊》〔台〕（美国期刊）

Astrophysical Journal Supplement Series （＝ApJS）《天体物理学报增刊》（美国期刊）

astrophysical maser 天体物理微波激射器 | 天文物理邁射儀〔台〕

astrophysical method 天体物理方法 | 天文物理方法〔台〕

Astrophysical Radiation with Ground-based Observatory at YangBaJing （＝ARGO-YBJ）羊八井 ARGO 实验（中国）

astrophysical technique 天体物理技术

Astrophysical Virtual Observatory （＝AVO）天体物理虚拟天文台（欧洲）

astrophysicist 天体物理学家 | 天文物理學家〔台〕

Astrophysics 《天体物理学》（亚美尼亚期刊）

astrophysics 天体物理学 | 天文物理學〔台〕

Astrophysics and Space Science （＝ApSS）《天体物理和空间科学》|《天文物理和太空科學》〔台〕

astropolarimeter 天体偏振计

astropolarimetry 天体偏振测量 | 天體偏振測量[法]〔台〕 又见：astronomical polarimetry

astro-position line 天文位置线

astrorelativity 宇宙相对论

Astrosat 天文号卫星（印度）

astroscope 天文仪, 星宿仪 | 天球儀, 星宿儀〔台〕

astroseismology 星震学 又见：asteroseismology; stellar seismology

astrospectrograph 天体摄谱仪 又见：astronomical spectrograph

astrospectrometer 天体分光计 | 天體分光計，天體光譜儀〔台〕

astrospectrometry 天体分光测量

astrospectroscope 天体分光镜 又见：astronomical spectroscope

astrospectroscopy ① 天体波谱学 ② 天体光谱学 又见：astronomical spectroscopy

astrostatistics 天文统计学

astrostereogram 天文立体照片

astrotelevision 天文电视

astrotomography 天体层析摄像

astrovelocimeter 天体视向速度仪

A subdwarf star A 型亚矮星 | A 型次矮星〔台〕

A subgiant star A 型亚巨星 | A 型次巨星〔台〕

A supergiant star A 型超巨星

asymmetrical galaxy 非对称星系

asymmetric drift 非对称星流 | 非對稱流〔台〕

asymmetric effect 非对称效应

asymmetric periodic orbit 非对称周期轨道

asymmetric radiation 非对称辐射

asymmetric top 非对称陀螺

asymmetry 非对称

asymmetry spectrum 非对称谱

asymptotic approximations 渐近近似

asymptotic branch 渐近支

asymptotic branch giant （＝AGB star）渐近支巨星，AGB 星 | 漸近支巨星〔台〕

asymptotic freedom 渐近自由

asymptotic giant branch （＝AGB）渐近巨星支
asymptotic model 渐近模型
asymptotic orbit 渐近轨道
asymptotic representation 渐近表示
asymptotic solution 渐近解
asymptotic stability 渐近稳定性
asynchronous correlator 非同步相关器
AT ①（＝Australia Telescope）澳大利亚巨型望远镜｜澳洲巨型望遠鏡〔台〕②（＝atomic time）原子时
At （＝astatine）砹｜砈〔台〕（85号元素）
ATA （＝Allen Telescope Array）艾伦望远镜阵
Atacama Cosmology Telescope （＝ACT）阿塔卡马宇宙学望远镜
Atacama Large Millimeter/sub-millimeter Array （＝ALMA）阿塔卡马大型毫米[/亚毫米]波阵｜阿塔卡瑪大型毫米[/次毫米]波陣〔台〕
Atacama Pathfinder Experiment telescope （＝APEX）阿塔卡马探路者实验望远镜
Atalante 驰神星（小行星36号）
atautun 阿托顿（玛雅历日，约 2.3×10^{12} 天）
ataxite 杂陨石｜無紋隕鐵〔台〕
ATC （＝Automatic Transit Circle）自动子午环
ATCA （＝Australia Telescope Compact Array）澳大利亚望远镜致密阵
AT-cut AT 切割
Ate 苟神星（小行星111号）
Aten 阿登（小行星2062号）
Aten asteroid 阿登型小行星
Aten group 阿登群
Ati 卷舌增七（英仙座 o） 又见：*Atik*
ATIC （＝Advanced Thin Ionization Calorimeter experiment）高新薄离子量能器实验
Atik 卷舌增七（英仙座 o） 又见：*Ati*
Atlantic Standard Time （＝AST）大西洋标准时间
Atlas ①土卫十五 ②昴宿七（金牛座27）
atlas ①天图 又见：*sky atlas; celestial chart; sky map* ②星图 又见：*star map; stellar map*
Atlas crater 阿特拉斯环形山（月球）
Atlas of Peculiar Galaxies 《特殊星系图集》
ATM （＝Apollo Telescope Mount）阿波罗望远镜装置
atmosphere 大气
atmospheric absorption 大气吸收
atmospheric agitation 大气抖动｜大氣攪動〔台〕
atmospheric attenuation 大气衰减

atmospheric Cherenkov telescope 大气切伦科夫望远镜
atmospheric composition 大气成分
atmospheric concentration 大气浓度
atmospheric correction 大气改正
atmospheric correlation length 大气相关长度
atmospheric correlation time 大气相关时间
atmospheric density 大气密度
atmospheric depth 大气深度
atmospheric dispersion 大气色散
atmospheric dispersion compensator 大气色散补偿器
atmospheric drag 大气阻力 又见：*air drag*
atmospheric eclipse 大气食
atmospheric effect 大气效应
atmospheric electricity 大气电学
atmospheric entry trajectory 大气进入轨道
atmospheric extinction 大气消光
atmospheric fluctuation 大气起伏
atmospheric fluorescence 大气荧光
atmospheric interference 天电干扰
 又见：*atmospherics; parasite*
atmospheric layer 大气层 又见：*aerosphere*
atmospheric line 大气谱线｜[地球]大氣譜線〔台〕
 又见：*telluric line*
atmospheric loading 大气负载
atmospheric mass density 大气质量密度
atmospheric model 大气模型
atmospheric noise 大气噪声｜大氣雜訊〔台〕
atmospheric optics 大气光学
atmospheric parameter 大气参量
atmospheric perturbation 大气扰动
atmospheric pressure 大气压｜[大]氣壓[力]〔台〕
atmospheric radiation 大气辐射
atmospheric refraction 大气折射
 又见：*astronomical refraction*
atmospheric refraction corrector 大气折射改正器
atmospherics ①天电干扰 又见：*atmospheric interference; parasite* ②天电 又见：*sferics; stray*
atmospheric scattering 大气散射
atmospheric scintillation 大气闪烁
atmospheric seeing 大气视宁度｜视相，大氣寧靜度〔台〕
atmospheric tide 大气潮｜大氣潮[汐]〔台〕
atmospheric transmission 大气透射
atmospheric transparency 大气透明度
atmospheric turbulence 大气湍流｜大氣亂流〔台〕

atmospheric window	大气窗｜大氣窗口〔台〕
atmospherium	云象仪
ATNF	（＝Australia Telescope National Facility）澳洲国立巨型望远镜
atomic absorption coefficient	原子吸收系数
atomic beam	原子束
atomic beam frequency standard	原子束频标
atomic beam magnetic resonance	原子束磁共振
atomic beam tube	原子束管
atomic clock	原子钟
atomic constant	原子常数
atomic density	原子密度
atomic diffusion	原子扩散
atomic energy level	原子能级　又见：*atomic level*
atomic frequency standard	原子频标
Atomichron	原子钟（第一款商业产品）
atomic hydrogen maser	氢原子激射器，氢原子脉泽
atomic kernel	原子实
atomic level	原子能级　又见：*atomic energy level*
atomic line	原子谱线
atomic mass unit	原子质量单位
atomic process	原子过程
atomic second	原子时秒
atomic selective absoption coefficient	原子选择吸收系数
atomic spectrum	原子光谱
atomic spin	原子自旋
atomic standard	原子标准
atomic state	原子能态
atomic time	（＝AT）原子时
Atomic Time 1	（＝A1）A1 原子时
atomic time scale	原子时标度
atomic time standard	原子时标准
atomic unit	原子单位
atomic unit of time interval	原子时单位
Atria	三角形三（南三角座 α）
ATST	（＝Advanced Technology Solar Telescope）高新技术太阳望远镜｜先進技術太陽望遠鏡〔台〕
attachment coefficient	附着系数
attenuation	衰减
attenuation constant	衰减常量
attenuation curve	衰减曲线
attenuation distance	衰减距离
attenuation factor	衰减因子｜減因素〔台〕
attenuator	衰减器
attitude control	姿态控制
attitude parameter	姿态参数
attractive force	引力　又见：*gravitation; gravitational force*
A-type asteroid	A 型小行星
A-type star	A 型星　又见：*A star*
AU	（＝astronomical unit）天文单位
Au	（＝gold）金（79 号元素）
au	（＝astronomical unit）天文单位
aubrite	无球粒顽辉陨石｜頑火無球隕石〔台〕
audiofrequency	声频
Auger effect	俄歇效应｜奧傑效應〔台〕
Auger shower	俄歇簇射
augmentation	增大
Augusta	奥古斯塔（小行星 243 号）
Augusta family	奥古斯塔族
August meteors	八月流星雨｜英仙 [座] 流星雨，八月流星群〔台〕
AU Microscopii	显微镜座 AU 型星
Aur	（＝Auriga）御夫座
AURA	（＝Association of Universities for Research in Astronomy）大学天文研究联合组织（美国）
Aura	①奥拉号（法国紫外卫星）②大气号（美国科学卫星）
Aureole	日晕号（苏联 - 法国科学卫星）又见：*Oreol*
aureole	日晕｜日暈，華蓋〔台〕
Auriga	（＝Aur）御夫座
Aurora	①彩神星（小行星 94 号）②极光号（美国通信卫星）
aurora	极光　又见：*polar glow; polar light*
aurora australis	南极光　又见：*southern light*
aurora borealis	北极光　又见：*northern light*
aurora brightness	极光亮度
Aurorae	曙光号（欧洲科学卫星）
auroral arc	极光弧
auroral corona	极光冕　又见：*auroral crown*
auroral crown	极光冕　又见：*auroral corona*
auroral electrojet	极光电喷流
auroral hiss	极光嘘声｜極光撕聲〔台〕
auroral jet	极光喷流
auroral latitude	极光纬区
auroral line	极光谱线
auroral oval	卵形极光
auroral proton flux	极光质子流量
auroral storm	极光暴
auroral substorm	极光亚暴
auroral zone	极光带

Ausonia 奥索尼亚（小行星 63 号）
auspicious star 瑞星
auspicious vapour 瑞气
Australian Astronomical Observatory （＝AAO）澳大利亚天文台（Anglo-Australian Observatory 的现用名）
Australian Astronomical Telescope （＝AAT）澳大利亚望远镜（Anglo-Australian Telescope 的现用名）
Australian Patrol Telescope （＝APT）澳洲巡天望远镜
Australian Research Council （＝ARC）澳大利亚研究委员会
Australian Square Kilometre Array Pathfinder （＝ASKAP）澳大利亚平方千米阵探路者
Australian Virtual Observatory （＝Aus-VO）澳大利亚虚拟天文台
Australia Telescope （＝AT）澳大利亚巨型望远镜 | 澳洲巨型望遠鏡〔台〕
Australia Telescope Compact Array （＝ATCA）澳大利亚望远镜致密阵
Australia Telescope National Facility （＝ATNF）澳洲国立巨型望远镜
australite 澳洲玻璃陨体 | 澳洲似曜石〔台〕
Aus-VO （＝Australian Virtual Observatory）澳大利亚虚拟天文台
autocoder 自动编码器
autocoding 自动编码 又见：*automatic coding*
autocollimation 自准直 | 自[動]準直〔台〕
autocollimation spectrograph 自准直摄谱仪
autocollimator 自准直管
autocontrol 自动控制 又见：*automatic control*
autocorrection 自动校正
autocorrelation 自相关 又见：*self-correlation*
autocorrelation coefficient 自相关系数
　　又见：*coefficient of autocorrelation*
autocorrelation function 自相关函数
autocorrelation radiometer 自相关型辐射计
autocorrelation spectrometer 自相关频谱计
autocorrelator 自相关器
autocovariance 自协方差
autoexcitation 自激发
autographic record 自动记录
autoguider 自动导星装置 | 自動導星儀〔台〕
　　又见：*automatic guider*
auto-guiding system 自动导星系统
autoionization 自电离 | 自游離〔台〕
Automated Astronomical Positioning System （＝AAPS）自动天文定位系统
Automated Plate Measuring Survey （＝APM Survey）底片自动测量巡天，APM 巡天
Automated Plate-Measuring System （＝APM）底片自动测量系统
automated telescope 自动化望远镜
automatic astronomical station （＝ASTRON）天体号（苏联天文卫星）
automatic celestial navigation （＝ACN）自动天文导航
automatic coding 自动编码 又见：*autocoding*
automatic comet-seeker 自动寻彗器
automatic control 自动控制 又见：*autocontrol*
automatic data processing （＝ADP）自动数据处理
automatic gain control 自动增益控制
automatic guider 自动导星装置 | 自動導星儀〔台〕 又见：*autoguider*
automatic guiding 自动导星
automatic interplanetary station （＝AIS）自动行星际站
automatic measuring machine 自动量度仪
Automatic Meridian Circle （＝AMC）自动子午环
automatic observatory 自动观测台
automatic patrol telescope 自动巡天望远镜
automatic phase corrector （＝APC）自动相位改正器
automatic photoelectric telescope （＝APT）自动光电测光望远镜
Automatic Plate Measuring galaxy catalogue （＝APM galaxy catalogue）底片自动测量星系表，APM 星系表
automatic plate-measuring machine （＝APM）底片自动测量仪
automatic tidemeter 自动潮汐计
automatic tracking 自动跟踪 | 自動追蹤〔台〕
Automatic Transit Circle （＝ATC）自动子午环
automatic zenith tube 自动天顶筒
automation system 自动系统
automorph 自守
Autonoe 木卫二十八
autonomous system 自控系统
autoregression 自回归
autoregressive coefficient 自回归系数
autoregressive-moving average （＝ARMA）自回归移动平均
autosyn 自整角机 又见：*selsyn*
Autumnal Equinox 秋分（节气）

autumnal equinox 秋分点　又见：*autumnal point; first point of Libra*
autumnal point 秋分点　又见：*autumnal equinox; first point of Libra*
Auva 东次相，太微左垣三（室女座δ）
auxiliary circle 辅助圆
auxiliary instrument 附属仪器
auxiliary lens 辅助透镜
auxiliary level 辅助水准
auxiliary mirror 辅助反射镜
auxiliary optics 辅助光学系统
auxiliary variable 辅助变量
available electron 可用电子
available energy 可用能
available photon 可用光子
available power 可用功率 | 可得功率〔台〕
available work 可用功
avalanche photodiode 雪崩光电二极管
average brightness 平均亮度
average clock 平均钟
average coherence 平均相干
average departure 平均偏差
average distance 平均距离
average error 平均误差
average life 平均寿命　又见：*mean life; mean lifetime*
average magnitude 平均星等　又见：*mean magnitude*
average power 平均功率
average rate 平均速率　又见：*average speed*
average sample 平均取样
average speed 平均速率　又见：*average rate*
average time 平均时间，平均时刻
average variance 平均方差
average velocity 平均速度
averaging method 平均法
averted hemisphere 背面半球
averted vision 眼角余光法
aviation astronomy 航空天文学
Avicenna crater 阿维森纳环形山（月球）
Avior 海石一（船底座ε）
AVO （＝Astrophysical Virtual Observatory）天体物理虚拟天文台（欧洲）
Avogadro crater 阿伏伽德罗环形山（月球）
avoided crossing 交叉过渡区
Awakening from Hibernation 惊蛰（节气）
　又见：*Waking of Insects*

AXAF （＝Advanced X-ray Astrophysics Facility）高新 X 射线天体物理台 | 先進 X 光天文物理觀測台〔台〕（Chandra X-ray Observatory 曾用名）
axial aberration 轴向像差
axial acceleration 轴向加速度
axial defocusing 轴向离焦
axial deformation 轴向形变
axial force 轴向力
axial period 自转周期　又见：*rotational period; rotation period*
axial ratio 轴比
axial ray 近轴光线，轴上光线 | 近軸 [光] 線〔台〕
axial rotation 自转　又见：*rotation*
axial symmetry 轴对称
axial tilt 轴倾角
axial vector coupling 轴矢量耦合
Axilla 斗宿六（人马座ζ）　又见：*Ascella*
axino 轴微子
axion 轴子
axis of angular momentum 角动量轴
axis of coordinate 坐标轴　又见：*coordinate axis*
axis of figure 形状轴
axis of inertia 惯量轴
axis of reference 参考轴
axis of rotation 自转轴　又见：*rotating axis; rotation axis*
axisymmetric accretion 轴对称吸积
axisymmetric system 轴对称系统
AXP （＝anomalous X-ray pulsar）反常 X 射线脉冲星
azel 方位和高度　又见：*azimuth-elevation*
Azelfafage 螣蛇四（天鹅座π）
Azha 天苑六（波江座η）
Azimech 角宿一（室女座α）　又见：*Epi; Spica*
azimuth 方位角，地平经度　又见：*azimuthal angle*
azimuthal action 周向作用量
azimuthal angle 方位角，地平经度　又见：*azimuth*
azimuthal control 方位控制
azimuthal equidistant projection 等距方位投影
azimuthal period 周向周期
azimuthal telescope 方位仪　又见：*azimuth telescope*
azimuthal velocity dispersion 周向速度弥散度
azimuth axis 方位轴，垂直轴
azimuth circle ① 方位圈，地平经圈 | 方位圈〔台〕② 地平经圈，垂直圈 | 地平經圈〔台〕
　又见：*vertical circle*
azimuth constant 方位差　又见：*azimuth equation*

azimuth correction 方位改正 | 方位修正〔台〕
azimuth-elevation 方位和高度　又见：*azel*
azimuth equation 方位差　又见：*azimuth constant*
azimuth error 方位误差
azimuth line 方位线
azimuth mark 方位标
azimuth mounting 地平装置，经纬仪式装置 | 地平 [式] 装置〔台〕　又见：*altazimuth mounting*
azimuth quadrant 地平象限仪　又见：*quadrant altazimuth*

azimuth-range 方位和距离　又见：*azran*
azimuth shift 方位变化
azimuth star 测方位星 | [测] 方位星〔台〕
azimuth surveying 方位测量
azimuth telescope 方位仪　又见：*azimuthal telescope*
Azmidiske 弧矢增十七（船尾座 ξ）
　　又见：*Asmidiske*
azran 方位和距离　又见：*azimuth-range*
Azur 蓝天号（西德 -美国科学卫星）
Azure Dragon 苍龙（四象）　又见：*Blue Dragon*

B

B　（＝boron）硼（5号元素）
Ba　（＝barium）钡（56号元素）
BAA　（＝British Astronomical Association）英国天文协会
Baade crater　巴德环形山（月球）
Baade's star　巴德星
Baade's Window　巴德窗
Baade-Wesselink analysis　巴德－韦塞林克分析
Baade-Wesselink mass　巴德－韦塞林克质量
Baade-Wesselink method　巴德－韦塞林克方法
Baade-Wesselink radius　巴德－韦塞林克半径
Babbage crater　巴贝奇环形山（月球）
Babcock crater　巴布科克环形山（月球）
Babcock magnetograph　巴布科克磁像仪
Babinet compensator　巴俾涅补偿器｜巴比内補償器〔台〕
Babylonian calendar　巴比伦历
BAC　（＝British Association Catalogue）英国天文协会星表
Bach crater　巴赫环形山（水星）
back end　后端
back feed factor　反馈因子　又见：feedback factor
back focus　后焦点
background brightness　背景亮度
background Compton scattering　背景康普顿散射
background continuum　背景连续谱
background count　背景计数
background elimination　背景消除
background fluctuation　背景起伏
background galaxy　背景星系
background galaxy cluster　背景星系团
background glow　背景光
background intensity　背景强度
background noise　背景噪声｜背景雜訊〔台〕
background-noise level　背景噪声电平
background radiation　背景辐射
background radiation intensity　背景辐射强度
background radio radiation　背景射电
background rejection　本底抑制
background source　背景源
background source function　背景源函数
background star　背景星
background subtraction　背景减除
background temperature　背景温度
backlash　齿隙，空回｜齒隙〔台〕
backlit view　背照视像
back lobe　后瓣
Backlund crater　巴克伦德环形山（月球）
back-off　补偿
Back-Paschen effect　贝克－帕邢效应
backscatter　反向散射｜後向散射〔台〕
backscatter cross-section　反向散射截面
backscatter peak　反向散射峰
backside-illuminated CCD　背照式 CCD
backside-illuminated CMOS　背照式 CMOS
back surface　后表面（陨星）　又见：rear surface
backup structure　背架
backward difference　反向较差
backward movement　逆行　又见：retrograde motion; retrogression; antecedence
back warming　反向致热｜逆向致熱〔台〕
backwarming　返回加热
back wave　回波，反向波　又见：echo wave; return wave
Baco crater　培根环形山（月球）
Badger's equation　巴德格方程
bad pixel　坏像元
baffle　遮光罩
Baghdad Sulcus　巴格达沟（土卫二）
Baham　危宿二（飞马座θ）　又见：Biham
Ba II star　电离钡星
Baikal Deep Underwater Neutrino Telescope　（＝BDUNT）贝加尔湖深水中微子望远镜
Bailey type variable　贝利型变星

Baillaud crater 巴约环形山（月球）
Bailly crater 巴伊环形山（月球）
Baily's Beads 贝利珠 | 倍里珠〔台〕（日食）
Bajin 巴金（小行星 8315 号）
Baker-Nunn camera 贝克－纳恩照相机 | 貝克－儂人造衛星照相機〔台〕
Baker-Schmidt mirror system 贝克－施密特反射镜系统
Baker-Schmidt telescope 贝克－施密特望远镜
Baktun 白克顿（玛雅历，144000 天）
BAL （=broad absorption line）宽吸收线
balance force 平衡力
balancer 平衡器
balance wheel 摆轮
Balboa crater 巴尔沃亚环形山（月球）
Balch crater 鲍尔奇环形山（金星）
Baldet crater 巴尔代环形山（月球）
Baldet-Johnson band 巴尔德特－约翰逊谱带
Baldwin effect 鲍德温效应
Ballik-Ramsay band 巴利克－拉姆塞谱带
ballistic camera 弹道照相机
ballistic curve 弹道 又见: *ballistic trajectory*
ballistic trajectory 弹道 又见: *ballistic curve*
Ball of wool 昴宿（二十八宿） 又见: *Stopping-place*
balloon astronomy 球载天文学 | 氣球天文學〔台〕 又见: *balloon-based astronomy*
balloon-based astronomy 球载天文学 | 氣球天文學〔台〕 又见: *balloon astronomy*
balloon-based IR astronomy 球载红外天文学
balloon-based UV astronomy 球载紫外天文学
balloon-based X-ray astronomy 球载 X 射线天文学
balloon-based γ-ray astronomy 球载 γ 射线天文学
balloon-borne astronomy 球载天文学
balloon-borne instrument 球载仪器
balloon-borne IR telescope 球载红外望远镜
balloon-borne photometer 球载光度计
balloon-borne spectroscope 球载分光镜
balloon-borne telescope 球载望远镜 又见: *balloon telescope*
balloon-borne UV telescope 球载紫外望远镜
balloon-borne X-ray telescope 球载 X 射线望远镜
balloon-borne γ-ray telescope 球载 γ 射线望远镜
Balloon Observations of Millimetric Extragalactic Radiation and Geophysics telescope （=Boomerang telescope）飞镖球载望远镜
balloon photography 气球照相
balloon platform 球载平台
balloon satellite 球载卫星 又见: *satelloon*
balloon telescope 球载望远镜 又见: *balloon-borne telescope*
balloon theodolite 球载经纬仪
Balmer continuum 巴耳末连续区 | 巴耳麥連續譜區〔台〕
Balmer crater 巴尔末环形山（月球）
Balmer decrement 巴耳末减幅 | 巴耳麥減幅〔台〕
Balmer discontinuity 巴耳末跳跃，巴耳末跃变 | 巴耳麥陡變〔台〕 又见: *Balmer jump*
Balmer jump 巴耳末跳跃，巴耳末跃变 | 巴耳麥陡變〔台〕 又见: *Balmer discontinuity*
Balmer limit 巴耳末系限 | 巴耳麥系限〔台〕
Balmer line 巴耳末谱线 | 巴耳麥譜線〔台〕
Balmer progression 巴耳末渐进 | 巴耳麥漸增〔台〕
Balmer series 巴耳末线系 | 巴耳麥譜線系〔台〕
BAL quasar （=broad absorption-line quasar）宽吸收线类星体
Baltic Astronomy 《波罗的海天文学》（立陶宛期刊）
balun 平衡－不平衡变换器
Bamberga 班贝格（小行星 324 号）
Bamberg variable （=BV）班贝格变星
Banachiewicz crater 巴纳赫维奇环形山（月球）
Banach space 巴纳赫空间
banana orbit 香蕉形轨道
band 带，谱带 | [譜] 帶〔台〕
band constant 谱带常数
band envelope 谱带包络
band filter 带通滤波器 又见: *bandpass filter*
band-gap energy 带隙能
band half-width 通带半宽
band head 谱带头 | [譜] 帶頭〔台〕
band lamp 条形灯
band-limit coronagraph 带限星冕仪
band-limit mask 带限掩模
band of rotation 转动谱带
band of rotation-vibration 转振谱带
band of the Milky Way 银河带
band origin 谱带基线
bandpass 带通
bandpass filter 带通滤波器 又见: *band filter*
Bandpowers 带功率
band profile 谱带轮廓
band series 谱带系
band spectrum 带光谱

band splitting　谱带分裂
band width　带宽 | [频] 带宽 [度]〔台〕
　又见：*bandwidth*
bandwidth　带宽 | [频] 带宽 [度]〔台〕　又见：*band width*
bandwidth effect　带宽效应
Banting crater　班廷环形山（月球）
BAO　① （＝Baryon Acoustic Oscillation）重子声学振荡　② （＝Beijing Astronomical Observatory）北京天文台
bar　巴（压强单位）
Barbier crater　巴尔比耶环形山（月球）
Bardeen-Bond-Kaiser-Szalay transfer function　（＝BBKS transfer function）BBKS 转移函数
bare mass　裸质量
bare nucleon　裸核子
bare nucleus　裸核　又见：*stripped nucleus*
bare particle　裸粒子
bar instability　棒不稳定性（星系）
barium　（＝Ba）钡（56 号元素）
barium star　钡星　又见：*Ba star*
Barlow lens　巴罗透镜
bar mode　棒模式
barn　靶，靶恩 | 邦〔台〕（核反应截面单位，10^{-28} 平方米）
Barnard crater　巴纳德环形山（月球）
Barnard Gap　巴纳德环缝（土星）
Barnard satellite　巴纳德木卫
Barnard's Galaxy　巴纳德星系（NGC 6822）
Barnard's Loop　巴纳德环，巴纳德圈 | 巴納得環星雲〔台〕（Sh 2-276）　又见：*Barnard's Ring*
Barnard's Merope Nebula　巴纳德昴宿五星云
Barnard's Ring　巴纳德环，巴纳德圈 | 巴納得環星雲〔台〕（Sh 2-276）　又见：*Barnard's Loop*
Barnard's star　巴纳德星
Barnett effect　巴涅特效应 | 巴內特效應〔台〕
barometer　气压计
barometric disturbance　气压扰动
barometric pressure　气压　又见：*air pressure*
barothermograph　气压－温度记录仪
barotropic equation of state　正压状态方程
barotropic gas　正压气体
barotropic star　正压恒星
barred galaxy　有棒星系
barred spiral galaxy　（＝SB galaxy）棒旋星系，SB 型星系
barrel distortion　桶形畸变，负畸变
barrel shutter　筒状快门

barrier film　阻挡膜
barrier penetration　势垒穿透
Barringer crater　巴林杰陨星坑 | 巴林杰陨石坑〔台〕（地球）
Barrow crater　巴罗环形山（月球）
Bartels crater　巴特尔斯环形山（月球）
Bartel's diagram　巴特尔图
barycenter　质心 | 引力中心，質 [量中] 心〔台〕
　又见：*center of mass*
Barycentric Celestial Reference System　（＝BCRS）质心天球参考系（太阳系）
barycentric coordinate　质心坐标
Barycentric Coordinate Time　（＝TCB）质心坐标时
Barycentric Dynamical Time　（＝TDB）质心力学时
barye　微巴（压强单位，10^{-6} 巴）
baryogenesis　重子合成
baryon　重子
Baryon Acoustic Oscillation　（＝BAO）重子声学振荡
baryon asymmetry　重子不对称性
baryon catastrophe　重子灾难
baryon conservation　重子守恒
baryon density　重子密度
baryon density parameter　重子密度参数
baryonic dark matter　重子暗物质
baryonic matter　重子物质
baryon number　重子数
baryon oscillation　重子振荡
baryon oscillation spectroscopic survey　（＝BOSS）重子振荡光谱巡天
baryon-photon fluid　重子－光子流体
baryon-radiation coupling　重子－辐射耦合
baryon-radiation plasma　重子－辐射等离子体
baryon star　重子星
baryon-to-entropy ratio　重子－熵比
baryon-to-photon ratio　重子－光子比
baryton　介子　又见：*meson; mesotron*
barytron　重电子
basal flux　基流量
basal plane　底平面
basalt　玄武岩
basaltic achondrite　玄武岩无球粒陨石
basaltic layer　玄武岩层
baseband　基带
base field　基域

base level 背景辐射电平，宁静辐射电平
baseline 基线
baseline bootstrapping 基线自举法
baseline curvature 基线弯曲
baseline fitting 基线拟合
baseline ripple 基线波纹
baseline vector 基线矢量
basis transformation 基变换
basis vector 基矢量，基矢
Basket 箕宿（二十八宿） 又见：*Winnowing-basket; Dustpan*
basket weaving 织篮法
Ba star 钡星 又见：*barium star*
BAT （＝γ-ray Burst Alert Telescope）γ暴预警望远镜
Baten Kaitos 天仓四（鲸鱼座ζ）
batholite 岩基
batholith 岩盘
bathyseism 深源地震
Bauhinia 洋紫荆（小行星 151997 号）
Baumbach corona 鲍姆巴赫日冕
Baumbach formula 鲍姆巴赫公式
Bautz-Morgan type 鲍茨—摩根形态型
Bavaria 巴伐利亚（小行星 301 号）
Bayer constellation 拜尔星座|拜耳星座〔台〕
Bayer letter 拜尔星名字母
Bayer name 拜尔星名|拜耳星名〔台〕
Bayesian evidence （＝BE）贝叶斯证据
Bayesian information criterion （＝BIC）贝叶斯信息判据
Bayesian network 贝叶斯网 又见：*Bayes network*
Bayesian statistics 贝叶斯統計
Bayes network 贝叶斯网 又见：*Bayesian network*
Bayes theorem 贝叶斯定理
Bay of Astronauts 宇航员湾（月球）
B band B 谱带|B 波段〔台〕（夫琅和费线）
BBGKY hierarchy （＝Bogoliubov-Born-Green-Kirkwood-Yvon hierarchy）BBGKY 方程组
BBKS transfer function （＝Bardeen-Bond-Kaiser-Szalay transfer function）BBKS 转移函数
BBN （＝Big Bang nucleosynthesis）大爆炸核合成
BC （＝bolometric correction）热改正|熱 [星等] 修正〔台〕
BCA model （＝Bilderberg continuum atmosphere model）比德伯格连续大气模型
BCDG （＝blue compact dwarf galaxy）蓝致密矮星系
BCG （＝blue compact galaxy）蓝致密星系
B coefficient B 系数
BCRS （＝Barycentric Celestial Reference System）质心天球参考系（太阳系）
BD （＝Bonner Durchmusterung）【德】波恩星表，BD 星表|波昂星表，波昂星圖〔台〕
BDS （＝Burnham's General Catalogue of Double Stars）伯纳姆双星总表
BDUNT （＝Baikal Deep Underwater Neutrino Telescope）贝加尔湖深水中微子望远镜
BE （＝Bayesian evidence）贝叶斯证据
Be （＝beryllium）铍（4 号元素）
beaded structure 带珠结构
Beagle 2 Mars lander 猎兔犬 2 火星着陆器
Beal's classification 比尔分类
beam angle of scattering 散射束锥角
beam antenna 定向天线 又见：*directional antenna*
beam aperture 波束孔径
beam area 波束面积
beam broadening 波束展宽|波束變寬〔台〕
beam combiner 光束合束器
beam current convective instability 束流对流不稳定性
beam diaphragm 束光阑
beam dilution 波束稀化
beamed radiation 成束辐射
beam efficiency 波束效率
beamformer 波束形成器
beaming effect 束流效应，聚束效应（脉冲星）
beaming in AGN 活动星系核集束效应 又见：*AGN beaming*
beam instability 波束不稳定性
beam intensity 束强度
beam leakage 波束渗漏
beam-limiting aperture 束限制孔径
beam line 光束线
beam maser 束微波激射器，束脉泽
beam model 束流模型
beam optics 束光学
beam pattern 波束方向图
beam-plasma interaction 束等离子体相互作用
beam reducer 光束缩束器
beam-reducing telescope 光束缩束望远镜
beam reversal 束反向
beam solid angle 束立体角

beam splitter 分束器，光束分离器 | 光束分離器〔台〕 又见: *beamsplitter*
beamsplitter 分束器，光束分离器 | 光束分離器〔台〕 又见: *beam splitter*
beam stabilization 束稳定化
beam steering 波束控制
beam switching 波束切换
beam thermal velocity 束热速度
beam tube 束管 又见: *storage beam tube*
beam waveguide 束波导
beamwidth 波束宽度
beard 向阳彗尾
bearing mark 象限角
bearing strain 承压应变
bearing stress 支承应力
beat Cepheid 拍频造父变星 | 差频造父[型]變星〔台〕
beat effect 拍频效应
beat frequency 拍频
beat mass 拍频质量
beat period 拍频周期 | 拍合周期〔台〕
Beatrix 欣女星（小行星83号）
Beaumont crater 博蒙环形山（月球）
Beaverhead crater 比弗黑德陨星坑（地球）
Bebhionn 土卫三十七
Becklin-Neugebauer object （=BN object）BN天体，贝克林－诺伊格鲍尔天体 | BN 天體〔台〕
Becrux 十字架三（南十字座β） 又见: *Mimosa*
Bečvář crater 贝奇瓦日环形山（月球）
bediasite 贝迪阿熔融石
Be dwarf Be 矮星
Beehive Cluster 蜂巢星团（M44） 即: *Praesepe*
Beemim 九州殊口四（波江座ν） 又见: *Theemin*
Beer-Lambert law 比尔－朗伯定律
Beethoven region 贝多芬地区（水星）
beginning height 出现点高度 又见: *appearance height*
Beginning of Autumn 立秋（节气）
beginning of lunation 朔，新月 又见: *new moon*
beginning of morning twilight 晨光始 | 曙光始〔台〕
beginning of partial eclipse 初亏，偏食始 | 偏食始〔台〕
Beginning of Spring 立春（节气）
Beginning of Summer 立夏（节气）
beginning of totality 食既，全食始 又见: *total eclipse beginning*
Beginning of Winter 立冬（节气）

beginning of year 岁首
beginning point 出现点 又见: *appearance point*
Behaim crater 贝海姆环形山（月球）
Beid 九州殊口二（波江座o）
Beiguan 北京天文馆（小行星59000号）
Beijerinck crater 拜耶林克环形山（月球）
Beijing Ancient Observatory 北京古观象台
Beijingaoyun 北京奥运（小行星23408号）
Beijing Astronomical Observatory （=BAO）北京天文台
Beijingdaxue 北京大学（小行星7072号）
Beishida 北师大（小行星8050号）
Beishizhang 贝时璋（小行星31065号）
Bekenstein-Hawking entropy of black hole 贝肯斯坦－霍金黑洞熵
Belinda 天卫十四 | 天[王]衛十四〔台〕
Bel'kovich crater 别利科维奇环形山（月球）
Bellatrix 参宿五（猎户座γ）
Bell crater 贝尔环形山（月球）
Bellingshausen crater 别林斯高晋环形山（月球） 又见: *Bellinsgauzen crater*
Bellinsgauzen crater 别林斯高晋环形山（月球） 又见: *Bellingshausen crater*
Bellona 战神星（小行星28号）
Bell's theorem 贝尔定理
bell tower 钟楼
Belopol'skiy crater 别洛波利斯基环形山（月球）
Belt of Orion 猎户腰带（星组） 又见: *Orion's belt*
belt of totality 全食带 又见: *zone of totality; path of total eclipse; path of totality*
Belyaev crater 别利亚耶夫环形山（月球）
bench mark 水准点
bending instability 弯曲不稳定性
bending of light 光线弯曲
bending wave 弯曲波
Benetnash 摇光，北斗七（大熊座η） 又见: *Alkaid*
Benett comet 贝内特彗星
bent double radio source 弧形双射电源
bent-pillar mounting 弧形基架
BepiColombo 贝比科隆博水星探测计划
BeppoSAX （=Satellite per Astronomia X）【意】贝波 X 射线天文卫星
Bergedorfer Spectral Durchmusterung （=BSD）贝格多夫恒星光谱表
Bergelmir 土卫三十八
Bergman series 伯格曼线系

Berkeley-Illinois-Maryland Association array （＝BIMA array）伯克利－伊利诺斯－马里兰联合[射电]阵，BIMA 射电望远镜阵
berkelium （＝Bk）锫 | 鉳〔台〕（97 号元素）
Berkner crater 伯克纳环形山（月球）
Berlage crater 伯尔拉赫环形山（月球）
Bernoulli probability 伯努利概率 | 白努利機率〔台〕
Bernoulli's equation 伯努利方程
Berosus crater 贝罗索斯环形山（月球）
beryllium （＝Be）铍（4 号元素）
beryllium mirror 铍镜
Berzelius crater 贝尔塞柳斯环形山（月球）
Bessel crater 贝塞尔环形山（月球）
Bessel equation 贝塞尔方程 | 白塞耳方程〔台〕
Bessel Gap 贝塞尔环缝（土星）
Besselian date 贝塞尔日期 | 白塞耳日期〔台〕
Besselian day number 贝塞尔日数 | 白塞耳日數〔台〕
Besselian elements 贝塞尔根数 | 白塞耳要數，白塞耳根數〔台〕
Besselian epoch 贝塞尔历元
Besselian solar year 贝塞尔太阳年
Besselian star constant 贝塞尔恒星常数 | 白塞耳恆星常數〔台〕
Besselian star number 贝塞尔星数 | 白塞耳星數〔台〕
Besselian year 贝塞尔年 | 白塞耳年〔台〕
Be star Be 星 | Be 型星，B 型發射[譜線]星〔台〕
Bestia 土卫三十九
Beta Regio 贝塔区（金星）
Beta Taurid meteor shower 金牛座 β 流星雨 又见：Beta Taurids
Beta Taurids 金牛座 β 流星雨 又见：Beta Taurid meteor shower
betatron acceleration 电子感应加速
betatron effect 电子感应加速效应
betatron mechanism 电子感应加速机制
betatron process 电子感应加速过程
Betelgeuse 参宿四（猎户座 α）
Bethe cycle 贝蒂循环
Bethe-Weizsäcker cycle 贝蒂－魏茨泽克循环 | 貝特－魏茨澤克循環〔台〕 即：CNO cycle
Bettina 贝蒂（小行星 250 号）
Be/X-ray binary Be/X 射线双星 | Be/X 光雙星〔台〕
B galaxy B 星系
BGO scintillator 锗酸铋闪烁体

BH （＝black hole）黑洞
Bhabha crater 巴巴环形山（月球）
Bi （＝bismuth）铋（83 号元素）
Bianca 天卫八 | 天[王]衛八〔台〕
Bianchi cosmology 比安基宇宙论
Bianchi identity 比安基恒等式
Biandepei 卞德培（小行星 6742 号）
bias ①偏向 ②偏差 ③本底
bias correction 偏离改正，本底改正
biased error 系统误差 | 系統[誤]差〔台〕 又见：systematic error
biased galaxy formation 星系偏袒形成
bias field 本底场
bias parameter 偏袒参数
bias uncertainty 偏离不定度
BIB detector （＝blocked impurity band detector）阻杂带探测器
Bibliographical Star Index （＝BSI）恒星文献索引
BIC （＝Bayesian information criterion）贝叶斯信息判据
biconcave lens 双凹透镜 又见：double concave glass; double concave lens
biconvex lens 双凸透镜 又见：double convex glass; double convex lens
bicrystal 双晶体
bidimensional spectrography 二维摄谱
bidimensional spectroscopy 二维分光 | 二維分光法〔台〕 又见：two-dimensional spectroscopy
Biela crater 比拉环形山（月球）
Biela's comet 比拉彗星 | 比拉彗[星]〔台〕
Bielid meteors 比拉流星群 | 比拉流星雨〔台〕 即：Andromedids; 又见：Bielids
Bielids 比拉流星群 | 比拉流星雨〔台〕 即：Andromedids; 又见：Bielid meteors
bifid 二叉彗尾
bifurcated E-layer 分叉 E 层（电离层）
bifurcation 分歧，分支
bifurcation point 歧点
Big Bang chronology 大爆炸年代学
Big Bang cosmology 大爆炸宇宙论 | 大爆炸宇宙論，霹靂說〔台〕
Big Bang model 大爆炸模型
Big Bang nucleosynthesis （＝BBN）大爆炸核合成
Big Bang singularity 大爆炸奇点
Big Bang theory 大爆炸理论
Big Bear Solar Observatory 大熊湖太阳观测台

big bounce 大反冲
big chill 冷寂
big crunch 大挤压
Big Dipper 北斗 [七星]（星组）　又见：*Triones; Northern Dipper; Charles' Wain; Wain; Plough; Plow*
big flare 大耀斑 | 大閃焰〔台〕　又见：*major flare*
big rip 大撕裂
big splash 大溅撞
BIH （＝Bureau International de l'Heure）【法】国际时间局
Biham 危宿二（飞马座 θ）　又见：*Baham*
bilateral observation 对向观测
Bilderberg continuum atmosphere model （＝BCA model）比德伯格连续大气模型
billiard-ball collision 弹性碰撞　又见：*elastic collision*
billitonite 勿里洞玻璃陨石
BIMA array （＝Berkeley-Illinois-Maryland Association array）伯克利-伊利诺斯-马里兰联合 [射电] 阵，BIMA 射电望远镜阵
bi-Maxwellian distribution 双麦克斯韦分布
bimolecular 双分子
bimorph mirror 双压电晶片镜
binarity 成双性
binary 双星 | [物理] 雙星〔台〕　即：*DS*；又见：*binary star*
binary asteroid 双小行星　又见：*double asteroid*
binary collision 二体碰撞
binary combination head 双复合谱带头
binary flare star 耀发双星 | 閃焰雙星〔台〕
binary frequency 双星出现率
binary galaxy 双重星系　又见：*double galaxy; twin galaxy*
binary mask 二元掩模
binary millisecond pulsar 毫秒脉冲双星
binary planet 双行星　又见：*double-planet*
binary protostar 原双星　又见：*proto-binary*
binary pulsar 脉冲双星
binary quasar 双类星体　又见：*twin quasar; double quasar*
binary radio pulsar 射电脉冲双星 | 電波脈衝雙星〔台〕
binary star 双星 | [物理] 雙星〔台〕　即：*DS*；又见：*binary*
binary system 双星系统，双重星系 | 雙星系統〔台〕
binary X-ray source （＝XRB）X 射线双星 | X 光雙星〔台〕　又见：*X-ray binary*

binding energy 结合能 | 結合能，束縛能〔台〕
binding force 结合力
bineutron 中子对
binoculars 双目望远镜，双筒望远镜 | 雙筒 [望遠] 鏡〔台〕　又见：*binocular telescope*
binocular telescope 双目望远镜，双筒望远镜 | 雙筒 [望遠] 鏡〔台〕　又见：*binoculars*
binomial array 二项式天线阵
binomial coefficient 二项式系数
binomial distribution 二项式分布 | 二項 [式] 分布〔台〕
binomial probability 二项式概率
binormal 副法线
bin size 分格尺寸
bioastronomy 生物天文学
biological effect 生物效应
biosphere 生物圈　又见：*ecosphere*
BIPM （＝Bureau International des Poids et Measures）【法】国际计量局
bipolar coordinate 双极坐标
bipolar flow 偶极流
bipolar galaxy 偶极星系 | 雙極星系〔台〕
bipolar group 双极群 | 雙極 [黑子] 群〔台〕
bipolar jet 双极喷流
bipolar magnetic region （＝BMR）双极磁区
bipolar nebula 偶极星云 | 雙極星雲〔台〕
bipolar outflow 偶极外向流
bipolar planetary nebula 双极行星状星云
bipolar sunspot 双极太阳黑子 | 雙極黑子〔台〕
biprism 双棱镜
birefringence 双折射　又见：*double refraction*
birefringent filter 双折射滤光器 | 雙折射濾 [光] 鏡〔台〕
birefringent interferometer 双折射干涉仪
birefringent monochrometer 双折射单色计
birefringent prism 双折射棱镜
Birkeland crater 伯克兰环形山（月球）
Birkhoff crater 伯克霍夫环形山（月球）
Birkhoff regularization 伯克霍夫正规化
Birkhoff theorem 伯克霍夫定理
Birmingham crater 伯明翰环形山（月球）
bisection 平分
bisection error 平分误差
bisector 平分线，等分线
bismuth （＝Bi）铋（83 号元素）
bispectrum 双谱

bissextile day 闰日　又见：*leap day; epagomenal day; intercalary day*

bissextile year 闰年　又见：*leap year; embolismic year; intercalary year*

bistatic radar 双站雷达

bivariate distribution 双变量分布

bizarre variable 奇异变星

Bk （=berkelium）锫｜鉳〔台〕（97号元素）

Blaauw mechanism 布劳机制

blackbody photocell 黑体光电管

blackbody radiation 黑体辐射　又见：*black radiation*

blackbody temperature 黑体温度

black brane 黑膜

black drop 黑滴（金星凌日）

black dwarf 黑矮星

blackening 致黑｜變黑，黑化度〔台〕

Blackett crater 布莱克特环形山（月球）

Black-eye Galaxy 黑眼睛星系（NGC 4826）

black hole （=BH）黑洞

black hole accretion 黑洞吸积　又见：*accretion by black hole*

black hole binary 黑洞双星

black hole dynamics 黑洞动力学

black hole entropy 黑洞熵

black hole neighbourhood 黑洞邻域

black hole spin 黑洞自旋

blackout 中断｜中斷〔無線電通訊〕〔台〕

black radiation 黑体辐射　又见：*blackbody radiation*

Black Tortoise 玄武（四象）

Black Widow Pulsar 毒蜘蛛脉冲星（PSR 1957+20）

black widow pulsar 毒蜘蛛型脉冲星

blade shutter 叶片快门

Blanford-Payne mechanism of jet formation （=BP mechanism）BP[喷流形成]机制

Blanford-Znajek mechanism of jet formation （=BZ mechanism）BZ[喷流形成]机制

blanketing effect 覆盖效应｜覆蓋效應〔台〕

blanketing factor 覆盖因子

blank field 星系际巨洞　又见：*galaxy void*

blanking 间歇，断路

Blashko effect 布拉什科效应

blast wave 爆震波

B layer B层

blazar 耀变体｜蠍虎BL[型]類星體〔台〕

blazar-like activity 类耀活动｜耀變活動性〔台〕

blazar-like object 类耀变体

blaze angle 闪耀角｜炫耀角〔台〕

blazed grating 定向光栅｜炫耀光栅〔台〕

Blaze Star 闪耀星（T CrB）

blaze wavelength 闪耀波长｜炫耀波長〔台〕

Blazhko crater 布拉日科环形山（月球）

Blazhko effect 布拉日科效应

B-L conservation 重子轻子差守恒

blend line 混合谱线｜混合[譜]線〔台〕

BL Herculis star （=BL Her star）武仙BL型星｜武仙[座]BL型星〔台〕

BL Her star （=BL Herculis star）武仙BL型星｜武仙[座]BL型星〔台〕

blind spot 盲点

blind survey 盲巡天

b-lines b三重线

blink comparator 闪视仪，闪视比较仪｜閃視[比較]鏡〔台〕

blinking 闪视

Blinking Planetary Nebula 闪视行星状星云（NGC 6826）

blink microscope 闪视镜｜閃視[比較]鏡〔台〕

BL Lac （=BL Lacertid）蝎虎天体

BL Lacertae object 蝎虎天体　又见：*BL Lacertid; Lacertid*

BL Lacertid （=BL Lac）蝎虎天体　又见：*BL Lacertae object; Lacertid*

blob 小斑点，小云

Bloch-Siegert [frequency] shift 布洛赫－西格特频移

Bloch-Siegert effect 布洛赫－西格特效应

block adjustment 面积平差

block diagram 方框图

blocked impurity band detector （=BIB detector）阻杂带探测器

blocking filter 截止滤光片

block movement 断块运动

bloomed lens 消晕透镜

BLR （=broad-line region）宽线区

BLRG （=broad-line radio galaxy）宽线射电星系｜寬線電波星系〔台〕

blue band 蓝波段

blue branch 蓝分支

blue bump 蓝鼓包

blue clearing 蓝洁化

blue compact dwarf galaxy （=BCDG）蓝致密矮星系

blue compact galaxy （=BCG）蓝致密星系

Blue Dragon 苍龙（四象）　又见：*Azure Dragon*

blue dwarf 蓝矮星
blue edge 蓝界 | 藍限〔台〕
blue flash 蓝闪
blue galaxy 蓝星系
blue giant 蓝巨星
blue-green flame 蓝绿闪
blue halo star 蓝晕星
blue horizontal branch 蓝水平支
blue horizontal branch star 蓝水平支星
blue loop 蓝回绕
blue magnitude 蓝星等
blue moon 蓝月亮
blue object 蓝天体
Blue Planetary Nebula 蓝行星状星云（NGC 3918）
blue populous cluster 蓝族星团
blue-sensitive plate 蓝敏底片
blue shift 蓝移 | 藍[位]移〔台〕
Blue Snowball 蓝雪球（NGC 7662）
Blue Snowball Nebula 蓝雪球星云
blue star 蓝星
blue straggler 蓝离散星 | 藍掉隊星，藍脫序星〔台〕
blue supergiant 蓝超巨星
blue variable 蓝变星
B-magnitude B 星等
BM area BM 区（太阳）
B mode polarization B 模式偏振
BMR （=bipolar magnetic region）双极磁区
BN object （=Becklin-Neugebauer object）BN 天体，贝克林－诺伊格鲍尔天体 | BN 天體〔台〕
Bode's Galaxy 波德星系
Bode's law 波得定则 | 波德定律〔台〕
bodily wave 体波　又见: *body wave*
body-fixed coordinate system 地固坐标系
　又见: *earth-fixed coordinate system*
body force 彻体力
body tide 固体潮 | [物] 體潮〔台〕　又见: *solid tide*
body wave 体波　又见: *bodily wave*
Bogoliubov-Born-Green-Kirkwood-Yvon hierarchy （=BBGKY hierarchy）BBGKY 方程组
Bogomol'nyi-Prasad-Sommerfield bound （=BPS）博格莫尼－普拉萨德－萨默菲尔德限
Boguslawsky crater 博古斯瓦夫斯基环形山（月球）
Bohlin group method 波林群法
Bohr-Coster diagram 玻尔－科斯特图

Bohr crater 玻尔环形山（月球）
Bohr magneton 玻尔磁子 | 波耳磁元〔台〕
Bohr radius 玻尔半径
Bok globule 博克球状体 | 包克雲球〔台〕
bolide 火流星　又见: *fireball*
Bologna survey 博洛尼亚星表
bolograph 测辐射热仪
bolometer 测辐射热计 | 輻射熱 [測定] 計〔台〕
bolometer array 测辐射热计面阵
bolometric absolute magnitude 绝对热星等
　又见: *absolute bolometric magnitude*
bolometric albedo 热反照率
bolometric amplitude 热星等变幅 | 熱 [星等] 變幅〔台〕
bolometric correction （=BC）热改正 | 熱 [星等] 修正〔台〕
bolometric light curve 热光变曲线
bolometric luminosity 热光度
bolometric magnitude 热星等
bolometric radiation 热辐射 | 總輻射，熱輻射〔台〕　又见: *heat radiation*
bolometric temperature 热温度
Boltysh crater 波泰士陨星坑（地球）
Boltzmann brain paradox 玻尔兹曼大脑佯谬
Boltzmann constant 玻尔兹曼常量 | 波茲曼常數〔台〕
Boltzmann crater 玻尔兹曼环形山（月球）
Boltzmann distribution 玻尔兹曼分布
Boltzmann-Einstein equations 玻尔兹曼－爱因斯坦方程
Boltzmann-Einstein system of equations 玻尔兹曼－爱因斯坦方程组
Boltzmann equation 玻尔兹曼方程
Boltzmann equation of state 玻尔兹曼物态方程
Boltzmann excitation formula 玻尔兹曼激发公式 | 波茲曼激發公式〔台〕
Boltzmann factor 玻尔兹曼因子
Boltzmann formula 玻尔兹曼公式 | 波茲曼公式〔台〕
Boltzmann-Saha theory 玻尔兹曼－萨哈理论 | 波茲曼－沙哈理論〔台〕
Boltzmann statistics 玻尔兹曼统计
Bolyai crater 鲍耶环形山（月球）
Bond 娄宿（二十八宿）　又见: *Lasso*
Bond albedo 邦德反照率 | 邦德反照率，球面反照率〔台〕
Bond Gap 邦德环缝（土星）
Bondi accretion 邦迪吸积

Bonner Durchmusterung （=BD）【德】波恩星表，BD 星表 | 波昂星表，波昂星圖〔台〕
Bonnor-Ebert mass 邦纳-埃伯特质量
Bonn Telescope 波恩射电望远镜　即：*Effelsberg Telescope*; *Effelsberg Radio Telescope*
Bonpland crater 邦普朗环形山（月球）
Boo （=Boötes）牧夫座
Boole crater 布尔环形山（月球）
Boomerang Nebula 旋镖星云（ESO-172707）
Boomerang telescope （=Balloon Observations of Millimetric Extragalactic Radiation and Geophysics telescope）飞镖球载望远镜
booster 助推器，运载火箭 | 助推器〔台〕
Boötes （=Boo）牧夫座
Boötes void 牧夫巨洞
Bootids 牧夫流星群
boot strap 自举作用
bootstrap resampling 自举复采样
Borasisi 博拉西西（小行星 66652 号）
Boreal sign 北方宫（占星）
boresight 视轴，漏孔
Borman crater 博尔曼环形山（月球）
Born approximation 玻恩近似 | 波恩近似法〔台〕
Born-Oppenheimer approximation 玻恩-奥本海默近似 | 波恩-歐本海默近似法〔台〕
boron （=B）硼（5 号元素）
borosilicate glass 硅酸硼玻璃
Borrelly Comet 博雷利彗星
Bose crater 博斯环形山（月球）
Bose-Einstein distribution 玻色-爱因斯坦分布
Bose-Einstein nuclei 玻色-爱因斯坦核 | 玻司-愛因斯坦原子核〔台〕
Bose-Einstein statistics 玻色-爱因斯坦统计 | 玻司-愛因斯坦統計法〔台〕
boson 玻色子
Bosonic string 玻色弦
BOSS （=baryon oscillation spectroscopic survey）重子振荡光谱巡天
Boss General Catalogue （=GC）博斯总星表，GC 星表 | 博斯星表〔台〕
Botein 天阴四（白羊座 δ）
bottle stone 暗绿玻璃
Bottlinger diagram 玻特林格图
Bottlinger model 玻特林格模型
bottom-up galaxy formation ①自下而上星系形成②自下而上式星系形成
bottom-up scenario ①自下而上图景②自下而上式图景

Bouguer plot 布格图
bounce-back 反冲　又见：*recoil*
bounce resonance 反冲共振
bouncing model 弹跳模型
boundary layer 边界层
boundary plane 边界面
boundary temperature 边界温度
boundary value 边界值
bound-bound absorption 束缚-束缚吸收
bound-bound transition 束缚-束缚跃迁
bound electron 束缚电子
bound-free absorption 束缚-自由吸收
bound-free transition 束缚-自由跃迁
bound model 束缚模型
bound orbit 束缚轨道
Boussinesq approximation 博欣内斯克近似
Boussinesq convection 博欣内斯克反演
Boussinesq equation 博欣内斯克方程
Boussingault crater 布森戈环形山（月球）
Bouwers telescope 鲍维尔斯望远镜
Bowen compensator 鲍恩补偿器
Bowen fluorescence mechanism 鲍恩荧光机制 | 波文螢光機制〔台〕　又见：*Bowen mechanism*
Bowen image slicer 鲍恩像切分器
Bowen line 鲍恩谱线 | 鮑文譜線〔台〕
Bowen mechanism 鲍恩荧光机制 | 波文螢光機制〔台〕　又见：*Bowen fluorescence mechanism*
Bowen-Walraven image slicer 鲍恩-瓦尔拉文像切分器，B-W 像切分器
bow shock 弓形激波 | 弓形震波〔台〕
bow-shock nebula 弓形激波星云 | 弓形震波星雲〔台〕
bowtie antenna 领结天线
Boxhole crater 博克斯霍尔陨星坑（地球）
box orbit 盒轨道（恒星运动）
box photometry 方格测光法
boxy elliptical galaxy 盒型椭圆星系
Boyer-Lindquist coordinate 博耶-林德奎斯特坐标
Boyle crater 玻意耳环形山（月球）
BP mechanism （=Blanford-Payne mechanism of jet formation）BP[喷流形成]机制
BPS （=Bogomol'nyi-Prasad-Sommerfield bound）博格莫尼-普拉萨德-萨默菲尔德限
Bp star （=peculiar B star）B 型特殊星 | B 型特殊星，Bp 星〔台〕
Br （=bromine）溴（35 号元素）

Brachium 折威七（天秤座 σ，天蝎座 γ）
　　又见：*Zubenalgubi*
brachy-axis 短轴
Brackett continuum 布拉开连续区
Brackett limit 布拉开系限
Brackett series 布拉开线系 | 布拉克系〔台〕
Bragg angle 布拉格角
Bragg cell spectrometer 布拉格盒频谱仪
Bragg crater 布喇格环形山（月球）
Bragg crystal spectrometer 布拉格晶体分光计
Brahms crater 布拉姆斯环形山（水星）
braking absorption 阻尼吸收 | 制动吸收〔台〕
braking index 转慢指数，制动指数（脉冲星）
braking orbit 制动轨道 | 螺旋軌道〔台〕
braking radiation 韧致辐射，阻尼辐射 | 制動輻射〔台〕　　又见：*Bremsstrahlung*
branching network 分支网络
branching ratio 分支比（核裂变）
branch point 分支点
brane collision 膜碰撞
braneworld 膜世界
Brans-Dicke cosmology 布兰斯－迪克宇宙论 | 卜然斯－狄基宇宙論〔台〕
Brans-Dicke gravity 布兰斯－迪克引力
Brans-Dicke theory 布兰斯－迪克理论 | 卜然斯－狄基理論〔台〕
Brashear crater 布拉希尔环形山（月球）
breadth of spectral line 谱线宽度　　又见：*spectral line width*
breakdown potential 击穿电势
breakdown voltage 击穿电压
breaking point 断点
breccia 角砾岩
Bredikhin crater 布列季欣环形山（月球）
Breit-Wigner equation 布赖特－维格纳方程 | 布萊特－維格納公式〔台〕
Bremsstrahlung 韧致辐射，阻尼辐射 | 制動輻射〔台〕　　又见：*braking radiation*
Brent crater 布伦特陨星坑（地球）
Brianchon crater 布利安生环形山（月球）
brick-wall method 砖墙方法
Bridgman crater 布里奇曼环形山（月球）
bright band 亮带
bright bridge 亮桥　　又见：*light bridge*
bright component 亮子星
bright core 亮核
brightening 增亮

brightening towards the limb 临边增亮
　　又见：*limb brightening*
brightest cluster galaxy 最亮团星系
bright flocculus 亮谱斑
bright galaxy 亮星系
bright giant 亮巨星　　又见：*luminous giant*
bright grain 亮颗粒
bright hydrogen flocculus 亮氢谱斑
bright limb 亮边缘
bright line 明线
bright line spectrum 明线光谱
bright mottle 亮日芒
bright nebula 亮星云 | 光星雲〔台〕
　　又见：*luminous nebula*
bright nebulosity 亮星云状物质 | 亮星雲〔台〕
brightness 亮度
brightness coefficient 亮度系数
brightness contour 等亮度线
brightness distribution 亮度分布
brightness function 亮度函数
brightness ratio 亮度比
brightness temperature 亮温度 | 亮度溫度〔台〕
bright point 亮点　　又见：*bright spot*
bright rays 亮纹
bright rim structure 亮环结构　　又见：*elephant trunk*
bright ring 亮环
bright spot 亮点　　又见：*bright point*
bright star 亮星
BRight Target Explorer - Constellation （＝BRITE Constellation）亮目标探测器卫星网
brilliance 辉度　　又见：*brilliancy*
brilliancy 辉度　　又见：*brilliance*
Brillouin scattering 布里渊散射 | 布里元散射〔台〕
Brillouin spectrum 布里渊谱
Brillouin zone 布里渊区 | 布里元區〔台〕
BRITE Constellation （＝BRight Target Explorer - Constellation）亮目标探测器卫星网
British Association Catalogue （＝BAC）英国天文协会星表
British Astronomical Association （＝BAA）英国天文协会
broad absorption line （＝BAL）宽吸收线
broad absorption-line quasar （＝BAL quasar）宽吸收线类星体
broadband continuum 宽带连续区
broadband imaging 宽带成像
broadband photometry 宽带测光

broadband spectral measurement 宽带分光测量
broadband width 宽带宽度
broadcast ephemeris 广播星历表
broad emission line 宽发射线
broadening 致宽，展宽
broadening by damping 阻尼致宽　又见：*damping broadening*
broad halo 延伸晕
broad-line radio galaxy （=BLRG）宽线射电星系｜寬線電波星系〔台〕
broad-line region （=BLR）宽线区
broad resonance 宽共振
broadside array 垂射天线阵
Brocchi's Cluster 布罗基星团
broken-beam technique 截止束技术
broken transit 折轴中星仪　又见：*broken transit instrument*
broken transit instrument 折轴中星仪　又见：*broken transit*
bromine （=Br）溴（35 号元素）
Bronk crater 布朗克环形山（月球）
Brontë crater 勃朗特环形山（水星）
bronzite 古铜辉石
Brooks comet 布鲁克斯彗星
broom star 帚星
Brouwer crater 布劳威尔环形山（月球）
brown dwarf 褐矮星｜棕矮星〔台〕
Brown's lunar theory 布朗月离理论
Brown-Twiss interferometer 布朗—特威斯干涉仪
Brucia 布鲁斯（小行星 323 号）
Brunner crater 布隆内尔环形山（月球）
Bruns theorem 勃隆斯定理
BS （=The Bright Star Catalogue）亮星星表
BSD （=Bergedorfer Spectral Durchmusterung）贝格多夫恒星光谱表
BSI （=Bibliographical Star Index）恒星文献索引
B star B 型星　又见：*B-type star*
B subdwarf B 型亚矮星｜B 型次矮星〔台〕
B supergiant B 型超巨星
BTA （=Large Altazimuth Telescope）大型地平装置望远镜（俄罗斯 6 米望远镜）
B-type asteroid B 型小行星
B-type star B 型星　又见：*B star*
bubble model of reionization 再电离泡模型
Bubble Nebula 气泡星云（NGC 7635）
bubble nucleation 泡核化
bubble sextant 气泡六分仪
bubble tube 水准管　又见：*bubble vial*
bubble vial 水准管　又见：*bubble tube*
Buch crater 布赫环形山（月球）
buckling instability 翘曲不稳定性
Buffon crater 布丰环形山（月球）
Bug Nebula 小虫星云（NGC 6302）
build-up effect 积累效应
Buisson crater 比松环形山（月球）
bulge 核球　又见：*nuclear bulge*
bulge X-ray source 核球 X 射线源｜核球 X 光源〔台〕
bulk flow 体流
bulk motion 体运动
bulk property 体特性
bulk viscosity 体黏滞
Bullet Cluster 子弹星系团（1E 0657-56）
Bulletin of The Astronomical Society of India 《印度天文学会通报》（印度期刊）
bump Cepheid 驼峰造父变星
Bunch-Davies vacuum 邦奇—戴维斯真空
bunch of particles 粒子束　又见：*particle beam*
Bunda 天垒城一（宝瓶座 ξ）
bundle frame 标架丛
Buneman instability 布尼曼不稳定性
Bunsen crater 本生环形山（月球）
Bunsen photometer 本生光度计
Burckhardt crater 布尔克哈特环形山（月球）
Bureau International de l'Heure （=BIH）【法】国际时间局
Bureau International des Poids et Measures （=BIPM）【法】国际计量局
Bureau of Astronomy 钦天监，司天监｜欽天監〔台〕　又见：*Astronomical Bureau*
Burgers equation 伯格斯方程
buried-channel CCD 埋入沟道型 CCD
Burnham's General Catalogue of Double Stars （=BDS）伯纳姆双星总表
Burnham's Nebula 伯纳姆星云（Sh2-238）
Burns Cliff 伯恩斯峭壁（火星）
burst 暴｜爆發〔台〕
burster 暴源｜爆發源〔台〕　又见：*burst source*
bursting pulsar 暴态脉冲星（GRO J1744-28）
burst noise 暴噪
burst source 暴源｜爆發源〔台〕　又见：*burster*
Büsching crater 比兴环形山（月球）
Butcher-Oemler effect 布彻—厄姆勒效应
Butler matrix 巴特勒矩阵

butt 触
Butterfly Cluster 蝴蝶星团
butterfly diagram 蝴蝶图
Butterfly Nebula 蝴蝶星云（IC 2220） 又见：*Toby Jug Nebula*
Buys-Ballot crater 白贝罗环形山（月球）
BV （＝Bamberg variable）班贝格变星
B-V color index B－V 色指数
BV photometry BV 测光
Bw star Bw 型星，B 型弱氦线星

BY Draconis star 天龙座 BY 型星 即：*BY Draconis variable*
BY Draconis variable 天龙 BY 型变星 | 天龍[座]BY 型[變]星〔台〕
Byrd crater 伯德环形山（月球）
Byurakan Astrophysical Observatory 布拉堪天文台
Byzantine era 拜占庭纪年
BZ mechanism （＝Blanford-Znajek mechanism of jet formation）BZ[喷流形成] 机制

C

C ① （=Cambridge Catalogue of Radio Sources）剑桥射电源表｜劍橋電波源表〔台〕
② （=carbon）碳（6 号元素）
CA （=center of activity）活动中心
Ca （=calcium）钙（20 号元素）
Cabannes crater 卡巴纳环形山（月球）
Cabibbo angle 卡比博角
cadmium （=Cd）镉（48 号元素）
cadmium sulfide 硫化镉
Cae （=Caelum）雕具座
Caelum （=Cae）雕具座
caesium （=Cs）铯（55 号元素）
caesium antimonide 含锑铯
caesium atomic beam 铯原子束
caesium clock 铯钟｜鉙 [原子] 鐘〔台〕
又见：*cesium clock; cesium-beam clock*
caesium cycle 铯周期
caesium resonator 铯共振器
cage 观测笼
CAHA （=Centro Astronómico Hispano-Alemán）【西】德国－西班牙天文中心 即：*Calar Alto Observatory*
CaII emission line CaII 发射线，电离钙发射线
Cailleux crater 卡耶环形山（月球）
Cajori crater 卡约里环形山（月球）
Calabash Nebula 葫芦星云（OH 231.84 +4.22）
即：*Rotten Egg Nebula*
Calabi-Yau manifold 卡拉比－丘流形
Calabi-Yau spaces 卡拉比－丘空间
Calar Alto Observatory 卡拉阿托天文台（西班牙）
calbtun 卡勃顿（玛雅历日）
calcite 方解石
calcium （=Ca）钙（20 号元素）
calcium cloud 钙云
calcium flocculus 钙谱斑 又见：*calcium plage*
calcium line 钙线

calcium network 钙网络
calcium parallax 钙吸收视差
calcium plage 钙谱斑 又见：*calcium flocculus*
calcium prominence 钙日珥
calcium star 钙星
caldera 火山喷口
calendar clock 历钟
calendar date 历日期
calendar day 历日
calendarist 历算家
calendar month 历月
calendar reform 改历
calendar stone 历石
calendar year 历年
Calendrical Chapter 历志
Caliban 天卫十六
calibrating receiver 校准接收机
calibrating source 校准发生器
calibration 校准，定标｜定標〔台〕
calibration curve 定标曲线
calibration procedure 定标程序
calibration source 定标源
calibration spectrum 定标谱
calibration star 定标星
calibration system 定标系统
calibrator 校准器
California Extremely Large Telescope （=CELT）加州特大望远镜
California Nebula 加利福尼亚星云（IC 1499）
californium （=Cf）锎｜鉲〔台〕（98 号元素）
Callipic cycle 卡利普周（4 个默冬周或 76 年）
Callirrhoe 木卫十七
Callisto 木卫四｜木衛四，卡利斯多〔台〕
calm day 宁静日 又见：*quiet day*
calorimetric spectrometer 量能器能谱仪
calorimetry 量热学

Caloris Basin　卡路里盆地（水星）
Caltech Infrared Catalogue　（＝IRC）加州理工学院红外源表
Caltech Submillimeter Observatory　（＝CSO）加州理工学院亚毫米波天文台
Calypso　土卫十四
Cam　（＝Camelopardalis）鹿豹座
cam　凸轮
Cambridge Anisotropy Telescope　（＝CAT）剑桥各向异性望远镜
Cambridge Catalogue of Radio Sources　（＝C）剑桥射电源表 | 劍橋電波源表〔台〕
Cambridge Low-Frequency Synthesis Telescope　（＝CLFST）剑桥低频综合孔径望远镜 | 劍橋低頻孔徑綜合望遠鏡〔台〕
Cambridge Optical Aperture Synthesis Telescope　（＝COAST）剑桥光学综合孔径望远镜 | 劍橋光學孔徑合成望遠鏡〔台〕
Cambridge pulsar　（＝CP）剑桥脉冲星
CAMC　（＝Carlsberg Automatic Meridian Circle）卡尔斯伯格自动子午环
Camelopardalis　（＝Cam）鹿豹座
camera lens　照相透镜
Camilla　驭神星（小行星 107 号）
Campbell crater　坎贝尔环形山（月球）
Canada-France-Hawaii Telescope　（＝CFHT）加拿大－法国－夏威夷望远镜，CFH 望远镜 | 加法夏望遠鏡〔台〕
Canada-France redshift survey　（＝CFRS）加拿大－法国红移巡天，CFRS 巡天
Canadian Space Agency　（＝CSA）加拿大宇航局
Canadian Virtual Observatory　（＝CVO）加拿大虚拟天文台
canal of Mars　火星运河 | [火星] 運河〔台〕
cancellation　对消
Cancer　①（＝Cnc）巨蟹座 ②（＝Cnc）巨蟹宫，鹑首，未宫
candle-flame star　烛星
candle power　烛光
candoluminescence　烛发光率
Candor Chasma　堪德深谷（火星）
Canes Venatici　（＝CVn）猎犬座
Canicula　天狼 [星]（大犬座 α）　又见：*Sirius; Aschere; Dog Star*
canicular days　偕日升前后
Canis Major　（＝CMa）大犬座
Canis Major Dwarf Galaxy　大犬矮星系
Canis Minor　（＝CMi）小犬座

cannibalism　吞食
cannibalizing of galaxies　星系吞食　又见：*galactic cannibalism*
Cannizzaro crater　坎尼扎罗环形山（月球）
Cannonball　炮弹号（美国科学卫星）
Cannon crater　坎农环形山（月球）
Canon der Finsternisse　【德】《日月食典》
canonical assemblage　正则系综　又见：*canonical ensemble*
canonical Big Bang　典型大爆炸
canonical change　正则变化
canonical commutation relation　正则置换关系
canonical conjugate　正则共轭
canonical constants　正则常数
canonical coordinate　正则坐标
canonical distribution　正则分布
canonical elements　正则根数 | 正則要素〔台〕
canonical ensemble　正则系综　又见：*canonical assemblage*
canonical extension　正则扩充
canonical map　正则映射
canonical model　正则模型
canonical momentum　正则动量
canonical nonthermal source　典型非热源
canonical time unit　正则时间单位
canonical transformation　正则变换
Canopus　老人星（船底座 α）　又见：*Suhail*
cantaloupe terrain　甜瓜形地表
Cantor crater　康托尔环形山（月球）
canyon　深峡谷
Canyon Diablo meteorite　代阿布洛峡谷陨星
Cap　①（＝Capricornus）摩羯座 ②（＝Capricornus）摩羯宫，星纪，丑宫
Cape Canaveral　卡纳维拉尔角（美国地名）
Capella　五车二（御夫座 α）
Cape Photographic Atlas　好望角照相星图
Cape Photographic Catalogue　（＝CPC）好望角照相星表
Cape Photographic Durchmusterung　①（＝CPD）【德】好望角照相巡天 ② 好望角照相星表
Cape photometry　好望角测光
Cape RI photometry　好望角 RI 测光
Cape St Vincent　圣文森特角（火星）
Cape York meteorite　约克角陨石
Caph　王良一（仙后座 β）　又见：*Chaf; Kaff*
capping shutter　叠合快门
cap prominence　冠状日珥

Capricornid meteor shower 摩羯流星雨
Capricornids 摩羯流星群
Capricornus ①（=Cap）摩羯座 ②（=Cap）摩羯宫，星纪，丑宫
capsule 密封舱 又见：*airlock module; airlock*
capture 俘获｜捕獲〔台〕 又见：*entrapment*
capture cross-section 俘获截面
captured rotation 受俘自转，同步自转
capture event 俘获事件
capture hypothesis 俘获假说｜捕獲假說〔台〕
capture theory 俘获理论
capture time 俘获时间
Caput Trianguli 娄宿增六（三角座 α）
Car （=Carina）船底座
Carafe Galaxy 水瓶星系（PGC 15172）
carbon （=C）碳（6 号元素）
carbonaceous asteroid 碳质小行星
carbonaceous chondrite 碳粒陨星｜碳粒隕石〔台〕
carbon branch 碳分支
carbon burning 碳燃烧
carbon-carbon bond 碳一碳键
carbon cloud 碳云
carbon cycle 碳循环
carbon-deflagration model 碳暴燃模型
carbon detonation 碳爆轰｜碳引爆〔台〕
carbon dioxide 二氧化碳
carbon dwarf star 碳矮星 又见：*dwarf carbon star*
carbon fiber reinforced polymer mirror （=CFRP mirror）碳纤维复合材料镜
carbon flash 碳闪
carbon line 碳谱线
carbon monosulfide 一硫化碳（CS）
carbon monoxide 一氧化碳（CO）
carbon-nitrogen cycle （=CN cycle）碳氮循环｜碳氮循環，CN 循環〔台〕
carbon-nitrogen-oxygen cycle （=CNO cycle）碳氮氧循环
carbon-poor star 贫碳星
carbon-rich planet 富碳行星
carbon-rich star 富碳星
carbon sequence 碳序｜碳星序〔台〕
carbon star （=C star）碳星
carbonyl sulfide 氧硫化碳（COS）
Car-Cyg arm （=Carina-Cygnus Arm）船底一天鹅臂
card catalogue 卡片型星表
cardinal direction 基向

cardinal point 四方点，基点｜基點，基本方位〔台〕
cardinal signs 黄道带主宫
cargo bay 货舱
Carina （=Car）船底座
Carina arm 船底臂
Carina-Cygnus Arm （=Car-Cyg arm）船底一天鹅臂
Carina Dwarf Galaxy 船底座矮星系
Carina Nebula 船底星云（NGC 3372） 又见：*Eta Carina Nebula*
Carina OB 2 船底 OB 2 星协
Carina-Sagittarius arm 船底一人马臂
Carina system 船底星系
Carlsberg Automatic Meridian Circle （=CAMC）卡尔斯伯格自动子午环
CARMA （=Combined Array for Research in Millimeter-wave Astronomy）CARMA 毫米波组合阵（美国）
Carme 木卫十一｜木衛十一，卡米〔台〕
Carnegie Observatories 卡耐基天文台
Carnot crater 卡诺环形山（月球）
Carnot cycle 卡诺循环
Carpenter crater 卡彭特环形山（月球）
Carpo 木卫四十六
carrier frequency 载频
carrier signal 载波信号
carrier-to-noise ratio 载波噪声比
carrier vehicle 运载飞行器｜運載飛行器，運載火箭〔台〕
carrier wave 载波
Carrington coordinate 卡林顿坐标｜卡林吞坐標〔台〕
Carrington longitude 卡林顿经度｜卡林吞經度〔台〕
Carrington meridian 卡林顿子午线｜卡林吞子午圈〔台〕
Carrington rotation number 卡林顿自转序号｜卡林吞自轉序〔台〕
Carswell crater 卡斯韦尔陨星坑（地球）
Carte du Ciel （=CdC）【法】照相天图｜照相天圖星表〔台〕
Carter theorem 卡特定理
Cartesian coordinate 笛卡儿坐标
carte synoptique 日面综合图
cartographic projection 制图式投影
Cartwheel galaxy 车轮星系（A0035-335）
Carver crater 卡弗环形山（月球）

CAS （＝Chinese Astronomical Society）中国天文学会
Cas （＝Cassiopeia）仙后座
cascaded image converter 级联像转换器
cascade transition 级联跃迁
Casimir effect 卡西米尔效应
Cassegrain antenna 卡塞格林天线
Cassegrain beam 卡塞格林射束
Cassegrain configuration 卡塞格林系统
Cassegrain crater 卡塞格林环形山（月球）
Cassegrain focus 卡氏焦点，卡塞格林焦点｜卡塞格林焦點〔台〕
Cassegrain focus corrector 卡焦改正镜
Cassegrain reflector 卡氏反射望远镜｜卡塞格林[式]反射望遠鏡〔台〕
Cassegrain spectrograph 卡塞格林摄谱仪，卡焦摄谱仪｜卡塞格林焦攝譜儀〔台〕
Cassegrain telescope 卡塞格林望远镜｜卡塞格林[式]望遠鏡〔台〕　即：Cassegrain reflector
Cassini 卡西尼号土星探测器｜卡西尼號[土星探測器]〔台〕
Cassini crater 卡西尼环形山（月球）
Cassini division 卡西尼环缝（土星）
Cassini-Huygens spacecraft 卡西尼－惠更斯号土星探测器
Cassini Regio 卡西尼区（土卫八）
Cassini's law 卡西尼定律（月球运动）
Cassini spacecraft 卡西尼号土星探测器
Cassini state 卡西尼态
Cassiopeia （＝Cas）仙后座
Cassiopeia Nebula 仙后星云
Cassiopeids 仙后流星群｜仙后[座]流星雨〔台〕
Castor 北河二（双子座α）　又见：Apollo
Castula 造父五（仙王座ν）
CAT ①（＝Cambridge Anisotropy Telescope）剑桥各向异性望远镜 ②（＝Cosmic Anisotropy Telescope）宇宙各向异性望远镜
cataclysm 激变
cataclysmic binary 激变双星｜激變[雙]星〔台〕
cataclysmic event 激变事件
cataclysmic explosion 激变爆发
cataclysmic variable 激变变星｜激變[變]星〔台〕
catadioptric objective 折反射物镜
catadioptric system 折反射系统
catadioptric telescope 折反射望远镜
　又见：reflector-corrector
catalog place 星表位置
catalogue astronomy 星表天文学

catalogue equinox 星表分点
Catalogue for Stellar Identification （＝CSI）恒星证认表
catalogue number 星表编号
Catalogue of Bright Stars 亮星星表　又见：The Bright Star Catalogue
Catalogue of Cometary Orbits 彗星轨道表
Catalogue of Faint Stars （＝KSZ）暗星星表，暗星表｜暗星表〔台〕
Catalogue of Galaxies and Clusters of Galaxies （＝CGCG）星系和星系团表
Catalogue of Geodetical Stars 测地星表
Catalogue of Nearby Stars 近星星表
Catalogue of QSO and Active Nuclei 类星体和活动星系核表
catalogue of stars 星表　又见：star catalogue
Catalogue of Time Services （＝CTS）授时星表
catalogue parameter 星表参数
catarinite 镍铁陨星｜鎳鐵隕石〔台〕
catastrophe 灾变
catastrophe theory 突变理论
catastrophic collision 灾变碰撞
catastrophic event 灾变事件
catastrophic explosion 灾变爆发
catastrophic hypothesis 灾变假说，灾变说｜災變說〔台〕
catastrophic theory 灾变理论
catastrophic variable 灾变变星
catena 环形山串　又见：crater chain
Catena Abulfeda 【拉】艾布·菲达链（月球）
Catena Artamonov 【拉】阿尔塔莫诺夫坑链（月球）
Catena Davy 【拉】戴维坑链（月球）
Catena Dziewulski 【拉】杰武尔斯基坑链（月球）
Catena Humboldt 【拉】洪堡坑链（月球）
Catena Krafft 【拉】克拉夫特坑链（月球）
Catena Kurchatov 【拉】库尔恰托夫坑链（月球）
Catena Lucretius 【拉】卢克莱修坑链（月球）
Catena Mendeleev 【拉】门捷列夫坑链（月球）
Catena Sumner 【拉】萨姆纳坑链（月球）
Catena Sylvester 【拉】西尔维斯特坑链（月球）
cat-eye interferometer 猫眼干涉仪
cathodoluminescence 阴极发光
Cathysia 华夏古陆
cation 阳离子
catoptrics 反射光学

catoptric system 反射系统
catoptric telescope 反射望远镜　又见：*reflecting telescope; mirror telescope; reflector*
Cat's Eye Nebula 猫眼星云（NGC 6543）
Cat's Paw Nebula 猫爪星云（NGC 6334）
Cauchy crater 柯西环形山（月球）
Cauchy distribution 柯西分布
Cauchy horizon 柯西视界
Cauchy profile 柯西轮廓
Cauchy's dispersion formula 柯西色散公式
Cauchy sequence 柯西序列
causality 因果律　又见：*law of causation*
caustics 焦散线
Cavendish crater 卡文迪什环形山（月球）
Cave Nebula 洞穴星云
cavity 空腔
cavity clocking 空腔计时
cavity coupling 腔体耦合
cavity pulling 腔牵引
cavus 深坑
Cayley formation 凯利形成
C band C 谱带 | C 波段〔台〕（4-8 GHz）
CBI （＝Cosmic Background Imager）宇宙背景成像仪
CCD （＝charge-coupled device）电荷耦合器件 | 電荷耦合元件〔台〕
CCD array CCD 阵
CCD astrometry CCD 天体测量
CCD astronomy CCD 天文学
CCD camera CCD 照相机 | CCD 相機, 電荷耦合元件相機〔台〕
CCD detector CCD 探测器
CCD fringing CCD 条纹
CCD imaging CCD 成像
CCD meridian circle CCD 子午环
CCD mosaic CCD 拼接
CCD observation CCD 观测
CCD photometer CCD 光度计
CCD photometry CCD 测光
CCD polarimeter CCD 偏振计
CCD radiation damage CCD 辐射损伤
CCDS （＝Comité Consultatif pour la Definition de la Seconde）【法】秒定义咨询委员会
CCD spectrograph CCD 摄谱仪
CCD spectrometer CCD 分光计
CCD spectroscopy CCD 分光
CCD spectrum CCD 光谱

CCIR （＝International Radio Consultative Committee）国际电信咨询委员会
CCRS （＝conventional celestial reference system）习用天球参考系
CCTF （＝Comité Consultatif du Temps et des Fréquences）【法】时间频率咨询委员会
CD （＝Cordoba Durchmusterung）【德】科尔多瓦巡天星表
Cd （＝cadmium）镉（48 号元素）
CdC （＝Carte du Ciel）【法】照相天图 | 照相天圖星表〔台〕
CDF （＝Chandra Deep Field）钱德拉深空区
CDFN （＝Chandra Deep Field North）钱德拉北深空区
CDFS （＝Chandra Deep Field South）钱德拉南深空区
cD galaxy （＝central-Dominated galaxy）中央主导星系, cD 星系 | cD 星系〔台〕
CDM （＝cold dark matter）冷暗物质
CDM model （＝cold dark matter model）冷暗物质模型
CDMS （＝Cryogenic Dark Matter Search）暗物质低温搜寻计划
CDQ （＝core-dominated quasar）核主导类星体
CD-ROM （＝compact disc read-only memory）只读光盘
CDS ① （＝Centre de Donnees Stellaires）【法】恒星数据中心 ② （＝Centre de Données astronomiques de Strasbourg）【法】斯特拉斯堡天文数据中心
CdTe detector array 碲化镉面阵（探测器）
CDW （＝centrifugally driven wind）离心力驱星风
CdZnTe detector array （＝CZT array）碲锌镉面阵（探测器）
CE （＝color excess）色余
Ce （＝cerium）铈（58 号元素）
Celaeno 昴宿增十六（金牛座 16）
Celb-al-Rai 宗正一（蛇夫座 β）　又见：*Celbalrai; Cheleb*
Celbalrai 宗正一（蛇夫座 β）　又见：*Celb-al-Rai; Cheleb*
Celescope 天空巡视仪
Celescope Catalogue of UV Magnitude 天空巡视紫外星等表
celestial axis 天轴
celestial baton 天杵（星官）
celestial body 天体　又见：*aster; celestial object; orb*
celestial chart 天图　又见：*sky atlas; sky map; atlas*

celestial clock 天体钟
celestial coordinate 天球坐标
celestial coordinate system 天球坐标系
又见: *celestial system*
celestial cryptography 天体密码学
celestial dog 天狗
Celestial Ephemeris Origin （＝CEO）天球历书零点
celestial ephemeris pole （＝CEP）天球历书极
celestial equator 天赤道 | 天[球]赤道〔台〕
又见: *equinoctial*
celestial equator system of coordinate 天赤道坐标系
celestial flail 天棓（星官）
celestial globe 天球仪，浑象 | 天球仪〔台〕
又见: *celestial sphere*
celestial guidance 天文导航 | 太空导航〔台〕
又见: *astronavigation; celestial navigation*
celestial horizon 天球地平
celestial inertial guidance 天文惯性导航
Celestial Intermediate Origin （＝CIO）天球中间零点
Celestial Intermediate Pole （＝CIP）天球中间极
Celestial Intermediate Reference System （＝CIRS）天球中间参考系
celestial lance 天枪（星官）
celestial latitude 黄纬 又见: *ecliptic latitude*
celestial longitude 黄经 又见: *ecliptic longitude*
Celestial Market Enclosure 天市垣
celestial maser 天体微波激射，天体脉泽
celestial matter 宇宙物质
celestial mechanician 天体力学家
celestial mechanics 天体力学 又见: *astromechanics*
Celestial Mechanics & Dynamical Astronomy 《天体力学和动力天文学》（荷兰期刊）
celestial meridian 天球子午圈 又见: *principal vertical circle*
celestial navigation 天文导航 | 太空导航〔台〕
又见: *astronavigation; celestial guidance*
celestial object 天体 又见: *aster; celestial body; orb*
Celestial Observation Satellite （＝COS）COS天文卫星
celestial parallel 天球纬圈
celestial perimeter parts 周天分
celestial photograph 天体照片
celestial photography 天体照相 | 天體照相學〔台〕
celestial planisphere 平面星图
celestial polar distance 天极距

celestial pole 天极
celestial pole offsets 天极偏差
Celestial Reference System （＝CRS）天球参考系
celestial revolution 天周
celestial source 宇宙源 又见: *cosmic source*
celestial spear 天锋
celestial sphere ① 天球仪，浑象 | 天球仪〔台〕
又见: *celestial globe* ② 天球 又见: *coelosphere*
celestial stem 天干 又见: *heavenly stem*
celestial system 天球坐标系 又见: *celestial coordinate system*
celestial thermal background 天空热背景
celestial UV telescope 紫外空间望远镜
celestial X-ray source 宇宙X射线源 又见: *cosmic X-ray source*
celestial γ-ray source 宇宙γ射线源 又见: *cosmic γ-ray source*
cell field model 蜂窝场模型
cell summing 网格和
cellular structure 蜂窝状结构
cell variances 单格方差
CELT （＝California Extremely Large Telescope）加州特大望远镜
Cen （＝Centaurus）半人马座
Cen A （＝Centaurus A）半人马射电源A
Cen B （＝Centaurus B）半人马射电源B
Centaur ① 半人马型小行星 ② 半人马座火箭
Centaur group 半人马群
Centaurus （＝Cen）半人马座
Centaurus A （＝Cen A）半人马射电源A
Centaurus arm 半人马臂
Centaurus B （＝Cen B）半人马射电源B
Centaurus cluster 半人马星系团
centennial variation 百年变化
Center for High Angular Resolution Astronomy （＝CHARA）高角分辨率天文中心
center-limb variation 中心－边缘变化
center of activity （＝CA）活动中心 又见: *active center*
center of curvature 曲率中心
center of gravity 重心 又见: *gravity center*
center of gyration 回转中心
center of inertia 惯性中心
Center of land 地中
center of light 光心 又见: *optical center; photocenter*
center of mass 质心 | 引力中心，質[量中]心〔台〕
又见: *barycenter*

center of mass angle 质心角
center of mass frame 质心系
center of oscillation 振动中心
center of rotation 转动中心
centigrade scale 百分度
centigrade system 百分度制 | 百分 [度] 制〔台〕
centimeter-excess object 厘米波超天体
centimeter wave 厘米波
centimetric emission 厘米波辐射　又见: *centimetric radiation*
centimetric radiation 厘米波辐射
　　又见: *centimetric emission*
central angle 中心角
central axis 中心轴
central blockage 中心遮拦
Central Bureau for Astronomical Telegrams 天文电报中央局
central concentration 中心聚度 | 中心密集度, 中聚度〔台〕
central condensation 中心凝聚物 | 中心凝聚 [物], 中央密集體〔台〕
central configuration 中心构形　又见: *central figure*
central core 中心核
central date 中央日期
central-Dominated galaxy （＝cD galaxy）中央主导星系, cD 星系 | cD 星系〔台〕
central eclipse 中心食
central European time （＝CET）欧洲中部时间
central figure 中心构形　又见: *central configuration*
central force 有心力
central image 中心像
central intensity 线心强度
central line 中心线
central lobe 中心瓣
centrally directed field of force 中心力场
central meridian （＝CM）中央子午线, 日心子午线 | 中央子午圈〔台〕
central mountain 中央峰（环形山）　又见: *central peak*
central overlap technique 中心重叠法
central peak 中央峰（环形山）　又见: *central mountain*
central standard time （＝CST）美国中部标准时
central star 中央星
central temperature 中心温度
central time （＝CT）中部时
Centre de Données astronomiques de Strasbourg （＝CDS）【法】斯特拉斯堡天文数据中心

Centre de Donnees Stellaires （＝CDS）【法】恒星数据中心
Centre National de la Recherche Scientifique （＝CNRS）【法】国家科学研究中心（法国）
Centre National d'Etudes Spatiales （＝CNES）【法】法国国家空间研究中心
centrifugal acceleration 离心加速度
centrifugal force 离心力
centrifugally driven wind （＝CDW）离心力驱星风
centring error 对心误差
centripetal acceleration 向心加速度
centripetal force 向心力
Centro Astronómico Hispano-Alemán （＝CAHA）【西】德国—西班牙天文中心
　　即: *Calar Alto Observatory*
centroid of stars 恒星群形心
centrosphere 地心圈
centurial year 世纪年
century 世纪
CEO （＝Celestial Ephemeris Origin）天球历书零点
CEP （＝celestial ephemeris pole）天球历书极
Cep （＝Cepheus）仙王座
Cepheid ①造父一（仙王座δ）②造父变星
　　又见: *Cepheid variable*
Cepheid distance 造父距离
Cepheid distance scale 造父距离尺标
Cepheid instability strip 造父变星不稳定带
Cepheid parallax 造父视差
Cepheids 仙王流星群 | 仙王 [座] 流星群〔台〕
Cepheid variable 造父变星　又见: *Cepheid*
Cepheus （＝Cep）仙王座
Ceraski crater 采拉斯基环形山（月球）
　　又见: *Tseraskiy crater*
Cerberus 地狱犬座（现已不用）
Cerenkov counter 切伦科夫计数器 | 契忍可夫计數器〔台〕
Cerenkov effect 切伦科夫效应
Cerenkov radiation 切伦科夫辐射 | 契忍可夫辐射〔台〕
Ceres 谷神星（小行星 1 号, 矮行星）
CERI （＝connected-element radio interferometry）联线射电干涉测量
cerium （＝Ce）铈（58 号元素）
CERN （＝European Centre for Nuclear Research）欧洲核子研究中心

Cerro Tololo Inter-American Observatory（＝CTIO）托洛洛山美洲天文台
certain event 确定事件
cervit 微晶玻璃　又见：*zerodur; glass-ceramic; sytall*
cesium-beam clock 铯钟 | 銫 [原子] 鐘〔台〕
　　又见：*cesium clock; caesium clock*
cesium-beam resonator 铯束共振器
cesium-beam tube 铯束管
cesium clock 铯钟 | 銫 [原子] 鐘〔台〕
　　又见：*caesium clock; cesium-beam clock*
cesium frequency standard 铯频标
cesium iodide 碘化铯
cesium resonanc frequency 铯共振频率
cesium transition frequency 铯跃迁频率
CET （＝central European time）欧洲中部时间
Cet （＝Cetus）鲸鱼座
CETI （＝communication with extra-terrestrial intelligence）地外智能生物通信 | 地 [球] 外智慧生物通訊〔台〕
Cetids 鲸鱼流星群 | 鯨魚 [座] 流星雨〔台〕
Cetus （＝Cet）鲸鱼座
Cetus Arc 鲸鱼弧形星云
Cf （＝californium）锎 | 鉲〔台〕（98 号元素）
CfA （＝Harvard-Smithsonian Center for Astrophysics）哈佛史密松天体物理中心
CfA Redshift Survey CfA 红移巡天
CFHT （＝Canada-France-Hawaii Telescope）加拿大－法国－夏威夷望远镜，CFH 望远镜 | 加法夏望遠鏡〔台〕
C-field （＝creation field）创生场
CFRP mirror （＝carbon fiber reinforced polymer mirror）碳纤维复合材料镜
CFRS （＝Canada-France redshift survey）加拿大－法国红移巡天，CFRS 巡天
C galaxy C 星系
CGCG （＝Catalogue of Galaxies and Clusters of Galaxies）星系和星系团表
CGPM （＝Conférence Générale des Poids et Mesures）【法】国际计量大会
CGRO （＝Compton Gamma-Ray Observatory）康普顿伽马射线天文台 | 康卜吞 γ 射線天文台〔台〕（美国卫星）
Cha （＝Chamaeleon）蝘蜓座
Chacornac crater 沙科纳克环形山（月球）
Chaf 王良一（仙后座 β）　又见：*Caph; Kaff*
chain method 链锁法
chain of bursts 爆发链
chain of galaxies 星系链

chain of QSOs 类星体链
chain reaction 链式反应 | 鏈鎖 [式] 反應〔台〕
Chaldene 木卫二十一
Chaliubieju 查刘璧如（小行星 3960 号）
Challis crater 查理士环形山（月球）
Chamaeleon （＝Cha）蝘蜓座
Chamber 房宿（二十八宿）　又见：*Room*
Chamberlin crater 钱柏林环形山（月球）
chameleon field 变色龙场
Champollion crater 商博良环形山（月球）
Chandler crater 钱德勒环形山（月球）
Chandler motion 钱德勒运动
Chandler number 钱德勒数 | 張德勒數〔台〕（变星）
Chandler period 钱德勒周期 | 張德勒週期〔台〕（地球极移）
Chandler wobble 钱德勒摆动 | 張德勒搖轉〔台〕
Chandra Deep Field （＝CDF）钱德拉深空区
Chandra Deep Field North （＝CDFN）钱德拉北深空区
Chandra Deep Field South （＝CDFS）钱德拉南深空区
Chandrasekhar limit 钱德拉塞卡极限 | 錢卓極限〔台〕
Chandrasekhar mass 钱德拉塞卡质量
Chandrasekhar-Schoenberg limit 钱德拉塞卡－申贝格极限 | 錢卓－荀伯極限〔台〕
Chandrasekher dynamical friction 钱德拉塞卡动力摩擦
Chandrasekher potential energy tensor 钱德拉塞卡势能张量
Chandra X-ray Observatory 钱德拉 X 射线天文台 | 錢卓 X 光天文台〔台〕（美国卫星）
Chandrayaan 月船号（印度月球探测器）
Chang 张钰哲 | 張 [鈺哲]〔台〕（小行星 2051 号）
Changchun 长春（小行星 7485 号）
Chang'e 嫦娥（小行星 4047 号）
Chang'e 1 lunar probe 嫦娥 1 号月球探测器（中国）
Changjiangcun 长江村（小行星 5384 号）
Changshi 常熟（小行星 3221 号）
Chanmatchun 陈密珍（小行星 20760 号）
channel filter 通道滤波器，波道滤波器
channel plate 通道板
CH anomaly 碳氢反常
Chanwainam 陈伟南（小行星 8126 号）
Chanyikhei 陈易希（小行星 20780 号）
Chaokuangpiu 曹光彪（小行星 4566 号）

Chaos 卡俄斯（小行星19521号）
chaos theory 混沌理论 又见：*chaotic theory*
chaotic cosmology 混沌宇宙论
chaotic dynamics 混沌动力学
chaotic inflation 混沌暴胀
chaotic layer 混沌层
chaotic model 混沌模型
chaotic orbit 混沌轨道，不规则轨道 | 混沌軌道〔台〕 又见：*irregular orbit*
chaotic region 混沌区
chaotic spiral arm 混沌状旋臂
chaotic theory 混沌理论 又见：*chaos theory*
Chaoyichi 赵依祈（小行星21436号）
Chaozhou 潮州（小行星5217号）
Chaplygin crater 恰普雷金环形山（月球）
Chaplygin gas 恰普雷金气体
Chapman crater 查普曼环形山（月球）
Chapman equation 查普曼方程
Chapman-Jouguet detonation 查普曼－儒居特爆炸
Chapman layer 查普曼层
Chappe crater 沙佩环形山（月球）
Chappell crater 查普尔环形山（月球）
CHARA （＝Center for High Angular Resolution Astronomy）高角分辨率天文中心
Chara 常陈四（猎犬座β） 又见：*Asterion*
CHARA Array 高角分辨率天文中心望远镜阵（美国）
characteristic age 特征年龄
characteristic asteroid 特征小行星
characteristic constant 特征常数
characteristic curve 特征曲线 | 特性曲綫〔台〕
characteristic dyadic 特征并矢
characteristic envelope 特征包络
characteristic equation 特征方程
characteristic exponent 特征指数
characteristic frequency 特征频率
characteristic function 特征函数
characteristic index 特征指标
characteristic length 特征长度
characteristic number 特征数
characteristic radius 特征半径
characteristic root 特征根
characteristic temperature 特征温度
characteristic thickness 特征厚度
characteristic time 特征时间
characteristic time scale 特征时标

characteristic value 特征值
characteristic velocity 特征速度
CHaracterizing ExOPlanet Satellite （＝CHEOPS）基奥普斯系外行星表征卫星
charactron 字码管
charge conjugation 电荷共轭
charge conservation 电荷守恒
charge-coupled device （＝CCD）电荷耦合器件 | 電荷耦合元件〔台〕
charge density 电荷密度
charged particle 荷电粒子
charged particle track 荷电粒子迹
charge excess plasma 电荷过剩等离子体
charge exchange 电荷交换
charge injected device （＝CID）电荷注入器件
charge multiplet 电荷多重态
charge number 电荷数
Charge-Parity-Time reversal invariance （＝CPT invariance）电荷－宇称－时间反演不变性，CPT不变性
Charge-Parity violation （＝CP violation）电荷－宇称破坏，CP破坏
charge-space symmetry 荷空对称性
charge-to-mass ratio 荷质比
charge transfer device （＝CTD）电荷转移器件
Chariot cross-board 轸宿（二十八宿） 又见：*Chariot Platform; Cross-stick at back of chariot*
Chariot Platform 轸宿（二十八宿） 又见：*Chariot cross-board; Cross-stick at back of chariot*
Charitum Montes 查瑞腾山脉（火星）
Charles' law 查理定律
Charles' Wain 北斗[七星]（星组） 又见：*Triones; Big Dipper; Northern Dipper; Wain; Plough; Plow*
Charlevoix crater 夏洛瓦陨星坑（地球）
Charlier crater 沙利叶环形山（月球）
Charlier universe 沙利叶宇宙
Charon 冥卫一 | 冥[王]衛一，凱倫〔台〕
Chasma Boreale 北极深谷（火星）
chasma （复数：*chasmata*） 深谷
chassignite 纯橄无球粒陨石
Chauvenet crater 肖夫内环形山（月球）
CHDM （＝cold hot dark matter model）冷热暗物质混合模型
Chebyshev crater 切比雪夫环形山（月球）
check sum 检验和
Cheleb 宗正一（蛇夫座β） 又见：*Celb-al-Rai; Celbalrai*
chemical abundance 化学丰度

chemical composition 化学组成
chemical enrichment 化学增丰
chemical equilibrium 化学平衡
chemical evolution 化学演化
chemically peculiar star 化学组成特殊星
chemical potential 化学势
chemiluminiscence 化学发光
chemosphere 光化层
Chendonghua 陈栋华（小行星 19872 号）
Chenfangyun 陈芳允（小行星 10929 号）
Chengbruce 郑崇华（小行星 168126 号）
Chengdu 成都（小行星 2743 号）
Chengmaolan 程茂兰（小行星 47005 号）
Chengyuhsuan 陈永介（小行星 20879 号）
Chenhungjen 陈泓任（小行星 23279 号）
Chen Jiageng 陈嘉庚（小行星 2963 号）
Chenjiansheng 陈建生（小行星 33000 号）
Chenjingrun 陈景润（小行星 7681 号）
Chenkegong 陈克攻（小行星 151362 号）
Chenqian 陈骞（小行星 3560 号）
Chentao 陈韬（小行星 19873 号）
Chentsaiwei 蔡辰葳（小行星 21645 号）
CHEOPS （＝CHaracterizing ExOPlanet Satellite）基奥普斯系外行星表征卫星
Cherenkov Telescope Array （＝CTA）CTA 切伦科夫望远镜阵
Chern 陈省身（小行星 29552 号）
Chern-Simons interaction term 陈－西蒙斯相互作用项
Chernyshev crater 切尔内绍夫环形山（月球）
Chertan 西次相，太微右垣四（狮子座 θ）
 又见: Chort; Coxa
Chesapeake Bay crater 切萨皮克湾陨星坑（地球）
Chevallier crater 谢瓦利尔环形山（月球）
chevelure 云状包层
Chiayi 嘉义（小行星 147918 号）
Chicxulub crater ① 奇克苏鲁布陨星坑（地球）② 奇克苏鲁布陨星坑（地球）
China 中华（小行星 1125 号）
China-VO （＝Chinese Virtual Observatory）中国虚拟天文台
Chinese Astronomical Society （＝CAS）中国天文学会
Chinese Astronomy and Astrophysics 《天文学报》
Chinese Journal of Astronomy and Astrophysics （＝ChJAA）《中国天文和天体物理学报》（中国期刊）

Chinese Virtual Observatory （＝China-VO）中国虚拟天文台
Chinese VLBI network （＝CVN）中国甚长基线干涉网
Ching-Sung Yu 余青松（小行星 3797 号）
CHIPS （＝Cosmic Hot Interstellar Plasma Spectrometer Satellite）星际热等离子体光谱卫星
chiral anomaly 手征反常
chirality 手征性
chirality operator 手征算符
chiral symmetry 手征对称性
Chiron 喀戎 | 開朗〔台〕（小行星 2060 号）
Chiron-like object 类喀戎型天体 | 類開朗型天體〔台〕
Chiron-type object 喀戎型天体 | 開朗型天體〔台〕
chirp transform spectrometer 线性调频变换频谱仪，啁啾变换频谱仪
ChJAA （＝Chinese Journal of Astronomy and Astrophysics）《中国天文和天体物理学报》（中国期刊）
chladnite 顽辉石陨星 | 頑火無球陨石〔台〕
chlorine （＝Cl）氯（17 号元素）
Choikaiyau 蔡继有（小行星 5389 号）
chondrite 球粒陨星 | [球] 粒陨石〔台〕
chondrule 粒状体 | [球] 粒〔台〕
Chongqing 重庆（小行星 3011 号）
Choo 杵二（天坛座 α）
chopper 斩波器
chopper wheel method 斩波轮法
chopping 斩波法
chopping angle 切角
chopping primary mirror 斩波主镜
chopping secondary mirror 斩波副镜
chopping shutter 断口快门
chopping technique 斩波技术
Chort 西次相，太微右垣四（狮子座 θ）
 又见: Chertan; Coxa
Chow 天右五（巨蛇座 β）
Chowmeeyee 周美儿（小行星 71461 号）
Chrétien crater 克雷蒂安环形山（月球）
Christiansen cross 克里斯琴森十字
Christiansen interferometer 克里斯琴森干涉仪
Christmas Tree Cluster 圣诞树星团（NGC 2264）
Christmas tree model 圣诞树模型
Christoffel symbol 克里斯托弗尔符号
chromatic aberration 色差 又见: chromatism
chromatic curve 色差曲线
chromatic difference of magnification 倍率色差

chromatic dispersion 色散
chromatic image 色差像
chromatic refraction 色折射
chromatic resolution 色分解
chromatism 色差 又见：*chromatic aberration*
chromite 陨星铬铁
chromium （＝Cr）铬（24 号元素）
chromium star 铬星
chromosphere 色球 | 色球 [層]〔台〕
chromosphere-corona transition region 色球－日冕过渡区 | 色球日冕過渡區〔台〕
  又见：*chromosphere-corona transition zone*
chromosphere-corona transition zone 色球－日冕过渡区 | 色球日冕過渡區〔台〕
  又见：*chromosphere-corona transition region*
chromospheric ablation 色球蒸发
  又见：*chromospheric evaporation*
chromospheric activity 色球活动
chromospherically active binary star 色球活动双星
chromospherically active star 色球活动星 | 色球活躍星〔台〕
chromospheric bubble 色球泡
chromospheric condensation 色球压缩区 | 色球壓縮區，色球凝聚區〔台〕
chromospheric ejection 色球抛射
chromospheric eruption 色球爆发 | 耀斑，色球爆發〔台〕
chromospheric evaporation 色球蒸发
  又见：*chromospheric ablation*
chromospheric facula 色球光斑
chromospheric fine structure 色球精细结构
chromospheric flare 色球耀斑
chromospheric flocculus 色球谱斑
  又见：*chromospheric plage*
chromospheric knot 色球结
chromospheric line 色球谱线
chromospheric material 色球物质
chromospheric mottling 色球日芒
chromospheric network 色球网络
chromospheric plage 色球谱斑
  又见：*chromospheric flocculus*
chromospheric spectrum 色球光谱
chromospheric spicule 色球针状体
  又见：*chromospheric spike*
chromospheric spike 色球针状体
  又见：*chromospheric spicule*
chromospheric telescope 色球望远镜

chromospheric temperature 色球温度
chromospheric whirl 色球旋涡
chronicle 编年史
chronogeometry 计时几何学
chronograph 记时仪 又见：*time keeper*
chronological table 年表 | 年 [代] 表〔台〕
chronology 纪年法 | 年代學〔台〕
chronometer 时计 又见：*horologe; time piece*
chronometer correction 时计改正 | 時計差，時計改正 [量]〔台〕
chronometer rate 时计日速
chronometric invariant 时计不变量
chronometry 计时学
chronon 双时元
chronoscope 记时镜 | 計時鏡〔台〕
Chryseis 赫露斯（小行星 202 号）
Chryse Planitia 克律塞平原（火星）
CH star 碳氢星，CH 星 | CH 星〔台〕
Chubb crater 丘布陨星坑（地球）
Chungchiyung 钟期荣（小行星 34779 号）
Churyumov-Gerasimenko Comet 丘留莫夫－格拉西缅科彗星
Chushuho 朱树豪（小行星 4988 号）
Chuwinghung 朱永鸿（小行星 38962 号）
CI （＝color index）色指数
CID （＝charge injected device）电荷注入器件
CI emission line CI 发射线，中性碳发射线
Cigar Galaxy 雪茄星系（M82, NGC 3034）
CII region CII 区，电离碳区
cinematography 电影照相
cine theodolite 电影经纬仪 又见：*kinotheodolite*
CIO ①（＝Celestial Intermediate Origin）天球中间零点 ②（＝Conventional International Origin）国际协议原点 | 國際慣用 [極] 原點〔台〕
CIO locator CIO 定位角
CIP （＝Celestial Intermediate Pole）天球中间极
CIPM （＝Comité International des Poids et Mesures）【法】国际计量委员会
Cir （＝Circinus）圆规座
Circe 巫神星（小行星 34 号）
Circinus （＝Cir）圆规座
Circinus Galaxy 圆规座星系（ESO 97-G13）
circle division 度盘分划
circle left 盘左，正镜
circle of altitude 地平纬圈，平行圈 | 地平緯圈，等高圈〔台〕 又见：*altitude circle; parallel of altitude; almucantar*

circle of confusion 模糊圈
circle of declination 赤纬圈 又见: *declination circle; parallel of declination; declination parallel*
circle of equal altitudes 等位圈 | 等高圈〔台〕
circle of geodesic curvature 大地曲率圈
circle of longitude ① 经度圈 ② 黄经圈
　　又见: *longitude circle*
circle of perpetual apparition 上规, 恒显圈 | 恆顯圈〔台〕 又见: *upper circle*
circle of perpetual occultation 恒隐圈, 下规 | 恆隱圈〔台〕 又见: *lower circle*
circle of position 位置圈 | 方位圈〔台〕
　　又见: *position circle*
circle of right ascension 赤经圈
circle reading 度盘读数
circle right 盘右, 倒镜
circle scanner 度盘扫描器
Circlet of Pisces 双鱼座小环（星组）
circular aperture 圆形孔径
circular array 环形阵
circular dichroism 圆振二向色性
circular frequency 圆频率
circular Gaussian brightness distribution 圆高斯亮度分布
circularity 正圆性
circular motion 圆周运动
circular orbit 圆轨道
circular polarization 圆偏振
circular polarized radiation 圆偏振辐射
circular repolarization 再生圆偏振
circular restricted three-body problem 圆型限制性三体问题 | 圓型設限三體問題〔台〕
circular solution 圆轨解
circular variable filter （=CVF）圆形可变滤光片
circular velocity 环绕速度 | 圓周速度, 環繞速度〔台〕
circulation cell 环流圈（木星）
circulator 环形器
circulatory motion 环形运动
circumgalactic medium 星系周介质
circumhorizontal arc 环地平弧, 日承 | 日承, 環地平弧〔台〕
circumlunar flight 环月飞行
circumlunar orbit 环月轨道 又见: *circumlunar trajectory*
circumlunar satellite 环月卫星
circumlunar space 月周空间

circumlunar trajectory 环月轨道
　　又见: *circumlunar orbit*
circummeridian altitude 拱子午线高度
Circumnuclear Disc （=CND）[银河] 核周盘
circumnuclear disk （=CND）核周盘
circumnuclear star formation 核周产星
circumnuclear star-forming ring 核周产星环
circumplanetary flight 环行星飞行
circumplanetary matter 行星周物质
circumplanetary orbit 环行星轨道
　　又见: *circumplanetary trajectory*
circumplanetary satellite 环行星卫星
circumplanetary space 行星周空间
circumplanetary trajectory 环行星轨道
　　又见: *circumplanetary orbit*
circumpolar constellation 拱极星座
circumpolar region 拱极区 又见: *circumpolar zone*
circumpolar star 拱极星
circumpolar zone 拱极区 又见: *circumpolar region*
circumscribed halo 外接日晕
circumsolar flight 环日飞行
circumsolar orbit 环日轨道 又见: *circumsolar trajectory*
circumsolar satellite 环日卫星
circumsolar space 日周空间
circumsolar trajectory 环日轨道 又见: *circumsolar orbit*
circumstances of eclipse 交食概况
circumstances of meteorite fall 陨降概况
circumstellar astrophysics 星周天体物理
circumstellar cloud 星周云 | 拱星星雲〔台〕
　　又见: *circumstellar nebula*
circumstellar debris disk 星周碎屑盘
circumstellar disc 星周盘 | 拱星盤〔台〕
　　又见: *circumstellar disk*
circumstellar disk 星周盘 | 拱星盤〔台〕
　　又见: *circumstellar disc*
circumstellar dust 星周尘 | 拱星塵〔台〕
circumstellar dust disk 星周尘盘 | 拱星塵盤〔台〕
circumstellar dust shell 星周尘壳
circumstellar envelope 星周包层 | 拱星包層〔台〕
circumstellar gas 星周气体 | 拱星氣體〔台〕
circumstellar grain 星周粒子
circumstellar line 星周谱线 | 拱星譜線〔台〕
circumstellar maser 星周脉泽
circumstellar material 星周物质 | 拱星物質〔台〕
　　又见: *circumstellar matter*

circumstellar matter 星周物质 | 拱星物質〔台〕
  又见: *circumstellar material*
circumstellar nebula 星周云 | 拱星星雲〔台〕
  又见: *circumstellar cloud*
circumstellar shell 星周壳
circumsystem material 双星周物质
circumterrestrial flight 环地飞行
circumterrestrial orbit 环地轨道
  又见: *circumterrestrial trajectory*
circumterrestrial satellite 环地卫星
circumterrestrial space 地周空间
circumterrestrial trajectory 环地轨道
  又见: *circumterrestrial orbit*
circumzenithal 拱天顶仪
circumzenithal arc 环天顶弧,日载 | 日戴,環天頂弧〔台〕
circus 圆谷
Cirrus Nebula 卷毛星云 (NGC 6960)
CIRS (=Celestial Intermediate Reference System) 天球中间参考系
CIS (=conventional inertial system) 习用惯性系
cis-cytherean space 金地空间
cislunar space 月地空间
cis-martian space 火地空间
cis-planetary space 行星内空间
civil calendar 民用历
civil date 民用日期
civil day 民用日
civilization 文明
civil time 民用时
civil twilight 民用晨昏蒙影 | 民用曙暮光,民用晨昏蒙影〔台〕
civil year 民用年
Cl (=chlorine) 氯 (17 号元素)
Clairaut's theorem 克莱罗定理
clamp screw 制动螺旋
Clapeyron equation 克拉珀龙方程
Claritas Fossae 克拉利塔堑沟 (火星)
CLASS (=Cosmic Lens All-Sky Survey) 宇宙透镜巡天
classical Algol system 经典大陵双星
Classical and Quantum Gravity 《经典引力和量子引力》(英国期刊)
classical astronomy 经典天文学 | 古典天文學〔台〕
classical Cepheid 经典造父变星 | 古典造父變星〔台〕
classical Compton scattering 经典康普顿散射
classical cosmology 经典宇宙学

classical damping constant 经典阻尼常数
classical integral 经典积分 | 古典積分〔台〕
classical Kuiper belt object 经典柯伊伯带天体
  又见: *cubewano*
classical nova 经典新星 即: *nova*
classical planet 经典行星
classical quasar 经典类星体
classical R CrB star 经典北冕 R 型星
classical theory 经典理论
classical transport 经典输运
classical T Tauri star 经典金牛 T 型星
Classic of the Stars 《星经》
classification 分类
classification criterion 分类判据
class S spectra S 类频谱
class T spectra T 类频谱
clathrate 包合物
Clausius-Clapeyron relation 克劳修斯－克拉珀龙方程 即: *Clapeyron equation*
Clayden effect 克莱登效应
clean beam 洁束
cleaned map 洁化图
clean method 洁化方法
Clear and Bright 清明 (节气) 又见: *Fresh Green*; *Pure Brightness*
Clearwater Lake crater 清水湖陨星坑 (地球)
Clementine 克莱芒蒂娜 | 克萊芒蒂娜 [月球探测器]〔台〕(美国月球探测器)
Cleomedes crater 克莱奥迈季斯环形山 (月球)
Cleopatra crater 克莱奥帕特拉环形山 (金星)
clepsydra 漏壶,漏刻 | 漏壺〔台〕
CLFST (=Cambridge Low-Frequency Synthesis Telescope) 剑桥低频综合孔径望远镜 | 劍橋低頻孔徑綜合望遠鏡〔台〕
cliff interferometer 海岸干涉仪 又见: *sea interferometer*
climatology 气候学
C line C 谱线
clinopyroxene 单斜辉石
clipped signal 限幅信号
clock aging 钟老化
clock comparison 时钟比对 | 時鐘比對,對鐘〔台〕
clock coordination 时钟协调
clock correction 时钟改正 | 鐘差,時鐘校正 [量]〔台〕
clock dial 钟面
clock drift 时钟漂移

clock drive 转仪钟 又见: *drive clock; driving clock; sidereal drive; sidereal driving clock*
clock ensemble 钟组
clock error 钟差
clock hand 指针 又见: *pointer*
clock indication 钟面读数
clock malfunction 时钟失调
clock mechanism 时钟机构
clock rate 钟速 又见: *rate of clock*
clock room 钟房
clock star 测时星｜[测] 時星〔台〕
clock transportation 时钟搬运｜搬鐘〔台〕
clock weight 钟锤
clockwise 顺时针方向
close binary 密近双星 又见: *close binary star*
close binary galaxy 密近双重星系
close binary star 密近双星 又见: *close binary*
close binary system 密近双重天体
close configuration 闭合位形
closed-box model 闭箱模型
closed fork mounting 闭合型叉式装置
closed long-axis orbit 闭合长轴轨道
closed loop orbit 闭合圈形轨道
closed model 闭模型
closed orbit 闭合轨道
closed string 闭弦
closed system 闭合系
closed timelike curve （=CTC）闭合类时曲线
closed universe 闭宇宙｜封闭宇宙〔台〕
close encounter 密近交会
closely coiled arm 紧卷旋臂 又见: *tightly wound arm*
closely coupled state 密耦态
close satellite 密近卫星
closest approach 最接近态, 最接近时刻｜最接近態〔台〕
closest approach point 近站点
closing error 闭合误差｜閉合 [誤] 差〔台〕
closing sum 闭合和
closure 闭合, 闭合差
closure amplitude 闭合幅度
closure density 闭合密度
closure phase 闭合相位
closure phase imaging technique 锁相成像技术
cloud chamber 云室｜雾室〔台〕
cloud cover 云量
cloud-in-cloud problem 云中云问题
clouding 云蔽
cloudlet 小云
cloud structure 星云结构
Cloverleaf quasar 四叶形类星体
Clown Face Nebula 小丑脸星云（NGC 2392）
  即: *Eskimo Nebula*
clumping 簇聚
Cluster 簇群空间探测器
cluster ① 星系团 又见: *galaxy cluster; cluster of galaxies* ② 星团 又见: *star cluster; stellar cluster; cluster of stars*
cluster center 团中心
cluster Cepheid 星团造父变星 又见: *cluster-type Cepheid*
cluster galaxy 团星系｜屬團星系〔台〕
cluster infall [星系] 团沉降
clustering 成团
clustering evolution 成团性演化
clustering of galaxies 星系成团 又见: *galaxy clustering*
clustering of peaks 峰值成团性
cluster member ① 星团成员 ② 星系团成员
cluster model 束状模型
cluster nebula 团星云｜屬團星雲〔台〕
cluster normalization [星系] 团归一化
cluster of galaxies 星系团 又见: *galaxy cluster; cluster*
cluster of nebulae 星云团
cluster of stars 星团 又见: *star cluster; stellar cluster; cluster*
cluster parallax ① 星团视差 ② 星群视差｜星群视差, 星團視差〔台〕 又见: *group parallax*
cluster point 丛点
cluster rotation ① 星团自转 ② 星系团自转
cluster star 团星｜屬團星〔台〕
cluster-type Cepheid 星团造父变星 又见: *cluster Cepheid*
cluster-type variable 星团变星 又见: *cluster variable*
cluster variable 星团变星 又见: *cluster-type variable*
CM ① （=central meridian）中央子午线, 日心子午线｜中央子午圈〔台〕② （=command module）指令舱
Cm （=curium）锔（96 号元素）
CMa （=Canis Major）大犬座
CMB （=cosmic microwave background）宇宙微波背景
CMB anisotropy （=cosmic microwave background anisotropy）宇宙微波背景各向异性

CMB experiment （＝cosmi microwave background experiment）宇宙微波背景实验装置

CMB polarization （＝cosmic microwave background polarization）宇宙微波背景偏振，宇宙微波背景极化

CMBR （＝cosmic microwave background radiation）宇宙微波背景辐射 | [宇宙] 微波背景辐射〔台〕

CMB spectral distortion （＝cosmic microwave background spectral distortion）宇宙微波背景谱畸变

CMD （＝color-magnitude diagram）颜色－星等图

c-m diagram （＝color-magnitude diagram）颜色－星等图

CME （＝coronal mass ejection）日冕物质抛射 | 日冕物質噴發〔台〕

CMi （＝Canis Minor）小犬座

CMOS detector CMOS 探测器

CMR （＝cosmic microwave radiation）宇宙微波辐射

c-m relation （＝color-magnitude relation）颜色－星等关系

CN anomaly 碳氮反常

CN band CN 谱带

Cnc ① （＝Cancer）巨蟹座 ② （＝Cancer）巨蟹宫，鹑首，未宫

CN cycle （＝carbon-nitrogen cycle）碳氮循环 | 碳氮循環，CN 循環〔台〕

CND ① （＝circumnuclear disk）核周盘 ② （＝Circumnuclear Disc）[银河] 核周盘

CNES （＝Centre National d'Etudes Spatiales）【法】法国国家空间研究中心

CN index 碳氮指数

CNO anomaly 碳氮氧反常

CNO bi-cycle 碳氮氧双循环

CNO cycle （＝carbon-nitrogen-oxygen cycle）碳氮氧循环

C/NOFS （＝Communication/Navigation Outage Forecasting System）通信/导航中断预报系统

CNO tri-cycle 碳氮氧三循环

CNRS （＝Centre National de la Recherche Scientifique）【法】国家科学研究中心（法国）

CN star CN 星

Co （＝cobalt）钴（27 号元素）

coacervate 凝聚层

coaction 公共作用

coalesced star 并合星

Coalsack Dark Nebula 煤袋星云 又见：Coalsack Nebula; Southern Coalsack

Coalsack Nebula 煤袋星云 又见：Coalsack Dark Nebula; Southern Coalsack

co-altitude 余高度

coarse grain 粗粒

coarse-grained distribution function 粗粒分布函数

coarse motion 粗动

coarse mottle 粗日芒

COAST （＝Cambridge Optical Aperture Synthesis Telescope）剑桥光学综合孔径望远镜 | 劍橋光學孔徑合成望遠鏡〔台〕

Coathanger Cluster 衣架星团

coating 镀膜 | 鍍鏡膜〔台〕

cobalt （＝Co）钴（27 号元素）

cobalt star 钴星

COBE （＝Cosmic Background Explorer）宇宙背景探测器 | 宇宙背景探測衛星〔台〕（美国卫星）

Cockcroft crater 考克饶夫环形山（月球）

Cocoon Nebula 蚕茧星云，茧状星云（IC 5146）

cocoon star 茧星

CoD （＝Cordoba Durchmusterung）【德】科尔多瓦巡天星表

codeclination 余赤纬，极距，去极度 | 去極度〔台〕 又见：polar distance degrees

coded aperture imaging 遮幅孔径成像

coded aperture telescope 编码孔径望远镜

coded CW waveform 编码连续波波形

coded disk 码盘

coded mask 编码 [孔] 板，编码掩模，编码遮罩

coded mask imaging 编码掩模成像

coded mask telescope 编码孔罩望远镜

coder 编码器 又见：encoder

code translator 译码器 又见：decipherer; decoder

coefficient of absorption 吸收系数 又见：absorption coefficient

coefficient of amplification 放大系数

coefficient of autocorrelation 自相关系数 又见：autocorrelation coefficient

coefficient of correlation 相关系数 又见：correlation coefficient

coefficient of elasticity 弹性系数

coefficient of expansion 膨胀系数 又见：expansion coefficient

coefficient of opacity 不透明系数 又见：opacity coefficient

coefficient of reflection 反射系数 又见: *reflection coefficient*
coefficient of refraction 折射系数 又见: *refraction coefficient*
coefficient of selective absorption 选择吸收系数
又见: *selective absorption coefficient*
coefficient of true selective absorption 真选择吸收系数
coefficient of viscosity 黏性系数
coelosphere 天球 又见: *celestial sphere*
coelostat 定天镜
coesite 联合址
coherence 相干性 | 相干性, 同调性〔台〕
又见: *coherency*
coherence bandwidth 相干带宽
coherence function 相干函数
coherence length 相干长度
coherence time 相干时间
coherence width 相干宽度
coherency 相干性 | 相干性, 同調性〔台〕
又见: *coherence*
coherent averager 相干平均器
coherent communication system 相干通信系统
coherent detector 相干检测器
coherent emission 相干发射
coherent light 相干光
coherent object 相干天体
coherent radiometer 相干式接收机 又见: *coherent receiver*
coherent receiver 相干式接收机 又见: *coherent radiometer*
coherent reflection 相干反射
coherent scattering 相干散射 | 相干散射, 同調散射〔台〕
coherent synchrotron radiation 相干同步加速辐射
cohesion 内聚力 又见: *cohesive force*
cohesive force 内聚力 又见: *cohesion*
coincidence argument 巧合论点
coincidence problem 巧合问题
Col (=Columba) 天鸽座
colatitude 余纬度, 余黄纬
cold atom clock 冷原子钟
Cold Bokkeveld meteorite 冷伯克费尔德陨星
cold camera 冷相机
cold dark matter (=CDM) 冷暗物质
cold dark matter model (=CDM model) 冷暗物质模型

Cold Dew 寒露 (节气)
cold emission 冷辐射
cold-gas approximation 冷气体近似
cold hot dark matter model (=CHDM) 冷热暗物质混合模型
cold intergalactic medium 冷星系际介质
cold phase 冷相
cold plasma 冷等离子体 | 冷離子體, 冷電漿〔台〕
cold relic particle 冷遗迹粒子
cold sky 无源天区
cold stop 冷光阑
cold supergiant 冷超巨星 又见: *cool supergiant*
cold universe 冷宇宙
Coleman-Weinberg potential 科尔曼－温伯格势
collapsar 坍缩星 | 塌縮星〔台〕
collapse 坍缩 | 塌縮〔台〕
collapsed object 坍缩天体
collapsed star 坍缩星 | 塌縮星〔台〕
又见: *collapsing star*
collapse fraction 坍缩比例
collapse simulation 坍缩模拟
collapse time 坍缩时间
collapsing cloud 坍缩云 | 塌縮雲〔台〕
collapsing instability 坍缩不稳定性
collapsing star 坍缩星 | 塌縮星〔台〕
又见: *collapsed star*
collecting area 接收面积 又见: *receiving area*
collecting lens 聚光透镜 又见: *convergent lens; converging lens; condensing lens*
collective interaction 集聚相互作用
collective plasma 集合等离子体
colliding galaxy 碰撞星系
colliding-wind binary 星风碰撞双星
collimated γ-ray scintillation spectrometer 准直γ射线闪烁谱仪
collimating device 准直器 又见: *collimator*
collimating lens 准直透镜 又见: *collimation lens*
collimating telescope 准直望远镜
collimation 准直
collimation axis 准直轴
collimation constant 准直常数
collimation error 准直误差 | 準直[誤]差〔台〕
collimation lens 准直透镜 又见: *collimating lens*
collimation plane 准直面
collimator 准直器 又见: *collimating device*
collinear point 共线点
collis 矮丘

collisional bremsstrahlung 碰撞韧致辐射 | 碰撞制动辐射〔台〕
collisional broadening 碰撞致宽 又见: impact broadening; collision broadening
collisional damping 碰撞阻尼 | 碰撞制动〔台〕
collisional excitation 碰撞激发 又见: impact excitation
collisional ionization 碰撞电离 | 碰撞游离〔台〕 又见: collision ionization; impact ionization; ionization by collision
collisional plasma 碰撞等离子体
collisional radio source 碰撞射电源
collisional shock wave 碰撞激波
collisional turbulence 碰撞湍流
collision broadening 碰撞致宽 又见: collisional broadening; impact broadening
collision cross-section 碰撞截面
collision ejection hypothesis 碰撞抛射假说
collision-free bow shock 无碰撞弓形激波
collision frequency 碰撞频率
collision hypothesis 碰撞假说 又见: impact hypothesis
collision induced instability 碰撞感生不稳定性
collision induced spectrum 碰撞感生谱
collision induced transition 碰撞感生跃迁
collision ionization 碰撞电离 | 碰撞游離〔台〕 又见: collisional ionization; impact ionization; ionization by collision
collisionless Boltzmann equation 无碰撞玻尔兹曼方程
collisionless damping 无碰撞阻尼
collisionless gas 无碰撞气体
collisionless magnetosonic shock wave 无碰撞磁声激波
collisionless plasma 无碰撞等离子体
collisionless shock wave 无碰撞激波
collisionless system 无碰撞系统
collisionless tearing instability 无碰撞撕裂不稳定性
collision lifetime 碰撞寿命
collision line broadening 谱线碰撞致宽
collision loss 碰撞损失
collision narrowing 碰撞致窄
collision of the first kind 第一类碰撞
collision of the second kind 第二类碰撞
collision orbit 碰撞轨道
collision parameter 碰撞参数 又见: impact parameter

collision probability 碰撞概率
collision process 碰撞过程
collision radiation 碰撞辐射
collision rate 碰撞速率
collision strength 碰撞强度
collision term 碰撞项
collision time 碰撞时间
collumn diameter 迹厚度
colocational observation 并址观测
Colombo crater 哥伦布环形山（月球）
Colombo Gap 科隆博环缝（土星）
colongitude 余经度，余黄经
color-apparent magnitude diagram 颜色－视星等图
color coding 色码
color-color diagram 两色图 | 兩色圖，色[指数]－色[指数]圖〔台〕 又见: two-color diagram; color-color plot
color-color plot 两色图 | 兩色圖，色[指数]－色[指数]圖〔台〕 又见: two-color diagram; color-color diagram
color contrast 色衬度 | 色对比〔台〕
color correction 色改正
color curve 颜色曲线 | [颜]色曲线〔台〕
color dependent error 色相关误差
color development 彩色显影
color difference photography 色较差照相
color distortion 色畸变
colored glass filter 颜色玻璃滤光片
color equivalent 色当量
color excess （=CE）色余
color factor 色因子
color filter 滤色器
color force 色力
color gradient 色梯度
color image 彩色图像
colorimeter 色度计
colorimetry 色度测量
color index （=CI）色指数
color-luminosity correlation 颜色－光度相关
color-luminosity diagram 颜色－光度图
color-luminosity relation 颜色－光度关系
color magnification error 色放大率误差
color-magnitude diagram （=c-m diagram; CMD）颜色－星等图
color-magnitude effect 颜色－星等效应

color-magnitude relation （＝c-m relation）颜色－星等关系
color-redshift diagram 颜色－红移图
color-redshift relation 颜色－红移关系
color singlet 色单态
color standard 色标准
color subcarrier 彩色副载波
color superconductivity 色超导
color television 彩色电视
color temperature 色温度
Columba （＝Col）天鸽座
column abundance 柱丰度
column density 柱密度
colure 分至圈｜二分圈，二至圈〔台〕
Com （＝Coma Berenices）后发座
coma ① 彗发 ② 彗差｜彗形像差〔台〕
　　又见：*comatic aberration*
Coma Berenices （＝Com）后发座
Coma cluster 后发星系团
coma corrector 彗差改正器
coma sidelobe 彗形旁瓣
Comas Solá Comet 科马斯－索拉彗星
Coma Star Cluster 后发星团
comatic aberration 彗差｜彗形像差〔台〕
　　又见：*coma*
Coma-Virgo Cluster 后发－室女星系团
comb dekker 德克梳
combination band 组合谱带
combination line 组合谱线
combination scattering 组合散射
combination variable 共生变星　又见：*symbiotic variable*
Combined Array for Research in Millimeter-wave Astronomy （＝CARMA）CARMA 毫米波组合阵（美国）
combined magnitude 合成星等
comb-like structure 梳状结构
combustion clock 火钟
comes（复数：**comites**）【拉】伴星
　　即：*companion star*; *partner star*
comet 彗星
cometary astronomy 彗星天文学｜彗星天文[学]〔台〕
cometary burst 彗暴
cometary dust 彗尘｜彗[星]尘[埃]〔台〕
cometary dust tail 尘彗尾
cometary dynamics 彗星动力学
cometary flare 彗耀

cometary gas 彗星气体
cometary gas tail 气体彗尾　又见：*gas tail*
cometary globule 彗形球状体
cometary halo 彗晕　又见：*comet halo*
cometary head 彗头　又见：*comet head*; *head of comet*
cometary HII region 彗状电离氢区
cometary hydrogen cloud 彗状氢云
cometary ion 彗星离子
cometary meteor 彗生流星
cometary molecule 彗星分子
cometary nebula 彗状星云　又见：*comet-shaped nebula*
cometary neutral tail 中性彗尾
cometary nucleus 彗核　又见：*nucleus of comet*
cometary orbit 彗星轨道
cometary outburst 彗星爆发
cometary pause 彗顶
cometary physics 彗星物理学｜彗星物理[学]〔台〕
cometary plasma 彗星等离子体
cometary proplyd 彗状原行星盘
cometary stream 彗生流星雨
cometary tail 彗尾　又见：*comet tail*; *tail of comet*
cometary zone 彗星区
Comet Bennett 本内特彗星
comet catalogue 彗星星表
comet cloud 彗云｜彗星雲〔台〕
Comet de Chéseaux 德塞瑟彗星　又见：*de Cheseaux's comet*
comet designation 彗星命名　又见：*designation of comets*
Comet Donati 多纳蒂彗星
Comet Encke 恩克彗星｜恩克彗[星]〔台〕
　　又见：*Encke's comet*
cometesimal 彗星子
comet family 彗星族　又见：*family of comets*
comet finder 寻彗镜　又见：*comet seeker*
comet group 彗星群　又见：*group of comets*
Comet Hale-Bopp 海尔－波普彗星
comet halo 彗晕　又见：*cometary halo*
comet head 彗头　又见：*cometary head*; *head of comet*
comet hunter 寻彗者
Comet Hyakutake 百武彗星
comet intercept mission 彗星拦截器
Comet ISON 艾森彗星
comet-like activity 类彗活动
comet-like dust tail 彗状尘尾
comet-like galaxy 彗状星系

comet-like gas tail 彗状气尾
comet-like object 类彗天体
comet-like tail 彗状尾
Comet Machholz 麦克霍尔茨彗星
Comet Nucleus Tour （＝CONTOUR）CONTOUR 彗核探测器
cometocentric coordinate 彗心坐标
comet of Jupiter family 木族彗星 | 木 [星] 族彗星〔台〕 又见：*Jupiter's comet family; Jupiter's family of comets; Jupiter-type comet*
comet of Neptune family 海王族彗星 | 海 [王星] 族彗 [星]〔台〕
comet of Saturn family 土族彗星 | 土 [星] 族彗 [星]〔台〕
comet of Uranus family 天王族彗星 | 天 [王星] 族彗 [星]〔台〕
cometography 彗星志
cometoid 小彗星
Comet Pan-STARRS 泛星彗星（C/2011 L4）
comet radio radiation 彗星射电
comet seeker 寻彗镜 又见：*comet finder*
comet-shaped nebula 彗状星云 又见：*cometary nebula*
Comet Shoemaker-Levy 9 （＝SL9）舒梅克－列维 9 号彗星 | 舒梅克－李維 9 號彗星〔台〕
comet shower 彗星雨
comet space probe 彗星空间探测器
comet tail 彗尾 又见：*cometary tail; tail of comet*
Comité Consultatif du Temps et des Fréquences （＝CCTF）【法】时间频率咨询委员会
Comité Consultatif pour la Definition de la Seconde （＝CCDS）【法】秒定义咨询委员会
Comité International des Poids et Mesures （＝CIPM）【法】国际计量委员会
comma-like nebula 逗点状星云
command and service module （＝CSM）指令－服务舱
command module （＝CM）指令舱
commensurability 通约 | 通約 [性]〔台〕
commensurable motion 通约运动
commensurable orbit 通约轨道 又见：*commensurate orbit*
commensurate orbit 通约轨道 又见：*commensurable orbit*
commercial satellite 商用卫星
Committee on Space Research （＝COSPAR）国际空间研究委员会 | 太空研究委員會〔台〕
common center of gravity 公共重心
common envelope 共 [有] 包层 | 共有包層〔台〕
common envelope binary 共包层双星
common-envelope evolution 共包层演化
common-envelope star 共包层星
common establishment 常用潮候时差
common-mount interferometer 共机架干涉仪
common proper motion （＝c.p.m.）共自行
common proper-motion binary 共自行双星
common proper-motion pair 共自行星对
common proper-motion stars 共自行星
common sign 常用宫
common-view observation 共视观测
common year 平年（非闰年） 又见：*non-leap year*
Communication/Navigation Outage Forecasting System （＝C/NOFS）通信/导航中断预报系统
communication satellite 通信卫星
communication with extra-terrestrial intelligence （＝CETI）地外智能生物通信 | 地 [球] 外智慧生物通訊〔台〕
comoving coordinate 共动坐标
comoving coordinate system 共动坐标系
comoving distance 共动距离
comoving gauge 共动规范
comoving horizon 共动视界
comoving observer 共动观测者
comoving radius 共动半径
comoving separation 共动间隔
compact array 射电望远镜密集阵
compact binary 致密双星
compact binary galaxy 致密双重星系
compact cluster 致密星系团 又见：*compact cluster of galaxy*
compact cluster of galaxy 致密星系团 又见：*compact cluster*
compact disc read-only memory （＝CD-ROM）只读光盘
compact elliptical 致密椭圆星系
compact flare 致密耀斑 | 緻密閃焰〔台〕
compact galaxy 致密星系
compact galaxy nucleus 致密星系核
compact group of galaxies 致密星系群
compact HII region 致密电离氢区 | 緻密 HII 區〔台〕
compactification 紧致化
compact infrared source 致密红外源
compaction age 组成年龄
compact nebula 致密星云
compact nucleus 致密核

compact object 致密天体
compact radio source 致密射电源 | 緻密 [無線] 電波源〔台〕
compact source 致密源
compact star 致密星
compact X-ray source 致密 X 射线源 | 緻密 X 光源〔台〕
compact γ-ray source 致密 γ 射线源
companion galaxy 伴星系　又见：*satellite galaxy*
companion star 伴星　又见：*partner star*
comparative planetology 比较行星学
comparative sensitivity 相对灵敏度
comparator 比长仪　又见：*spectral comparator; spectrocomparator*
comparison band 比较谱带 | 比較 [頻] 帶〔台〕
comparison beam 比较束
comparison image device 像比较装置
comparison load 比较负载
comparison residual 比较残差
comparison spectrum 比较光谱
comparison star 比较星
compass dial 罗盘日晷
compensated microphotometer 补偿显微光度计 | 補償測微光度計〔台〕
compensated pendulum 补偿摆
compensated receiver 补偿接收机
compensation colorimeter 补偿色度计
compensator 补偿器
compilation catalogue 编纂星表
compiler 编译程序
complementary function 补函数
complete absorption 完全吸收
complete Baumbach corona 完全鲍姆巴赫日冕
complete catalogue 完整星表
complete degeneracy 完全简并性
complete eclipse 完整食
complete eclipsing binary 完整食双星
complete solution 全解
complete stability 完全稳定性
complete synchrone 全等时线
complex 复合体
complex fringe amplitude 复条纹幅度
complex group 复杂群 | 複雜 [黑子] 群〔台〕
complex of activity 活动复合体
Complex Orbital Observations Near-Earth of Activity of the Sun （＝Koronas）日冕系列卫星（俄罗斯）

complex singularity 复奇点
complex source 复合源
complex visibility function 复可见度函数
component of the four displacements 四游仪
　　又见：*movable sighting set; sighting-tube ring*
component of the six cardinal points 六合仪
component of the three arrangers of time 三辰仪
　　又见：*three luminary set*
component star 子星
component velocity 分速度
composite diagram method 复合图法
composite model 复合模型 | 複雜模型〔台〕
composite particle 合成粒子
composite spectrum 复合光谱
composite-spectrum binary 复谱双星
composition of force 力合成
composition of the earth 地球组成
compound eyepiece 复合目镜
compound field-flattener 复合平场镜
compound-grating antenna 复合栅天线
compound interferometer 复合干涉仪
compound lens 复合透镜
compound pendulum 复摆　又见：*physical pendulum*
comprehensive flare index 综合耀斑指数
compressibility 可压缩性 | 壓縮性，壓縮係數〔台〕
compressible fluid 可压缩流体
compressional Alfvén wave 可压缩阿尔文波
compression cap 压缩冠
compression of the earth 地球扁率　又见：*earth ellipticity; earth flattening; earth oblateness*
Comptes Rendus Physique 【法】《法国科学院报告：物理学》（法国期刊）
Compton cooling 康普顿冷却
Compton crater 康普顿环形山（月球）
Compton cross-section 康普顿截面
Compton effect 康普顿效应 | 康卜吞效應〔台〕
Compton frequency 康普顿频率
Compton Gamma-Ray Observatory （＝CGRO）康普顿伽马射线天文台 | 康卜吞 γ 射線天文台〔台〕（美国卫星）　又见：*Compton γ-Ray Observatory*
Compton heating 康普顿加热
Comptonization 康普顿化 | 康卜吞化〔台〕
Compton loss 康普顿耗损
Compton recoil particle 康普顿反冲粒子
Compton scattering 康普顿散射 | 康卜吞散射〔台〕
Compton scattering cross-section 康普顿散射截面

Compton telescope 康普顿望远镜
Compton wavelength 康普顿波长
Compton y-parameter 康普顿 y 参数
Compton γ-Ray Observatory 康普顿伽马射线天文台 | 康卜吞 γ 射線天文台〔台〕（美国卫星）
又见：*Compton Gamma-Ray Observatory*
computational astrophysics 计算天体物理学
computational celestial mechanics 计算天体力学
computer control 计算机控制
computer-controlled telescope 计算机控制望远镜
computerized simulation 计算机模拟
又见：*computer simulation*
computerized telescope 程控望远镜
computer model 计算机模型
computer program 计算机程序
computer simulation 计算机模拟
又见：*computerized simulation*
computing method 计算方法
computing technique 计算技术
computus 计算表册
Comrie crater 科姆里环形山（月球）
concave grating 凹光栅
concave lens 凹透镜
concave mirror 凹镜 | 凹面鏡〔台〕
concavity 凹度
concavo-convex lens 凹凸透镜
conceal 伏
concentration 聚集度 又见：*concentration class*
concentration class 聚集度 又见：*concentration*
concentration index 聚集指数
concentration of stress 应力集中
concentration parameter 聚集参数
concentric eclipse 同心食
concordance model 协调模型
Concordia 协神星（小行星 58 号）
Concordia family 协神星族
condensation energy 凝聚能
condensation point 聚点 又见：*accumulation*
condensation region 凝聚区
condensed star 凝聚星
condenser 聚光器
condensing lens 聚光透镜 又见：*convergent lens; converging lens; collecting lens*
conditional luminosity function 条件光度函数
conditionally periodic function 条件周期函数
conditional mass function 条件质量函数
conditional multiplicity function 条件多重度函数

conditional stability 条件稳定性
conditionary periodic motion 条件周期运动
condition equation 条件方程
Condorcet crater 孔多塞环形山（月球）
conductance 电导 又见：*electric conductance*
conduction band 导带
conduction flux 传导通量
conductivity 电导率 又见：*electroconductivity*
conductor 导体
Cone nebula 锥状星云（NGC 2264）
Conférence Générale des Poids et Mesures （＝CGPM）【法】国际计量大会
confidence 置信度 又见：*degree of confidence*
confidence level 置信级
confidence limit 置信限 又见：*fiducial limit*
configuration mixing 组态混合
configuration space 构形空间
confocal ellipsoidal coordinate 共焦椭球坐标
confocal lens 共焦透镜
conformal diagram 共形图
conformal Newtonian gauge 共形牛顿规范
conformal projection 正形投影
conformal structure 共形结构
conformal time 共形时间
conformal transformation 共形变换
Confucius 孔子（小行星 7853 号）
confusion level 致淆电平 | 致淆水平〔台〕
confusion noise 致淆噪声
Congreve crater 康格里夫环形山（月球）
conical nebula 锥形星云
conical point 锥点
conical spiral feed 圆锥螺旋馈源
conic section 圆锥截线
conjugate focus 共轭焦点
conjugate point 共轭点
conjunction 合
conjunction in ecliptical longitude 黄经合
conjunction in right ascension 赤经合
Connaissance des Temps 【法】《法国天文年历》 | 《[法國] 天文年曆》〔台〕
connected-element interferometer 联线干涉仪
connected-element radio interferometry （＝CERI）联线射电干涉测量
connected region 连通区域
conoid 劈锥曲面
consecutive collision orbit 相邻碰撞轨道
consecutive number 相邻数

conservation 守恒
conservation law 守恒定律
conservation of angular momentum 角动量守恒
conservation of energy 能量守恒　又见：*energy conservation*
conservation of mass 质量守恒　又见：*mass conservation*
conservation of mass-energy 质能守恒
conservation of momentum 动量守恒
conservation of vorticity 涡度守恒
conservative field 保守场
conservative force 保守力
conservative process 守恒过程
conservative scattering 守恒散射
conservative system 守恒系
conserved quantity 守恒量
constant 常数，恒量
constant force 恒力
constant of aberration 光行差常数
　　又见：*aberration constant*
constant of apsidal motion 拱线运动常数
constant of gravitation 引力常数 | 重力常數〔台〕
　　又见：*gravitational constant; gravitation constant*
constant of motion 运动常数
constant of nutation 章动常数　又见：*nutation constant; nutational constant*
constant of precession 岁差常数　又见：*precession constant; precessional constant*
constant of refraction 折射常数　又见：*refraction constant*
constant vector field 常矢量场
constellation 星座
constituent day 潮汐日
constraint model 约束模型
constructive interference 相长干涉
consumption rate 消耗率
contact binary 相接双星 | 密接雙星〔台〕
　　又见：*contact system*
contact chronometer 接触时计
contact copying 接触复制法
contact drum 接触鼓
contact micrometer 接触测微计
contact system 相接双星 | 密接雙星〔台〕
　　又见：*contact binary*
contact transformation 接触变换
contact width 接触宽度
continental block 大陆块

continental drift 大陆漂移　又见：*continental movement*
continental movement 大陆漂移　又见：*continental drift*
continental tide 陆潮　又见：*earth tide*
continuity condition 连续性条件
continuity equation 连续性方程 | 連續[性]方程〔台〕
continuous absorption 连续吸收
continuous absorption coefficient 连续吸收系数
continuous background 连续背景
continuous contour 连续轮廓图
continuous creation 连续创生
continuous distribution 连续分布
continuous emission 连续谱发射　又见：*continuum emission*
continuous filled-aperture 连续满面孔径
continuous filled-aperture array 连续满面天线阵
continuous medium 连续介质
continuous opacity 连续不透明度
continuous radiation 连续辐射
continuous spectrum 连续谱 | 連續[光]譜〔台〕
continuous wave （＝CW）连续波
continuum 连续区，连续谱 | 連續譜，連續區〔台〕
continuum burst 连续谱爆发
continuum emission 连续谱发射　又见：*continuous emission*
continuum radiation 连续谱辐射
continuum receiver 连续谱接收机
continuum source 连续谱源
CONTOUR （＝Comet Nucleus Tour）CONTOUR 彗核探测器
contour 轮廓，等高线，恒值线
contour diagram 轮廓图　又见：*contour map*
contour line 轮廓线，等高线，恒值线
contour map 轮廓图　又见：*contour diagram*
contour mapping 轮廓绘制
contracting model 收缩宇宙模型 | 收縮[宇宙]模型〔台〕
contracting universe 收缩宇宙
contraction 收缩
contraction age 收缩年龄
contraction hypothesis 收缩假说（太阳能源）
contraction of tensor 张量缩并
contraction phase 收缩阶段 | 收縮相，收縮階段〔台〕
contragradience 反步，逆步

contrast coefficient 反衬系数
contrast developer 强反差显影剂
contrast factor 反衬因子 | 對比因子〔台〕
contrast of fringes 条纹反衬度
contrast photometer 对比光度计
contravariant component 逆变分量
contravariant index 逆变指标
contravariant vector 逆变矢量
Contributions of The Astronomical Observatory Skalnate Pleso 《斯洛伐克天文台论文集》
control azimuth 控制方位
control device 控制裝置
control latitude 控制纬度
controller 控制器 又见: control unit
control levelling 校核水准测定
control range 控制范围
control system 控制系统
control unit 控制器 又见: controller
conus density 锥体密度
convection 对流
convection cells 对流元
convection current 对流气流 | 對流〔台〕
convection instability 对流不稳定性
Convection rotation and planetary transits (=CoRoT) 科罗系外行星探测器 | 科洛[系外行星探测器]〔台〕（法国卫星）
convection zone 对流层
convective cell 对流元
convective core 对流核 | 對流核[心]〔台〕
convective derivative 对流导数
convective envelope 对流包层
convective equilibrium 对流平衡
convective instability criterion 对流不稳定性判据
convective mixing 对流混合
convective operator 对流算子
convective overshooting 对流过冲
convective overturn time 对流翻转时间
convective region 对流区 | 對流帶〔台〕
  又见: convective zone
convective shell 对流壳
convective transfer 对流转移
convective zone 对流区 | 對流帶〔台〕
  又见: convective region
conventional celestial reference system (=CCRS) 习用天球参考系
conventional inertial system (=CIS) 习用惯性系

Conventional International Origin (=CIO) 国际协议原点 | 國際慣用[極]原點〔台〕
conventional terrestrial reference system (=CTRS) 习用地球参考系
convergence 会聚 又见: convergency
convergency 会聚 又见: convergence
convergent lens 聚光透镜 又见: converging lens; collecting lens; condensing lens
convergent mirror 聚光镜 又见: converging mirror
convergent point 会聚点 | 匯聚點〔台〕
convergent point method 汇聚点方法
converging beam 会聚波束
converging lens 聚光透镜 又见: convergent lens; collecting lens; condensing lens
converging magnetic mirror 会聚磁镜
converging mirror 聚光镜 又见: convergent mirror
conversion fraction 转换系数
conversion gain coefficient 变频增益系数
conversion of coordinates 坐标变换
  又见: coordinate transformation; transformation of coordinates
conversion of time 时间换算, 时间反演 | 時間換算〔台〕
conversion transconductance 变频跨导
convexity 凸度
convex lens 凸透镜
convexo-concave lens 凸凹透镜
convolution 卷积 | 摺積〔台〕
Cook crater 库克环形山（月球）
Cooke method 库克法
cool component 冷子星
cool dwarf 冷矮星
cooled camera 制冷相机
cooled parametric amplifier 致冷参量放大器
cooled Schottky-barrier mixer 致冷肖特基位垒混频器
cool giant 冷巨星
cooling catastrophe 冷却灾难
cooling effect 冷却效应
cooling flow 冷却流
cooling flow galaxy 冷流星系 | 冷卻流星系〔台〕
cooling frequency 冷却频率
cooling function 冷却函数
cooling rate 冷却率
cooling time 冷却时间
cool star 冷星
cool subdwarf 冷亚矮星
cool subgiant 冷亚巨星

cool supergiant 冷超巨星 又见: *cold supergiant*
co-orbit 共轨
coorbital satellite 共轨卫星
coordinate axis 坐标轴 又见: *axis of coordinate*
coordinate basis 坐标基
coordinate direction 坐标方向
Coordinated Universal Time (=UTC) 协调世界时
coordinate measuring instrument 坐标量度仪
coordinate of the pole 地极坐标
coordinate perturbation 坐标摄动
coordinate system 坐标系 又见: *system of coordinate*
coordinate time 坐标时
coordinate transformation 坐标变换
   又见: *conversion of coordinates; transformation of coordinates*
Copeland's Septet 科普兰七重星系
Copenhagen category 哥本哈根判据
Copenhagen classification 哥本哈根分类
Copenhagen problem 哥本哈根问题
Copernican argument 哥白尼论点
Copernican Principle 哥白尼原理
Copernican system 哥白尼体系
Copernicus 哥白尼天文卫星（美国）
   又见: *OAO-3*
Copernicus crater 哥白尼环形山（月球）
coplanarity 共面性
coplanar orbits 共面轨道
copper (=Cu) 铜（29号元素）
Coprates Chasma 科普莱特斯深谷（火星）
coprocessor 协处理器
CORAVEL spectrometer (=Correlation Radial Velocities spectrometer) 相关式视向速度仪
Cor Caroli 常陈一（猎犬座α）
Cordelia 天卫六 | 天[王]衛六〔台〕
Cordoba Durchmusterung (=CD; CoD)【德】科尔多瓦巡天星表
Cordoba Zone Catalogue (=CZ) 科尔多瓦分区星表
core collapse 核区坍缩
core-collapse progenitor 核坍缩前身天体
core collapse supernova 核心坍缩超新星
core-dominated quasar (=CDQ) 核主导类星体
core fitting method 核区拟合法
core-halo galaxy 核晕星系 | 核－暈星系〔台〕
core-halo model 核晕模型
core-halo source 核晕源

core of a line 线心 | [谱]線[中]心〔台〕
   又见: *line core; line center*
corequake 核震
core radius 核半径
Cor Hydrae 星宿一（长蛇座α） 又见: *Alfard; Alphard*
Cori crater 科里环形山（月球）
Coriolis coupling 科里奥利耦合
Coriolis crater 科里奥利环形山（月球）
Coriolis distortion 科里奥利畸变
Coriolis effect 科里奥利效应
Coriolis force 科里奥利力 | 柯氏力〔台〕
Cor Leonis 轩辕十四（狮子座α） 又见: *Regulus*
Cornu prism 考组棱镜
Corona Australids 南冕流星群
Corona Australis (=CrA) 南冕座
Corona Borealis (=CrB) 北冕座
Corona Borealis cluster 北冕星系团
coronagraph 日冕仪 又见: *coronograph*
coronagraphic camera 日冕照相机
   又见: *coronographic camera*
coronal abundance 日冕丰度
coronal active region 日冕活动区
coronal activity 日冕活动，星冕活动 | 星冕活動〔台〕
coronal arch 冕拱
coronal bright point 日冕亮点
coronal bubble 冕泡
coronal cavity 冕穴，冕腔 | 冕穴〔台〕
coronal cloud 冕云
coronal condensation 日冕凝区，日冕凝聚物 | 日冕凝聚物〔台〕
coronal continuum 日冕连续谱
coronal dividing line 星冕分界线
coronal electrophotometer 日冕光电光度计
coronal enhancement 日冕增强区
coronal equilibrium 日冕平衡 | 冕平衡〔台〕
coronal fan 冕扇
coronal forbidden line 日冕禁线
coronal gas 冕区气体 | 星冕氣體〔台〕
coronal green line 日冕绿线，星冕绿线
coronal heating 日冕加热
coronal helmet 冕盔
coronal hole 冕洞 | [日]冕洞〔台〕 又见: *solar coronal hole*
coronal light 冕光
coronal line 日冕谱线 | 日冕[谱]線〔台〕

coronal loop 冕环　又见: *solar coronal loop*
coronal magnetic energy 冕区磁能
coronal mass ejection （＝CME）日冕物质抛射 | 日冕物質噴發〔台〕
coronal optical polarization 日冕光学偏振
coronal oscillation 冕震 | 日冕振動〔台〕
coronal plasma 日冕等离子体
coronal plume 冕羽
coronal prominence 冕珥
coronal rain 冕雨
coronal ray 日冕射线
coronal red line 日冕红线，星冕红线
coronal streamer 冕流 | 日冕流〔台〕
coronal sunspot prominence 黑子冕珥
coronal transient 日冕瞬变
coronal whip 冕鞭
coronal wind 冕风
corona of galaxy 星系冕　又见: *galactic corona*
corona of the Galaxy 银河系冕
coronascope 日冕观测镜
Coronas-Photon 日冕质子号卫星（俄罗斯）
　　又见: *Koronas-Foton*
coronium 氢（miǎn，假想元素）
coronograph 日冕仪　又见: *coronagraph*
coronographic camera 日冕照相机
　　又见: *coronagraphic camera*
CoRoT （＝Convection rotation and planetary transits）科罗系外行星探测器 | 科洛[系外行星探测器]〔台〕（法国卫星）
corotating enclosure 共转围罩
corotation 共转，正转，顺转 | 共轉〔台〕
corotational departure 顺自转方向发射
corotational torque 共旋力矩
corotation circle 共转圈
corotation electric field 正转电场
corotation radius 共转半径
corotation resonance 共转共振
corotation zone 共转带
corpuscle 微粒　又见: *particulate*
corpuscular beam 微粒束
corpuscular cloud 微粒云
corpuscular eclipse 微粒食
corpuscular emission 微粒发射
corpuscular radiation 微粒辐射
corpuscular stream 微粒流
corrected area 改正面积
corrected establishment 改正潮候时差

correcting lens 改正透镜 | 修正透鏡〔台〕
correcting plate 改正片，改正板 | 修正鏡片〔台〕
　　又见: *corrector plate*
correction coefficient 改正系数 | 修改係數〔台〕
correction screw 改正螺旋 | 校正螺絲〔台〕
correction to time signal 时号改正数
corrector 改正镜，改正器 | 修正鏡〔台〕
corrector cell 改正镜室
corrector plate 改正片，改正板 | 修正鏡片〔台〕
　　又见: *correcting plate*
correlated double-sampling 相关双采样
correlation 相关　又见: *dependence*
correlation analysis 相关分析
correlation coefficient 相关系数　又见: *coefficient of correlation*
correlation detection 相关探测
correlation function 相关函数
correlation function of clusters 星系团相关函数
correlation function of galaxies 星系相关函数
correlation function of halo 晕相关函数
correlation function of quasars 类星体相关函数
correlation interferometer 相关干涉仪
correlation length 相关长度
Correlation Radial Velocities spectrometer （＝CORAVEL spectrometer）相关式视向速度仪
correlation receiver 相关接收机
correlator 相关器
correlogram 相关图
corrugated horn 波纹喇叭
corrugation 褶皱
Corsa-B 天鹅号（日本X射线天文卫星）
　　又见: *Hakucho*
Cor Scorpii 心宿二，大火 | 大火〔台〕（天蝎座 α）　又见: *Antares*
Cor Serpentis 天市右垣七，蜀（巨蛇座 α）
　　又见: *Unuk al Hai; Unukalhai*
Cor Tauri 毕宿五（金牛座 α）　又见: *Aldebaran*
Corvus （＝Crv）乌鸦座
COS （＝Celestial Observation Satellite）COS天文卫星
COS-B COS－B天文卫星（欧洲 γ 射线卫星）
COS-B Catalogue COS－B伽马源表
cosecant law 余割律 | 餘割定律〔台〕
cosine law 余弦律　又见: *cosine rule*
cosine rule 余弦律　又见: *cosine law*
cosine wave 余弦波
cosmic absorption 宇宙吸收
cosmic abundance 宇宙丰度

cosmic acceleration 宇宙加速
cosmic aerodynamics 宇宙气体动力学
  又见: *cosmic gasdynamics*
cosmic age 宇宙年龄 又见: *age of the universe*
cosmicality 宇宙性
Cosmic Anisotropy Telescope （=CAT）宇宙各向异性望远镜
cosmic background 宇宙背景 又见: *universal background*
Cosmic Background Explorer （=COBE）宇宙背景探测器 | 宇宙背景探测卫星〔台〕（美国卫星）
Cosmic Background Imager （=CBI）宇宙背景成像仪
cosmic background radiation 宇宙背景辐射
cosmic brightness 宇宙亮度
cosmic catastrophe 宇宙灾变
cosmic censorship 宇宙监察 | 宇宙監察說〔台〕
cosmic chemistry 宇宙化学
cosmic cloud 宇宙云
cosmic constant 宇宙学常数 | 宇宙常数〔台〕
  又见: *cosmological constant*
cosmic deceleration 宇宙减速
cosmic density field 宇宙密度场
cosmic density parameter 宇宙密度参数
cosmic distance 宇宙距离
cosmic distance ladder 宇宙距离阶梯
cosmic dust 宇宙尘 又见: *cosmic spherule*
cosmic electrodynamics 宇宙电动力学
cosmic emission 宇宙辐射 又见: *cosmic radiation*
Cosmic Evolution Survey （=COSMOS）宇宙演化巡天
cosmic expansion 宇宙膨胀 又见: *expansion of the universe*
cosmic experiment 宇宙实验
cosmic frame 宇宙坐标架
cosmic gasdynamics 宇宙气体动力学
  又见: *cosmic aerodynamics*
cosmic gusher 宇宙喷射源
cosmic hole 宇宙空洞
cosmic horizon 宇宙视界
Cosmic Hot Interstellar Plasma Spectrometer Satellite （=CHIPS）星际热等离子体光谱卫星
cosmic jet 宇宙喷流
Cosmic Lens All-Sky Survey （=CLASS）宇宙透镜巡天
cosmic light 宇宙光 | 宇宙背景 [夜] 光〔台〕

cosmic light horizon 宇宙光子视界
cosmic Mach number 宇宙马赫数
cosmic magnetic field 宇宙磁场
cosmic magnetohydrodynamics 宇宙磁流体力学
cosmic mean 宇宙均值
cosmic mean density 宇宙平均密度
cosmic microwave background （=CMB）宇宙微波背景
cosmic microwave background anisotropy （=CMB anisotropy）宇宙微波背景各向异性
cosmic microwave background polarization （=CMB polarization）宇宙微波背景偏振, 宇宙微波背景极化
cosmic microwave background radiation （=CMBR）宇宙微波背景辐射 | [宇宙] 微波背景辐射〔台〕
cosmic microwave background spectral distortion （=CMB spectral distortion）宇宙微波背景谱畸变
cosmic microwave radiation （=CMR）宇宙微波辐射
cosmic noise 宇宙噪声 又见: *cosmic static*
cosmic physics 宇宙物理学 又见: *cosmophysics*
cosmic plasma 宇宙等离子体
cosmic pressure 宇宙压力
cosmic radiation 宇宙辐射 又见: *cosmic emission*
cosmic radio astronomy 宇宙射电天文学
cosmic radio radiation 宇宙射电 | 宇宙 [無線] 電波〔台〕
cosmic radio source 宇宙射电源
cosmic radio wave 宇宙射电波
cosmic ray 宇宙线 又见: *ultraray*
cosmic ray abundance 宇宙线丰度
cosmic ray age 宇宙线年龄
cosmic ray anisotropy 宇宙线各向异性
cosmic ray astronomy 宇宙线天文学
cosmic ray astrophysics 宇宙线天体物理
cosmic ray background 宇宙线背景
cosmic ray burst 宇宙线暴
cosmic ray event 宇宙线事件
cosmic ray exposure age 宇宙线曝射法年龄
cosmic ray exposure time 宇宙线曝射时间
cosmic ray flare 宇宙线耀斑
cosmic ray shower 宇宙线簇射
cosmic ray source 宇宙线源
cosmic ray spectrum 宇宙线谱
cosmic ray telescope 宇宙线望远镜
cosmic ray trajectory 宇宙线轨迹
cosmic redshift 宇宙红移

cosmic refraction 宇宙大气折射
Cosmic Research 《宇宙研究》（俄罗斯期刊）
cosmic rocket 宇宙火箭
cosmic scale factor 宇宙标度因子 又见：*scale factor of the universe*
cosmic shear 宇宙切变
cosmic singularity 宇宙奇点 又见：*singularity of the universe*
cosmic source 宇宙源 又见：*celestial source*
cosmic spherule 宇宙尘 又见：*cosmic dust*
cosmic static 宇宙噪声 又见：*cosmic noise*
cosmic string 宇宙弦
cosmic texture 宇宙纹形
cosmic time 宇宙时
cosmic turbulence 宇宙湍动
cosmic variance 宇宙方差
cosmic velocity field 宇宙速度场
cosmic velocity stage 宇宙速度级
cosmic virial theorem 宇宙位力定理
cosmic void 巨洞 又见：*void*
cosmic web 宇宙网
cosmic X-ray background 宇宙 X 射线背景
cosmic X-ray burst 宇宙 X 射线暴
cosmic X-ray burster 宇宙 X 射线暴源
cosmic X-ray source 宇宙 X 射线源 又见：*celestial X-ray source*
cosmic yardstick 量天尺 又见：*sky measuring scale*
cosmic year 宇宙年 即：*Galactic year*
cosmic γ-ray background 宇宙 γ 射线背景
cosmic γ-ray burst 宇宙 γ 射线暴
cosmic γ-ray burster 宇宙 γ 射线暴源
cosmic γ-ray source 宇宙 γ 射线源 又见：*celestial γ-ray source*
cosmi microwave background experiment （＝CMB experiment）宇宙微波背景实验装置
cosmobiology 宇宙生物学，天体生物学
cosmochemistry 宇宙化学，天体化学 | 宇宙化學〔台〕
cosmochronological method 宇宙纪年法
cosmochronology 宇宙纪年学
cosmogenic activation 宇宙线激活
cosmogonal theory 天体演化理论
cosmogoner 天体演化学家 又见：*cosmogonist*
cosmogonist 天体演化学家 又见：*cosmogoner*
cosmogony 天体演化学 | 天文演化學〔台〕
cosmographer 宇宙学家 又见：*cosmographist; cosmologist*
cosmographist 宇宙学家 又见：*cosmographer; cosmologist*
cosmography 宇宙志
cosmolabe 星盘式测角仪
cosmological constant 宇宙学常数 | 宇宙常數〔台〕 又见：*cosmic constant*
cosmological dimming 宇宙学昏暗
cosmological dipole 宇宙学偶极矩
cosmological distance 宇宙学距离 | 宇宙距離〔台〕
cosmological distance scale 宇宙学距离尺度
cosmological effect 宇宙学效应
cosmological hypothesis 宇宙学假设
cosmological infall 宇宙学沉降
cosmological interpretation 宇宙学解释
cosmological model 宇宙学模型 | 宇宙模型〔台〕
cosmological nucleosynthesis 宇宙核合成
cosmological paradox 宇宙学佯谬 | 宇宙論佯繆〔台〕
cosmological parameter 宇宙学参数
cosmological phase transition 宇宙学相变
cosmological principle 宇宙学原理 | 宇宙論原則〔台〕
cosmological redshift 宇宙学红移 | 宇宙紅位移〔台〕
cosmological simulation 宇宙学模拟
cosmological space-time 宇宙学时空
cosmological time 宇宙学时间
cosmological time scale 宇宙学时标
cosmologist 宇宙学家 又见：*cosmographer; cosmographist*
cosmology 宇宙学，宇宙论
cosmonaut 宇航员 | 太空人〔台〕 又见：*astronaut*
cosmonautics 宇航学 | 宇[宙]航[行]學，太空航行學〔台〕 又见：*astronautics*
cosmophysics 宇宙物理学 又见：*cosmic physics*
COSMOS （＝Cosmic Evolution Survey）宇宙演化巡天
Cosmos 宇宙系列卫星 | 宇宙號 [天文衛星]〔台〕（苏联）
cosmos 宇宙 又见：*universe*
COSMOS machine COSMOS 多功能底片测量仪
cosmosophy 宇宙说
cosmosphere 天球仪
cosmos remote sensing 宇宙遥感
cosmozoic theory 宇宙生命说 又见：*cosmozoism*
cosmozoism 宇宙生命说 又见：*cosmozoic theory*
cospace 共空间

COSPAR (＝Committee on Space Research) 国际空间研究委员会 | 太空研究委員會〔台〕
cost-aperture analysis 造价一口径分析
COSTAR (＝The Corrective Optics Space Telescope Axial Replacement) COSTAR 光学改正系统
Coster-Kronig transition 科斯特一克勒尼希跃迁
cotangent bundle 余切丛
cotidal chart 等潮图 又见：*cotidal map*
cotidal hour 等潮时
cotidal map 等潮图 又见：*cotidal chart*
coude focus 折轴焦点
coude mounting 折轴装置
coude reflector 折轴反射望远镜
coude refractor 折轴折射望远镜 又见：*elbow refractor*
coude spectrograph 折轴摄谱仪
coude spectrum 折轴光谱
coude telescope 折轴望远镜
Coulomb attraction 库仑引力
Coulomb barrier 库仑势垒
Coulomb-Born approximation 库仑一玻恩近似
Coulomb collision 库仑碰撞
Coulomb coupling 库仑耦合
Coulomb crater 库仑环形山（月球）
Coulomb drag 库仑阻力
Coulomb field 库仑场
Coulomb force 库仑力
Coulomb interaction 库仑互作用
Coulomb law 库仑定律
Coulomb logarithm 库仑对数
Coulomb scattering 库仑散射
Coulomb wave function 库仑波函数
count 计数 又见：*counting; number counting*
countdown 倒计数
counter 计数器 又见：*numerator*
counteraction 反作用
counterbalance 平衡锤 又见：*counterpoise*
counter clockwise 逆时针方向
counterglow 对日照 又见：*counter-twilight; zodiacal counterglow*
counter-Jupiter 太岁，岁阴 又见：*Shadow-Planet*
counterpart 对应体
counterpoise ① 平衡锤 又见：*counterbalance* ② 地网
counter-rotating core 反旋核
counter-rotational departure 反自转发射

counter sun 反日
counter telescope 计数望远镜
counter-twilight 对日照 又见：*counterglow; zodiacal counterglow*
counter weight 平衡重
counter weight arm 平衡臂
counting 计数 又见：*count; number counting*
counting rate 计数率
count-magnitude relation 计数一星等关系
counts in cells 分格计数
coupling constant 耦合常数
Courant condition 柯朗条件
Cousins color system 卡曾斯颜色系统
covalent compound 共价化合物
covariance 协变性，协方差
covariance function 协变函数
covariance matrix 协方差矩阵
covariance spectrum 协方差谱
covariant component 协变分量
covariant derivative 协变导数
covariant divergence 协变散度
covariant index 协变指标
coverage 覆盖度
coverage function 覆盖函数
Cowell method 科威尔方法
C-O white dwarf 碳氧白矮星
Cowling approximation 柯林近似
Cowling mechanism 柯林机制
Coxa 西次相，太微右垣四（狮子座 θ）
 又见：*Chertan; Chort*
CP (＝Cambridge pulsar) 剑桥脉冲星
CPC (＝Cape Photographic Catalogue) 好望角照相星表
CPD (＝Cape Photographic Durchmusterung)【德】好望角照相巡天
c.p.m. (＝common proper motion) 共自行
CPT invariance (＝Charge-Parity-Time reversal invariance) 电荷一宇称一时间反演不变性，CPT 不变性
CP violation (＝Charge-Parity violation) 电荷一宇称破坏，CP 破坏
Cr (＝chromium) 铬（24 号元素）
CrA (＝Corona Australis) 南冕座
Crablike source 类蟹源
Crab Nebula 蟹状星云（NGC 1952）
Crab pulsar 蟹云脉冲星 | 蟹狀星雲衝星〔台〕（NP 0532）

CrAO （=Crimean Astrophysical Observatory）克里米亚天体物理台
crape ring 暗环　又见：*dusky ring*
Crater （=Crt）巨爵座
crater ① 环形山｜環形山，[陨石] 坑洞，火山口〔台〕　又见：*ring mountain* ② 陨击坑（行星地貌）　又见：*impact crater*
crater chain 环形山串　又见：*catena*
crater cluster 环形山群
crater counting 环形山计数
crater floor 环形山底｜坑洞底〔台〕
cratering 陨击　又见：*meteorite impact*
craterlet 小环形山｜小坑洞〔台〕
CrB （=Corona Borealis）北冕座
creation field （=C-field）创生场
creation operator 产生算符
creep 频率漂移，蠕变｜潜移〔台〕
Cremona crater 克雷莫纳环形山（月球）
Crepe ring C 环（土星）
crepuscular arch 曙暮辉弧
crepuscular rays 曙暮辉
crescent moon ① 上蛾眉月　又见：*waxing crescent* ② 蛾眉月
Crescent Nebula 蛾眉星云（NGC 6888）
crescent sun 蛾眉日
Cressida 天卫九｜天[王]卫九〔台〕
crest 波峰　又见：*wave crest*
crest value 峰值　又见：*peak value*
crevasse 双峰共振，裂缝
crew module 乘员舱
Crimean Astrophysical Observatory （=CrAO）克里米亚天体物理台
crisp image 匀边像
cristobalite 方英石
criterion 判据｜判據，判别準則〔台〕
critical angle 临界角
critical argument 临界幅角
critical constant 临界常数
critical current 临界电流
critical density 临界密度
critical equatorial velocity 临界赤道速度
critical equipotential lobe 临界等位瓣
critical equipotential surface 临界等位面
critical field 临界场
critical frequency 临界频率
critical inclination 临界倾角
critical layer 临界层

critical line 临界线
critical mass 临界质量
critical mass-ratio 临界质量比
critical mass-to-light ratio 临界质光比
critical model 临界模型
critical point 临界点
critical radius for accretion 临界吸积半径
critical Rayleigh number 临界瑞利数
critical refraction 临界折射
critical surface 临界面｜臨界表面〔台〕
critical surface density 临界面密度
critical temperature 临界温度
critical term 临界项
critical value 临界值
critical velocity 临界速度
Crocco crater 克罗科环形山（月球）
crochet 磁钩　又见：*magnetic crochet*
Crommelin crater 克罗姆林环形山（月球）
Crommelin's comet 克罗姆林彗星
cross antenna 十字天线
cross-axis mounting 十字轴基架
cross-correlation 互相关
cross-correlation function 互相关函数
cross-correlation interferometer 互相关干涉仪
cross-correlation method 互相关法　又见：*cross-correlation technique*
cross-correlation receiver 互相关接收器
cross-correlation spectrometer 互相关频谱仪
cross-correlation technique 互相关法　又见：*cross-correlation method*
cross correlator 互相关器
cross-disperser 横向色散器
cross-disperser prism 横向色散棱镜
crossed-Dragone telescope 德拉戈相交式望远镜
crossed-field 交叉场
crossed-grating antenna 十字栅天线
crossed lens 最小球差单透镜
crossed Nicols 正交尼科尔棱镜
cross hair 叉丝｜十字絲，叉絲〔台〕　又见：*cross wire*
crossing time 跨越时间｜穿越时间〔台〕　又见：*crossover time*
cross interferometer 十字干涉仪
crossover effect 跨越效应
crossover time 跨越时间｜穿越时间〔台〕　又见：*crossing time*
cross polarization 交叉偏振

cross-power spectrum 互功率谱
cross section 截面 又见: *surface of section*
cross spectrum 互谱
cross-spectrum analyzer 互谱分析器
cross-staff 直角照准仪，十字杆
Cross-stick at back of chariot 轸宿（二十八宿）
 又见: *Chariot cross-board; Chariot Platform*
cross-talk 串扰
cross-track component 垂迹分量
cross-track error 垂迹误差
cross wire 叉丝 | 十字絲，叉絲〔台〕 又见: *cross hair*
cross-wire micrometer 十字丝测微计
crown flint glass 冕火石玻璃
crown glass 冕牌玻璃
CRS （＝Celestial Reference System）天球参考系
Crt （＝Crater）巨爵座
Cru （＝Crux）南十字座
crustal block 地壳块
crustal deformation 地壳变形
crustal dynamics 地动力学，星壳动力学
crustal spreading 地壳扩张
crust-mantle boundary 地壳－地幔边界
crust-mantle system 地壳－地幔系统
crust movement 地壳运动
crust of the earth 地壳 又见: *earth crust; earth shell*
Crux （＝Cru）南十字座
Crux-Scutum arm 南十字－盾牌臂
Crv （＝Corvus）乌鸦座
cryocooler 低温制冷机
cryogen 制冷剂，冷却剂
cryogenically-cooled parametric amplifier 低温致冷参量放大器
cryogenic camera 致冷照相机
cryogenic cooling 致冷
Cryogenic Dark Matter Search （＝CDMS）暗物质低温搜寻计划
cryogenics 低温学
cryogenic system 低温制冷系统
cryogenic telescope 低温望远镜
cryostat 低温恒温器
cryovolcano 冰火山
crystal controlled oscillator 晶控振荡器
crystal lattice 晶格
crystalline heaven 水晶天
crystallite 微晶
crystallographic axis 晶轴

crystal mixer 晶体混频器
Cs （＝caesium）铯（55号元素）
CSA （＝Canadian Space Agency）加拿大宇航局
CSI （＝Catalogue for Stellar Identification）恒星证认表
CSM （＝command and service module）指令－服务舱
CSO （＝Caltech Submillimeter Observatory）加州理工学院亚毫米波天文台
C-S star C－S型星，强氰线星
CST （＝central standard time）美国中部标准时
C star （＝carbon star）碳星
CT （＝central time）中部时
CTA （＝Cherenkov Telescope Array）CTA切伦科夫望远镜阵
CTC （＝closed timelike curve）闭合类时曲线
CTD （＝charge transfer device）电荷转移器件
CTIO （＝Cerro Tololo Inter-American Observatory）托洛洛山美洲天文台
CTRS （＝conventional terrestrial reference system）习用地球参考系
CTS （＝Catalogue of Time Services）授时星表
C-type asteroid C型小行星
Cu （＝copper）铜（29号元素）
cubewano 经典柯伊伯带天体 又见: *classical Kuiper belt object*
cubic distortion 三次畸变
Cubic Kilometre Neutrino Telescope （＝KM3NeT）立方千米中微子望远镜
cubic parsec 立方秒差距
Cujam 斗一（武仙座ω）
Culann Patera 库兰托边火山（木卫一）
culminant star 中天星，中星
culmination 中天 又见: *meridian transit; meridian passage*
cumulative error 累积误差
cumulative shift 累积位移
cumulative stress 累积应力
Cunitz crater 库尼茨环形山（金星）
Cupid 天卫二十七
cupola 圆顶，圆顶室 又见: *dome*
Curie crater 居里环形山（月球）
Curie temperature 居里温度
Curiosity rover 好奇号火星车
curium （＝Cm）锔（96号元素）
curl 旋度
current density 流密度
current-free field 无流场

current-sheath model　流鞘模型
current sheet　电流片，中性片　又见：*neutral sheet*
Cursa　玉井三（波江座 β）　又见：*Nath*
curtate distance　黄道面投影距离
Curtius crater　库尔提乌斯环形山（月球）
curvaton　曲率子
curvature　曲率
curvature constant　曲率常数
curvature correction　曲率改正 | 星徑曲率改正〔台〕
curvature index　曲率指数
curvature of field　场曲 | 视野弯曲像差，[像] 场 [弯] 曲〔台〕　又见：*field curvature*
curvature of parallel　纬线弯曲
curvature of space　空间曲率
curvature of space-time　时空曲率　又见：*space-time curvature*
curvature of the universe　宇宙曲率
curvature perturbation　曲率扰动
curvature radiation　曲率辐射
curvature scalar　曲率标量
curvature wavefront sensor　曲率波前传感器
curved fan　曲扇状流
curved field　弯曲场
curved grating　曲面光栅
curved jet　弯曲喷流 | 曲線噴流〔台〕
curved space　弯曲空间
curved spectrum　弯曲频谱
curved surface　曲面
curved tail　弯曲彗尾 | 曲彗尾〔台〕
curve fitting　曲线拟合 | 曲線擬合，曲線求律〔台〕
curve of growth　生长曲线　又见：*growth curve*
curve of zero velocity　零速度线　又见：*zero velocity curve*
curvilinear coordinate　曲线坐标
curvilinear motion　曲线运动
cushion distortion　枕形畸变
cusp　月角，尖点 | 尖角，尖點〔台〕
cusp cap　金星角
cusp catastrophe　尖峰突变
cusp-core degeneracy　尖峰－核简并性
cut-off error　截断误差　又见：*truncation error*
cut-off frequency　截止频率
cut-off rigidity　截断刚度
Cuvier crater　居维叶环形山（月球）
CVF　（＝circular variable filter）圆形可变滤光片

CVN　（＝Chinese VLBI network）中国甚长基线干涉网
CVn　（＝Canes Venatici）猎犬座
CVO　（＝Canadian Virtual Observatory）加拿大虚拟天文台
CW　（＝continuous wave）连续波
cyanoacetylene　丙炔腈（HC_3N）
cyanoethylene　丙烯腈（C_3H_3N）　又见：*vinyl cyanide*
cyanogen band　氰带
cyanogen line　氰线
cyanogen radical　氰基（CN）
cyanometry　天色测量
Cybele　原神星（小行星 65 号）
Cybele group　库柏勒群
cycle-amplitude relation　周幅关系
cycle of the sun　太阳周　又见：*heliacal cycle*
cycle theorem　循环定理
cyclical variability　周期性变化
cyclicity　周期性　又见：*periodicity*
cyclic model　循环模型
cyclic period　循环周期
cyclic transition　循环跃迁
cyclic universe　循环宇宙
cyclone　气旋
Cyclops project　独眼神计划 | 獨眼神計畫〔台〕
cyclotron absorption　回旋加速吸收
cyclotron damping　回旋加速阻尼
cyclotron emission　回旋加速发射　又见：*gyroemission*
cyclotron frequency　回旋加速频率 | 迴旋频率〔台〕　又见：*gyrofrequency*
cyclotron harmonic wave　回旋加速谐波
cyclotron instability　回旋加速不稳定性
cyclotron maser　回旋加速微波激射
cyclotron radiation　回旋加速辐射
cyclotron resonance　回旋共振　又见：*gyroresonance*
cyclotron scattering　回旋散射
cyclotron turnover　回旋频谱反转
Cyg　（＝Cygnus）天鹅座
Cygnids　天鹅流星群 | 天鹅 [座] 流星群〔台〕
Cygnus　（＝Cyg）天鹅座
Cygnus A　天鹅射电源 A
Cygnus Cloud　天鹅暗云
Cygnus Loop　天鹅圈（NGC 6960）　又见：*Network Nebula*
Cygnus Rift　天鹅座暗隙

Cygnus Star Cloud　天鹅恒星云
cylindrical coordinate　柱坐标 | 柱[面]坐標〔台〕
cylindrical coordinate system　柱坐标系
cylindrical equilibrium　柱体平衡
cylindrical harmonics　圆柱调和函数 | 圆柱[調和]函數〔台〕
cylindrical lens　柱面透镜
cylindrical paraboloid radio telescope　柱形抛物面射电望远镜
cylindrical wave　柱面波
Cyllene　木卫四十八

Cyrano crater　西哈诺环形山（月球）
Cyrillids　西里尔流星雨　又见：*Cyrillid Shower*
Cyrillid Shower　西里尔流星雨　又见：*Cyrillids*
Cyrillus crater　西里尔环形山（月球）
Cytherean atmosphere　金星大气
CZ　（＝Cordoba Zone Catalogue）科尔多瓦分区星表
Czerny-Turner spectrometer　切尔尼－特纳光谱仪
CZT array　（＝CdZnTe detector array）碲锌镉面阵（探测器）

D

Dabih Major 牛宿一（摩羯座 β）
Dabih Minor 牛宿增十二（摩羯座 16）
Dabu 大埔（小行星 3611 号）
D abundance 氘丰度
Dactyl 艾卫（小行星 Ida 的卫星）
Daedalia Planum 代达利亚高原（火星）
Daedalus crater 代达罗斯环形山（月球）
Dahlgren Polar Monitoring Service （＝DPMS）达尔格伦地极监测服务 | 達爾格連 [地] 極監測服務〔台〕
daily mean 日平均
daily motion 日运动 | 周日運動〔台〕
daily range 日差程
daily rate 日速
Daiwensai 戴文赛（小行星 3405 号）
d'Alembert characteristic 达朗贝尔性质
d'Alembert crater 达朗贝尔环形山（月球）
d'Alembert's principle 达朗贝尔原理
Dalgaranga crater 达尔加兰加陨星坑（地球）
Dalian 大连（小行星 3187 号）
Dali Chasma 达利深谷（金星）
Dall-Kirkham Cassegrain telescope D－K 式卡塞格林望远镜
Dall-Kirkham telescope 多尔－柯卡姆望远镜
Dalton crater 道尔顿环形山（月球）
DAMA/NaI dark matter search experiment DAMA 碘化钠暗物质探测实验
DAMPE （＝Dark Matter Particle Explorer）暗物质粒子探测器（中国卫星）
damped harmonic motion 阻尼谐振动
damped Lyman α system （＝DLA）阻尼莱曼 α 系统
damped oscillation 阻尼振荡 | 阻尼振動，阻尼振蕩〔台〕
damped wave 阻尼波
damping 阻尼，衰减 | 阻尼〔台〕

damping broadening 阻尼致宽 又见：broadening by damping
damping coefficient 阻尼系数
damping constant 阻尼常数
damping decrement 阻尼减幅
damping radiation 阻尼辐射
damping rate 阻尼速率
damping term 阻尼项
damping time 阻尼时间
damping wing 阻尼翼
Damsel 女宿（二十八宿） 又见：Maid; Girl; Woman
Danaë 囚神星（小行星 61 号）
Danieltsui 崔琦（小行星 77318 号）
Danilova crater 丹尼洛娃环形山（金星）
Danjon astrolabe 丹戎等高仪 | 丹容等高儀〔台〕
Danjon crater 丹戎环形山（月球）
Danjon scale 丹戎标度
Dante crater 但丁环形山（月球）
DAO （＝Dominion Astrophysical Observatory）自治领天体物理台 | 自治領天文物理台〔台〕（加拿大）
DAP （＝digital area photometer）数字面积光度计
Daphne 桂神星（小行星 41 号）
Daphnis 土卫三十五
dark adaptation 暗适应
dark age 黑暗时期
dark burst 暗暴
dark cloud 暗云
dark companion 暗伴天体，暗伴星，暗伴星系 | 暗伴星〔台〕 又见：faint companion
dark component 暗成分
dark current 暗 [电] 流
dark diffuse nebula 暗弥漫星云
dark dome 暗拱
dark dust lane 暗尘带
dark dust nebula 暗尘云

dark-eclipsing variable　暗食变星
Darke equation　达克方程
dark energy　暗能量 | 暗能 [量]〔台〕
dark energy model　暗能量模型
darkening towards the limb　临边昏暗 | 周边减光，临邊昏暗〔台〕　又见：*limb darkening*
dark field　暗 [视] 场
dark flocculus　暗谱斑
dark Fraunhofer line　夫琅和费暗线
dark Galactic nebula　银河暗星云
dark galaxy　暗星系
dark halo　暗晕
dark halo crater　暗晕环形山
dark lane　暗带 | 暗帶，暗綫〔台〕
dark limb　暗边缘
dark matter　暗物质 | [黑] 暗物質，暗質〔台〕
dark matter annihilation　暗物质湮灭
dark matter candidate　暗物质候选体
dark matter decay　暗物质衰变
dark matter halo　暗物质晕
dark matter indirect detection　暗物质间接探测
Dark Matter Particle Explorer　（=DAMPE）暗物质粒子探测器（中国卫星）
dark matter search　暗物质搜寻
dark matter star　暗物质星
dark mottle　暗日芒
dark nebula　暗星云
darkness　黑度
dark of the moon　新月暗期，无月光期
darkroom　暗室
Dark Shadow　暗虚
Dark S Nebula　蛇形暗云（Barnard 72）又见：*Snake Nebula*
dark star　暗星　又见：*faint star*
dark γ-ray burst　暗 γ 暴
Darwin crater　达尔文环形山（月球）
Darwin ellipsoid　达尔文椭球体
Darwin Space Interferometer　达尔文空间干涉仪 | 達爾文太空干涉儀〔台〕
DASI　（=Degree Angular Scale Interferometer）度角尺度干涉仪（美国）
data acquisition　数据采集
data analysis　数据分析
database　数据库
data capacity　信息容量　又见：*information capacity*
data center　信息中心
data compression　数据压缩

data cube　数据立方
data display　数据显示　又见：*data presentation*
data filtering　数据滤波
data handling　数据处理　又见：*treatment of data; data processing*
data intensive astronomy　数据密集型天文学
data inversion　数据反演
data presentation　数据显示　又见：*data display*
data processing　数据处理　又见：*treatment of data; data handling*
data processor　数据处理机
data recording　数据记录
data reduction　数据归算
data-relay satellite　（=DRS）数据转播卫星
data retrieval　数据回收
data simulation　数据模拟
data storage　数据存储
date line　日界线
dating　纪年，计年 | 定年〔台〕
dating method　纪年法
DATS　（=Despun Antenna Test Satellite）消旋天线试验卫星（美国）
datum　基准面　又见：*datum level*
datum level　基准面　又见：*datum*
datum of elevation　高程基准面
daughter isotope　子 [体] 同位素
daughter substance　子系物质
Davida　戴维（小行星 511 号）
David Dunlap Observatory　（=DDO）大卫·邓拉普天文台，DDO 天文台
David Dunlap Observatory system　（=DDO system）DDO 分类系统
Davies-Cotton Cherenkov telescope　戴维斯－科顿型切伦科夫望远镜
da Vinci crater　达·芬奇环形山（月球）
Davis-Greenstein effect　戴维斯－格林斯坦效应
Davisson crater　戴维孙环形山（月球）
Davy Chain　戴维环形山链
Dawes Gap　道斯环缝（土星）
Dawes limit　道斯极限
DA white dwarf　DA 型白矮星
dawn side　黎明侧
Dawn Spacecraft　曙光号小行星探测器
day　日，昼 | 日〔台〕
day arc　昼弧
day break　拂晓
day glow　日辉 | 晝輝〔台〕

day light　日光　又见：*sunlight*
daylight fireball　白昼火流星
daylight meteor　白昼流星　又见：*daytime meteor*
daylight saving meridian　夏令子午线
daylight saving noon　夏令正午
daylight saving time　（＝DST）夏令时｜日光節約時間〔台〕　又见：*summer time*
daylight stream　白昼流星群　又见：*daytime stream*
day number　日数
day of autumnal equinox　秋分日
day of summer solstice　夏至日
day of vernal equinox　春分日
day of winter solstice　冬至日
day of year　积日
day range　日变化　又见：*diurnal change; diurnal variation*
day side　白昼侧
day star　晨星，启明星
daytime meteor　白昼流星　又见：*daylight meteor*
daytime stream　白昼流星群　又见：*daylight stream*
daytime train　白昼流星余迹｜白晝[流星]餘跡〔台〕
daytime transparency　白昼天空透明度
day value　白天值
db galaxy　（＝dumbbell galaxy）哑铃状星系
DBI action　（＝Dirac-Born-Infeld action）狄拉克－玻恩－因费尔德作用量
D bottleneck　氘瓶颈效应
D-brane　D－膜
DB white dwarf　DB 型白矮星
DC　（＝dispersion constant）色散常数
d.c. compensation　直流补偿
D-class asteroid　D 型小行星
DC white dwarf　DC 型白矮星
DDO　（＝David Dunlap Observatory）大卫·邓拉普天文台，DDO 天文台
DDO classification　DDO 分类
DDO color system　DDO 颜色系统
DDO photometry　DDO 测光
DDO system　（＝David Dunlap Observatory system）DDO 分类系统
deactivation constant　钝化常数｜鈍化係數〔台〕
dead reckoning　盲航法｜推测航行〔台〕
dead time　死区时间
death date　坠落时间
debris　残骸
de Broglie frequency　德布罗意频率

de Broglie wave　德布罗意波
de Broglie wavelength　德布罗意波长
Debye crater　德拜环形山（月球）
Debye frequency　德拜频率
Debye-Hückel model　德拜－许克尔模型
Debye length　德拜长度
Debye potential　德拜势
Debye radius　德拜半径
Debye shielding distance　德拜屏蔽距离
Debye sphere　德拜球
Debye unit　德拜单位
decadent wave　减幅波　又见：*decaying wave*
decameter activity　十米波活动
decameter continuum　十米波连续谱
decameter wave　十米波　又见：*decametric wave*
decametric radiation　十米波辐射
decametric wave　十米波　又见：*decameter wave*
decan　黄道十度分度
decans　旬星　又见：*ten-day star*
decay coefficient　衰变系数
decay constant　衰变常数
decay curve　衰变曲线
decaying dark matter　衰变暗物质
decaying wave　减幅波　又见：*decadent wave*
decay instability　衰变不稳定性
decay mode　衰变模式
decay phase　衰变阶段
decay process　衰变过程
decay product　衰变产物
decay rate　衰变率
decay scheme　衰变图
decay time　衰变时间
deceleration parameter　减速参数，减速因子｜減速參數〔台〕
deceleration radiation　减速辐射
December Solstice　冬至（节气）　又见：*Winter Solstice*
December solstice　冬至点　又见：*winter solstice; first point of Capricornus*
decentering distortion　偏心畸变
de Cheseaux's comet　德塞瑟彗星　又见：*Comet de Chéseaux*
decimal counter　十进位计数器｜十進[位]計數器〔台〕
decimal exponent　（＝dex）对数增量｜岱，十进制数〔台〕
decimeter activity　分米波活动

decimeter continuum 分米波连续谱
decimeter wave 分米波 又见: *decimetric wave*
decimetric radiation 分米波辐射
decimetric wave 分米波 又见: *decimeter wave*
decipherer 译码器 又见: *decoder; code translator*
declination 赤纬
declination axis 赤纬轴
declination circle ① 赤纬度盘 又见: *declination setting circle* ② 赤纬圈 又见: *circle of declination; parallel of declination; declination parallel*
declination compass 赤纬计, 磁偏仪 又见: *declinometer*
declination parallel ① 赤纬圈 又见: *declination circle; circle of declination; parallel of declination* ② 等纬圈
declination setting circle 赤纬度盘 又见: *declination circle*
decline phase 下降阶段
declining phase 衰减阶段
declinometer 赤纬计, 磁偏仪 又见: *declination compass*
decoder 译码器 又见: *decipherer; code translator*
decoherence 退相干
decoloration 消色 又见: *discoloration*
decomposition 分解
decompression wave 减压波
deconvolution 消卷积, 解卷积, 退卷积
deconvolution image 消旋图像
decoupling 退耦
decoupling epoch 退耦期 | [物质－辐射] 退耦时间〔台〕 又见: *decoupling era*
decoupling era 退耦期 | [物质－辐射] 退耦时间〔台〕 又见: *decoupling epoch*
decrement 减幅 | 减幅, 减缩量〔台〕
decremental arc 渐缩环形山弧
decremental chain 渐缩环形山链
decrescent 亏月 又见: *waning moon*
decuplet 十重线
Dedekind ellipsoid 迪特金椭球体
dedisperser 消色散器, 消频散器
Deep Bay crater 深湾陨星坑（地球）
deep-depletion CCD 深耗尽型 CCD
deep earthquake 深震
deep-field observation 深天区观测, 深场观测 | 深空觀測〔台〕
deep focus 深震源
deep focus earthquake 深源震 又见: *deep focus shock*
deep focus shock 深源震 又见: *deep focus earthquake*
Deep Impact space probe 深度撞击空间探测器
deep inelastic scattering 深度非弹性散射
Deep Lens Survey （＝DLS）引力透镜深空巡天
deeply embedded infrared source （＝DEIS）深埋红外源
DEep Near Infrared Survey of the Southern Sky （＝DENIS）南天近红外深度巡天
deep sky 深空 又见: *deep space*
deep-sky object 深空天体
deep-sky phenomena 深空天象
deep space 深空 又见: *deep sky*
Deep Space 1 （＝DS1）深空 1 号（美国探测器）
Deep Space Network （＝DSN）深空探测网
deexcitation 去激发, 退激 [发]
deexcitation cross-section 去激发截面
defect of illumination 圆面未照亮区
deferent 均轮
deferred charge 滞后电荷
deficit angle 缺陷角
defining constant 定义常数
definite designation 正式命名, 正式编号 | 正式命名〔台〕
definition of image 像清晰度
definitive orbit 既定轨道
definitive time 确定时
definitive weight 确定权
deflagration 暴燃
deflagrationwave 暴燃波
deflecting force 偏转力
deflecting magnet 致偏磁体
deflection angle 偏转角 又见: *angle of deflection*
deflection of light 光线偏折 | 光線偏轉〔台〕
deflection of vertical 垂线偏差 又见: *deviation of the vertical; plumb-line deviation*
defocused reflector 离焦反射镜
defocussing 离焦
De Forest crater 德福雷斯特环形山（月球）
deformable body 可变形体
deformable mirror 变形镜
deformation 形变
deformation tensor 形变张量
dE galaxy （＝dwarf elliptical galaxy）矮椭圆星系
Degas crater 德加环形山（水星）
degeneracy 简并 | 簡併性, 簡併度〔台〕
degeneracy collapse 简并坍缩 | 簡併塌縮〔台〕

degeneracy pressure　简并压　又见：degenerate pressure
degenerate configuration　简并组态
degenerate dwarf　简并矮星
degenerate electron gas　简并电子气
degenerate electron pressure　电子简并压　又见：electron degeneracy pressure
degenerate gas　简并气体
degenerate matter　简并物质
degenerate neutron pressure　简并中子压力
degenerate parametric amplifier　简并参量放大器
degenerate plasma　简并等离子体
degenerate pressure　简并压　又见：degeneracy pressure
degenerate star　简并星
degenerate state　简并态
degenerate stellar configuration　简并恒星组态
degenerate system　简并系
degeneration　简并化
degree　度　又见：degree of arc; arc degree
Degree Angular Scale Interferometer　（＝DASI）度角尺度干涉仪（美国）
degree of accuracy　精确度
degree of arc　度　又见：degree; arc degree
degree of concentration　中聚度 | 集中度〔台〕
degree of confidence　置信度　又见：confidence
degree of dispersion　弥散度
degree of excitation　激发度
degree of freedom　自由度
degree of ionization　电离度 | 游離度〔台〕　又见：ionicity; ionizability; ionization degree
degree of obscuration　食分　又见：magnitude of eclipse; eclipse factor
degree of polarization　偏振度
degree of vacuum　真空度
degree of wetness　湿度　又见：humidity
Deimos　火卫二
deionization　去电离
DEIS　（＝deeply embedded infrared source）深埋红外源
Dekker mask　德科掩模
Del　（＝Delphinus）海豚座
Delambre crater　德朗布尔环形山（月球）
De la Rue crater　德拉鲁环形山（月球）
Delaunay element　德洛奈根数
Delaunay theory　德洛奈理论
Delaunay variable　德洛奈变量 | 德洛內變數〔台〕

de Laval nozzle　拉瓦尔型喷嘴，渐缩渐阔喷嘴
delay ambiguity function　延迟模糊度函数
delay beam　延迟波束
delay-Doppler mapping　延迟多普勒成图
delay fringe frequency　延迟条纹率
delay function　延迟函数
delay line　延迟线
delay resolution　延迟分辨率
delay signal　滞后信号 | 遲滯信號〔台〕
delay time　时延 | 延遲時間〔台〕
delay tracker　延迟跟踪器
delay tracking　延迟跟踪
Dellinger crater　德林杰环形山（月球）
Delphinus　（＝Del）海豚座
Delta　奎宿五（仙女座δ）
Delta Aquarid meteor shower　宝瓶δ流星雨
Demeter　木卫十（曾用名）　即：Lysithea
demodulation　解调制
demodulator　解调制器
Demonax crater　泽莫纳克斯环形山（月球）
Demon Star　恶魔星（英仙座β）　即：Algol
Deneb　天津四（天鹅座α）　又见：Deneb Cygni
Deneb Alased　五帝座一（狮子座β）　又见：Deneb Aleet; Denebola
Deneb Aleet　五帝座一（狮子座β）　又见：Deneb Alased; Denebola
Deneb Algedi　垒壁阵四（摩羯座δ）　又见：Deneb Algiedi
Deneb Algenubi　天仓二（鲸鱼座η）
Deneb Algiedi　垒壁阵四（摩羯座δ）　又见：Deneb Algedi
Deneb Cygni　天津四（天鹅座α）　又见：Deneb
Deneb Dulfim　败瓜一（海豚座ε）
Deneb Kaitos　①天仓一（鲸鱼座ι）②土司空（鲸鱼座β）　又见：Difda al Thani; Diphda
Deneb Okab　①吴越（天鹰座ζ）②吴越增一（天鹰座ε）
Denebola　五帝座一（狮子座β）　又见：Deneb Alased; Deneb Aleet
Dengfeng Observatory　登封观星台
DENIS　（＝DEep Near Infrared Survey of the Southern Sky）南天近红外深度巡天
dense array　密集阵列
dense cloud　稠密云
dense core　稠密云核
dense flint　重火石玻璃
dense interstellar dust cloud　（＝DIDC）稠密星际尘云

dense matter　稠密物质
dense plasma　稠密等离子体
dense star cluster　稠密星团
densified pupil imaging　密瞳成像法
densimeter　密度计　又：densitometer
densitometer　密度计　又：densimeter
density arm　密度臂
density correlation function　密度相关函数
density cusp　密度尖峰
density distribution　密度分布
density drift instability　密度漂移不稳定性
density evolution　密度演化
density excess　密度超　又：excess density
density-exposure curve　密度—曝光曲线
density fluctuation　密度起伏
density function　密度函数
density hole　密度空洞
density matrix　密度矩阵
density maxima　密度最大值
density of ionization　电离密度
density parameter　密度参数 | 密度參數，封閉參數〔台〕
density peak　密度峰
density perturbation　密度扰动
density profile　密度轮廓
density threshold　密度阈值
density variation　密度变化
density wave　密度波
density-wave model　密度波模型
density-wave theory　密度波理论
denudation　剥蚀
departure point　起标点
dependence　相关　又：correlation
dependence method　依数法　又：method of dependences
dependent equatorial coordinate　时角赤道坐标
dependent equatorial coordinate system　时角赤道坐标系
DEPFET pixel sensor　（=depleted p-channel field effect transistor pixel sensor）DEPFET 像元传感器
depleted p-channel field effect transistor pixel sensor　（=DEPFET pixel sensor）DEPFET 像元传感器
depolarization　消偏振
depolarizer　消偏振镜
depressed pole　下天极
depth equation　深度方程

depth of focus　焦深 | 焦深 [度]〔台〕
depth of penetration　贯穿深度　又见：penetration depth
depth relation　深度关系
dereddening　红化改正
De revolutionibus orbium coelestium　【拉】《天体运行论》（哥白尼著作）
derived constant　推导常数
derotator　消转器
Desargues crater　德扎尔格环形山（月球）
Descartes crater　笛卡儿环形山（月球）
descending branch　下降支 | 下降部分〔台〕
descending node　降交点　又见：south-bound node
Descent of Hoar Frost　霜降（节气）　又见：First Frost; Frost's Descent
descriptive astronomy　通俗天文学
Desdemona　天卫十 | 天 [王] 衛十〔台〕
designation of asteroids　小行星命名，小行星编号
designation of comets　彗星命名　又见：comet designation
designation of novae　新星命名，新星编号
designation of supernovae　超新星命名，超新星编号
designation of variable stars　变星命名，变星编号
design directivity　设计方向性
de Sitter model　德西特模型
de Sitter space　（=dS）德西特空间
de Sitter universe　德西特宇宙
Deslandres crater　德朗达尔环形山（月球）
desorption　退吸 | 退吸，脱附〔台〕
Despina　海卫五
despinning　自旋减速
Despun Antenna Test Satellite　（=DATS）消旋天线试验卫星（美国）
destabilizing effect　减稳效应
destructive interference　相消干涉
detached binary　不接双星 | 分離雙星〔台〕　又见：detached system
detached system　不接双星 | 分離雙星〔台〕　又见：detached binary
detailed balance　细致平衡
detailed balancing　细致平衡
detection　探测，检测，检波
detection limit　探测极限
detective quantum efficiency　（=DQE）探测量子效率
detectivity　探测能力，探测率
detector　探测器，检波器 | 探測器〔台〕

detector array　探测器阵
determination of orbit　定轨，轨道测定
　　又见：*orbital determination*
determinative star　距星
determinative star distance　入宿度
　　又见：*equatorial lodge degrees; lunar lodge degrees*
determining function　生成函数　又见：*generating function*
detonating fireball　发声火流星
　　又见：*sound-emitting fireball*
detonation　爆震
detonation wave　爆震波
Deucalion　丢卡利翁（小行星53311号）
deuterium　氘（^{2}H，氢同位素）
deuterium bottleneck　氘瓶颈
deuterium era　氘时期
deuterium line radiation　氘线辐射
deuteron　氘核　又见：*deuton; diplon*
deuton　氘核　又见：*deuteron; diplon*
Deutsch-Spanisches Astronomisches Zentrum　【德】德国－西班牙天文中心　即：*Calar Alto Observatory*
Devana Chasma　德瓦娜深谷（金星）
de Vaucouleurs classification　德沃古勒分类
de Vaucouleurs' law　德沃古勒定律
de Vaucouleurs radius　德沃古勒半径
de Vaucouleurs-Sandage classification　德沃古勒－桑德奇分类
developer　显影剂　又见：*developing agent; photographic developer*
developing agent　显影剂　又见：*developer; photographic developer*
deviant　偏差值
deviation　偏差，偏离
deviation of light　光偏差
deviation of the vertical　垂线偏差　又见：*plumb-line deviation; deflection of vertical*
De Vries crater　德弗里斯环形山（月球）
Dewar crater　杜瓦环形山（月球）
dew-cap　露罩　又见：*dew-shield*
dew-shield　露罩　又见：*dew-cap*
dex　（＝decimal exponent）对数增量｜岱，十进制数〔台〕
dextrogyrate component　右旋子线
dextrorotation　右旋
Deyoung-Axford model　德扬－阿克斯福德模型
DFC　（＝discrete Fourier convolution）离散傅里叶卷积
DFT　（＝discrete Fourier transform）离散傅里叶变换
D galaxy　D星系
DGP gravity　（＝Dvali-Gabadadz-Porrati gravity）DGP引力
Dhur　西上相，太微右垣五（狮子座δ）
　　又见：*Duhr*
Diabolo Nebula　哑铃星云（NGC 6853）
　　又见：*Dumbbell Nebula*
Diadem　太微左垣五，东上将（后发座α）
diagenesis　成岩作用
diagnostic diagram　诊断图
diagnostics　诊断法
diagonal eyepiece　对角目镜
diagonal horn　对角喇叭
diagonal mirror　对角镜
diagonal tensor　对角张量
Dial　日晷号（法－德科学卫星）
dial　①度盘｜[刻]度盤〔台〕②日晷｜日晷，日规〔台〕　又见：*sundial*
diameter　直径
diamond antenna　菱形天线　又见：*rhombic antenna*
diamond ring　金刚石环｜鑽石環〔台〕
diamond-ring effect　金刚石环效应
Diana　月神星（小行星78号）
diaphaneity　透明度　又见：*transparency*
diaphragm　光阑｜光闌，膜片〔台〕
diaphragm aperture　光阑孔径
diapositive　反底片｜複製底片，反底片〔台〕
diastrophism　地壳变动
diatomic molecule　双原子分子
DIB　（＝diffuse interstellar band）弥漫星际带（恒星光谱）
dichotomy　弦，半月
dichroic beam-splitter　双色分束器
dichroic-cross image divider　十字形双色分像器
dichroic extinction　双色消光
dichroic filter　双色滤光器
dichroic mirror　双色镜
Dicke radiometer　迪克辐射计
Dicke receiver　迪克接收机
Dicke switch　迪克开关
Dicke-switched receiver　迪克开关接收机
DIDC　（＝dense interstellar dust cloud）稠密星际尘云
Dido crater　迪多环形山（土卫四）
dielectric constant　介电常数

dielectronic recombination 双电子复合
Difda al Thani 土司空（鲸鱼座 β） 又见：*Deneb Kaitos; Diphda*
difference frequency 差频
difference pattern 差值方向图
differentiable manifold 微分流形
differential aberration 较差光行差 | 光行差較差〔台〕
differential astrometry 较差天体测量
differential atmosphere absorption 较差大气吸收 | 大氣吸收較差〔台〕
differential catalogue 较差星表 又见：*differential star catalogue*
differential correction 较差改正
differential cross-section 微分截面
differential delay 较差时延
differential determination 较差测定
differential Doppler 较差多普勒
differential energy distribution 微分能量分布
differential energy flux 较差能流
differential flexure 较差弯沉
differential galactic rotation 银河系较差自转
differential measurement 较差测量
differential microwave radiometer （＝DMR）较差微波辐射计
differential nutation 较差章动
differential observation 较差观测
differential photographic method 较差照相方法
differential photometry 较差测光
differential precession 较差岁差
differential radiometer 差分辐射计
differential reddening 较差红化
differential refraction 较差折射 | 較差〔大氣〕折射〔台〕
differential rotation 较差自转
differential scattering 较差散射
differential scattering cross-section 较差散射截面
differential spectrum 微分能谱
differential star catalogue 较差星表 又见：*differential catalogue*
differential visibility 微分能见度函数
differentiated object 层化天体
differentiation 分化
differentiator 差示器
differentional image motion monitor （＝DIMM）差分图像运动测量仪
diffracted wave 衍射波
diffraction 衍射 | 繞射〔台〕

diffraction angle 衍射角 又见：*angle of diffraction*
diffraction disk 衍射圆面 | 繞射盤〔台〕
diffraction grating 衍射光栅 | 繞射光柵〔台〕
diffraction halo 衍射晕
diffraction image 衍射像
diffraction limit 衍射极限 | 繞射限制〔台〕
diffraction pattern 衍射图样 | 繞射圖樣〔台〕
diffraction ring 衍射环 | 繞射環〔台〕
diffraction scattering 衍射散射
diffraction spectrum 衍射光谱 | 繞射光譜〔台〕
diffuse absorption band 漫吸收谱带
diffuse cloud 漫射云
diffuse density 漫射密度
diffused light 漫射光
diffuse dwarf galaxy 弥漫矮星系
diffuse-enhanced spectrum 漫强谱 | 擴散增強譜〔台〕
diffuse equilibrium 弥散平衡
diffuse Galactic infrared radiation 银河漫射红外辐射
diffuse Galactic light 银河漫射光
diffuse Galactic X-ray emission 银河漫射 X 射线辐射
diffuse Hα radiation 弥漫 Hα 辐射
Diffuse Infrared Background Experiment （＝DIRBE）弥漫红外背景实验
diffuse interstellar band （＝DIB）弥漫星际带（恒星光谱）
diffuse interstellar medium 弥漫星际介质
diffuse matter 弥漫物质
diffuse nebula 弥漫星云
diffuse nebulosity 弥漫状星云物质
diffuseness 弥漫态，漫射
diffuse pinch effect 漫箍缩效应
diffuse radiation 漫辐射
diffuse radiation field 漫辐射场
diffuse radio emission 漫射电
diffuse reflecting power 漫反射率
diffuse reflection 漫反射 又见：*scattered reflection*
diffuse scattering 漫散射
diffuse series 漫线系
diffuse X-ray 弥漫 X 射线
diffuse X-ray background 弥漫 X 射线背景 | 瀰漫 X 光背景〔台〕
diffuse X-ray emission 弥漫 X 射线辐射 | 瀰漫 X 光輻射〔台〕
diffuse γ-ray emission 弥漫 γ 射线

diffusion acceleration 扩散加速
diffusion approximation 扩散近似
diffusion coefficient 扩散系数
diffusion constant 扩散常数
diffusion damping 扩散阻尼
diffusion equation 扩散方程
diffusion of magnetic field line 磁力线扩散
diffusivity 扩散率　又见: *rate of diffusion*
dI galaxy （＝dwarf irregular galaxy）矮不规则星系
digicon 数字像管　又见: *digital image tube*
digital analyser 数字分析器 | 數位分析器〔台〕
digital area photometer （＝DAP）数字面积光度计
digital astrophotography 数字天体摄影
digital correlator 数字相关器
digital filtering 数字滤波
digital image 数字像
digital image recorder 数字图像记录仪
digital image tube 数字像管　又见: *digicon*
digital optical sky survey 数字化光学巡天
digital recording 数字记录
digital sky survey 数字化巡天 | 數位巡天〔台〕
digital time dissemination 数字时间发播
digitiser 数字化装置　又见: *digitizer*
Digitized Sky Survey （＝DSS）数字化巡天
digitizer 数字化装置　又见: *digitiser*
Dike 泰神星（小行星 99 号）
dilatation effect 钟慢效应　又见: *dilation effect*
dilation effect 钟慢效应　又见: *dilatation effect*
dilaton 伸缩子
dilute aperture 稀疏孔径　又见: *sparse aperture*
dilute array 稀化天线阵
diluted radiation 稀化辐射
dilution factor 稀化因子 | 稀釋因子〔台〕
dimensional orientation 空间方位
dimensionless notation 无量纲符号
dimensionless number 无量纲数　又见: *non-dimensional number*
dimensionless quantity 无量纲量
dimensionless unit 无量纲单位
dimethyl ether 二甲醚（CH_3OCH_3）
DIMM （＝differentional image motion monitor）差分图像运动测量仪
dineutron 双中子
diode array 二极管阵
diode detector 二极管探测器，二极管检波器

diode noise source 二极管噪声源
Diogenite 奥长古铜无球粒陨石
Diomedes 迪奥梅黛斯（小行星 1437 号）
Dione ① 土卫四 ② 坤神星（小行星 106 号）
Dione B 土卫十二（曾用名）即: Helene
Diophantus crater 丢番图环形山（月球）
diopside 透辉石
diopter 折光度，照准仪
dioptra 窥管，望筒　又见: *sighting-tube*
dioptric system 折射系统
dioptric telescope 折射望远镜　又见: *refractor; refracting telescope*
Diphda 土司空（鲸鱼座 β）　又见: *Deneb Kaitos; Difda al Thani*
diplon 氘核　又见: *deuteron; deuton*
dip of the horizon 地平俯角　又见: *horizon dip*
dipole anisotropy 偶极各向异性
dipole antenna 偶极天线
dipole array 偶极天线阵
dipole feed 偶极馈源
dipole magnetic field 偶极磁场
dipole moment 偶极矩　又见: *moment of dipole*
dipole radiation 偶极辐射
dipole term 偶极项
dipole transition 偶极跃迁
Dipper 斗宿（二十八宿）
Dirac-Born-Infeld action （＝DBI action）狄拉克－玻恩－因费尔德作用量
Dirac cosmology 狄拉克宇宙学
Dirac hypothesis 狄拉克假说
Dirac-Jordan cosmology 狄拉克－乔丹宇宙学
Dirac mass term 狄拉克质量项
Dirac neutrino 狄拉克中微子
Dirac's large-number hypothesis （＝LNH）狄拉克大数假说
Dirac string 狄拉克弦
DIRBE （＝Diffuse Infrared Background Experiment）弥漫红外背景实验
direct ascent 直接上升
directional antenna 定向天线　又见: *beam antenna*
directional coupler 定向耦合器
directional diagram 方向图　又见: *directional pattern*
directional distribution 按方向分布
directional pattern 方向图　又见: *directional diagram*
direction angle 方向角
direction cosine 方向余弦
direction-determining board 正方案

directive gain	定向增益 | 指向增益〔台〕
directivity	方向性 | 指向性〔台〕
directivity factor	方向性因子 | 指向性因子〔台〕
directivity function	方向性函数
directivity pattern	方向图样
direct motion	顺行 又见：*prograde motion; prograde*
director	导向器，控制仪 | 導向器〔台〕
Directorate of Astronomy and Calendar	司天监
direct orbit	顺行轨道 又见：*prograde orbit*
director of imperial observatory	钦天监监正（明清官职）又见：*imperial astronomer*
direct plate	正片
direct stationary	顺留
direct vision objective prism	直视物端棱镜
direct vision prism	直视棱镜
direct vision spectroscope	直视分光镜
dirty beam	不洁射束
dirty ice	脏冰
dirty map	脏图
dirty snowball	脏雪球理论
dirty-snowball model	脏雪球模型
dirty-snowball theory	脏雪球理论
disaggregation	解集
disappearance	掩始 又见：*immersion*
disappearance point	消失点 又见：*end point*
discoloration	消色 又见：*decoloration*
disconnection event	断尾事件 又见：*tail-disconnection event*
discontinuity	跃变，不连续性
discontinuous spectrum	不连续光谱
Discourse on the Conformation of the Heavens	安天论 又见：*Theory of Stable Heavens*
Discourse on the Diurnal Revolution of the Heavens	昕天论 又见：*Theory of Bright Heavens*
Discourse on the Vastness of the Heavens	穹天论 又见：*Theory of Vaulting Heavens*
Discoverer	发现者号（美国军事卫星）
Discovery Channel Telescope	发现频道望远镜
Discovery Rupes	发现号峭壁（水星）
discrepance	差异 又见：*discrepancy*
discrepancy	差异 又见：*discrepance*
discrepant redshift	差异红移
discrete band	分立谱带
discrete energy state	分立能态
discrete error	个别误差
discrete Fourier convolution	（=DFC）离散傅里叶卷积
discrete Fourier transform	（=DFT）离散傅里叶变换
discrete interval theorem	分立间隔定理
discrete radio source	分立射电源 | 分立電波源〔台〕
discrete space-time	离散时空
discriminator	鉴别器
dish antenna	碟形天线
dish sensitivity	天线灵敏度
disintegration	蜕变
disk accretion	盘吸积
disk cluster	盘族星团
disk galaxy	盘星系
disk globular cluster	盘族球状星团
disk-halo degeneracy	盘－晕简并性
disk instability	盘不稳定性
disk-like structure	盘状结构
disk of diffusion	弥散盘
disk population	盘族，薄盘族 | [星系] 盤 [星] 族，薄盤族〔台〕 又见：*thin-disk population*
disk population star	盘族星
disk potential	盘势
disk radio emission	盘族射电
disk scale height	盘标高
disk scale length	盘标长
disk-shaped galaxy	盘状星系
disk shocking	盘冲击
disk simulation	盘模拟
disk-spheroid model	盘－椭球子系模型
disk star	盘族恒星 | [星系] 盤族星〔台〕
disk temperature	圆面温度
disky elliptical	盘状椭圆星系
disordered field	无序场
disparation brusque	突异
dispersed fringe sensor	色散条纹传感器
dispersion constant	（=DC）色散常数
dispersion curve	色散曲线
dispersion ellipse	弥散椭圆
dispersion equation	色散方程
dispersion force	色散力
dispersion index	色散率 又见：*dispersion power*
dispersion measure	（=DM）频散量 | 色散量〔台〕
dispersion of velocities	速度弥散 | 速度瀰散 [度]〔台〕
dispersion orbit	弥散轨道
dispersion power	色散率 又见：*dispersion index*

dispersion relation　色散关系
dispersion removal　消色散，消频散，消色散器，消频散器
dispersion ring　弥散环
dispersion velocity　弥散速度
displacement actuator　位移触动器
displacement sensor　位移传感器
disruption rate　瓦解率
disruption time　瓦解时间
dissector　析像管　又见：*image dissector*
dissemination　弥散
dissipation　耗散 | 耗散，消散〔台〕
dissipational collapse　耗散坍缩
dissipation interaction　耗散相互作用
dissipationless collapse　无耗散坍缩
dissipation mass　耗散质量
dissipation of energy　能量耗散　又见：*energy dissipation*
dissipation scale　耗散尺度
dissipative force　耗散力
dissipative model　耗散模型
dissociation　离解 | 解離〔台〕
dissociation energy　离解能
dissociation equilibrium　离解平衡 | 解離平衡〔台〕
dissociation potential　离解电势 | 解離電位〔台〕
dissociation process　离解过程　又见：*dissociative process*
dissociation recombination　离解复合　又见：*dissociative recombination*
dissociation temperature　离解温度 | 解離溫度〔台〕
dissociation time　解体时间 | 瓦解時間〔台〕
dissociative process　离解过程　又见：*dissociation process*
dissociative recombination　离解复合　又见：*dissociation recombination*
dissymmetric ejection　不对称抛射
dissymmetric expansion　不对称膨胀
dissymmetric mode　不对称模式
distance control　遥控　又见：*remote control; telecontrol*
distance determination　距离测定
distance estimator　估距关系
distance gap　距离间隙
distance gauge　测距仪　又见：*range finder; telemeter*
distance indicator　①示距参数 ②示距天体
distance ladder　距离阶梯
distance-luminosity relation　距离－光度关系

distance measurement　距离测量
distance-measuring equipment　（＝DME）测距装置
distance-measuring satellite　测距卫星
distance modulus　距离模数　又见：*modulus of distance*
distance-redshift relation　距离－红移关系
distance scale　距离尺度
distant encounter　远距交会
distant observer approximation　远距观测者近似
distended magnetic field　延展磁场
distorted image　畸变像
distortion　畸变 | 畸變，失真〔台〕
distortionless image　无畸变像
distortion shift　畸变位移
distortion tensor　畸变张量
distribution curve　分布曲线
distribution function　分布函数
distribution in space　空间分布　又见：*spatial distribution; space distribution*
distribution parameter　分布参数
disturbance　摄动，微扰　又见：*perturbation*
disturbed area　扰动区
disturbed body　受摄体　又见：*perturbed body; perturbed object*
disturbed coordinate　受摄坐标　又见：*perturbed coordinate*
disturbed element　受摄根数 | 受攝要素，受攝根數〔台〕　又见：*perturbed element*
disturbed galaxy　受扰星系
disturbed motion　受摄运动　又见：*perturbed motion*
disturbed orbit　受摄轨道　又见：*perturbed orbit*
disturbed sun　扰动太阳
disturbing body　摄动体　又见：*perturbing body; perturbing object*
disturbing effect　摄动效应
disturbing force　摄动力　又见：*perturbative force*
disturbing function　摄动函数　又见：*perturbation function*
disturbing galaxy　扰动星系
dithering　抖动法
diurnal aberration　周日光行差
diurnal arc　日间弧 | 周日弧〔台〕
diurnal change　日变化　又见：*diurnal variation; day range*
diurnal circle　周日圈 | 周日[平]圈〔台〕
diurnal clock rate　周日钟速
diurnal inequality　日差

diurnal libration　周日天平动
diurnal motion　周日运动
diurnal nutation　周日章动
diurnal parallax　周日视差
diurnal phenomenon　周日现象
diurnal sign　奇数宫
diurnal variation　日变化　又见：*diurnal change; day range*
divergence　散度
divergence theorem　散度定理
divergent star cluster　散开星团
diverging lens　发散透镜
diversity technique　分集技术
divided circle　刻度盘　又见：*graduated circle*
dividing line　分界线
division line　分划线
DLA　（＝damped Lyman α system）阻尼莱曼 α 系统
D layer　D 层
D line　D 线
DLS　（＝Deep Lens Survey）引力透镜深空巡天
DM　（＝dispersion measure）频散量 | 色散量〔台〕
DME　（＝distance-measuring equipment）测距装置
dMe star　（＝dwarf Me star）Me 型矮星
DMR　（＝differential microwave radiometer）较差微波辐射计
Dobsonian mounting　多布森装置
Dobsonian reflector　多布森反射望远镜
Dobsonian telescope　多布森望远镜
Dobsonian telescope mount　多布森望远镜支架
docking　对接
docking adapter　对接接口
Dodecahedron　十二面体号（美国卫星）
　又见：*Dodecapole; Porcupine*
Dodecapole　十二面体号（美国卫星）
　又见：*Dodecahedron; Porcupine*
Doerfel crater　德费尔环形山（月球）
dog days　伏日 | 天狼星東升日〔台〕
Dog Star　天狼[星]（大犬座 α）　又见：*Sirius; Canicula; Aschere*
Dollfus polarimeter　多尔富斯偏振计
domain wall　畴壁
dome　圆顶，圆顶室　又见：*cupola*
dome flat　圆顶平场
domeless reflector　无圆顶反射望远镜
domeless refractor　无圆顶折射望远镜

domeless solar telescope　无圆顶太阳望远镜
domeless telescope　无圆顶望远镜
dome seeing　圆顶视宁度
dome servo　圆顶随动
dominant galaxy　主星系
Dominical Letter　主日字母
Dominion Astrophysical Observatory　（＝DAO）自治领天体物理台 | 自治领天文物理台〔台〕（加拿大）
Dominion Radio Astrophysical Observatory　（＝DRAO）自治领射电天体物理台（加拿大）
Domovoy crater　多莫伏伊环形山（天卫一）
Donati's comet　多纳提彗星 | 多納提彗[星]〔台〕
Dongguan　东莞（小行星 3476 号）
donkey effect　犟驴效应
Donner crater　唐纳环形山（月球）
donor　输质星
Doppelmayer crater　多佩尔迈尔环形山（月球）
Doppler boosting factor　多普勒提升因子
Doppler broadening　多普勒致宽 | 都卜勒致宽〔台〕
Doppler contour　多普勒轮廓 | 都卜勒周線〔台〕
　又见：*Doppler profile*
Doppler core　多普勒线核 | 都卜勒核心〔台〕
Doppler counting　多普勒计数
Doppler crater　多普勒环形山（月球）
Doppler data　多普勒资料
Doppler displacement　多普勒位移 | 都卜勒位移〔台〕
Doppler effect　多普勒效应 | 都卜勒效應〔台〕
Doppler equation　多普勒方程
Doppler-Fizeau effect　多普勒－菲佐效应 | 都卜勒－斐索效應〔台〕
Dopplergram　多普勒图 | 都卜勒圖〔台〕
Doppler mapping　多普勒测绘
Doppler motion　多普勒运动 | 都卜勒運動〔台〕
Doppler parameter　多普勒参数
Doppler peak　多普勒峰
Doppler Polar Monitoring Service　（＝DPMS）多普勒地极监测服务
Doppler positioning　多普勒定位　又见：*Doppler translocation*
Doppler profile　多普勒轮廓 | 都卜勒周線〔台〕
　又见：*Doppler contour*
Doppler ranging　多普勒测距 | 都卜勒測距〔台〕
Doppler receiver　多普勒接收机
Doppler relation　多普勒关系
Doppler shift　多普勒频移 | 都卜勒位移〔台〕

Doppler-shift measurement 多普勒频移测量
Doppler spread 多普勒频散
Doppler system with double frequency 双频多普勒
Doppler system with single frequency 单频多普勒
Doppler translocation 多普勒定位 又见：*Doppler positioning*
Doppler width 多普勒宽度 | 都卜勒寬度〔台〕
Dor （＝Dorado）剑鱼座
Dorado （＝Dor）剑鱼座
Doris 昏神星（小行星48号）
dormant volcano 休眠火山
Dorsa Aldrovandi 【拉】阿尔德罗万迪山脊（月球）
Dorsa Andrusov 【拉】安德鲁索夫山脊（月球）
Dorsa Argand 【拉】阿尔甘山脊（月球）
Dorsa Barlow 【拉】巴洛山脊（月球）
Dorsa Burnet 【拉】伯内特山脊（月球）
Dorsa Cato 【拉】加图山脊（月球）
Dorsa Dana 【拉】达纳山脊（月球）
Dorsa Ewing 【拉】尤因山脊（月球）
Dorsa Geikie 【拉】盖基山脊（月球）
Dorsa Harker 【拉】哈克山脊（月球）
Dorsa Lister 【拉】利斯特山脊（月球）
Dorsa Mawson 【拉】莫森山脊（月球）
Dorsa Rubey 【拉】鲁比山脊（月球）
Dorsa Smirnov 【拉】斯米尔诺夫山脊（月球）
Dorsa Sorby 【拉】索比山脊（月球）
Dorsa Stille 【拉】施蒂勒山脊（月球）
Dorsa Whiston 【拉】惠斯顿山脊（月球）
Dorsum Arduino 【拉】阿尔杜伊诺山脊（月球）
Dorsum Azara 【拉】阿萨拉山脊（月球）
Dorsum Bucher 【拉】布赫山脊（月球）
Dorsum Buckland 【拉】巴克兰山脊（月球）
Dorsum Cayeux 【拉】卡耶山脊（月球）
Dorsum Cloos 【拉】克洛斯山脊（月球）
Dorsum Cushman 【拉】库什曼山脊（月球）
Dorsum Gast 【拉】加斯特山脊（月球）
Dorsum Grabau 【拉】葛利普山脊（月球）
Dorsum Heim 【拉】海姆山脊（月球）
Dorsum Nicol 【拉】尼科尔山脊（月球）
Dorsum Niggli 【拉】尼格利山脊（月球）
Dorsum Oppel 【拉】奥佩尔山脊（月球）
Dorsum Owen 【拉】欧文山脊（月球）
Dorsum Scilla 【拉】希拉山脊（月球）
Dorsum Termier 【拉】泰尔米埃山脊（月球）
Dorsum Von Cotta 【拉】冯·科塔山脊（月球）
Dorsum Zirkel 【拉】齐克尔山脊（月球）
dorsum（复数：dorsa） 【拉】环形山脊
dose 剂量
dosimetry 剂量测定
dot image 点像
double asteroid 双小行星 又见：*binary asteroid*
double astrograph 双筒天体照相仪 | 雙筒攝星儀〔台〕
double-beam photometer 双束光度计
double-beam polarimeter 双光束偏振计
double-beam system 双束系统
double cluster 双重星团 | 雙[重]星團〔台〕
double-Compton process 双康普顿过程
double concave glass 双凹透镜 又见：*biconcave lens; double concave lens*
double concave lens 双凹透镜 又见：*biconcave lens; double concave glass*
double convex glass 双凸透镜 又见：*biconvex lens; double convex lens*
double convex lens 双凸透镜 又见：*biconvex lens; double convex glass*
double-counting method 双计数法
double-double radio source 双双射电源
double drift 二星流
double elliptical galaxy 双重椭圆星系
double equatorial 双筒赤道仪
double exponential disk 双指数盘
double extragalactic radio source 河外射电双源
double galaxy 双重星系 又见：*binary galaxy; twin galaxy*
double-hours 辰
double-humped profile 双峰轮廓
double interferometer 双元干涉仪
又见：*two-element interferometer*
double-lensed quasar 引力透镜效应双像类星体
double-lined [spectroscopic] binary （＝SB 2）双谱[分光]双星 | 復綫[分光]雙星〔台〕
又见：*double-spectrum binary; two-spectrum binary*
double lobed structure 双瓣结构（射电星系）
double meteor 双流星
double-mode Cepheid 双模造父变星 | 雙頻造父變星〔台〕
double-mode pulsation 双模脉动
double-mode pulsator 双模脉动星
double-mode RR Lyrae star 双模天琴RR型星 | 雙模天琴[座]RR型星〔台〕
double-mode variable 双模变星
double-mode variable star 双模变星

double nebula 双星云

double-pass echelle spectrometer 双通阶梯光栅光谱仪

double-planet 双行星　又见：*binary planet*

double-prism spectrograph 双棱镜摄谱仪
　　又见：*2-prism spectrograph*

double quasar 双类星体　又见：*twin quasar; binary quasar*

double radio source 双射电源 | 雙電波源〔台〕

double reflection 双反射

double refraction 双折射　又见：*birefringence*

double reversal 双重自食

double-ring galaxy 双环星系

double sideband receiver 双边带接收机

double slit 双缝

double source 双源

double spectrograph 双路摄谱仪

double spectroheliograph 双筒太阳单色光照相仪

double spectroprojector 双光束光谱投射仪

double-spectrum binary 双谱 [分光] 双星 | 復綫 [分光] 雙星〔台〕　又见：*double-lined [spectroscopic] binary; two-spectrum binary*

double star （=DS）双星 | [物理] 雙星〔台〕

double system 双重天体系统

doublet ① 双重线 ② 双重态

doublet antenna 偶极天线

doublet lens 双合透镜

double transit 二次中天

doublet structure 双重结构

double wave 双波

Dove prism 道威棱镜

DO white dwarf DO 型白矮星

down-Comptonization 康普顿软化

down-conversion 下变频

downdraft 下沉气流

down-leg light time 下行光行时

downsizing 降序，瘦身

downthrust 下冲板块

downward transition 向下跃迁

DPMS ①（=Dahlgren Polar Monitoring Service）达尔格伦地极监测服务 | 達爾格連 [地] 極監測服務〔台〕②（=Doppler Polar Monitoring Service）多普勒地极监测服务

DQE （=detective quantum efficiency）探测量子效率

DQ Herculis star （=DQ Her star）武仙 DQ 型星 | 武仙 [座]DQ 型星〔台〕

DQ Her star （=DQ Herculis star）武仙 DQ 型星 | 武仙 [座]DQ 型星〔台〕

DQ white dwarf DQ 型白矮星

Dra （=Draco）天龙座

Draco （=Dra）天龙座

Draco Dwarf Galaxy 天龙矮星系

Draco galaxy 天龙星系

draconic month 交点月　又见：*nodical month*

draconic revolution 交点周 | 交點轉動，交點周〔台〕

draconic year 交点年　又见：*nodical year*

Draconid meteor shower 天龙流星雨

Draconids 天龙流星群 | 天龍 [座] 流星雨〔台〕

drag effect 曳引效应 | 拖曳效應〔台〕

Dragon Nebula 飞龙星云

drain diffusion 泄漏扩散

DRAO （=Dominion Radio Astrophysical Observatory）自治领射电天体物理台（加拿大）

Draper Catalogue 德雷伯星表

Draper classification 德雷伯分类 | 杜雷伯分類〔台〕

draw tube 目镜接筒

dredge-up 上翻

D-region D 区

drift beam instability 漂移束不稳定性

drift curve 漂描曲线 | 漂移曲線〔台〕

drift cyclotron wave 漂移回旋波

drift frequency 漂移频率

drifting pulse 漂移脉冲　又见：*drift pulse; marching pulse*

drifting subpulse 漂移次脉冲　又见：*drift subpulse; marching subpulse*

drift of stars 星流　又见：*star drift; star streaming*

drift pulse 漂移脉冲　又见：*drifting pulse; marching pulse*

drift rate 漂移率

drift scan 漂移扫描　又见：*drift scanning*

drift-scan measurement 漂描测量

drift scanning 漂移扫描　又见：*drift scan*

drift subpulse 漂移次脉冲　又见：*drifting subpulse; marching subpulse*

drift turbulence 漂移湍流

drift velocity 漂移速度

drift wave 漂移波

Dritter Fundamental Katalog （=FK3）【德】FK3 星表，第三基本星表

drive clock 转仪钟　又见：*driving clock; clock drive; sidereal drive; sidereal driving clock*

driving clock 转仪钟　又见：*drive clock; clock drive; sidereal drive; sidereal driving clock*
driving mechanism 转仪装置
driving system 驱动系统
droplet 微滴
DRS (＝data-relay satellite) 数据转播卫星
drum tower 鼓楼
Dryden crater 德赖登环形山（月球）
Drygalski crater 德里加尔斯基环形山（月球）
dry ice 干冰
drying agent 干燥剂
dry merger 干并合，贫气体并合
DS (＝double star) 双星 | [物理] 雙星〔台〕
dS (＝de Sitter space) 德西特空间
DS1 (＝Deep Space 1) 深空 1 号（美国探测器）
Dschubba 房宿三（天蝎座 δ）
Dsiban 女史增一（天龙座 ψ）　又见：*Al Dhibain*
DSN (＝Deep Space Network) 深空探测网
DSS (＝Digitized Sky Survey) 数字化巡天
DST (＝daylight saving time) 夏令时 | 日光節約時間〔台〕
D star D 型星
D-type asteroid D 型小行星
dual-beam observation 双束观测
dual-beam technique 双束技术
dualistic nature 二象性　又见：*duality*
duality ① 二象性　又见：*dualistic nature* ② 对偶性
dual of tangent space 对偶切空间
dual photometer 双光度计
dual-rate moon camera 双速月球照相机
Dubhe 天枢，北斗一（大熊座 α）
Dubyago crater 杜比亚戈环形山（月球）
Duck Bay 杜克湾（火星）
Dufu 杜甫（小行星 110289 号）
Dugan crater 杜根环形山（月球）
Duhr 西上相，太微右垣五（狮子座 δ）
　　又见：*Dhur*
dumbbell galaxy (＝db galaxy) 哑铃状星系
Dumbbell Nebula 哑铃星云（NGC 6853）
　　又见：*Diabolo Nebula*
dumbbell radio galaxy 哑铃状射电星系
　　又见：*dumbbell-shaped radio galaxy*
dumbbell-shaped curve of zero velocity 哑铃形零速度线
dumbbell-shaped radio galaxy 哑铃状射电星系
　　又见：*dumbbell radio galaxy*
dummy antenna 仿真天线

Dunér crater 杜奈尔环形山（月球）
Dunhuang 敦煌（小行星 4273 号）
Dunhuang star chart 敦煌星图
dunite 纯橄榄岩
duodenary series 十二次　又见：*12 Jupiter-stations; twelve Jupiter-stations*
duplexer 双工器
duplicity 二重性，成双性 | 二重性〔台〕
duration of annular phase 环食时间
duration of eclipse 掩食时间 | 交食時間〔台〕
　　又见：*eclipse duration*
duration of pulse 脉冲宽度　又见：*pulse width*
duration of totality 全食时间
Durchmusterung 【德】巡天星表
dusk side 黄昏侧
dusky belt 暗带
dusky ring 暗环　又见：*crape ring*
dusky veil 暗纱
dust-ball 尘埃流星
dust cloud 尘云
dust disk 尘[埃]盘
dust driven wind 星尘风
dust globule 尘埃球状体
dust grain 尘粒
dust jet 尘埃喷流
dust lane 尘埃带
dustless galaxy 无尘星系
dust nebula 尘埃星云
Dustpan 箕宿（二十八宿）
　　又见：*Winnowing-basket; Basket*
dust ring 尘环
dust shell 尘壳
dust storm 尘暴
dust tail 尘埃彗尾
dust train 尘埃余迹 | 塵埃[流星]餘跡〔台〕
dusty primaeval galaxy 富尘埃原初星系
duty cycle 负载循环
duty factor 负载因子，占空因子
Dvali-Gabadadz-Porrati gravity (＝DGP gravity) DGP 引力
DW (＝Dwingeloo list) 德文格洛射电源表
dwarf carbon star 碳矮星　又见：*carbon dwarf star*
dwarf Cepheid 矮造父变星
dwarf elliptical galaxy (＝dE galaxy) 矮椭圆星系
dwarf galaxy 矮星系

dwarf irregular galaxy （=dI galaxy）矮不规则星系
dwarf Me star （=dMe star）Me 型矮星
dwarf M flare star M 型矮耀星
dwarf nova 矮新星
dwarf planet 矮行星
dwarf sequence 矮星序
dwarf spherical galaxy 矮球状星系
dwarf spheroidal galaxy 矮椭球星系
　　又见：*spheroidal dwarf galaxy*
dwarf spiral galaxy 矮旋涡星系 | 矮螺旋星系〔台〕
dwarf star 矮星
dwarf variable 矮变星
Dwingeloo galaxy 德温格洛星系
Dwingeloo list （=DW）德文格洛射电源表
Dwingeloo Radio Observatory 德文格洛射电天文台
Dy （=dysprosium）镝（66 号元素）
dyad 并矢
dynamical age 动力学年龄
dynamical astronomy 动力天文学
dynamical axis 动力轴
dynamical cosmology 动力学宇宙学
dynamical dissipation 动力耗散
dynamical ellipticity 力学椭率
dynamical equilibrium 动态平衡 | 動力學平衡〔台〕
dynamical equinox 动力学春分点 | 力學分點〔台〕
dynamical evolution 动力学演化
dynamical flattening 力学扁率 | 動態扁率〔台〕
　　又见：*dynamical oblateness*
dynamical friction 动力摩擦
dynamical gravity measurement 动态重力测量
dynamical instability 动力不稳定性
　　又见：*dynamic instability*
dynamical libration 动力学天平动，动力学秤动 | 力學天平動〔台〕

dynamical mass 动力学质量
dynamical mean equinox 动力学平春分点
dynamical method 动力学方法
dynamical oblateness 力学扁率 | 動態扁率〔台〕
　　又见：*dynamical flattening*
dynamical parallax 力学视差 | 動力視差〔台〕
dynamical pinch effect 动力箍缩效应
dynamical reference system 动力学参考系
dynamical relaxation 动力弛豫 | 動力鬆弛〔台〕
dynamical satellite geodesy 卫星动力测地学
dynamical spectrum 运动频谱
dynamical stability 动力稳定性 | 動態穩[定]度〔台〕　又见：*dynamic stability*
dynamical state 动态
dynamical symmetry breaking 动力学对称破缺
dynamical time 力学时
dynamical time-scale 力学时标
dynamical viscosity 动力黏性 | 動力粘度〔台〕
　　又见：*dynamic viscosity*
dynamic characteristic 动态特征曲线
dynamic instability 动力不稳定性　又见：*dynamical instability*
dynamic range 动态范围
dynamics of orbits 轨道动力学
dynamics of stellar systems 恒星系统动力学
dynamic stability 动力稳定性 | 動態穩[定]度〔台〕　又见：*dynamical stability*
dynamic viscosity 动力黏性 | 動力粘度〔台〕
　　又见：*dynamical viscosity*
dynamo geomagnetics theory 地磁发电机理论
dynamo theory 发电机理论
dynamo wave 发电机波
Dysnomia 阋卫（小行星 Eris 的卫星）
Dyson crater 戴森环形山（月球）
dysprosium （=Dy）镝（66 号元素）
DZ white dwarf DZ 型白矮星

E

E+A galaxy　E+A 星系
Eagle crater　老鹰环形山（火星）
Eagle mounting　伊格尔装置
Eagle Nebula　鹰状星云，鹰鹫星云（NGC 6611）
Eagle spectrometer　伊格尔光谱仪
earliest state　极早期状态
early decline　初降
early earth　早期地球
early planet　早期行星
early spectral type　早光谱型
early stage star　早期演化星
early stellar evolution　恒星早期演化
early sun　早期太阳
early-type cluster　早型星系团
early-type emission-line star　（=ETELS）早型发射线星
early-type emission star　早型发射星
early-type galaxy　早型星系
early-type spiral galaxy　早型旋涡星系 | 早型螺旋星系〔台〕
early-type star　早型星
early-type supergiant　早型超巨星
early-type variable　早型变星
early universe　早期宇宙
Earth　地球　又见：*terrestrial globe; globe*
earth-approaching asteroid　近地小行星
　又见：*near-earth asteroid*
earth-approaching comet　近地彗星
　又见：*near-earth comet*
earth-approaching object　近地天体
　又见：*near-earth object*
earth atmosphere　地球大气　又见：*terrestrial atmosphere*
earth attraction　地球引力　又见：*terrestrial gravitation*
earth axis　地轴

Earth-based observation　地基观测
　又见：*ground-based observation*
Earth-based telescope　地基望远镜
　又见：*ground-based telescope*
earth central angle　地球中心角
earth core　地核　又见：*earth nucleus*
Earth-crosser　越地小天体
earth-crossing asteroid　越地小行星
earth-crossing comet　越地彗星
earth-crossing object　越地天体
earth crust　地壳　又见：*crust of the earth; earth shell*
earth curvature correction　地球曲率改正
earth day　地球日
earth deformation　地形变
earth dynamic ellipticity　地球力学椭率
earth dynamics　地球动力学　又见：*geodynamics*
earth elasticity　地球弹性
earth ellipsoid　地球椭球体　又见：*terrestrial ellipsoid*
earth ellipticity　地球扁率　又见：*compression of the earth; earth flattening; earth oblateness*
earth-fixed coordinate system　地固坐标系
　又见：*body-fixed coordinate system*
earth flattening　地球扁率　又见：*compression of the earth; earth ellipticity; earth oblateness*
earth-grazer　掠地小天体
earth-grazing asteroid　掠地小行星
earth-light　地照 | 地[球反]照，地暉〔台〕
　又见：*earthshine*
Earth-like exoplanet　系外类地行星 | [太阳]系外类地行星〔台〕　又见：*terrestrial exoplanet; exo-Earth*
earth magnetic field　地球磁场
earth magnetosphere　地球磁层
earth mantle　地幔 | 地函〔台〕
earth model　地球模型
Earth Moon and Planets　《地球、月球与行星》（荷兰期刊）
Earth-Moon mass ratio　地月质量比

earth-moon space 地月空间
earth-moon system 地月系统
earth nucleus 地核 又见：earth core
earth oblateness 地球扁率 又见：compression of the earth; earth ellipticity; earth flattening
earth orbit 地球轨道
Earth-orbiting observatory 地球轨道天文台
Earth-orbiting probe 地球轨道探测器
Earth-orbiting satellite 地球轨道卫星
earth-orbit rendezvous 环地轨道会合
Earth orbit telescope 环地轨道望远镜
earth orientation parameter （＝EOP）地球定向参数
earth penumbra 地球半影
earth plasmasphere 地球等离子体层
earth point 陨星着地点
earth pole 地极 又见：terrestrial pole
earth potential field 地引力势场
earthquake 地震 又见：seism
earth radio radiation 地球射电
earth rate 地球自转速率 又见：earth rotation rate
earth-rate unit 地球速率单位
earthrise 地出
earth rotation 地球自转 又见：rotation of the earth
Earth Rotation Angle （＝ERA）地球自转角
earth-rotation aperture synthesis 地球自转孔径综合
earth rotation parameter （＝ERP）地球自转参数
earth rotation rate 地球自转速率 又见：earth rate
earth-rotation synthesis array 地球自转综合天线阵
earth satellite 地球卫星
Earth shadow 地影
earth shell 地壳 又见：earth crust; crust of the earth
earthshine 地照 | 地[球反]照，地暉〔台〕 又见：earth-light
earth spheroid 地球扁球体 | 地球球形體〔台〕
earth tide 陆潮 又见：continental tide
earth umbra 地球本影
earth year 地球年
east 东
East African Time （＝EAT）东非时间
eastern elongation ①东距角 ②东大距 又见：greatest eastern elongation; eastern greatest elongation
eastern greatest elongation 东大距 又见：greatest eastern elongation; eastern elongation

eastern hemisphere 东半球
eastern quadrature 东方照
Eastern Standard Time （＝EST）东部标准时间
Eastern Time （＝ET）东部时间
Eastern Wall 东壁
Eastern Well 东井
East European Time （＝EET）东欧时间
east longitude 东经
east point 东点 | 東方〔台〕
east-west asymmetry 东西不对称性
east-west effect 东西效应
east-west line 东西线
EAT （＝East African Time）东非时间
ebb 落潮 又见：ebb tide; fall of the sea
ebb tide 落潮 又见：ebb; fall of the sea
EBCCD （＝electron bombard CCD）电子轰击CCD
Eberhard effect 埃伯哈德效应
Ebert-Fastie spectrometer 艾勃特－法斯梯光谱仪
eccentric angle 偏心角 又见：angle of eccentricity
eccentric anomaly 偏近点角
eccentric dipole 偏心偶极
eccentric-disk model 偏心盘模型
eccentricity 偏心率 | 偏心率，離心率〔台〕
eccentricity resonance 偏心率共振
eccentric latitude 偏心纬度
eccentric orbit 偏心轨道 | 扁軌〔台〕
ecclesiatical calendar 教会历
echelette 小阶梯光栅 又见：echelette grating
echelette grating 小阶梯光栅 又见：echelette
echelette spectrograph 小阶梯光栅摄谱仪
echelle grating 中阶梯光栅
echelle spectrograph 中阶梯光栅摄谱仪
echelle spectrometer 中阶梯光栅分光仪
echelle spectroscopy 中阶梯光栅分光
echelle spectrum 中阶梯光谱
echelogram 阶梯光谱片
echelon grating 阶梯光栅
echelon prism 阶梯棱镜
Echidna 台卫（小行星 Typhon 的卫星）
Echo 司音星（小行星60号）
echo wave 回波，反向波 又见：return wave; back wave
E-class asteroid E型小行星
eclipse ①交食 ②食
eclipse beginning 食始
eclipse boundary 食界 | 初界〔台〕

eclipse comet　交食彗 | 日食彗〔台〕
eclipse cycle　食周 | 交食週期〔台〕
eclipsed body　被食天体
eclipsed component star　被食子星
eclipse depth　食深
eclipsed star　被食星
eclipse duration　掩食时间 | 交食時間〔台〕
　　又见：*duration of eclipse*
eclipse end　食终 | 復元〔台〕
eclipse factor　食分　又见：*magnitude of eclipse; degree of obscuration*
eclipse function　食函数
eclipse limit　食限
eclipse map　日食界限图
eclipse number　食数
eclipse of satellite　卫星食　又见：*satellite eclipse*
eclipse of the moon　月食　又见：*lunar eclipse*
eclipse of the sun　日食　又见：*solar eclipse*
eclipse path　食径
eclipse season　食季
eclipse series　食系
eclipse theory　食论 | 交食理論〔台〕
eclipse year　食年 | 食年，交點年〔台〕
eclipsing binary　食双星　又见：*eclipsing double star*
eclipsing component star　主食子星　又见：*eclipsing star*
eclipsing double star　食双星　又见：*eclipsing binary*
eclipsing infrared source　红外食变星
eclipsing star　① 主食子星　又见：*eclipsing component star* ② 食变星　又见：*eclipsing variable*
eclipsing variable　食变星　又见：*eclipsing star*
eclipsing X-ray source　X射线食变源　即：*eclipsing X-ray star; X-ray eclipsing star*
eclipsing X-ray star　X射线食变星 | X光食變星〔台〕　又见：*X-ray eclipsing star*
ecliptic　黄道
ecliptic armillary sphere　黄道经纬仪　又见：*simple ecliptic armilary*
ecliptic coordinate　黄道坐标
ecliptic coordinate system　黄道坐标系
　　又见：*ecliptic system of coordinate*
ecliptic diagram　黄道图
ecliptic latitude　黄纬　又见：*celestial latitude*
ecliptic limit　黄道限
ecliptic longitude　黄经　又见：*celestial longitude*
ecliptic map　黄道星图
ecliptic meteor　黄道流星

ecliptic obliquity　黄赤交角　又见：*obliquity of the ecliptic; obliquity*
ecliptic of date　瞬时黄道
ecliptic plane　黄道面
ecliptic pole　黄极　又见：*pole of the ecliptic*
ecliptic ring　黄道环
ecliptic stream　黄道流星雨
ecliptic system of coordinate　黄道坐标系
　　又见：*ecliptic coordinate system*
E component　E成分 | E分量〔台〕
E corona　E冕（太阳）
ecosphere　生物圈　又见：*biosphere*
Ed Asich　紫微左垣一，左枢（天龙座ι）　即：*Edasich; Eldsich*
Edasich　紫微左垣一，左枢（天龙座ι）　又见：*Ed Asich; Eldsich*
Eddington approximation　爱丁顿近似
Eddington-Barbier method　爱丁顿－巴比叶方法
Eddington crater　爱丁顿环形山（月球）
Eddington-Lemaître model　爱丁顿－勒梅特模型
Eddington limit　爱丁顿极限 | 艾丁吞極限〔台〕
Eddington luminosity　爱丁顿光度 | 艾丁吞光度〔台〕
Eddington's formula　爱丁顿公式
Eddington standard model　爱丁顿标准模型 | 艾丁吞標準模型〔台〕
Eddington time　爱丁顿时间
eddy　涡流 | 旋渦〔台〕　又见：*swirl; vortex flow*
eddy diffusion coefficient　涡流扩散系数
edge effect　边缘效应
edge-on galaxy　侧向星系
edge-on object　侧向天体
edge-on spiral galaxy　侧向旋涡星系 | 侧向螺旋星系〔台〕
edge sensor　边缘传感器
Edgeworth-Kuiper Belt　埃奇沃斯－柯伊伯带
Edison crater　爱迪生环形山（月球）
editing criterion　取舍判据
EEI　（＝Exo-Earth Imager）系外类地行星成像器
EELT　（＝European Extremely Large Telescope）欧洲特大望远镜 | 歐洲超大望遠鏡〔台〕
EET　（＝East European Time）东欧时间
effective absorption coefficient　有效吸收系数
effective acceleration of gravity　有效重力加速度
effective aperture　有效口径，有效孔径
effective area　有效面积
effective collecting area　有效接收面积
　　又见：*effective receiving area*

effective collision frequency 有效碰撞频率
effective cross-section 有效截面
effective cross-section of collision 有效碰撞截面
effective cross-section of ionization 有效电离截面
effective cross-section of recombination 有效复合截面
effective damping constant 有效阻尼常数
effective distance 有效距离
effective filter response 有效滤波器响应
effective focal length 有效焦距
effective height 有效高度
effective ion Larmor radius 有效离子拉莫尔半径
effective Lande g factor 有效朗德 g 因子
effective length 有效长度
effective line width 有效线宽 | 有效[譜]線寬[度]〔台〕
effective molecular weight 有效分子量
effective optical depth 有效光深
effective potential 有效势
effective pulse width 脉冲有效宽度
effective radiation 有效辐射
effective radius 有效半径
effective receiving area 有效接收面积
又见: *effective collecting area*
effective reflecting area 有效反射面积
effective temperature 有效温度
effective wavelength 有效波长
effective width 有效宽度
effective yield 有效产额
effect of evolution 演化效应
effect of relaxation 弛豫效应 | 鬆弛效應〔台〕
又见: *relaxation effect*
Effelsberg Radio Telescope 埃菲尔斯伯格望远镜 | 埃費爾斯貝格電波望遠鏡〔台〕 又见: *Effelsberg Telescope*
Effelsberg Telescope 埃菲尔斯伯格望远镜 | 埃費爾斯貝格電波望遠鏡〔台〕 又见: *Effelsberg Radio Telescope*
efflux 射流
e-folding time e 倍变化时间，e 倍衰减时间
e-fold length e 倍变化长度，e 倍衰减长度
e-folds e－叠数
EFR （＝emerging flux region）射流区
E galaxy （＝elliptical galaxy）椭圆星系
Egalité arc 平等环弧（海王星）
Egeria 芙女星（小行星13号）
EGG （＝evaporating gaseous globule）蒸发气态球狀体

Egg Nebula 卵形星云 | 蛋星雲〔台〕（CRL 2688）
egress 出食，出凌 | 出凌，終切〔台〕
EGRET （＝Energetic Gamma Ray Experiment Telescope）高能伽马射线实验望远镜
EGRET catalog EGRET 星表
Egyptian ancient astronomy 埃及古代天文学
Egyptian calendar 埃及历
eigenfunction 本征函数 又见: *proper function*
eigen mode 本征模
eigenperiod 本征周期
eigentemperature 本征温度
eigenvalue 本征值 又见: *proper value*
eigenvector 本征矢
eigenvibration 本征振动
Eight Burst Nebula 双环星云
eightfold way 八重法
eight trigrams 八卦
Eijkman crater 艾克曼环形山（月球）
Einasto profile 埃纳斯托轮廓
Einstein arc 爱因斯坦弧
Einstein-Bose statistics 爱因斯坦－玻色统计
Einstein-Cartan gravity 爱因斯坦－嘉当引力
Einstein coefficient 爱因斯坦系数
Einstein crater 爱因斯坦环形山（月球）
Einstein cross 爱因斯坦十字
Einstein-de Sitter cosmological model 爱因斯坦－德西特宇宙模型 | 愛因斯坦－迪西特宇宙模型〔台〕
Einstein-de Sitter model 爱因斯坦－德西特模型 | 愛因斯坦－迪西特模型〔台〕
Einstein-de Sitter universe 爱因斯坦－德西特宇宙 | 愛因斯坦－迪西特宇宙〔台〕
Einstein effect 爱因斯坦效应
Einstein equation 爱因斯坦方程
Einstein field equation 爱因斯坦场方程
Einstein-Hilbert Lagrangian 爱因斯坦－希尔伯特拉格朗日量
einsteinium （＝Es）锿 | 鎄〔台〕（99号元素）
Einstein Observatory （＝HEAO-2）爱因斯坦天文台
Einstein Observatory Catalogue 爱因斯坦天文台 X 射线源表
Einstein probability coefficient 爱因斯坦概率系数
Einstein radius 爱因斯坦半径
Einstein ring 爱因斯坦环
Einstein shift 爱因斯坦红移 | 愛因斯坦位移〔台〕
Einstein's model 爱因斯坦模型
Einstein tensor 爱因斯坦张量

Einstein universe 爱因斯坦宇宙
Einthoven crater 埃因托芬环形山（月球）
Eistla Regio 艾斯特拉区（金星）
ejecta 喷出物
ejecta blanket 喷出覆盖物
ejection 抛射
ejection rate 抛射率
ejection time 抛射时标
ejection velocity 抛射速度
Ekman layer 埃克曼层
Ekpyrotic universe model 火劫宇宙模型
Elaborate Equatorial Armillary Sphere 玑衡抚辰仪
Elara 木卫七 | 木衛七，依來拉〔台〕
elastic collision 弹性碰撞　又见：*billiard-ball collision*
elasticity 弹性
elastic strain 弹性应变
elastic stress 弹性应力
elastic yielding 弹性屈服
E layer E 层（地球）
elbow refractor 折轴折射望远镜　又见：*coude refractor*
ELDO （＝European Launcher Development Organization）欧洲发射器开发组织
Eldsich 紫微左垣一，左枢（天龙座 ι）　又见：*Ed Asich; Edasich*
Electra 昴宿一（金牛座 17）
electric charge 电荷
electric conductance 电导　又见：*conductance*
electric current helicity 电流螺度
electric dipole 电偶极子　又见：*electric doublet*
electric dipole moment 电偶极矩
electric dipole radiation 电偶极辐射
electric displacement 电位移 | 電[位]移〔台〕
electric doublet 电偶极子　又见：*electric dipole*
electric flux 电流量
electric moment 电矩
electric multipole radiation 电多极辐射
electric neutrality 电中性
electric quadrupole 电四极子 | 電四極〔台〕
electric quadrupole moment 电四极矩
electric quadrupole radiation 电四极辐射
electric resonance 电共振
electric spectrophotometry 光电分光光度测量　又见：*photoelectric spectrophotometry*
electric susceptibility 电极化率
electric vector 电矢

electrified body 带电体
electroacoustic wave 电声波
electrochronograph 电子记时仪 | 電子計時儀〔台〕
electrocolorimeter 电色度计
electrocolorimetry 电色度测量 | 電色度測量 [術]〔台〕
electroconductivity 电导率　又见：*conductivity*
electrode 电极
electrograph 电场仪
electrographic camera 电子照相机　又见：*electronographic camera; electron camera; electronic camera*
electrographic image tube 电子像管　又见：*electronic image-tube*
electrojet 电喷流
electrolysis 电解
electromagnetic acceleration 电磁加速
electromagnetic extraction 电磁提取
electromagnetic field 电磁场
electromagnetic force 电磁力
electromagnetic interaction 电磁相互作用
electromagnetic mass 电磁质量
electromagnetic moment 电磁矩
electromagnetic momentum 电磁动量
electromagnetic pulse 电磁脉冲
electromagnetic radiation 电磁辐射
electromagnetic spectrum 电磁波谱
electromagnetic viscosity 电磁黏度 | 電磁粘度〔台〕
electromagnetic wave 电磁波
electromagnetic wave scattering 电磁波散射
electromagnetism 电磁性
electrometer 静电计
electromotive force 电动势
Electron 电子号（苏联科学卫星）
electron affinity 电子亲和性
electron-atom bremsstrahlung 电子原子韧致辐射
electron attachment coefficient 电子附着系数
electron beam 电子束　又见：*electron jet*
electron beam parameter amplifier 电子束参量放大器
electron bombard CCD （＝EBCCD）电子轰击 CCD
electron camera 电子照相机　又见：*electrographic camera; electronographic camera; electronic camera*
electron charge 电子电荷
electron cloud 电子云

electron concentration 电子浓度
electron conduction 电子传导 | 電子傳熱〔台〕
electron configuration 电子组态
electron cyclotron oscillation 电子回旋振荡
electron cyclotron resonance 电子回旋共振
electron cyclotron wave 电子回旋波
electron degeneracy 电子简并
electron degeneracy pressure 电子简并压
 又见：*degenerate electron pressure*
electron density 电子密度
electron detachment coefficient 电子脱离系数
electron diffraction 电子衍射
electronegativity 电阴性
electron-electron bremsstrahlung 电子－电子韧致辐射
electron emission 电子发射　又见：*electronic emission*
electron energy spectrum 电子能谱　又见：*electron spectrum*
electron event 电子事件
electron flare 电子耀斑 | 電子閃焰〔台〕
electron gas 电子气
electron gyro-frequency 电子回旋频率
electron gyro-radiation 电子回旋辐射
electronic band spectrum 电子带光谱
electronic camera 电子照相机　又见：*electrographic camera; electronographic camera; electron camera*
electronic chronometer 电子时计
electronic emission 电子发射　又见：*electron emission*
electronic image-tube 电子像管
 又见：*electrographic image tube*
electronic imaging 电子成像
electronic state 电子态
electron jet 电子束　又见：*electron beam*
electron level 电子能级
electron mode 电子模式
electron multiplier tube 电子倍增管
electron-multiplying CCD （＝EMCCD）电子倍增CCD
electron neutrino 电子中微子　又见：*e-neutrino*
electronographic camera 电子照相机
 又见：*electrographic camera; electron camera; electronic camera*
electronographic detector 电子照相探测器
electronographic photometry 电子照相测光
electronography 电子照相
electron oscillation 电子振荡
electron pair 电子对

electron-photon scattering 电子－光子散射
electron-plasma frequency 电子－等离子体频率
electron-positron annihilation 电子－正电子湮灭
electron-positron pair 电子－正电子对
electron-positron pair annihilation 电子－正电子对湮灭
electron-positron pair creation 电子－正电子对产生
electron pressure 电子压力
electron pumping 电子抽运
electron rest frame 电子静止坐标系
electron rest mass 电子静止质量
electron scattering 电子散射
electron scattering continuum 电子散射连续谱
electron scattering opacity 电子散射不透明度
electron spectrum 电子能谱　又见：*electron energy spectrum*
electron spin 电子自旋
electron temperature 电子温度
electron transition 电子跃迁
electronvolt 电子伏
electron wave 电子波
electro-optical modulator 电光调制器
electro-optic crystal 电光晶体
electro-optic spectrograph 电光频谱仪
electrophonic meteor sound 电声变换流星声
electrophonic sound 电声响
electrophotometer 光电光度计　又见：*photoelectric photometer*
electrophotometry 光电测光 | 光電測光術〔台〕
 又见：*photoelectric photometry*
electrophotonic detector 光电探测器
electrophotonic imaging 光电成像
 又见：*photoelectronic imaging*
electrosphere 电子层
electrostatic bremsstrahlung 静电韧致辐射 | 靜電制動輻射〔台〕
electrostatic field 静电场　又见：*static electric field*
electrostatic influence 静电影响
electrostatic interaction 静电相互作用
electrostatic magnetic force 静电磁力
electrostatic state seperator 静电能态分离器
electrostatic wave 静电波
electroweak era 弱电时期
electroweak force 弱电力
electroweak interaction 弱电相互作用
electroweak phase transition 弱电相变

electroweak theory　弱电理论
element abundance　元素丰度　又见：*abundance of element; elemental abundance*
elemental abundance　元素丰度　又见：*element abundance; abundance of element*
Elementary Astronomical Instrument　简平仪
elementary excitation　元激发
elementary particle　基本粒子
elementary particle scattering　元粒子散射
element formation　元素形成　又见：*formation of elements*
element group　元素族
element of eclipse　交食要素
element of orbit　轨道根数 | 軌道要素，軌道根數〔台〕　又见：*orbital element*
elements of light variation　光变要素
Elephant's Trunk Nebula　象鼻星云
elephant trunk　亮环结构　又见：*bright rim structure*
elevated pole　上天极
elevation angle　仰角
elevation axis　高度轴
elimination date　灭日
E line　E 谱线
Elkhiffa Australis　氐宿一（天秤座 α）　又见：*Zuben Elgenubi*
Ellerman bomb　埃勒曼炸弹
ellipse of inertia　惯量椭圆　又见：*momental ellipse*
ellipse of zero velocity　零速度椭圆
ellipsoid　椭球
ellipsoidal binary　椭球双星
ellipsoidal collapse　椭球坍缩
ellipsoidal coordinate　椭球坐标　又见：*spheroidal coordinate*
ellipsoidal distribution　椭球分布
ellipsoidal distribution of velocities　速度椭球分布
ellipsoidal mirror　椭球面反射镜
ellipsoidal nebula　椭球星云
ellipsoidal telescope　椭球面望远镜
ellipsoidal variable　椭球变星 | 椭球 [狀] 變星〔台〕
ellipsoid of gyration　旋转椭球　又见：*spheroid of revolution*
ellipsoid of inertia　惯量椭球　又见：*momental ellipsoid*
ellipsoid of revolution　旋转椭球体
elliptic aberration　椭圆光行差
elliptical　椭圆星系　又见：*elliptical galaxy; elliptical system*

elliptical anagalactic nebula　椭圆河外星云　又见：*elliptical extragalactic nebula*
elliptical extragalactic nebula　椭圆河外星云　又见：*elliptical anagalactic nebula*
elliptical galaxy　（＝E galaxy）椭圆星系　又见：*elliptical; elliptical system*
elliptically polarized light　椭圆偏振光
elliptically polarized radiation　椭圆偏振辐射
elliptical motion　椭圆运动
elliptical nebula　椭圆星云
elliptical orbit　椭圆轨道
elliptical radio galaxy　椭圆射电星系
elliptical space　椭圆空间　又见：*elliptic space*
elliptical subsystem　椭圆次系
elliptical system　椭圆星系　又见：*elliptical galaxy; elliptical*
elliptic comet　椭圆轨道彗星 | 椭圆 [轨道] 彗星〔台〕
elliptic coordinate　椭圆坐标
ellipticity　椭率 | 橢圓率〔台〕
ellipticity effect　椭球状效应
elliptic restricted problem　椭圆型限制性问题
elliptic restricted three-body problem　椭圆型限制性三体问题
elliptic space　椭圆空间　又见：*elliptical space*
El Nath　娄宿三（白羊座 α）　又见：*Hamal*
Elnath　五车五（金牛座 β）　又见：*Alnath*
elongated orbit　椭长轨道
elongation　① 距角　又见：*angular elongation* ② 大距　又见：*greatest elongation*
elongation of circumpolar stars　拱极星大距
Elpis　乾神星（小行星 59 号）
ELT　（＝Extremely Large Telescope）特大望远镜
Eltanin　天棓四（天龙座 γ）　又见：*Etamin*
Elvey crater　埃尔维环形山（月球）
Elysium Planitia　埃律西昂平原（火星）
EM　（＝emission measure）发射量度 | 發射 [計] 量〔台〕
emanator　辐射器，辐射源
embedded cluster　内埋星团
embolismic year　闰年　又见：*leap year; bissextile year; intercalary year*
embryo of a planet　行星胎　又见：*planetary embryo*
embryo of a star　恒星胎　又见：*stellar embryo*
EMCCD　（＝electron-multiplying CCD）电子倍增 CCD
Emden crater　埃姆登环形山（月球）
Emden equation　艾姆登方程

Emden function 艾姆登函数
emergent pupil 出射光瞳 | 出射 [光] 瞳〔台〕
 又见: *exit pupil; eye-circle; ocular circle*
emerging flux region （＝EFR）射流区
emerging magnetic flux 浮现磁流
emersion 复现 | 復明〔台〕
emersion time 复现时刻
EMF （＝evolving magnetic feature）演化磁特征
emission 发射，辐射
emission area 发射区 又见: *emitting area*
emission band 发射带
emission coefficient 发射系数
emission component 发射子线
emission efficiency 发射效率
emission frequency 发射频率
emission line 发射线 | 發射 [譜] 線〔台〕
emission-line galaxy 发射线星系
emission-line nebula 发射线星云
emission-line object 发射线天体
emission-line star 发射线星
emission measure （＝EM）发射量度 | 發射 [計] 量〔台〕
emission mechanism 发射机制
emission nebula 发射星云
emission nebulosity 发射星云状物质 | 發射雲氣〔台〕
emission point 发射点
emission probability 发射概率
emission process 发射过程
emission spectrum 发射光谱
emission variable 发射线变星
emissive power 发射本领，发射强度 | 發射強度〔台〕
emissivity 发射率
emitron 光电摄像管
emittance 发射度
emitter 发射体
emitting area 发射区 又见: *emission area*
emitting atmosphere 发光大气
emitting dust mass 发光尘埃质量
E mode polarization E 模式偏振
EMP （＝Ephemeris of the Minor Planets）小行星星历表
emphasized second marker 加重秒信号
empirical coefficient 经验系数
empirical formula 经验公式 | 經驗 [公] 式〔台〕
empirical term 经验项

Emptiness 虚宿（二十八宿） 又见: *Void*
empty model 空模型 | 虛空模型，零質量模型〔台〕
emulated data 仿真数据
emulation 仿真
emulsion 乳胶
emulsion carrier 乳胶载体
emulsion defect 乳胶污点
emulsion photometer 乳胶光度计
emulsion shift 乳胶位移
Encampment 营室，室宿 | 室宿〔台〕（二十八宿） 又见: *House*
Enceladus 土卫二
encircled energy 能量集中度
Encke crater 恩克环形山（月球）
Encke gap 恩克环缝（土星）
Encke's comet 恩克彗星 | 恩克彗 [星]〔台〕
 又见: *Comet Encke*
Encke's method 恩克方法
enclosure 围罩
encoder 编码器 又见: *coder*
encounter 交会
encounter hypothesis 偶遇假说
encounter theory 偶遇理论
encounter-type orbit 交会型轨道
encroachment 侵
end-fire array 端射天线阵
end height 消失点高度 又见: *height of disappearance*
ending of evening twilight 昏影终 | 暮光終，昏影終〔台〕
endocrater 巨坑
endoergic process 吸能过程
endoergic reaction 吸能反应
End of Heat 处暑（节气） 又见: *Limit of Heat*
end of totality 全食终 | 生光〔台〕 又见: *total eclipse end*
end-on 端向
end-on object 端向天体
endothermic process 吸热过程
endothermic reaction 吸热反应
end point 消失点 又见: *disappearance point*
Endurance crater 坚忍环形山（火星）
Endymion crater 恩底弥昂环形山（月球）
energetic encounter 高能碰撞
Energetic Gamma Ray Experiment Telescope （＝EGRET）高能伽马射线实验望远镜

energetic particle 高能粒子 又见: *high-energy particle*
energetic particle burst 高能粒子暴
energetic particle event 高能粒子事件
energetic plasma 高能等离子体
energetic recoil particle 高能反冲粒子
energetics 力能学 | 吸能說〔台〕
energetic X-ray burst 高能 X 射线暴
energetic γ-ray burst 高能 γ 射线暴
energy absorption 能量吸收
energy attenuation 能量衰减
energy balance 能量平衡 又见: *energy equilibrium*
energy condition 能量条件
energy conservation 能量守恒 又见: *conservation of energy*
energy density 能量密度 | 能[量]密度〔台〕
energy density of radiation 辐射能量密度
energy density of the vacuum 真空能量密度
energy detectivity 能量探测率
energy dissipation 能量耗散 又见: *dissipation of energy*
energy distribution 能量分布
energy-driven wind 能量驱动风
energy equation 能量方程
energy equilibrium 能量平衡 又见: *energy balance*
energy equipartition 能量均分 又见: *equipartition of energy*
energy flux 能流量
energy-flux density 能流密度
energy gap 能隙
energy interval 能量间隔,能量区间
energy jump 能量跳变
energy level 能级 | 能階〔台〕
energy-level diagram 能级图 | 能階圖〔台〕
energy liberation 能量释放 又见: *energy release*
energy loss 能量损失
energy-momentum tensor 能动张量
energy of absolute zero 零点能量
energy of magnetization 磁化能
energy of nucleus 核能
energy of position 位能
energy of rotation 转动能 又见: *rotational energy*
energy of vibration 振动能 又见: *vibrational energy*
energy range 能幅
energy release 能量释放 又见: *energy liberation*
energy source 能源 又见: *source of energy*
energy spectral density 能谱密度

energy spectrum 能谱
energy state 能态
energy storage 能量储存
energy transfer 能量转移
energy transfer time 能量转移时间
energy transport 能量输运
energy trap 能量[陷]阱
e-neutrino 电子中微子 又见: *electron neutrino*
English mounting 英国式装置 | 英[國]式装置〔台〕
enhanced emission 增强发射
enhanced line 增强谱线
enhanced network 增强网络
enhanced radiation 增强辐射
Enif 危宿三（飞马座 ε）
Enki Catena 恩基环形山链（木卫三）
enlarger 放大机
enregistreur des vitesses 分光速度记录仪 又见: *spectro-enregistreur des vitesses*
ensemble 系综 又见: *assemblage*
ensemble average 集平均
ensemble of electrons 电子集
Ensisheim meteorite 恩希塞姆陨星
enstatite 顽辉石 | 顽火辉石〔台〕
enstatite chondrite 顽辉球粒陨石
enthalpy 热函,焓 | 焓〔台〕
entrance pupil 入射光瞳 | 入射[光]瞳〔台〕
entrapment 俘获 | 捕獲〔台〕 又见: *capture*
entropy bound 熵限
entropy density 熵密度
entropy equation 熵方程
entropy perturbation 熵扰动
entropy wave 熵波
envelope 包层,包络 | 包層,外殼〔台〕
envelope pattern 包络方向图
envelope star 气壳星 又见: *shell star*
EOP (＝earth orientation parameter）地球定向参数
EoR (＝epoch of reionization）再电离时期
EoS (＝equation of state）物态方程 | 狀態方程〔台〕
Eos 曙神星（小行星 221 号）
Eos Chasma 厄俄斯深谷（火星）
Eos family 曙神星族
Eötvös crater 厄缶环形山（月球）
epact 闰余,岁首月龄 | 元旦月龄〔台〕

epagomenal day 闰日 又见：*leap day; bissextile day; intercalary day*
Ep galaxy Ep 星系
ephemeral active region 瞬现区，瞬现活动区
　　又见：*ephemeral region*
ephemeral region 瞬现区，瞬现活动区
　　又见：*ephemeral active region*
ephemeris day 历书日
ephemeris hour angle 历书时角
ephemeris longitude 历书经度
ephemeris meridian 历书子午线
Ephemeris of the Minor Planets （=EMP）小行星星历表
ephemeris reference frame （=ERF）历书参考系
ephemeris second 历书秒
ephemeris sidereal time 历书恒星时
ephemerist 制历者
ephemeris time （=ET）历书时
ephemeris transit 历书中天
ephemeris（复数：ephemerides） 历表
ephemeris（复数：ephemerides） 星历表
Epi 角宿一（室女座 α） 又见：*Azimech; Spica*
epicenter 本轮中心，震中 | 震央〔台〕
epicycle 本轮 | 本輪，周轉圓〔台〕
epicycle energy 本轮能量
epicycle oscillation 本轮振荡
epicycle theory 本轮说
epicyclic approximation 本轮近似
epicyclic frequency 本轮频率
epicyclic motion 本轮运动
epicyclic orbit 本轮轨道
Epimetheus 土卫十一
epiplasma 超等离子体
epoch 历元，时期 | 曆元〔台〕
epoch of matter-radiation equality 物质-辐射等密度时期
epoch of neutralization 中和纪
epoch of observation 观测历元
epoch of orientation 定向历元
epoch of place 位置历元
epoch of reionization （=EoR）再电离时期
EPOXI （=Extrasolar Planet Observation and Deep Impact Extended Investigation）EPOXI 任务
e-process e 过程 | e-過程〔台〕
Equ （=Equuleus）小马座
equal altitude circle 等高圈

equal altitude method 等高法
equal brightness photometer 等亮度光度计
equality-of-contrast photometer 等反差光度计
equantequation 均衡点，均衡法
equation clock 时差钟
equation of equal altitude 等高差
equation of equipartition time 均时差
equation of light 光行时差 | 光 [行時] 差〔台〕
　　又见：*light equation*
equation of motion 运动方程
equation of radiative transfer 辐射转移方程
equation of state （=EoS）物态方程 | 狀態方程〔台〕
equation of the center 中心差 又见：*great inequality*
equation of the equinoxes 二分差
equation of the origins 零点差
equation of time 时差
equator 赤道
equator correction 赤道改正
equator determination 赤道测定
equatorial acceleration 赤道加速度 | 赤道加速度，赤道加速现象〔台〕
equatorial armillary sphere 赤道经纬仪
　　又见：*simple equatorial armilary*
equatorial band 赤道带 又见：*equatorial zone*
equatorial bulge 赤道隆起 | 赤道隆起 [部份]〔台〕
equatorial circumference 赤道圈
equatorial colure 二分圈
equatorial coordinate 赤道坐标
equatorial coordinate system 赤道坐标系
　　又见：*equatorial system of coordinate; equatorial system*
equatorial diameter 赤道直径
equatorial ejection 赤道抛射
equatorial horizontal parallax 赤道地平视差
equatorial instrument 赤道仪 又见：*equatorial telescope*
equatorial line 赤道线
equatorial lodge degrees 入宿度
　　又见：*determinative star distance; lunar lodge degrees*
equatorial mounting 赤道装置 | 赤道 [式] 装置〔台〕
equatorial orbit 赤道轨道
equatorial parallax 赤道视差
equatorial plane 赤道面
equatorial prominence 赤道日珥
equatorial radius 赤道半径
equatorial rectangular coordinate 赤道直角坐标
equatorial rotational velocity 赤道自转速度

equatorial satellite 赤道卫星
equatorial sundial 赤道日晷 | 赤道 [式] 日晷〔台〕
equatorial system 赤道坐标系 又见: *equatorial coordinate system; equatorial system of coordinate*
equatorial system of coordinate 赤道坐标系 又见: *equatorial coordinate system; equatorial system*
equatorial telescope 赤道仪 又见: *equatorial instrument*
Equatorial Torquetum 简仪 又见: *abridged armilla*
equatorial zone 赤道带 又见: *equatorial band*
equatorium 行星定位仪
equator of date 瞬时赤道
equator of epoch 历元赤道
equator of illumination 照明赤道
equator of position 位置赤道
equator ring 赤道环
equiareal mapping 保积映射 又见: *area-preserving map*
equidensite 等密度线
equidensitometry 等密度测量
equidistance motion 等距运动
equilateral point 等边点
equilateral triangle point 等边三角形点
equilibrium 平衡
equilibrium condition 平衡条件
equilibrium distribution 平衡分布
equilibrium figure 平衡形态
equilibrium point 平衡点
equilibrium position 平衡位置
equilibrium radiation 平衡辐射
equilibrium ratio 平衡比率 | 平衡比 [率]〔台〕
equilibrium solution 平衡解
equilibrium state 平衡态
equilibrium temperature 平衡温度
equilibrium tide 平衡潮
equinoctial 天赤道 | 天 [球] 赤道〔台〕 又见: *celestial equator*
equinoctial colure 二分圈
equinoctial day 恒星日 又见: *sidereal day*
equinoctial element 分点根数
equinoctial points 二分点 又见: *equinoxes*
equinoctial system of coordinate 分至坐标系
equinoctial year 分至年
equinox 分点
equinox correction 春分点改正
equinoxes 二分点 又见: *equinoctial points*
equinox motion 春分点运动

equinox position 春分点位置
equipartition 均分
equipartition instability 能均分不稳定性
equipartition of energy 能量均分 又见: *energy equipartition*
equipartition of kinetic energy 动能均分
equipartition parameter 均分参数
equipartition time 均分时间
equipotential surface 等势面，等位面
equivalence principle 等效原理 又见: *equivalent principle; principle of equivalence*
equivalent antenna 等效天线
equivalent breadth 等值宽度 又见: *equivalent width*
equivalent collecting area 等效面积
equivalent focal distance 等值焦距 | 等效焦距〔台〕 又见: *equivalent focal length*
equivalent focal length 等值焦距 | 等效焦距〔台〕 又见: *equivalent focal distance*
equivalent focus 等值焦点 | 等效焦點〔台〕
equivalent pendulum 等值摆
equivalent principle 等效原理 又见: *equivalence principle; principle of equivalence*
equivalent temperature 等效温度
equivalent thickness 等值厚度
equivalent width 等值宽度 又见: *equivalent breadth*
Equuleus （＝Equ）小马座
Er （＝erbium）铒（68 号元素）
ERA （＝Earth Rotation Angle）地球自转角
era 纪元，时代 | 紀元，代，時代〔台〕
era divisor 纪法
Erakis 造父四（仙王座 μ） 即: *Garnet star*
Erato 效神星（小行星 62 号）
Eratosthenes crater 厄拉多塞环形山（月球）
erbium （＝Er）铒（68 号元素）
erect image 正像
erecting eyepiece 正像目镜
erecting prism 正像棱镜
E region E 区（地球） 即: *E layer*
ERF （＝ephemeris reference frame）历书参考系
Erfle eyepiece 尔弗利目镜
ergodic density fields 各态历经密度场
ergodic hypothesis 遍历假说
ergodicity 遍历
ergodic motion 遍历运动 | 均曆運動〔台〕
ergodic theorem 各态历经定理
ergoregion 能层 | 動圈，動區〔台〕 又见: *ergosphere*

ergosphere 能层 | 動圈，動區〔台〕
    又见：*ergoregion*
ergosphere effect 能层效应
Eri （＝Eridanus）波江座
Eridalus Filaments 波江纤维
Eridalus Hot Spot 波江热斑
Eridalus Shell 波江壳
Eridanus （＝Eri）波江座
Erinome 木卫二十五
Eris 阋神星（小行星 136199 号，矮行星）
ERO （＝extremely red object）极红天体
Eros 爱神星（小行星 433 号）
erosion 侵蚀
erosive collision 侵蚀性碰撞
ERP （＝earth rotation parameter）地球自转参数
Er Rai 少卫增八（仙王座 γ）
Erriapus 土卫二十八
Erro crater 埃罗环形山（月球）
error accumulation 误差累积
error analysis 误差分析
error bar 误差棒
error beam 误差波束
error box 误差框
error distribution 误差分布
error ellipse 误差椭圆
error estimate 误差估计
error function 误差函数
error in pointing 指向误差   又见：*pointing error*
error of tilt 倾斜误差
error pattern 误差方向图
error propagation 误差传播
ERS （＝European Remote Sensing Satellite）欧洲遥感卫星
eruption 爆发   又见：*outbreak; outburst*
eruptive arch 爆发拱
eruptive binary 爆发双星
eruptive center 爆发中心
eruptive flare 爆发耀斑 | 爆發閃焰〔台〕
    又见：*eruptive solar flare; explosive flare*
eruptive galaxy 爆发星系   又见：*exploding galaxy; explosive galaxy*
eruptive period 爆发周期
eruptive prominence 爆发日珥   又见：*exploding prominence*
eruptive solar flare 爆发耀斑 | 爆發閃焰〔台〕
    又见：*eruptive flare; explosive flare*
eruptive star 爆发星   又见：*exploding star*

eruptive variable 爆发变星   又见：*explosive variable*
Es （＝einsteinium）锿 | 鑀〔台〕（99 号元素）
ESA （＝European Space Agency）欧洲空间局 | 歐[洲]空[間]局〔台〕
escape cone 逃逸锥   即：*loss cone*
escape energy 逃逸能量
escapement 擒纵系统 | 節擺件，擒縱件〔台〕
escape speed 逃逸速度 | 脫離速度〔台〕
    又见：*escape velocity; velocity of escape*
escape speed surface 逃逸速度面
escape trajectory 逸离轨道
escape velocity 逃逸速度 | 脫離速度〔台〕
    又见：*escape speed; velocity of escape*
Eskimo Nebula 爱斯基摩星云（NGC 2392）
Esnault-Pelterie crater 埃斯诺－佩尔蒂埃环形山（月球）
ESO （＝European Southern Observatory）欧南台 | 歐洲南天天文台〔台〕
ESO-MIDAS （＝European Southern Observatory - Munich Image Data Analysis System）欧洲南方天文台慕尼黑图像数据分析系统
ESO/SERC Southern Sky Survey ESO/SERC 南天天图
ESRO （＝European Space Research Organization）欧洲空间研究组织（欧洲空间局前身）
ESSENCE supernova survey ESSENCE 超新星巡天
essentially singular point 本性奇点
EST （＝Eastern Standard Time）东部标准时间
establishment of the port 潮候时差
estimated accuracy 估计精度
estimation 估计
estimation theory 估计理论
ET ① （＝Eastern Time）东部时间
  ② （＝ephemeris time）历书时
Eta Aquarid meteor shower 宝瓶 η 流星雨
Eta Carinae Nebula 船底 η 星云
Eta Carina Nebula 船底星云（NGC 3372）
    又见：*Carina Nebula*
etalon 标准，校准器 | 標準，標準具，法布立－柏洛干涉儀〔台〕
etalon frequency 标准频率   又见：*standard frequency*
etalon time 标准时   又见：*standard time*
Etamin 天棓四（天龙座 γ）   又见：*Eltanin*
ETELS （＝early-type emission-line star）早型发射线星
etendue 集光率
eternal inflation 永恒暴胀

ethane 乙烷（C$_2$H$_6$）
ethanol 乙醇（CH$_3$CH$_2$OH） 又见：*ethyl alcohol*
ether 以太 又见：*aether*
ethereal ring 薄环
ethyl alcohol 乙醇（CH$_3$CH$_2$OH） 又见：*ethanol*
ethyl cyanide 丙腈（C$_3$H$_5$N）
ethylene 乙烯（C$_2$H$_4$）
ethynyl radical 乙炔基（HCC）
E-type asteroid E 型小行星
Eu （=europium）铕（63 号元素）
Euanthe 木卫三十三
Euclidean continuation 欧几里得延拓
Euclidean source count 欧几里得源计数
Euclidean space 欧几里得空间
Euclidean time 欧几里得时间
Euclidean universe 欧几里得宇宙
Euclides crater 欧几里得环形山（月球）
Euclid space telescope 欧几里得空间望远镜
eucrite 钙长辉长无球粒陨石
Eudoxus crater 欧多克斯环形山（月球）
Eugenia 香女星（小行星 45 号）
Eukelade 木卫四十七
Euler angle 欧拉角
Euler crater 欧拉环形山（月球）
Euler equation 欧拉方程
Eulerian coordinate 欧拉坐标
Eulerian derivative 欧拉导数
Eulerian motion 欧拉运动
Eulerian period 欧拉周期
Euler-Lagrange equation 欧拉－拉格朗日方程
Euler-Poincare characteristic 欧拉－庞加莱特征
Eunomia 司法星（小行星 15 号）
Eunomia family 司法星族
Euphrosyne 丽神星（小行星 31 号）
Euporie 木卫三十四
Eurasia plate 欧亚板块
EURECA （=European Retrievable Carrier）尤里卡（欧洲可回收卫星）
Eureka 尤里卡（小行星 5261 号）
Europa ①木卫二 ②欧女星，欧罗巴（小行星 52 号）
Europa Orbiter 木卫二环轨探测器
European Centre for Nuclear Research （=CERN）欧洲核子研究中心
European Extremely Large Telescope （=EELT）欧洲特大望远镜｜欧洲超大望遠鏡〔台〕
European Launcher Development Organization （=ELDO）欧洲发射器开发组织
European Remote Sensing Satellite （=ERS）欧洲遥感卫星
European Retrievable Carrier （=EURECA）尤里卡（欧洲可回收卫星）
European Southern Observatory （=ESO）欧南台｜歐洲南天天文台〔台〕
European Southern Observatory - Munich Image Data Analysis System （=ESO-MIDAS）欧洲南方天文台慕尼黑图像数据分析系统
European Space Agency （=ESA）欧洲空间局｜歐[洲]空[間]局〔台〕
European Space Research Organization （=ESRO）欧洲空间研究组织（欧洲空间局前身）
European Virtual Observatory （=EURO-VO）欧洲虚拟天文台
European VLBI Network （=EVN）欧洲甚长基线干涉网｜歐洲特長基線干涉網〔台〕
European X-ray Observatory Satellite （=EXOSAT）EXOSAT 欧洲 X 射线天文卫星｜歐洲 X 光天文衛星〔台〕
europium （=Eu）铕（63 号元素）
EURO-VO （=European Virtual Observatory）欧洲虚拟天文台
Eurydike 狱神星（小行星 75 号）
Eurydome 木卫三十二
Eurynome 配女星（小行星 79 号）
EUSO （=Extreme Universe Space Observatory）极端宇宙空间天文台
Euterpe 司箫星（小行星 27 号）
EUV （=extreme-ultraviolet）极紫外
EUVE （=Extreme Ultra-Violet Explorer）极紫外探测器｜極紫外探測衛星〔台〕（美国）
EUV light （=Extreme ultraviolet light）极紫外射线｜超紫外線〔台〕
EVA （=extravehicular activity）舱外活动
evaluable primary frequency standard 可计值原始频标
evanescent wave 瞬即消失波
Evans crater 埃文斯环形山（月球）
evaporating gaseous globule （=EGG）蒸发气态球状体
evaporation of the black hole 黑洞蒸发
evaporation rate 蒸发率
evaporation time 蒸发时标
Evdokimov crater 叶夫多基莫夫环形山（月球）
evection 出差 又见：*moon's evection*

evection in latitude 黄纬出差
even coupling 偶耦合
even cycle 偶数周
even-even nuclei 偶－偶核
evening glow 暮辉
evening group 昏星组 | 黃昏星組〔台〕
evening star 昏星
evening twilight 暮光
even-odd nuclei 偶－奇核
even state 偶态
event horizon 视界 | [事件] 視界〔台〕
Evershed crater 埃弗谢德环形山（月球）
Evershed effect 埃弗谢德效应
EVLA （＝Expanded Very Large Array）增容甚大阵
EVN （＝European VLBI Network）欧洲甚长基线干涉网 | 歐洲特長基線干涉網〔台〕
evolutionary age 演化年龄
evolutionary cosmology 演化宇宙学
evolutionary mass 演化质量 又见：evolution mass
evolutionary phase 演化阶段 | 演化相, 演化階段〔台〕
evolutionary stage 演化期
evolutionary state 演化态
evolutionary time 演化时间
evolutionary time-scale 演化时标
evolutionary track 演化程 | 演化軌跡〔台〕
evolution diagram 演化图
evolution mass 演化质量 又见：evolutionary mass
evolution of clustering 成团性演化
evolution of luminosity function 光度函数演化
evolved object 晚期演化天体
evolved star 主序后星 又见：post-main-sequence star
evolving magnetic feature （＝EMF）演化磁特征
evolving object 早期演化天体
evolving star 零龄主序前星
ExAO （＝extreme adaptive optics）极端自适应光学
excess absorption 过剩吸收
excess density 密度超 又见：density excess
excess emission 过剩发射
excess energy 过剩能量
excessive star 超星
excess noise temperature 过剩噪声温度
excess redshift 剩余红移
exchange correlation 交换相关
exchange energy 交换能

exchange of mass 质量交换 又见：mass exchange
excitation cross-section 激发截面
excitation energy 激发能
excitation mechanism 激发机制
excitation of turbulence 湍动激发
excitation potential 激发势 | 激發電位, 激發電勢〔台〕
excitation temperature 激发温度
excited atom 受激原子
excited level 激发能级
excited nebula 受激星云
excited object 受激天体
excited star 受激星
excited state 受激态，激发态
exciting collision 激发碰撞
exciting object 激发天体
exciting star 激发星
exciton 激子
Exclamation Mark galaxy 惊叹号星系
exclusion energy 不相容能量
exclusion principle 不相容原理
excursion set 漫游集
excursion set model 漫游集模型
exfoliation 剥落
exhaust velocity 排气速度
exit cone 出射锥
exit pupil 出射光瞳 | 出射 [光] 瞳〔台〕
  又见：eye-circle; ocular circle; emergent pupil
exmeridian altitude 近子午线高度
  又见：extra-meridian altitude
exmeridian observation 近子午线观测
  又见：extra-meridian observation
ex-nova 爆后新星 又见：faded nova; postnova
exo-asteroid belt [太阳] 系外小行星带
exobase 外大气层底
exobiology 地外生物学 | 地 [球] 外生物學〔台〕
Exobiology on Mars （＝ExoMars）火星生命探测计划
exocomet ① 外星彗星 又见：exotic comet 太阳 系外彗星 又见：extrasolar comet; exotic comet
exo-Earth 系外类地行星 | [太陽] 系外類地行星〔台〕 又见：terrestrial exoplanet; Earth-like exoplanet
Exo-Earth Imager （＝EEI）系外类地行星成像器
exoergic process 放能过程
exogenic force 外力 又见：external force
exo-Jupiter 系外类木行星 | [太陽] 系外類木行星〔台〕 又见：Jovian exoplanet; Jupiter-like exoplanet

ExoMars （=Exobiology on Mars）火星生命探测计划

exoplanet 系外行星 | [太陽] 系外行星〔台〕
又见：*extrasolar planet; exotic planet*

exoplanet system 系外行星系 | [太陽] 系外行星系〔台〕 又见：*extrasolar planetary system*

exoplanet transit 系外行星凌星 | [太陽] 系外行星凌星〔台〕

EXOSAT （=European X-ray Observatory Satellite）EXOSAT 欧洲 X 射线天文卫星 | 歐洲 X 光天文衛星〔台〕

exo-solar wind [太阳] 系外星风

exosphere 外 [逸] 层，外大气层 | 外氣層〔台〕

exothermic process 产热过程

exothermic reaction 产热反应

exotic comet ① 外星彗星 又见：*exocomet* 太阳系外彗星 又见：*extrasolar comet; exocomet*

exotic particle 奇异粒子 又见：*strange particle*

exotic planet 系外行星 | [太陽] 系外行星〔台〕
又见：*extrasolar planet; exoplanet*

exotic star 奇异星 又见：*strange star*

exozodiacal dust 外星黄道尘

Expanded Very Large Array （=EVLA）增容甚大阵

expanding arm 膨胀臂 | 擴張旋臂〔台〕

expanding disk 膨胀盘

expanding envelope 膨胀包层

expanding model 膨胀模型

expanding nebula 膨胀星云

expanding phase 膨胀相

expanding plasmon 膨胀等离子体激元

expanding ring 膨胀环

expanding shell 膨胀壳

expanding universe 膨胀宇宙

expansion age 膨胀年龄 | 膨脹 [年] 齡〔台〕

expansion coefficient 膨胀系数 又见：*coefficient of expansion*

expansion lag 滞涨

expansion of the universe 宇宙膨胀 又见：*cosmic expansion*

expansion parameter 膨胀参数

expansion rate 膨胀率

expansion time-scale 膨胀时标

expectation value 期望值 又见：*expected value*

expected value 期望值 又见：*expectation value*

Experimental Astronomy 《实验天文学》（荷兰期刊）

experimental astronomy 实验天文学

explicit symmetry breaking 明显对称性破缺

exploding galaxy 爆发星系 又见：*eruptive galaxy; explosive galaxy*

exploding granule 爆发米粒

exploding prominence 爆发日珥 又见：*eruptive prominence*

exploding star 爆发星 又见：*eruptive star*

Explorer 探险者号（美国科学卫星和深空探测器系列）

Explorer 11 探险者 11 号（美国首颗高能天文卫星）

explosive era 爆发期

explosive flare 爆发耀斑 | 爆發閃焰〔台〕
又见：*eruptive flare; eruptive solar flare*

explosive galaxy 爆发星系 又见：*eruptive galaxy; exploding galaxy*

explosive instability 爆发不稳定性

explosive nucleosynthesis 爆发核合成

explosive phase 爆发相 | 爆發相，爆發階段〔台〕

explosive shower 爆炸簇射

explosive variable 爆发变星 又见：*eruptive variable*

expometer 曝光计 又见：*exponometer; exposure-meter*

exponential atmosphere 指数式大气

exponential disk 指数盘

exponentially stable 指数式稳定

exponential potential 指数势

exponential prediction 指数预测

exponential rigidity spectrum 指数硬度谱

exponential spectrum 指数谱

exponential spheroid 指数椭球体

exponometer 曝光计 又见：*expometer; exposure-meter*

exposure 曝光 | 曝光，露光〔台〕

exposure age 曝光时限 | 曝光範圍〔台〕
又见：*exposure latitude; latitude of exposure*

exposure latitude 曝光时限 | 曝光範圍〔台〕
又见：*exposure age; latitude of exposure*

exposure-meter 曝光计 又见：*expometer; exponometer*

exposure time 曝光时间 | 曝光時間，露光時間〔台〕

ex-supernova 爆后超新星 又见：*post-supernova*

extar X 射线星 | X 光星〔台〕 又见：*X-ray star*

extended atmosphere 厚大气，延伸大气 | 厚大氣〔台〕 即：*stellar envelope*

extended envelope 厚包层，延伸包层 | 厚外殼，延伸包層〔台〕

extended filamentary nebula 延展纤维星云

extended filamentary radio nebula 延展纤维射电星云
extended inflation 扩展暴胀
extended infrared source 红外展源 | 紅外 [線] 展源〔台〕
extended Kalman filter 扩充卡尔曼滤波器
extended low-surface brightness source 低面亮度展源
Extended net 张宿（二十八宿） 又见：*Extension*
extended object 延展天体
extended-phase space 扩充相空间
extended photosphere 厚光球，延伸光球 | 延伸光球〔台〕
extended point transformation 扩充点变换
extended quintessence 扩展精质（宇宙学）
extended radio source 射电展源 | 非點 [狀] 電波源，廣延電波源〔台〕
extended ROentgen Survey with an Imaging Telescope Array eROSITA 巡天
extended scattering medium 延展散射介质
extended source 展源 | 延展源，非點狀源〔台〕
extended supergravity 扩展超引力
extended X-ray source X 射线展源 | X 光展源〔台〕
extended γ-ray source γ 射线展源
Extension 张宿（二十八宿） 又见：*Extended net*
extensive air shower 广延大气簇射
extensive air shower array 广延大气簇射阵
exterior contact 外切
exterior ingress 外初切
external field 外场
external force 外力 又见：*exogenic force*
external galaxy 河外星系 又见：*extragalactic system*
externally dispersed interferometer 外部色散干涉仪
external occulter 外遮星器
external occulter coronagraph 外遮星冕仪
external potential 外引力势
external pressure confinement 外压力约束
external resonance 外共振
external shock 外激波
external source hypothesis 外源说
external weight 外部权
extinction 消光
extinction coefficient 消光系数
extinction cross-section 消光截面
extinction curve 消光曲线
extinction efficiency 消光效率

extinct nuclide 死核素
extinct volcano 死火山
extracentral telescope 偏侧望远镜
extra dimension model 额外维模型
extrafocal image 焦外像
extrafocal photometer 焦外光度计
extrafocal photometry 焦外测光 | 焦外光度测量，焦外測光〔台〕
extragalactic astronomy 河外天文学
extragalactic astrophysics 河外天体物理
extragalactic background 河外背景
extragalactic background light 河外背景光
extragalactic background radiation 河外背景辐射
extragalactic binary 河外双星
extragalactic Cepheid 河外造父变星
extragalactic HII region 河外电离氢区
extragalactic infrared astronomy 河外红外天文学
extragalactic infrared background 河外红外背景
extragalactic infrared source 河外红外源
extragalactic matter 河外物质
extragalactic medium 河外介质
extragalactic nebula 河外星云
extragalactic nova 河外新星
extragalactic object 河外天体
extragalactic planetary nebula 河外行星状星云
extragalactic radio astronomy 河外射电天文学
extragalactic radio background 河外射电背景
extragalactic radio radiation 河外射电
extragalactic radio source 河外射电源 | 河外電波源〔台〕
extragalactic source 河外源
extragalactic space 河外空间
extragalactic supernova 河外超新星
extragalactic system 河外星系 又见：*external galaxy*
extragalactic X-ray astronomy 河外 X 射线天文学
extragalactic X-ray background 河外 X 射线背景
extragalactic X-ray source 河外 X 射线源 | 河外 X 光源〔台〕
extragalactic γ-ray astronomy 河外 γ 射线天文学
extragalactic γ-ray background 河外 γ 射线背景
extragalactic γ-ray source 河外 γ 射线源
extragalectic jet 河外喷流
extrahead 反常谱带头
extra-meridian altitude 近子午线高度 又见：*exmeridian altitude*

extra-meridian observation　近子午线观测
　　又见：*exmeridian observation*
extraordinary ray　非常射线 | 非常射線，異常射線〔台〕
extraordinary wave　非常波 | 非常波，異常波〔台〕
extraplanetary space　行星区外空间
extrasensitivity　特高灵敏度
extrasolar comet　[太阳]系外彗星　又见：*exocomet; exotic comet*
extrasolar life　[太阳]系外生命
extrasolar planet　[太阳]系外行星　又见：*exoplanet; exotic planet*
extrasolar planetary system　[太阳]系外行星系
　　又见：*exoplanet system*
Extrasolar Planet Observation and Deep Impact Extended Investigation　（＝EPOXI）EPOXI任务
extrasolar radio source　[太阳]系外射电源
extrasolar X-ray source　太阳外 X 射线源
extrasolar γ-ray source　[太阳]系外 γ 射线源
extra spatial dimension　额外空间维度
extraterrestrial body　地外天体
　　又见：*extraterrestrial object*
extraterrestrial civilization　地外文明 | 地[球]外文明〔台〕
extraterrestrial dust　地外尘
extraterrestrial intelligence　地外智慧生物 | 地[球]外智慧生物〔台〕
extraterrestrial life　地外生命 | 地[球]外生命〔台〕
extraterrestrial matter　地外物质
extraterrestrial object　地外天体
　　又见：*extraterrestrial body*
extraterrestrial radiation　地外辐射
extraterrestrial source　地外源
extravehicular activity　（＝EVA）舱外活动
extremal curve　极值曲线
extremal field　致极场
extreme adaptive optics　（＝ExAO）极端自适应光学
extreme carbon star　极端碳星
extreme helium star　极端氦星
extreme Kerr black hole　极端克尔黑洞
Extremely Large Telescope　（＝ELT）特大望远镜
extremely red object　（＝ERO）极红天体
extreme metal-poor star　极贫金属星
extreme metal-rich star　极富金属星

extreme population I　臂族，极端星族 I | 旋臂星族，極端星族 I〔台〕　又见：*arm population*
extreme population II　晕族，极端星族 II | [銀]暈星族，極端第二星族〔台〕　又见：*halo population*
extreme Population I star　极端星族 I 恒星
extreme population I star　极端星族 I 恒星
extreme-ultraviolet　（＝XUV; EUV）极紫外
extreme ultraviolet astronomy　（＝XUV astronomy）极紫外天文学 | 超紫外天文學〔台〕
extreme-ultraviolet background　极紫外背景
extreme-ultraviolet background radiation　极紫外背景辐射
Extreme Ultra-Violet Explorer　（＝EUVE）极紫外探测器 | 極紫外探測衛星〔台〕（美国）
extreme-ultraviolet galaxy　极紫外星系
Extreme ultraviolet light　（＝XUV light; EUV light）极紫外射线 | 超紫外線〔台〕
extreme-ultraviolet radiation　极紫外辐射
extreme-ultraviolet source　极紫外源
extreme-ultraviolet star　极紫外星
Extreme Universe Space Observatory　（＝EUSO）极端宇宙空间天文台
extreme value　极值　又见：*extremum*
extremum　极值　又见：*extreme value*
extrinsic curvature　外在曲率
extrinsic variable　外因变星
extrusion　喷出
eye and ear method　耳目法
eye and key method　目键法
eye-circle　出射光瞳 | 出射[光]瞳〔台〕　又见：*exit pupil; ocular circle; emergent pupil*
eye-end　目端
eye estimate　目视估计
eyelens　接目镜
eye observation　目测
eyepiece　目镜　又见：*ocular*
eyepiece grid　目镜网络
eyepiece micrometer　目镜测微计　又见：*ocular micrometer*
eyepiece scale　目镜标度
eyepiece slide　目镜筒
eye-point distance　眼点距 | 適眼距〔台〕　即：*eye relief*
eye relief　适瞳距
eye sensitivity curve　目视灵敏曲线
eyesight tester　视力检测星

F

F　（＝fluorine）氟（9 号元素）
Fa　伐三（猎户座 ι）
Faber-Jackson law　费伯－杰克逊规律
　　即：*Faber-Jackson relation*
Faber-Jackson relation　费伯－杰克逊关系
fabrication technique　光纤技术
Fabricius crater　法布里休斯环形山（月球）
Fabry crater　法布里环形山（月球）
Fabry lens　法布里透镜 | 法布立透鏡〔台〕
Fabry-Perot etalon　法布里－珀罗标准具
Fabry-Perot imaging spectrograph　法布里－珀罗成像摄谱仪
Fabry-Perot interferometer　法布里－珀罗干涉仪 | 法布立－拍若干涉計〔台〕
Fabry-Perot interferometry　法布里－珀罗干涉测量
Fabry-Perot spectrograph　法布里－珀罗摄谱仪
Fabry-Perot spectrometer　法布里－珀罗分光计
face-on galaxy　正向星系
face-on object　正向天体
face-on spiral galaxy　正向旋涡星系
facility seeing　人为视宁度
facsimile　传真
factorability filter　可分因子滤波器，可分因子滤光片
factorability property　可分因子特性
factor of dilution of radiation　辐射稀化因子
facular area　光斑区　又见：*facular region*
facular granule　光斑米粒
facular point　光斑亮点
facular region　光斑区　又见：*facular area*
facula（复数：**faculae**）　光斑
faded nova　爆后新星　又见：*ex-nova; postnova*
fade-out　衰弱，消失
fading time　衰老时间
faint blue galaxy　暗蓝星系
faint blue object　暗蓝天体

faint blue star　暗蓝星
faint companion　暗伴天体，暗伴星，暗伴星系 | 暗伴星〔台〕　又见：*dark companion*
faint galaxy　暗星系 | 昏暗星系〔台〕
Faint Images of the Radio Sky at Twenty-Centimeters catalogue　（＝FIRST）FIRST 星表
faint meteor　暗流星
faint object　暗天体
faint source　暗源
faint star　暗星　又见：*dark star*
fair sample hypothesis　合理样本假说
fake Zeeman splitting　赝塞曼分裂
fall　① 落，没 ② 见落陨石
falling star　流星　又见：*shooting star; meteor*
fall of the sea　落潮　又见：*ebb; ebb tide*
false-color image　假彩色像 | 假色像〔台〕
false-color imagery　假彩色成像
false color images　假彩色图像
false-color photometry　假彩色测光
False Cross　赝十字（南天星组）
false dawn　假曙光
false image　误像
false vacuum　伪真空
false zodiacal light　假黄道光
Falx Italica　玄戈增二（牧夫座 38）　又见：*Merga*
FAME　（＝Full-sky Astrometric Mapping Explorer）全天天体测量卫星
Family Mountain　家庭山（月球）
family of asteroids　小行星族 | 小行星[家]族〔台〕　又见：*asteroid family; minor-planet family*
family of comets　彗星族　又见：*comet family*
family of nonclosed orbit　非闭合轨道族
Fanaroff-Riley class　法纳洛夫－里雷类型
Fanaroff-Riley type I　法纳洛夫－里雷 I 型射电星系，FRI 型射电星系

Fanaroff-Riley type II　法纳洛夫－里雷 II 型射电星系，FRII 型射电星系
fan-beam　扇束
Fangfen　方芬（小行星 5306 号）
fan in　输入端数
fan jet　扇形喷流
fan-like structure　扇形结构　又见：*sector structure*
fan out　输出端数
fan ray　扇形射线 | 扇狀射線〔台〕
fan-shaped nebula　扇状星云
fan-shaped tail　扇狀彗尾
Fanyang　杨帆（小行星 21815 号）
Faraday crater　法拉第环形山（月球）
Faraday cup　法拉第筒
Faraday depolarization　法拉第消偏振
Faraday effect　法拉第效应
Faraday pulsation　法拉第脉动
Faraday rotation　法拉第旋转
Farbauti　土卫四十
far end infrared frequency　远红外频
far-field pattern　远场方向图
far infrared　（＝far IR; FIR）远红外
Far Infrared Absolute Spectrophotometer　（＝FIRAS）远红外绝对分光光度计
Far-Infrared and Submillimeter Space Telescope　（＝FIRST）FIRST 空间望远镜，远红外和亚毫米波空间望远镜 | 遠紅外和次毫米波太空望遠鏡〔台〕（Herschel Space Observatory 曾用名）
far-infrared astronomy　远红外天文学
far-infrared object　远红外天体
far-infrared photometry　远红外测光
far-infrared radiation　远红外辐射
far-infrared spectroscopy　远红外分光
far-infrared spectrum　远红外光谱
far IR　（＝far infrared）远红外
far side　背面
far side of the Moon　月球背面
far ultraviolet　（＝far UV; FUV）远紫外
far-ultraviolet astronomy　远紫外天文学
Far Ultraviolet Camera/Spectrograph　远紫外相机/摄谱仪
far-ultraviolet object　远紫外天体
far-ultraviolet photometry　远紫外测光
far-ultraviolet radiation　远紫外辐射
Far Ultraviolet Spectroscopic Explorer　（＝FUSE）远紫外分光探测器（美国）
far-ultraviolet spectroscopy　远紫外分光
far-ultraviolet spectrum　远紫外光谱
far UV　（＝far ultraviolet）远紫外
FAST　①（＝Five-hundred-meter Aperture Spherical radio Telescope）500 米口径球面射电望远镜 | 500 米口徑球面電波望遠鏡〔台〕（中国设备，位于贵州）②（＝Fast Auroral Snapshot Explorer）极光快照探测者
fast angular variable　快变角变量
fast astrocamera　强光力天体照相机
Fast Auroral Snapshot Explorer　（＝FAST）极光快照探测者
fast drift burst　快漂暴 | 速漂爆發〔台〕
fast ejection　快速抛射
fast emulsion　快速乳胶
fast evolving star　快速演化星
fast fine structure　（＝FFS）快速精细结构
fast Fourier transform　（＝FFT）快速傅里叶变换 | 快速傅立葉變換〔台〕
fast fringe　快条纹
fast magnetosonic-wave speed　（＝FMS）快磁声波速率
fast mode wave　快模式波
fast moving object　（＝FMO）快动天体
fast-moving star　快速星
fast nova　快新星　又见：*rapid nova*
fast pulsar　快转脉冲星
fast-rotating star　快转星 | 快[速自]轉星〔台〕
fast-slewing telescope　快动望远镜
fast-spinning black hole　快自旋黑洞
fast steering mirror　快摆镜
fast supernova　快超新星
fast telescope　强光力望远镜　又见：*high-speed telescope*
fast wave　快波
fat zero　胖零
fault　断层
favorable opposition　大冲
fayalite　铁橄榄石
Faye's comet　费伊彗星
F band　F 谱带
F-class asteroid　F 型小行星
FCN　（＝free core nutation）自由核章动
F component　F 成分 | F 分量〔台〕
F corona　F 冕
F dwarf　F 型矮星
Fe　（＝iron）铁（26 号元素）
Feather Ridge Observatory　费瑟山射电天文台
Fechner crater　费希纳环形山（月球）

feed 馈源，照明器 | 饋源〔台〕
feedback 反馈
feedback circuit 反馈回路
feedback factor 反馈因子　又见：*back feed factor*
feedback loop 反馈环路
feed defocusing 馈源偏焦
feeder 馈线 | 饋[電]源〔台〕
feeder network 馈线网络
feed horn 喇叭馈源/号角形馈电源 | 號角形饋電器〔台〕
feeding zone 引力俘获区
Feijunlong 费俊龙（小行星9512号）
feldspar 长石
Felicitas 祥神星（小行星109号）
Fellgett advantage 费尔盖特增益
Fenrir 土卫四十一
Ferdinand 天卫二十四
Fermat principle 费马原理
Fermi acceleration mechanism 费米加速机制
Fermi crater 费米环形山（月球）
Fermi-Dirac distribution 费米－狄拉克分布
Fermi-Dirac nucleus 费米－狄拉克核
Fermi-Dirac statistics 费米－狄拉克统计
Fermi distribution 费米分布
Fermi energy 费米能
Fermi Gamma-ray Space Telescope （＝FGST）费米γ射线空间望远镜 | 費米γ射線太空望遠鏡〔台〕（美国）
Fermi gas 费米气体
Fermi interaction 费米相互作用
Fermi level 费米能级
Fermi mechanism 费米机制
Fermi momentum 费米动量
Fermi normal coordinate system 费米标准坐标系
fermion 费米子
Fermi pressure 费米压力
Fermi sphere 费米球
Fermi statistics 费米统计
Fermi surface 费米面
Fermi temperature 费米温度
Fermi threshold 费米阈
fermium （＝Fm）镄（100号元素）
Fermi-Walker transportation 费米－沃克移动
Fermi-Yang model 费米－杨模型
Feronia 期女星（小行星72号）
ferroelectric liquid crystal modulator （＝FLC modulator）铁电液晶调制器

Fersman crater 费斯曼环形山（月球）
few-body problem 少体问题
Feynman diagram 费曼图
Feynman-Gell-Mann hypothesis 费曼－盖尔－曼假设
Feynman-Gell-Mann universal formula 费曼－盖尔－曼普适公式
Feynman path integral 费曼路径积分
Feynman rule 费曼规则
FFS （＝fast fine structure）快速精细结构
FFT （＝fast Fourier transform）快速傅里叶变换 | 快速傅立葉變換〔台〕
F giant F型巨星
FG Sagittae star 天箭FG型星 | 天箭[座]FG型星〔台〕
FG Sagittae variable 天箭FG型变星
FGST （＝Fermi Gamma-ray Space Telescope）费米γ射线空间望远镜 | 費米γ射線太空望遠鏡〔台〕（美国）
fiber bundle 光纤束
fiber optics 纤维光学
fiber-optic spectrograph 光纤摄谱仪
fiber-optic spectroscopy 光纤分光
fiber plug plate 光纤插接板
fiber positioning system 光纤定位系统
fiber spectroscopy 纤维分光
fibril 小纤维　又见：*fibrille*
fibrille 小纤维　又见：*fibril*
fibrous nebula 纤维状星云 | 絲狀星雲〔台〕
　　又见：*filamentary nebula*
fictitious fluid 虚拟流体
fictitious force 虚拟力
fictitious mean sun 假平太阳
fictitious star 假星　又见：*artificial star*
fictitious sun 假太阳
fictitious time 虚时
fictitious year 假年
Fides 忠神星（小行星37号）
fiducial confidence bar 置信棒
fiducial confidence box 置信框
fiducial confidence circle 置信圆
fiducial confidence ellipse 置信椭圆
fiducial limit 置信限　又见：*confidence limit*
fiducial mark 基准标
fiducial point 基准点
field astronomy 野外天文学
field center 场中心

field correction 像场改正
field corrector 场改正镜
Field criterion for thermal instability 费尔德热不稳定性判据
field curvature 场曲 | 视野弯曲像差，[像] 场 [弯] 曲 〔台〕 又见: curvature of field
field distortion 场畸变
field effect 场效应
field equation 场方程
field flattener 平场器
field flattening 平像场
field flattening correction 平像场校正
field flattening lens 平场镜
field galaxy 场星系 | 视野星系，视场星系 〔台〕
field galaxy luminosity function 场星系光度函数
field intensity 场强 又见: field strength
field lens 场 [透] 镜 | 像场 [透] 镜 〔台〕
field line annihilation 磁力线湮灭
field line reconnection 磁力线重联
field nebula 场星云 | 视野星云，视场星云 〔台〕
field of curvature 曲率场
field of force 力场
field of prominences 日珥场
field of regard 能视域
field of view （=FoV）视场 | 视野 〔台〕
    又见: viewing field
field optics 场致光学，场镜系统，像场光学
field pattern 场方向图
field pulsar 场脉冲星
field reversal 场反转
field star 场星 | 视野星 〔台〕 又见: general field star
field stop 场止
field strength 场强 又见: field intensity
field strengthmeter 场强计
field transition arch 场贯联弧
figure of merit 优值，灵敏值，品质因素
figure of the earth 地球形状
figuring 修磨
figuring of mirror 镜面修磨
filament ① 暗条 ② 纤维，丝
filament activation 暗条激活
filamentary nebula 纤维状星云 | 丝状星云 〔台〕
    又见: fibrous nebula
filamentary nebulosity 纤维星云状物质
filamentary structure 纤维状结构 | 纖維絲狀結構 〔台〕
filament channel 暗条沟

filament foot 暗条足
filament of chromosphere 色球暗条，色球纤维 | 色球暗條，色球絲狀體 〔台〕
filament oscillation 暗条振荡 | 暗條振動 〔台〕
filament sudden disappearance 暗条突逝
    又见: sudden disappearance of filament
filar micrometer 动丝测微计 | 動絲測微器 〔台〕
filigree 网斑，光球细链 | 細鏈 〔台〕
filled aperture 连续孔径
filled-aperture dish 连续孔径碟形天线
filled-aperture radio telescope 连续孔径射电望远镜 | 連續孔徑電波望遠鏡 〔台〕
filled-aperture reflector 连续孔径反射面
filled array 连续天线阵
filled-center supernova remnant 云心超新星遗迹
filled disk 连续碟式天线
filled supernova remnant 云斑超新星遗迹
filling factor 填充因子
film distortion 底片变形 又见: plate distortion
film shift 底片药膜位移
filter 滤光片，滤波器 | 濾波器 〔台〕
filter attenuation band 滤光片减光带
filter-bank spectrometer 滤波频谱仪
filtergram 单色像
filtering 滤光，滤波
filter photography 滤光片照相观测
filter photometry 滤光片测光
filter stop band 滤光片不透明带
filter transmission band 滤光片透射带
filter wheel 滤光片转盘，滤光片转轮
final decline 终降
final orbit 终轨，既定轨道 | 既定軌道 〔台〕
final rise 终升
final state 终态
final velocity 末速度
find 寻获陨石
finder 寻星镜 又见: finderscope; viewfinder; star finder
finder chart 寻星图 又见: finding chart
finderscope 寻星镜 又见: viewfinder; finder; star finder
finding chart 寻星图 又见: finder chart
finding list 寻星星表
fine analysis 精细分析
fine-grain development 微粒显影
fine-grained distribution function 细粒分布函数
fine-grained image 微粒显像
fine guiding sensor 精确导星传感器

fine motion　微动
fine-motion screw　微动螺旋 | 微調螺旋〔台〕
fine mottle　细日芒
finesse　等强干涉束有效数
fine steering mirror　精密转向镜
fine structure　精细结构
fine-structure constant　精细结构常数
fine tuning problem　精细调节问题
finger of God　上帝手指
finite difference　有限差分
finite singularity　可去奇点
finite-temperature field theory　有限温度场论
finite wave-train　有限波列
Finsen crater　芬森环形山（月球）
FIR　（=far infrared）远红外
FIRAS　（=Far Infrared Absolute Spectrophotometer）远红外绝对分光光度计
fireball　火流星　又见：*bolide*
firehose instability　水龙带不稳定性 |「水龍帶」不穩定性〔台〕
firmament　天穹
firmware　固件
Firsov crater　菲尔索夫环形山（月球）
FIRST　①（=Far-Infrared and Submillimeter Space Telescope）FIRST 空间望远镜，远红外和亚毫米波空间望远镜 | 遠紅外和次毫米波太空望遠鏡〔台〕（Herschel Space Observatory 曾用名）②（=Faint Images of the Radio Sky at Twenty-Centimeters catalogue）FIRST 星表
first approximation　一级近似
first-ascent giant branch　初升巨星支　又见：*first giant branch*
first contact　初亏，第一切 | 初虧〔台〕
first cosmic velocity　第一宇宙速度
first day of lunar month　朔日
first difference　一次差
First Frost　霜降（节气）　又见：*Descent of Hoar Frost; Frost's Descent*
first galaxies　第一代星系
first giant branch　初升巨星支　又见：*first-ascent giant branch*
first ion cyclotron wave　基频离子回旋波
first Lagrangian point　第一拉格朗日点
first meridian　本初子午线，本初子午圈 | 本初子午線〔台〕　又见：*prime meridian*
first objects　第一代天体
first order phase transition　一级相变
first order spectrum　一级光谱
first order theory　一级近似理论
first phase　初相
first point of Aries　春分点　又见：*vernal equinox; vernal point; spring equinox*
first point of Cancer　夏至点　又见：*summer solstice; June solstice*
first point of Capricornus　冬至点　又见：*December solstice; winter solstice*
first point of Libra　秋分点　又见：*autumnal equinox; autumnal point*
first quarter　上弦
first-rank galaxy　一级星系
first stars　第一代恒星
first summation　一次累加
first type surface　主表面
Fisher information　费希尔信息
Fisher matrix　费希尔矩阵
Fisher-Tropsch reaction　费希尔－特罗普什反应
Fish on the Platter　盘鱼星云（Barnard 144）
fish orbit　鱼形轨道
fission　裂变 | 分裂，裂變〔台〕
fission fragment　分裂碎块
fission hypothesis　分裂假说
fission rocket　核裂变火箭
fission track　分裂轨迹
FITS　（=Flexible Image Transport System）普适图像传输系统
FITS-IDI　（=FITS Interferometry Data Interchange Format）干涉数据交换 FITS 格式
FITS Interferometry Data Interchange Format　（=FITS-IDI）干涉数据交换 FITS 格式
fitting　拟合
fitting condition　拟合条件
fitting curve　拟合曲线
FitzGerald contraction　菲茨杰拉德收缩
Fitzgerald crater　菲茨杰拉德环形山（月球）
Five College Observatory　五院校天文台
five elements　五行
Five-hundred-meter Aperture Spherical radio Telescope　（=FAST）500 米口径球面射电望远镜 | 500 米口徑球面電波望遠鏡〔台〕（中国设备，位于贵州）
Five-Kilometer Telescope　五千米射电望远镜 | 五千米電波望遠鏡〔台〕　又见：*5-Kilometer Telescope*
five-minute oscillation　五分钟振荡 | 五分[鐘]振蕩〔台〕　又见：*5-minute oscillation*
five palaces　五宫
fixation　定影　又见：*fixing*

fixed altitude mounting　固定高度式装置
fixed antenna　固定式天线
fixed mean pole　固定平极
fixed mounting　固定式装置
fixed observatory　定点天文台
fixed parabolic mirror　固定式抛物面反射镜
　　又见：*fixed parabolic reflector*
fixed parabolic radio telescope　固定式抛物面射电望远镜
fixed parabolic reflector　固定式抛物面反射镜
　　又见：*fixed parabolic mirror*
fixed-phase wave　固定相波
fixed point　定点，不动点
fixed point method　不动点方法
fixed point theorem　不动点定理
fixed pole　固定极
fixed radio telescope　固定式射电望远镜
fixed satellite　静止卫星
fixed spherical mirror　固定式球面反射镜
　　又见：*fixed spherical reflector*
fixed spherical radio telescope　固定式球面射电望远镜
fixed spherical reflector　固定式球面反射镜
　　又见：*fixed spherical mirror*
fixed star　恒星
fixed thread　定丝
fixed-tiltable mirror　半可动式反射镜
　　又见：*fixed-tiltable reflector*
fixed-tiltable reflector　半可动式反射镜
　　又见：*fixed-tiltable mirror*
fixer　定影剂　又见：*fixing agent*
fixing　定影　又见：*fixation*
fixing agent　定影剂　又见：*fixer*
fixing solution　定影液 | 定影液，定像液〔台〕
Fizeau crater　菲佐环形山（月球）
Fizeau interferometer　菲佐干涉仪
FK　① (=Fundamental Catalogue) 基本星表，FK 星表 | 基本星表〔台〕② (=Fundamental Katalog)【德】FK 星表，基本星表 | 基本星表〔台〕
FK3　(=Dritter Fundamental Katalog)【德】FK3 星表，第三基本星表
FK4　(=Vierter Fundamental Katalog)【德】FK4 星表，第四基本星表
FK5　(=Fuenfter Fundamental Katalog)【德】FK5 星表，第五基本星表
FK catalogue system　FK 星表系统
FK Comae Berenices star　后发 FK 型星

FK Comae Berenices variable　后发 FK 型变星
FLA　(=Florida Large Array) 佛罗里达大型天线阵，FLA 射电阵
Flame Nebula　火焰星云（NGC 2024）
flame spectrum　火焰光谱
Flaming Star Nebula　烽火恒星云（IC 405）
Flammarion crater　弗拉马里翁环形山（月球）
Flammarion Observatory　弗拉马里翁天文台
Flamsteed Catalogue　弗兰斯蒂德星表
Flamsteed number　弗兰斯蒂德星号 | 佛氏星號〔台〕
flare　① 耀发 | 閃焰〔台〕② 耀斑 | 閃焰，耀斑〔台〕
flare arc　耀弧
flare class　耀斑级，耀斑级别　又见：*flare importance*
flare flash　耀斑闪光
flare importance　耀斑级，耀斑级别　又见：*flare class*
flare indicator　耀斑指示器
flare kernel　耀斑核 | 閃焰核〔台〕
flare-like brightening　类耀斑增亮
flare-like phenomenon　类耀斑现象 | 似閃耀現象〔台〕
flare loop　耀斑环
flare nimbus　耀斑暗晕
flare onset　耀斑激发
flare particle emission　耀斑粒子发射
flare physics　耀斑物理
flare puff　耀斑喷焰 | 閃焰噴焰〔台〕
flare ribbon　耀斑带，耀斑亮带 | 閃焰亮條〔台〕
flare spectrum　耀斑光谱
flare star　耀星 | [閃] 焰星〔台〕
flare surge　耀斑日浪 | 閃焰噴流〔台〕
flare variable　耀发变星 | 突亮變星〔台〕
flare wave　耀斑波 | 閃焰波〔台〕
flaring　临边增厚
flaring chromosphere　耀现色球
flash　闪光　又见：*flashing light; lightening flash; lightening; shimmer*
flashing light　闪光　又见：*flash; lightening flash; lightening; shimmer*
flashing-light satellite　闪光卫星
flash phase　闪耀相，闪相 | 閃光相〔台〕
flash spectrum　闪光光谱 | 閃光譜〔台〕
flash star　闪星
flat　平场　又见：*flat field*
flat field　平场　又见：*flat*
flat field correction　平场改正

flat fielding 平场处理
flat field photometry 平场测光
flatness problem 平直性问题
flat rotation curve 平坦自转曲线
flat space 平直空间 | 平坦空间〔台〕
flat spectrum 平谱
flat spectrum radio quasar （＝FSRQ）平谱射电类星体 | 平譜電波類星體〔台〕
flat spectrum source 平谱源
flat spiral feed 扁平螺旋馈源
flattening factor 扁率 又见: oblateness
flat universe 平直宇宙，平坦宇宙
flavour 味
F layer F 层（大气电离层） 即: Appleton layer
FLC modulator （＝ferroelectric liquid crystal modulator）铁电液晶调制器
Fleming crater 弗莱明环形山（月球）
Fleurs Radio Observatory 弗勒尔射电天文台
Flexible Image Transport System （＝FITS）普适图像传输系统
flexion 拐变
flexure 弯沉
flexure of the tube 镜筒弯沉 | 鏡筒彎曲〔台〕
flexus 弯脊结构
flicker effect 闪变效应
flicker frequency 闪变频率
flickering 闪变
flickering binary 闪变双星
flickering meteor 闪变流星 | 光變不規則流星〔台〕
flickering star 闪变星
flickering white dwarf 闪变白矮星
flicker noise 闪变噪声
flicker phase 闪变相
flicker photometer 闪变光度计
flint glass 火石玻璃
floating zenith telescope （＝FZT）浮动天顶筒 | 浮動天頂儀〔台〕
flocculent spiral 絮状旋涡结构
flocculent spiral arm 絮状旋臂
flocculus 谱斑 又见: plage
flood tide 涨潮
Flora 花神星（小行星 8 号）
Flora group 花神星群
Florey crater 弗洛里环形山（月球）
Florida Large Array （＝FLA）佛罗里达大型天线阵，FLA 射电阵

flow chart 流程图
flow instability 流态不稳定性
flow pattern 流动图案
FLRW model （＝Friedmann-Lemaitre-Robertson-Walker model）弗里德曼－勒梅特－罗伯逊－沃克模型，FLRW 模型
fluctuating field 起伏场
fluctuation spectrum 起伏谱
fluence 注量，能流 | 累積通量〔台〕
fluid dynamics 流体动力学 又见: hydrodynamics
fluid mechanics 流体力学
fluorescence 荧光 | 螢光〔台〕
fluorescence astronomy 荧光天文学
fluorescence spectrum 荧光光谱
fluorescent line 荧光谱线
fluorescent radiation 荧光辐射 | 螢光輻射〔台〕
fluorine （＝F）氟（9 号元素）
fluorite 萤石
flute instability 槽形不稳定性
flux 流量 | 通量，流量〔台〕
flux collector 聚流器
flux deficit 流量亏损
flux density 流量密度 | 通量密度〔台〕
flux density scale 流量密度标
flux density spectrum 流量密度谱
flux depression 流量衰减
flux-gravity diagram 流量－重力图 | 通量－重力圖〔台〕
flux limited redshift survey 流量极限红移巡天
fluxmeter 磁流计
flux of radiation 辐射流量 | 辐射通[量]〔台〕
    又见: radiant flux; radiation flux
flux of signal particles 信号粒子流量
flux quantization 磁流量子化
flux quantum 磁流量子
flux scale 流量标
flux standard 流量标准，流量标准星
flux-tube dynamics 磁流管动力学
flux unit （＝f.u.）流量单位 | 通量單位〔台〕
flyby 飞掠
flyby interplanetary orbit 行星际飞掠轨道
flyby interplanetary trajectory 行星际飞掠轨迹
flyby orbit 飞掠轨道
flyby trajectory 飞掠轨迹
flying clock 飞行钟
flying clock measurement 飞行钟测量

Flying Sandbank model　飞沙堆模型
flying saucer　飞碟
FM　（＝frequency modulation）调频
Fm　（＝fermium）镄（100 号元素）
FME　（＝frequency measuring equipment）测频装置
FMO　（＝fast moving object）快动天体
f-mode　f 模，基本模 | f 模式，基本模〔台〕
FMS　（＝fast magnetosonic-wave speed）快磁声波速率
Fm star　Fm 型星
f number　f 数，焦比数
focal distance　焦距　又见：*focal length*
focal extender　延焦器
focal image　焦面像
focal length　焦距　又见：*focal distance*
focal line　焦线
focal plane　焦面 | 焦 [平] 面〔台〕　又见：*focal surface*
focal-plane array　焦面阵
focal plane assembly　（＝FPA）焦面组件
focal-plane instrumentation　焦面附属仪器
focal-plane modulator　焦面调制器
focal-plane spectrometer　焦面分光计
focal point　焦点　又见：*focus*
focal power　焦度
focal ratio　焦比
focal ratio degradation　（＝FRD）焦比衰退
focal reducer　缩焦器
focal surface　焦面 | 焦 [平] 面〔台〕　又见：*focal plane*
focus　焦点　又见：*focal point*
focused reflector　焦正反射面
focuser　聚焦器
focusing glass　聚焦透镜
focusing magnet　聚焦磁体
focusing ring　调焦环
focusing screw　调焦螺旋
focusing X-ray telescope　聚焦 X 射线望远镜 | 聚焦 X 光望遠鏡〔台〕
focusing γ-ray telescope　聚焦成像 γ 射线望远镜
Fokker-Planck approximation　福克－普朗克近似
Fokker-Planck equation　福克－普朗克方程 | 福克－卜朗克方程〔台〕
fold catastrophe　折叠突变
folded camera　折叠式照相机
folded dipole　折叠偶极

folded refractor　折叠式折射望远镜
folded Schmidt camera　折叠式施密特照相机
following arm　曳臂 | 尾隨旋臂〔台〕
　又见：*trailing arm*
following edge　东边缘，后随边缘　又见：*following limb*
following limb　东边缘，后随边缘　又见：*following edge*
following member　后随成员
following spot　后随黑子，f 黑子 | 尾隨黑子〔台〕
　又见：*following sunspot; trailer sunspot; trailer spot*
following sunspot　（＝f-spot）后随黑子，f 黑子 | 尾隨黑子〔台〕　又见：*following spot; trailer sunspot; trailer spot*
follow-up　随动
Fomalhaut　北落师门（南鱼座 α）
Fongyunwah　方润华（小行星 5198 号）
footpoint　地脚点
footprint　覆盖区
Footprint Nebula　脚印星云（M1-29）
foot screw　地脚螺旋 | [地] 腳螺旋〔台〕
For　（＝Fornax）天炉座
forbidden line　禁线 | 禁 [谱] 線〔台〕
forbidden transition　禁戒跃迁 | 禁制躍遷〔台〕
Forbush decrease　福布希型下降 | 福布殊衰減〔台〕
Forbush effect　福布希效应
force actuator　力型致动器
force-carrier particles　载体粒子
forced absorption　受迫吸收 | 強迫吸收〔台〕
　又见：*induced absorption*
forced emission　受迫发射 | 強迫發射，誘發發射〔台〕　又见：*induced emission*
force density　力密度
forced nutation　受迫章动 | 強迫章動〔台〕
forced oscillation　受迫振荡 | 強迫振動，強迫振盪〔台〕
forced transition　受迫跃迁 | 強迫躍遷，誘發躍遷〔台〕　又见：*induced transition*
force due to viscosity　黏性力
forced vibration　受迫振动
force-free field　无力场
force-free magnetic field　无力磁场
force function　力函数
force softening　力软化
foreground galaxy　前景星系
foreground galaxy cluster　前景星系团
foreground radiation　前景辐射

foreground removal 前景减除　又见: *foreground subtraction*
foreground rotation 前景旋转
foreground star 前景星
foreground subtraction 前景减除　又见: *foreground removal*
fore optics system 输入光学系统
forescattering 向前散射
foreshortening 投影缩减 | 缩减〔台〕
foreshortening effect 投影缩减效应
forked mounting 叉式装置　又见: *fork mounting*
fork mounting 叉式装置　又见: *forked mounting*
fork telescope mount 叉式望远镜装置
formal accuracy 形式精度
formaldehyde 甲醛（H_2CO）　又见: *methanal*
formally convergent 形式收敛
formal stability 形式稳定性
formal transformation 形式变换
formamide 甲酰胺（$HCONH_2$）
formation of elements 元素形成　又见: *element formation*
formation of galaxy 星系形成　又见: *galaxy formation; galactic formation*
formation of image 成像　又见: *image formation; imagery; imaging*
formation of stars 恒星形成　又见: *star formation; stellar formation*
form factor 形状因子，波形因子
formyl radical 甲酰基（HCO）
Fornacis 天苑增三，天苑增四（天炉座 α）
Fornax （=For）天炉座
Fornax cluster 天炉星系团 | 天爐[座]星系團〔台〕
Fornax Dwarf Galaxy 天炉矮星系
Fornax system 天炉星系 | 天爐[座]星系〔台〕
Fornjot 土卫四十二
forsterite 镁橄榄石
Fortuna 命神星（小行星 19 号）
Fortuna Fortunarum 虚宿一（宝瓶座 β）　又见: *Alshain*
Fortuna Tessera 命运女神镶嵌地块（金星）
forward difference 前向差分
forward scattering 前向散射
forward shock 正向激波
Foshan 佛山（小行星 2789 号）
fossa 堑沟（行星地貌）
fossil crater 化石环形山
fossil field model 化石场模型

fossil group 化石星系群
fossil Ströemgren sphere 古斯特龙根球
Foucaultgram 傅科检验图样
Foucault knife-edge test 傅科刀口检验 | 富可刀口檢驗〔台〕
Foucault's pendulum 傅科摆 | 富可擺〔台〕
Foucault test 傅科检验
fountain clock 喷泉钟
fountain model 喷泉模型
four-axis mounting 四轴装置　又见: *4-axis mounting; quadraxial mounting*
four-body problem 四体问题　又见: *4-body problem*
four celestial images 四象　又见: *four symbolic animals*
four-color photometry 四色测光　又见: *4-color photometry*
four-dimensional universe 四维宇宙　又见: *4-dimensional universe*
Fourier analysis 傅里叶分析 | 傅立葉分析〔台〕
Fourier coefficient 傅里叶系数
Fourier component 傅里叶分量
Fourier crater 傅里叶环形山（月球）
Fourier decomposition 傅里叶分解
Fourier frequency 傅里叶频率
Fourier series 傅里叶级数 | 傅立葉級數〔台〕
Fourier spectrometer 傅里叶分光仪，傅里叶频谱仪
Fourier spectroscopy 傅里叶分光
Fourier tachometer 傅里叶转速计
Fourier transform 傅里叶变换
Fourier transformer 傅里叶变换计
Fourier transform spectrometer ①（=FTS）傅里叶变换分光仪 | 傅立葉變換分光儀〔台〕②（=FTS）傅里叶变换频谱仪 | 傅立葉變換頻譜儀〔台〕
Fourier transform spectroscopy 傅里叶变换分光
four-momentum 四维动量
four-point correlation function 四点相关函数
four symbolic animals 四象　又见: *four celestial images*
Fourth Cambridge Survey catalogue （=4C catalogue）第 4 剑桥巡天星表，4C 星表
fourth contact 复圆　又见: *last contact of umbra; last contact*
fourth cosmic velocity 第四宇宙速度
FoV （=field of view）视场 | 視野〔台〕
Fowler crater 福勒环形山（月球）
Fowler sampling 福勒采样

FPA （＝focal plane assembly）焦面组件
Fp star F 型特殊星
Fr （＝francium）钫｜鈁〔台〕（87 号元素）
Fracastorius crater 弗拉卡斯托罗环形山（月球）
fractal structure 分形结构
fractal Universe 分形宇宙
fractional-bit-shift corrector 分数比特移位改正器
fractional frequency 相对频率
fractional gain error 相对增益误差
fractional method 部分化方法
fractional polarization 部分偏振 又见：*partially polarization*
fractiona of the year 年分
fractionation 部分化｜分馏〔台〕
fragmentation 碎裂｜分裂〔台〕
fragmentation limit 碎裂极限
Fra Mauro crater 弗拉·毛罗环形山（月球）
frame bias 参考架偏差
frame of axes 坐标架
frame of reference 参考架，参考系 又见：*reference frame*
frame transfer 帧转移
frame transfer CCD 帧转移 CCD
frame work 构架
Francais 法兰西号（美法科学卫星）
Francisco 天卫二十二
francium （＝Fr）钫｜鈁〔台〕（87 号元素）
Franck-Condon factor 富兰克－康登因子
Franck-Condon principle 富兰克－康登原理
Franklin-Adams chart 富兰克林－亚当斯星图
Franklin-Adams Survey 富兰克林－亚当斯巡天
Franklin crater 富兰克林环形山（月球）
Frankshu 徐遐生（小行星 18238 号）
Franzia 法兰西（小行星 862 号）
Fraternité arc 友谊环弧（海王星）
Fraunhofer component 夫琅和费分量｜夫朗和斐部分〔台〕
Fraunhofer corona 夫琅和费日冕
Fraunhofer crater 夫琅和费环形山（月球）
Fraunhofer diffraction 夫琅和费衍射
Fraunhofer line 夫琅和费谱线｜夫朗和斐[谱]線〔台〕
Fraunhofer spectrum 夫琅和费光谱｜夫朗和斐光譜〔台〕
FRD （＝focal ratio degradation）焦比衰退
Fred Lawrence Whipple Observatory 惠普尔天文台 又见：*Whipple Observatory*

free-bound absorption 自由－束缚吸收
free-bound emission 自由－束缚发射
free-bound transition 自由－束缚跃迁
free core nutation （＝FCN）自由核章动
free electron 自由电子
free energy 自由能
free fall 自由落体
free-fall time 自由下落时间
free-flying telescope 空间望远镜｜太空望遠鏡〔台〕 又见：*space telescope*
free-free absorption 自由－自由吸收｜自由態間吸收〔台〕
free-free emission 自由－自由发射｜自由態間發射〔台〕
free-free transition 自由－自由跃迁｜自由態間躍遷〔台〕
freely falling body 自由下落物体
Freeman's bar 弗里曼棒
free neutron 自由中子
free nutation 自由章动
free path 自由程
free pendulum 自由摆
free period 自由周期
free radical 自由基
free streaming 自由流动
free-streaming damping 自由流动阻尼
free vibration 自由振动
F region F 区（大气电离层） 即：*Appleton layer*
Freia 舒女星（小行星 76 号）
frequency allocation 频率分配 又见：*allocation of frequency*
frequency analysis 频率分析
frequency band 频带
frequency calibration 频率校准
frequency comparision 频率比对
frequency conversion 频率转换
frequency converter 变频器
frequency cutoff 频率截止
frequency dispersion 频散
frequency distribution 频率分布
frequency domain 频域
frequency drift 频率漂移
frequency locking 锁频
frequency measuring equipment （＝FME）测频装置
frequency modulation （＝FM）调频
frequency multiplier 倍频器

frequency offset 频率偏置
frequency resolution 频率分辨率
frequency scale 频标
frequency sensitivity 频率灵敏度
frequency shift 频移
frequency stability 频率稳定度
frequency standard 频率标准
frequency step 频率阶跃
frequency switching 频率开关
frequency transition 频率跃迁
Fresh Green 清明（节气） 又见：Clear and Bright; Pure Brightness
Fresnel diffraction 菲涅尔衍射
Fresnel lens 菲涅尔透镜
Fresnel pattern 菲涅尔图样
Fresnel rhomb 菲涅尔菱形
Fresnel scale 菲涅尔尺度
Fresnel X-ray telescope 菲涅尔 X 射线望远镜
Fresnel zone 菲涅尔区
Fresnel zone plate 菲涅尔环板
Fresnel zone plate telescope 菲涅尔波带片望远镜
Fresnel γ-ray telescope 菲涅尔 γ 射线望远镜
fretted terrain 侵蚀地形（火星）
fretum （复数：freta） 峡
Freud crater 弗洛伊德环形山（月球）
Freundlich crater 弗罗因德利希环形山（月球）
friction 摩擦
frictional force 摩擦力
Fridman crater 弗里德曼环形山（月球）
　　又见：Friedmann crater
Fried constant 弗里德常数
Fried Egg Galaxy 煎蛋星系（NGC 7742）
Friedmann cosmological model 弗里德曼宇宙学模型 | 弗里德曼宇宙模型〔台〕
Friedmann crater 弗里德曼环形山（月球）
　　又见：Fridman crater
Friedmann equation 弗里德曼方程
Friedmann-Lemaitre-Robertson-Walker model （＝FLRW model）弗里德曼－勒梅特－罗伯逊－沃克模型，FLRW 模型
Friedmann-Lemaitre universe 弗里德曼－勒梅特宇宙
Friedmann model 弗里德曼模型
Friedmann-Robertson-Walker cosmology 弗里德曼－罗伯逊－沃克宇宙学
Friedmann-Robertson-Walker model （＝FRW model）弗里德曼－罗伯逊－沃克模型，FRW 模型

Friedmann time 弗里德曼时间
Friedmann universe 弗里德曼宇宙
Fried parameter 弗里德参数
friends-of-friends cluster finder algorithm 二度好友寻团算法
Frigga 寒神星（小行星 77 号）
frigid zone 寒带
fringe amplitude 条纹幅度
fringe envelope 条纹包络
fringe-frequency spectrum 条纹－频率谱
fringe identification 条纹证认
fringe order 条纹级
fringe pattern 条纹图样
fringe phase 条纹相位
fringe rate 条纹率
fringe separation 条纹间距 又见：fringe spacing
fringe spacing 条纹间距 又见：fringe separation
fringe stopper 条纹驻留器
fringe stopping 条纹驻留
fringe stopping center 条纹驻留中心
fringe stopping system 条纹驻留系统
fringe tracking 条纹跟踪
fringe trackor 条纹跟踪器
fringe visibility 条纹可见度 | 條紋能見度〔台〕
fringe-visibility spectrum 条纹可见度谱
Froelich crater 弗勒利希环形山（月球）
front end 前端
front focus 前焦点
frontside-illuminated CCD 前照式 CCD
frontside-illuminated CMOS 前照式 CMOS
front surface 前表面
frosting model 结霜模型
frost point 霜点
Frost's Descent 霜降（节气） 又见：First Frost; Descent of Hoar Frost
Froude number 弗劳德数
frozen field 冻结场
frozen-in 冻结
frozen-in element 冻结元素
frozen-in magnetic field 冻结磁场
frozen star 冻结星
frustum of a cone 截锥 又见：truncated cone
FRW model （＝Friedmann-Robertson-Walker model）弗里德曼－罗伯逊－沃克模型，FRW 模型
f-spot （＝following sunspot）后随黑子，f 黑子 | 尾隨黑子〔台〕

FSRQ （=flat spectrum radio quasar）平谱射电类星体 | 平譜電波類星體〔台〕
F star F 型星 又见：*F-type star*
F subdwarf F 型亚矮星
F subgiant F 型亚巨星
f-sum rule f 求和定则
F supergiant F 型超巨星
FTS ①（=Fourier transform spectrometer）傅里叶变换分光仪 | 傅立葉變換分光儀〔台〕
②（=Fourier transform spectrometer）傅里叶变换频谱仪 | 傅立葉變換頻譜儀〔台〕
F-type asteroid F 型小行星
F-type star F 型星 又见：*F star*
f.u. （=flux unit）流量单位 | 通量單位〔台〕
fuel ratio 燃料比
Fuenfter Fundamental Katalog （=FK5）【德】FK5 星表，第五基本星表
Fujian 福建（小行星 2184 号）
full earth 满地
full-earth brightness 满地亮度
full-earth illumination 满地照明
full English mounting 全英国式装置
full-frame CCD 全帧式 CCD
full load 满载
full moon 望，满月
full-moon brightness 满月亮度
full-moon illumination 满月照明
full phase 满相
Full-sky Astrometric Mapping Explorer （=FAME）全天天体测量卫星
full-wave dipole 全波偶极
full width at half-maximum （=FWHM）半峰全宽 | 半峰全幅值〔台〕（谱线）
fully coded field of view 全编码视场
fully depleted CCD 全耗尽型 CCD
fully depleted pn-junction CCD （=pnCCD）全耗尽 pn 结 CCD
fully ionized gas 完全电离气体
fully ionized plasma 完全电离等离子体
fully steerable dish 全动碟形天线
fully steerable radio telescope 全动射电望远镜
fully steerable reflector 全动反射面
Fum al Samakah 霹雳一（双鱼座 β）
fundamental astrometry 基本天体测量学 | 基本天體測量[術]〔台〕
fundamental astronomical constant 基本天文常数
fundamental astronomy 基本天文学

Fundamental Catalogue （=FK）基本星表，FK 星表 | 基本星表〔台〕
fundamental circle 基本圈 | 基本大圓〔台〕
又见：*reference circle*
fundamental component 基本分量
fundamental constant 基本常数
fundamental coordinate system 基本坐标系
fundamental force 基本力
fundamental frequency 基频
fundamental function 基本函数
fundamental harmonic 基波 又见：*fundamental wave*
fundamental-harmonic pair 基波对
Fundamental Katalog （=FK）【德】FK 星表，基本星表 | 基本星表〔台〕
fundamental mode 基本模
fundamental observer 基本观测者
fundamental particle 基本粒子
fundamental plane 基面 | 基[本]面〔台〕
fundamental reference frame 基本参考架
fundamental reference system 基本参考系
fundamental series 基线系
fundamental star 基本星
fundamental string 基本弦
fundamental system 基本系统
fundamental unit 基本单位
fundamental wave 基波 又见：*fundamental harmonic*
funneling 漏斗
funneling effect 漏斗效应
funnel prominence 漏斗状日珥
funnel-shaped nebula 漏斗状星云
Fuor （=FU Orionis star）猎户 FU 型星
FU Orionis object 猎户 FU 型天体
FU Orionis star （=Fuor）猎户 FU 型星
FU Orionis variable 猎户 FU 型变星
furnace spectrum 电炉光谱
Furud 孙增一（大犬座 ζ）
FUSE （=Far Ultraviolet Spectroscopic Explorer）远紫外分光探测器（美国）
fused quartz 熔石英
fusee 均力圆锥轮 又见：*fuzee*
fusion 聚变
fusion crust 熔凝壳
fusion of wave 多波反应
fusion reaction 聚变反应
fusion rocket 热核火箭
future light cone 未来光锥

FUV （＝far ultraviolet）远紫外
fuzee 均力圆锥轮 又见：*fusee*
Fuzhougezhi 福州格致（小行星 55892 号）
fuzz 展云

FWHM （＝full width at half-maximum）半峰全宽｜半峰全幅值〔台〕（谱线）
FX correlator FX 相关器
FZT （＝floating zenith telescope）浮动天顶筒｜浮動天頂儀〔台〕

G

Ga （＝gallium）镓（31号元素）
Gabi 加比（小行星1665号）
Gacobini comet 贾科比尼彗星
Gacrux 十字架一（南十字座γ）
gadolinium （＝Gd）钆（64号元素）
Gadomski crater 加多姆斯基环形山（月球）
Gaea crater 盖亚环形山（木卫五）
Gagarin crater 加加林环形山（月球）
Gaia （＝Global Astrometric Interferometer for Astrophysics）全天天体测量干涉仪（全称已不用）
Gaia Astrometry Satellite 盖亚天文卫星 即：Gaia
gain 增益
gain error 增益误差
gain factor 增益因子
gain modulation 增益调制
Galactic absorbing layer 银河吸光层 | 銀河吸收層〔台〕
galactic absorbing layer 星系吸光层 | 星系吸收層〔台〕
Galactic absorption 银河吸光，河内吸收 | 銀河吸收〔台〕
galactic absorption 星系吸光 | 星系吸收〔台〕
galactic age 星系年龄
Galactic aggregate 银河星集
Galactic anticenter 反银心，反银心方向 | 反銀心〔台〕 又见：anticenter
Galactic arm 银河臂
galactic arm 星系臂
Galactic arm-population 银河系臂族
Galactic astronomy 银河系天文学
galactic astronomy 星系天文学 又见：galaxy astronomy
Galactic astrophysics 银河系天体物理
galactic astrophysics 星系天体物理
Galactic background 银河背景

Galactic background infrared radiation 银河红外背景辐射
Galactic background radiation 银河背景辐射
Galactic background radio radiation 银河射电背景辐射
Galactic background X-ray radiation 银河X射线背景辐射
Galactic background γ-ray radiation 银河γ射线背景辐射
Galactic bar 银棒
galactic bar 星系棒 又见：galaxy bar
galactic bridge 星系桥
Galactic brightness 银河亮度
galactic brightness 星系亮度
galactic bubble 星系泡
Galactic bulge 银河系核球 | 銀河核球〔台〕
galactic bulge 星系核球
galactic cannibalism 星系吞食 又见：cannibalizing of galaxies
Galactic center 银心 | 銀[河系中]心〔台〕
Galactic center region 银心区
Galactic center source 银心辐射源
Galactic Cepheid 银河造父变星
Galactic chimney 银河系通道
galactic chimney 星系通道
Galactic circle 银道圈
Galactic cloud 银河云
Galactic cluster 疏散星团 | 疏散星團，銀河星團〔台〕 又见：open cluster of stars; open cluster; open star cluster
galactic collision 星系碰撞
Galactic component 银河系子系
galactic component 星系子源
Galactic concentration 银面聚度 | 銀[面]聚度〔台〕
galactic content 星系成分 又见：galaxy content
galactic coordinate 银道坐标 | 銀河坐標〔台〕

galactic coordinate system　银道坐标系
Galactic core　银核　又见：*Galactic nucleus*
galactic core　星系核心
Galactic core source　银核源
galactic core source　星系核心源
Galactic core star　银核星
galactic core star　星系核心星
Galactic corona　银冕
galactic corona　星系冕　又见：*corona of galaxy*
Galactic cosmic rays　（＝GCR）银河系宇宙线
Galactic dark halo　银河系暗晕
Galactic diffuse light　银河弥漫光
Galactic disk　银盘 | 銀[河]盤[面]〔台〕
galactic disk　星系盘 | 銀[河]盤[面]〔台〕
Galactic disk-population　银盘族
Galactic dynamics　银河系动力学
galactic dynamics　星系动力学
galactic environment　星系环境
galactic equator　银道 | 銀[河赤]道〔台〕
Galactic evolution　银河系演化
galactic evolution　星系演化　又见：*galaxy evolution*
Galactic field　银河场
galactic field　星系场 | 銀河場〔台〕
galactic filament　星系纤维
galactic formation　星系形成　又见：*galaxy formation; formation of galaxy*
Galactic fountain　银道喷流
Galactic globular cluster system　银河系球状星团系统
Galactic gusher　银河喷射源
galactic gusher　星系喷射源
Galactic halo　银晕
galactic halo　星系晕　又见：*halo of galaxy*
Galactic halo emission　银晕射电
Galactic halo-population　银晕族
Galactic high-energy astrophysics　银河系高能天体物理
galactic high-energy astrophysics　星系高能天体物理
Galactic HII disk　银河系电离氢盘
Galactic infrared astronomy　银河系红外天文学
galactic infrared astronomy　星系红外天文学
Galactic infrared emission　（＝Galactic IR emission）银河系红外辐射
Galactic IR emission　（＝Galactic infrared emission）银河系红外辐射
galactic jet　星系喷流

Galactic kinematics　银河系运动学
galactic kinematics　星系运动学
galactic latitude　银纬
Galactic light　银河光
galactic longitude　银经
Galactic luminosity　银河系光度
galactic luminosity　星系光度
Galactic magnetic cavity　银河磁穴
Galactic magnetic field　银河系磁场
galactic magnetic field　星系磁场
galactic merging　星系并合　又见：*merge of galaxy; galaxy merging*
Galactic model　银河系模型
galactic model　星系模型
Galactic nebula　银河星云
Galactic noise　银河系噪声 | 銀河雜訊〔台〕
Galactic nova　银河新星
galactic nova　星系新星
Galactic nucleus　银核　又见：*Galactic core*
galactic nucleus　星系核　又见：*nucleus of galaxy; galaxy nucleus*
Galactic orbit　银心轨道
galactic pericenter　近银心点　又见：*perigalacticon; perigalacticum*
Galactic plane　银道面 | 銀河盤面〔台〕
galactic pole　银极
Galactic potential　银河系势
Galactic radio astronomy　银河系射电天文学
galactic radio astronomy　星系射电天文学
Galactic radio astrophysics　银河系射电天体物理
galactic radio astrophysics　星系射电天体物理
Galactic radio emission　银河系射电　又见：*Galactic radio radiation*
galactic radio emission　星系射电　又见：*galactic radio radiation*
Galactic radio noise　银河系射电噪声
Galactic radio radiation　银河系射电　又见：*Galactic radio emission*
galactic radio radiation　星系射电　又见：*galactic radio emission*
Galactic radio spur　银河射电支 | 銀河電波支〔台〕　又见：*Galactic spur*
Galactic ridge　银脊
Galactic rotation　银河系自转　又见：*rotation of the Galaxy*
galactic rotation　星系自转　又见：*rotation of galaxy*
Galactic rotation curve　银河系自转曲线
galactic rotation curve　星系自转曲线

galactic sheet 星系片	**galaxy counting** 星系计数 又见：*galaxy count*
Galactic source 银河源	**galaxy distribution** 星系分布
galactic space 星系空间	**galaxy encounter** 星系交会
Galactic spiral structure 银河系旋涡结构	**galaxy evolution** 星系演化 又见：*galactic evolution*
Galactic spur 银河射电支｜銀河電波支〔台〕 又见：*Galactic radio spur*	**Galaxy Evolution Explorer** （=GALEX）GALEX 星系演化探测器｜星系演化探索者〔台〕
Galactic stellar disk 银河系恒星盘	**galaxy formation** 星系形成 又见：*formation of galaxy; galactic formation*
Galactic structure 银河系结构	**galaxy-galaxy lensing** 星系－星系引力透镜
galactic structure 星系结构	**galaxy harassment** 星系干扰
Galactic subsystem 银河次系｜銀河系次系〔台〕	**galaxy merger** 并合星系 又见：*merging galaxy*
galactic superbubble 星系超泡	**galaxy merging** 星系并合 又见：*merge of galaxy; galactic merging*
Galactic supernova 银河超新星	
galactic supernova 星系超新星	**Galaxy model** 银河系模型
Galactic supernova remnant 银河超新星遗迹	**galaxy morphology** 星系形态
Galactic System 银河系 又见：*Galaxy; Milky Way galaxy*	**galaxy nucleus** 星系核 又见：*galactic nucleus; nucleus of galaxy*
Galactic variable 银河变星	**galaxy power spectrum** 星系功率谱
galactic void 大尺度巨洞	**galaxy strangulation** 星系遏制
galactic walls 星系长城	**galaxy supercluster** 超星系团 又见：*supercluster*
Galactic warp 银河系翘曲	**galaxy survey** 星系巡天
galactic wind 星系风	**galaxy type** 星系型
Galactic window 银河系窗口｜星系窗孔〔台〕	**galaxy void** 星系际巨洞 又见：*blank field*
Galactic X-ray source （=GX）银河 X 射线源	**GALEX** （=Galaxy Evolution Explorer）GALEX 星系演化探测器｜星系演化探索者〔台〕
Galactic year 银河年（太阳系绕银心公转的周期，约 2 亿年）	**Galilaei crater** 伽利略环形山（月球）
Galactocentric concentration 银心聚度	**Galilean moon** 伽利略卫星（木星） 又见：*Galilean satellite*
Galactocentric distance 银心距	**Galilean satellite** 伽利略卫星（木星） 又见：*Galilean moon*
Galatea ① 巫女星（小行星 74 号）② 海卫六	
galaxoid 星系体	**Galilean telescope** 伽利略望远镜｜伽利略式望遠鏡〔台〕
GALAXY （=General Automatic Luminosity and X Y Measuring Engine）GALAXY 底片自动测量仪｜GALAXY[底片自動測量儀]〔台〕	**Galilean transformation** 伽利略变换
	Galileon 伽利略子
Galaxy 银河系 又见：*Milky Way galaxy; Galactic System*	**Galileo National Telescope** （=TNG）国立伽利略望远镜（意大利）
galaxy 星系	**Galileon field** 伽利略场
galaxy assembly 星系组装	**Galileo spacecraft** 伽利略木星探测器｜伽利略號[木星探測]太空船〔台〕
galaxy astronomy 星系天文学 又见：*galactic astronomy*	**Galle ring** 伽勒环（海王星）
galaxy bar 星系棒 又见：*galactic bar*	**gallium** （=Ga）镓（31 号元素）
galaxy bias 星系偏袒	**Galois crater** 伽罗瓦环形山（月球）
galaxy bimodality 双星族星系组态	**Galvani crater** 伽伐尼环形山（月球）
galaxy cannibalism 星系吞并	**Gamma** 伽马号（苏联高能天文卫星）
galaxy classification 星系分类	**Gamma Cassiopeiae star** 仙后 γ 型星
galaxy cluster 星系团 又见：*cluster of galaxies; cluster*	**Gamma Crucis** 南十字 γ
galaxy clustering 星系成团 又见：*clustering of galaxies*	**Gamma-ray burst** γ 射线暴｜γ 射線爆發〔台〕 又见：*γ-ray burst*
galaxy content 星系成分 又见：*galactic content*	
galaxy count 星系计数 又见：*galaxy counting*	

Gamma-ray Light Detector （＝AGILE）敏捷号 γ射线天文卫星
Gamow crater 伽莫夫环形山（月球）
Gansu 甘肃（小行星 2515 号）
Ganswindt crater 甘斯文特环形山（月球）
Ganymede 木卫三｜木衛三，甘尼米德〔台〕
Gaoshiqi 高士其（小行星 3704 号）
Gaoyaojie 高耀洁（小行星 38980 号）
gap center 缝心
gap of asteroid ring 小行星环缝｜小行星帶隙〔台〕
gap of asteroids 小行星带隙
gap of satellite ring 卫星环缝
Garavito crater 加拉维托环形山（月球）
gardening 表土混合
Garnet star 石榴石星（仙王座 μ）
Gärtner crater 格特纳环形山（月球）
gas accretion 气体吸积
gas cell 气泡，气室
gas cloud 气体云
gas content 气体成分
gas density 气体密度
gas distribution 气体分布
gas-dust cloud 气尘云｜氣[體]塵[埃]雲〔台〕
gas-dust complex 气尘复合体
gas-dust envelope 气尘包层
gas-dust nebula 气尘星云
gas dynamics 气体动力学　又见：*aerodynamics*
gas electron multiplier （＝GEM）气体电子倍增器
gas envelope 气体包层
gaseous collapse 气态坍缩
gaseous disk 气体盘
gaseous emission nebula 气体发射星云
gaseous halo 气态晕
gaseous mass 气团｜氣[體]團〔台〕
gaseous nebula 气体星云　又见：*gas nebula*
gaseous planet 气态行星　又见：*gas planet*
gaseous ring 气环
gaseous spectrum 气体光谱
gaseous sphere 气体球
gaseous train 气体余迹｜氣體[流星]遺跡〔台〕
gas giant planet 气态巨行星
gas hypersensitization 气体敏化
gas jet 气体喷流
gas kinematics 气体运动学
gas-kinetic diameter 气体动理学直径
gas kinetics 气体动理学

gas-kinetic theory 气体分子运动论，气体动理[学理]论　又见：*kinetic theory of gases*
gas laser 气体激光器
gas law 气体定律
gas nebula 气体星云　又见：*gaseous nebula*
gas planet 气态行星　又见：*gaseous planet*
gas-poor comet 贫气彗星
Gaspra 加斯普拉（小行星 951 号）
gas-pressure 气体压力
gas retention age 气体保留年龄
gas-rich asteroid 富气小行星
gas-rich meteorite 富气陨星｜富氣隕石〔台〕
gas-rich satellite 富气卫星
gas scintillation proportional counter （＝gas SPC）气体闪烁正比计数器
Gassendi crater 伽桑狄环形山（月球）
gas SPC （＝gas scintillation proportional counter）气体闪烁正比计数器
gas stream 气体流
gas tail 气体彗尾　又见：*cometary gas tail*
gas-to-dust ratio 气尘比
gated noise 选通噪声
gated noise source 选通噪声源
gauge boson 规范玻色子
gauge condition 规范条件
gauge field 规范场
gauge freedom 规范自由度
gauge group 规范群
gauge invariance 规范不变性
gauge invariant 规范不变量
gauge particle 规范粒子
gauge symmetry 规范对称性
gauge theory 规范理论
gauge transformation 规范变换
Gaunt factor 冈特因子
Gauss-Bonnet term 高斯－博内项
Gauss-Bonnet theorem 高斯－博内定理
Gauss-Codacci equation 高斯－科达奇方程
Gauss constant 高斯常数
Gauss crater 高斯环形山（月球）
Gauss distribution 高斯分布
Gauss equation 高斯方程
Gauss eyepiece 高斯目镜
Gaussia 高斯（小行星 1001 号）
Gaussian beam 高斯波束
Gaussian curvature 高斯曲率
Gaussian density perturbation 高斯密度扰动

Gaussian distribution 高斯分布
Gaussian filter 高斯滤波器
Gaussian gravitational constant 高斯引力常数 | 高斯[重力]常數〔台〕
Gaussian perturbation 高斯扰动
Gaussian random field 高斯随机场
Gaussian random variable 高斯随机变量
Gaussian year 高斯年
Gauss line profile 高斯谱线轮廓
Gauss method 高斯方法
Gauss noise 高斯噪声
Gauss normal coordinate 高斯正则坐标
Gauss pattern 高斯方向图
Gauss theorem 高斯定理
GAVO （=German Astrophysical Virtual Observatory）德国天体物理虚拟天文台
Gavrilov crater 加夫里洛夫环形山（月球）
G band G 谱带 | G 波段〔台〕
GBT （=Green Bank Telescope）绿堤射电望远镜 | 格林班克電波望遠鏡〔台〕
GC （=Boss General Catalogue）博斯总星表，GC 星表 | 博斯星表〔台〕
GCE （=ground cosmic rays enhancement）地面宇宙线增强
G-class asteroid G 型小行星
GCR （=Galactic cosmic rays）银河系宇宙线
GCRS （=Geocentric Celestial Reference System）地心天球参考系
GCT （=Greenwich civil time）格林尼治民用时 | 格林[威治]民用時〔台〕
GCVS （=General Catalogue of Variable Stars）变星总表
Gd （=gadolinium）钆（64 号元素）
GDS （=Great Dark Spot）大暗斑
G dwarf problem G 矮星问题
G-dwarf problem G 型矮星问题
GE （=greatest elongation）大距
Ge （=germanium）锗（32 号元素）
Ge:Ga infrared array 锗掺镓红外面阵
gegenschein 【德】对日照 即: *counterglow; counter-twilight; zodiacal counterglow*
Geiger-Muller counter 盖革－弥勒计数器 | 蓋革－繆勒計數器〔台〕
Geiger-tube telescope 盖革计数望远镜
GEM （=gas electron multiplier）气体电子倍增器
Gem ①（=Gemini）双子座 ②（=Gemini）双子宫，实沈，申宫

Geminga 杰敏卡伽马射线源（2CG 195+04）
Geminga Pulsar 杰敏卡脉冲星（SN 437）
Gemini ①（=Gem）双子座 ②（=Gem）双子宫，实沈，申宫
Geminid meteor shower 双子流星雨
Geminids 双子流星群 | 雙子[座]流星雨〔台〕
Gemini Nebula 双子星云（IC 443）
Gemini Planet Imager （=GPI）双子座行星成像仪
Gemini project 双子载人卫星计划
Gemini Telescope 双子望远镜
Gemma 贯索四（北冕座 α） 又见: *Alphecca*
GEMS （=Global Environmental Monitoring System）全球环境监测系统
Genam 天棓一（天龙座 ξ） 又见: *Genan; Grumium*
Genan 天棓一（天龙座 ξ） 又见: *Genam; Grumium*
general astronomy 普通天文学
general astrophysics 普通天体物理 | 普通天文物理學〔台〕
General Automatic Luminosity and X Y Measuring Engine （=GALAXY）GALAXY 底片自动测量仪 | GALAXY[底片自動測量儀]〔台〕
General Catalogue of Variable Stars （=GCVS）变星总表
general circulation 总环流
general electric synchrotron 广义电同步加速
general elliptic type 一般椭圆型
general field star 场星 | 視野星〔台〕 又见: *field star*
generalization of Hill problem 广义希尔问题
generalized coordinate 广义坐标
generalized displacement 广义位移
generalized ensemble 广义系综
generalized force 广义力
generalized main sequence 广义主序 | 廣義主星序〔台〕
generalized momentum 广义动量
generalized solution of Lagrange 广义拉格朗日解
general magnetic field 普遍磁场
general perturbation 普遍摄动
general precession 总岁差
general precession in longitude 黄经总岁差
general-relativistic effect 广义相对论效应
general relativity （=GR）广义相对论
General Relativity and Gravitation 《广义相对论与引力》（美国期刊）
general theory of relativity 广义相对论

generating function 生成函数 又见: *determining function*
generating periodic solution 生成周期解
generator 发生器
Genesis 起源号探测器
Genesis rock 创世岩
genetic connection 演化关联
Geneva color system 日内瓦颜色系统
Geneva photometric system 日内瓦测光系统
Geneva photometry 日内瓦测光
genus 亏格
genus of cosmic density field 宇宙密度场亏格
Geoalert (=Geophysical Alert Broadcast) 地球物理警报广播
geoastrophysics 地球天体物理 | 地球天文物理學〔台〕
geoceiver 大地接收机
geocentric angle 地心角
geocentric apparent motion 地心视动 | 地球視動〔台〕
Geocentric Celestial Reference System (=GCRS) 地心天球参考系
geocentric colatitude 地心余纬
geocentric conjunction 地心合
geocentric conjunction in right ascension 地心赤经合
geocentric constant 地心常数
geocentric coordinate 地心坐标 | 地球坐標〔台〕
Geocentric Coordinate Time (=TCG) 地心坐标时
geocentric distance 地心距离 | 地球距離〔台〕
geocentric ephemeris 地心历表
geocentric gravitational constant 地心引力常数
geocentric horizon 地心地平 又见: *rational horizon*
geocentric latitude 地心纬度 | 地球緯度〔台〕
geocentric longitude 地心经度 | 地球經度〔台〕
geocentric orbit 地心轨道 | 地球軌道〔台〕
geocentric parallax 地心视差 | 地球視差〔台〕
geocentric phenomena 地心天象
geocentric position 地心位置
geocentric radiant 地心辐射点 | 地球輻射點〔台〕
geocentric system 地心体系 | 地球[宇宙]體系〔台〕(宇宙体系)
Geocentric Terrestrial Reference System (=GTRS) 地心地球参考系
geocentric velocity 地心速度
geocentric zenith 地心天顶
geochemistry 地球化学

geocorona 地冕
geodesically complete connection 测地完备联络
geodesic completeness 测地完备性
geodesic coordinate 测地坐标
geodesic equations 测地线方程组
geodesic line 测地线 | 短程線，大地線〔台〕
geodesic nutation 测地章动 又见: *geodetic nutation*
geodesic precession 测地岁差 又见: *geodetic precession*
geodesic satellite 测地卫星 又见: *geodetic satellite*
geodesy 大地测量学 又见: *geodetic surveying*
geodetic astronomy 大地天文学 | 測地天文學〔台〕
geodetic azimuth 大地方位角
geodetic base line 大地测量基线
geodetic constant 大地常数
geodetic coordinate 大地坐标 | 測地坐標〔台〕
geodetic datum 大地基准点
Geodetic Earth-Orbiting Satellite (=GEOS) GEOS 测地卫星
geodetic latitude 大地纬度 | 測地緯度〔台〕
geodetic longitude 大地经度 | 測地經度〔台〕
geodetic nutation 测地章动 又见: *geodesic nutation*
geodetic precession 测地岁差 又见: *geodesic precession*
Geodetic Reference System 大地参考系 | 測地參考系[統]，大地參考坐標[系]〔台〕
geodetic refraction 地平大气折射 又见: *horizontal refraction*
geodetic satellite 测地卫星 又见: *geodesic satellite*
geodetic surveying 大地测量学 又见: *geodesy*
geodetic triangle 大地定位三角形
geodetic zenith 大地天顶
geodimeter 光速测距仪
Geodynamic Experimental Ocean Satellite (=GEOS) 海洋地球动力实验卫星
geodynamics 地球动力学 又见: *earth dynamics*
geographic coordinate 地理坐标
geographic equator 地理赤道
geographic latitude 地理纬度
geographic longitude 地理经度
geographic meridian 地理子午线
geographic position 地理位置
Geographos 地理星（小行星1620号）
geoid 地球体，大地水准面 | 大地水準面〔台〕
geoidal height 大地水准面高度
geoidal horizon 大地水准面地平面
geoid surface 大地水准面

geological age 地质年龄
geological dating 地质计年
geological era 地质代
geological time 地质时代
geological time scale 地质时标
geology 地质学
geomagnetic activity 地磁活动
geomagnetic anomaly 地磁异常
geomagnetic axis 地磁轴
geomagnetic boundary 地磁边界
geomagnetic cavity 地磁穴
geomagnetic chart 地磁图
geomagnetic coordinate 地磁坐标
geomagnetic cut-off latitude 地磁截止纬度
geomagnetic declination 地磁偏角
geomagnetic dipole 地磁偶极子
geomagnetic disturbance 地磁扰动
　　又见：*terrestrial magnetic disturbance*
geomagnetic equator 地磁赤道
geomagnetic field 地磁场
geomagnetic inclination 地磁倾角
geomagnetic micropulsation 地磁微脉动
geomagnetic parameter 地磁参数
geomagnetic pole 地磁极
geomagnetic pulsation 地磁脉动
geomagnetic shell 地磁壳层
geomagnetic spiral field 地磁螺旋场
geomagnetic storm 磁暴 | 地磁暴〔台〕
geomagnetic tail 地磁尾
geomagnetic tide 地磁潮
geomagnetism 地磁　又见：*terrestrial magnetism*
geomechanics 地球力学，地质力学
geometric aberration 几何像差　又见：*geometrical aberration*
geometrical aberration 几何像差　又见：*geometric aberration*
geometrical albedo 几何学反照率
geometrical libration 几何天平动　又见：*geometric libration*
geometrically thin lens 几何薄透镜
geometric area 几何面积
geometric cross-section 几何截面
geometric delay 几何延迟
geometric depth 几何深度
geometric effect 几何效应
geometric latitude 几何纬度

geometric libration 几何天平动　又见：*geometrical libration*
geometric longitude 几何经度
geometric path 几何路程
geometric position 几何位置
geometric satellite geodesy 卫星几何测地学
geometric variable 几何变星
geometrodynamics 四维几何动力学 | 時空幾何動力論〔台〕
geomorphology 地形学
geon 引力电磁子
geonomy 地学，地球学
Geophysical Alert Broadcast （＝Geoalert）地球物理警报广播
Geophysical and Astrophysical Fluid Dynamics 《地球物理与天体物理流体动力学》（英国期刊）
geophysical effect 地球物理效应
geophysical observatory 地球物理观测台
geophysical phenomena 地球物理现象
geophysical satellite 地球物理卫星
geophysical station 地球物理观测站
geophysical year 地球物理年
geophysics 地球物理 | 地球物理學〔台〕
Geophysics Research Satellite （＝GRS）地球物理研究卫星
geopotential 大地势
georgiaite 乔治亚陨石
GEOS ① （＝Geodetic Earth-Orbiting Satellite）GEOS 测地卫星 ② （＝Geodynamic Experimental Ocean Satellite）海洋地球动力实验卫星
geospace 近地空间 | 地球空間〔台〕
　　又见：*terrestrial space*
geosphere 陆圈，陆界 | 陸圈〔台〕
geostational satellite 静地卫星　又见：*geostationary satellite*
Geostationary Operational Environmental Satellite （＝GOES）环境应用静地卫星（美国）
geostationary orbit 静地轨道
geostationary satellite 静地卫星　又见：*geostational satellite*
geostrophic motion 地转性运动
geosynchronous orbit 同步轨道 | 地球同步軌道〔台〕
geosynchronous satellite 地球同步卫星
geotectonics 大地构造学　又见：*tectonics*
gerade 【德】偶态　即：*even state*
Gerard crater 杰勒德环形山（月球）
Gerasimovich crater 格拉西莫维奇环形山（月球）

German Astrophysical Virtual Observatory （＝GAVO）德国天体物理虚拟天文台
Germania 德意志（小行星241号）
germanium （＝Ge）锗（32号元素）
germanium bolometer 锗测辐射热计
germanium detector 锗探测器
German mounting 德国式装置｜德式装置〔台〕
German-Spanish Astronomical Centre 德国－西班牙天文中心 即：Calar Alto Observatory
Gertrude crater 格特鲁德环形山（天卫三）
Geschichte des Fixsternhimmels （＝GFH）【德】《星空史》（恒星历史位置资料汇编）
Geschichte und Literatur des Lichtwechsels der Veränderlichen Sterne （＝GuL）【德】《变星光变史和文献》（变星资料汇编）
Ge γ-ray detector 锗γ射线探测器
g-factor g因子
GFH （＝Geschichte des Fixsternhimmels）【德】《星空史》（恒星历史位置资料汇编）
gf-value 加权振子强度
G giant G型巨星
GGSE （＝Gravity Gradient Satellite Experiment）重力梯度稳定实验卫星
GGTS （＝Gravity Gradient Test Satellite）重力梯度实验卫星
GHA （＝Greenwich hour angle）格林尼治时角
Ghirlanda relation 吉兰达关系
Ghost 鬼宿（二十八宿）
ghost crater 假环形山
ghost field 鬼场
ghost image 鬼像｜鬼影〔台〕
ghost line 鬼线
Ghost Nebula 鬼魂星云（NGC 1977）
Ghost of Jupiter 木魂星云（NGC 3242）
 又见：*Jupiter's Ghost*
Ghost Stream 鬼流
Ghost Vehicle 舆鬼
Giacobinid meteor shower 贾科比尼流星雨
Giacobinids 贾科比尼流星群｜賈可比尼流星群〔台〕
Giacobini-Zinner comet 贾科比尼－津纳彗星｜賈可比尼－金納彗星〔台〕
Gianfar 上辅，紫微右垣三（天龙座λ）
 又见：*Giansar; Juza*
Giansar 上辅，紫微右垣三（天龙座λ）
 又见：*Gianfar; Juza*
giant arc 巨型光弧
giant branch 巨星支

giant-branch star 巨星分支星
giant cell 巨泡
giant donor 输质巨星
giant elliptical galaxy 巨椭圆星系
giant galaxy 巨星系
giant granulation 巨米粒组织
giant granule 巨米粒
giant impact hypothesis 大碰撞假说
Giant Magellan Telescope （＝GMT）巨麦镜，巨麦哲伦望远镜｜巨麥[哲倫望遠]鏡〔台〕
giant maximum 巨极大
Giant Metrewave Radio Telescope （＝GMRT）大型米波射电望远镜｜巨型米波電波望遠鏡〔台〕
giant minimum 巨极小
giant molecular cloud （＝GMC）巨分子云
giant molecular cloud complex 巨分子云复合体
giant planet 巨行星
giant pulse 巨脉冲
giant radio galaxy 巨射电星系｜巨電波星系〔台〕
giant radio pulse 巨射电脉冲
Giant Segmented Mirror Telescope （＝GSMT）巨型拼合镜面望远镜
giant spiral galaxy 巨旋涡星系｜巨螺旋星系〔台〕
giant star 巨星
gibbous moon 凸月
Gibbs crater 吉布斯环形山（月球）
Gibbs ensemble 吉布斯系综
Gibbs free energy 吉布斯自由能
GID （＝gradual ionospheric disturbance）电离层渐扰
Gienah ① 天津九（天鹅座ε） 又见：*Gienah Cygni; Gienar* ② 轸宿一（乌鸦座γ） 又见：*Gienah Ghurab*
Gienah Cygni 天津九（天鹅座ε） 又见：*Gienah; Gienar*
Gienah Ghurab 轸宿一（乌鸦座γ） 又见：*Gienah*
Gienar 天津九（天鹅座ε） 又见：*Gienah; Gienah Cygni*
Gilbert crater 吉尔伯特环形山（月球）
Gill crater 吉尔环形山（月球）
gimbal-mounted polarimeter 换向装置偏振计
gimbal mounting 换向装置
Ginga 银河号（日本X射线天文卫星）
 又见：*Astro-C*
Giordano Bruno crater 焦尔达诺·布鲁诺环形山（月球）
Giotto 乔托行星际探测器｜喬陶號[太空船]〔台〕
Girl 女宿（二十八宿） 又见：*Maid; Woman; Damsel*
Girtab 尾宿七（天蝎座κ）

glacial period 冰期　又见：*ice age*
glaciation 冰蚀
glaciology 冰川学
glancing incidence telescope 掠入成像望远镜
Glan-Foucault prism 格兰－傅科棱镜
Glan-Thompson prism 格兰－汤姆森棱镜
GLAO （=ground-layer adaptive optics）地面层自适应光学
glass-ceramic 微晶玻璃　又见：*cervit; zerodur; sytall*
glass circle 玻璃度盘
glass fiber 玻璃纤维
glass filter 玻璃滤光片
glass-lined crater 搪玻璃环形山
glass-lined microcrater 搪玻璃微型环形山
glass-prism spectrograph 玻璃棱镜摄谱仪
glass-prism spectroscope 玻璃棱镜分光镜
glass spherule 玻璃球粒
Glatton meteorite 格拉顿陨星
GLE （=ground level event）地面粒子事件
Gleissberg period 格莱斯伯格周期
Gliese Catalogue of Nearby Stars 格利泽近星星表
glitch 自转突变 | 頻率突變〔台〕
glitch activity 自转突变活动
Global Astrometric Interferometer for Astrophysics （=Gaia）全天天体测量干涉仪（全称已不用）
global change 全球变化
Global Environmental Monitoring System （=GEMS）全球环境监测系统
global error 全局误差
global helioseismology 整体日震学
Global Microlensing Alert Network 全球微引力透镜预警网络
global mode 整体模型
global oscillation 全球振荡
Global Oscillation Network Group （=GONG）全球[太阳]振荡监测网
Global Positioning System （=GPS）全球定位系统
global regularization 全局正规化
global sensitivity 总体灵敏度
global symmetry 全局对称性
global time synchronization 全球时间同步
globe 地球　又见：*Earth; terrestrial globe*
globular cluster 球状星团　又见：*globular star cluster*
globular galaxy 球状星系　又见：*spherical galaxy*
globular star cluster 球状星团　又见：*globular cluster*

globule 球状体 | 雲球〔台〕
Globus Aerostaticus 热气球座（现已不用）
gluon 胶子
GMAT （=Greenwich mean astronomical time）格林尼治平天文时
GMC （=giant molecular cloud）巨分子云
GMN （=Greenwich mean noon）格林尼治平午 | 格林[威治]平午〔台〕
g-mode g 模，重力模 | g 模式〔台〕
GMRT （=Giant Metrewave Radio Telescope）大型米波射电望远镜 | 巨型米波電波望遠鏡〔台〕
GMST （=Greenwich mean sidereal time）格林尼治平恒星时 | 格林[威治]平恆星時〔台〕
GMT ① （=Giant Magellan Telescope）巨麦镜，巨麦哲伦望远镜 | 巨麥[哲倫望遠]鏡〔台〕② （=Greenwich mean time）格林尼治平时 | 格林[威治]平時〔台〕
gnomon ① 圭表 | 日圭，圭表〔台〕② 表
gnomonic projection 心射切面投影 | 心射切面投影，日晷投影〔台〕
gnomonics 日晷法
gnomon shadow template 圭
Goddard crater 戈达德环形山（月球）
Goddard Space Flight Center （=GSFC）戈达德航天中心 | 哥達德太空飛行中心〔台〕（美国）
Gödel universe 哥德尔宇宙
GOES （=Geostationary Operational Environmental Satellite）环境应用静地卫星（美国）
Goethe Link Observatory 哥德林克天文台（美国）
goethite 针铁矿
Golay cell 高莱探测器
gold （=Au）金（79 号元素）
gold-doped germanium detector 锗掺金探测器
gold number 金数（默冬章周数）
Goldschmidt crater 戈尔德施米特环形山（月球）
gold spot disease 金斑病
Goldstack interferometer 戈德斯塔克干涉仪
Goldstino 戈德斯通微子
Goldstone boson 戈德斯通玻色子
Goldstone-Haystack interferometer 戈德斯通－海斯塔克干涉仪
Goldstone Radio Astronomy Station 金石射电天文站
Gomeisa 南河二（小犬座 β）　又见：*Algomeyla*
Gomez's Hamburger Nebula 戈麦斯汉堡星云（IRAS 18059-3211）

GONG (＝Global Oscillation Network Group）全球[太阳]振荡监测网
Gongyi 巩义（小行星 19258 号）
good quantum number 好量子数
Gordon equation 戈登方程
Gorgonea Quarta 大陵增十八（英仙座 ω）
Gorgonea Secunda 积尸（英仙座 π）
Gorgonea Tertia 大陵六（英仙座 ρ）
Gossamer Ring 薄纱环（木星）
Gosses Bluff crater 戈斯崖陨星坑（地球）
go-to telescopes 自动寻星望远镜
Göttingen University Observatory 哥廷根大学天文台（德国）
Gould belt 古德带 又见：*Gould's belt*
Gould's belt 古德带 又见：*Gould belt*
G-parity G 字称
GP-B （＝Gravity Probe B）引力探测器 B（美国科学卫星）
GPI （＝Gemini Planet Imager）双子座行星成像仪
GPS （＝Global Positioning System）全球定位系统
GR （＝general relativity）广义相对论
Gr. （＝Groombridge's Catalogue of Circumpolar Stars）格鲁姆布里奇拱极星表
graben 地堑
GRACE （＝Gravity Recovery and Climate Experiment）重力测量和气候实验（美国科学卫星）
graceful exit problem 优雅退出问题（暴胀）
gradation 渐变
gradient 梯度
grading 照明
grading function 照明函数
gradual burst 缓慢暴 | 緩慢爆發〔台〕
gradual ionospheric disturbance （＝GID）电离层渐扰
gradual production 缓慢产生
graduated arc 刻度弧
graduated circle 刻度盘 又见：*divided circle*
graduated leaker 刻漏
graduated scale 刻度尺
graduation 刻度 又见：*scale division*
graduation error 刻度误差
Graffias 房宿四（天蝎座 β） 又见：*Acrab*
Grafias 尾宿三（天蝎座 ζ）
GRAIL （＝Gravity Recovery and Interior Laboratory）圣杯号（美国月球探测器）

Grain Fills 小满（节气） 又见：*Lesser Fullness*
Grain in Ear 芒种（节气）
Grain Rain 谷雨（节气）
gram atom 克原子
gram weight 克重
Granat 石榴号（苏联高能天文卫星）
grand canonical ensemble 巨正则系综
grand design 宏观图像
grand design spiral 宏象旋涡结构
grand ensemble 巨系综
grand origin ①上元 又见：*superior epoch* ②上元积年
grand unified theory （＝GUT）大统一理论
granite 花岗岩
Gran Telescopio CANARIAS （＝GTC）【西】加那利大型望远镜
granulation 米粒组织（太阳） 又见：*rice grain*
granule 米粒（太阳）
graphic method 图解法，图示法
graphite 石墨
graphite flake 石墨片
graticule 栅网
grating 光栅
grating array 栅阵
grating constant 光栅常数
grating image 光栅星像
grating lobe 栅瓣
grating response 栅阵响应
grating ring 栅环
grating space 栅线间距
grating spectrograph 光栅摄谱仪
grating spectrometer 光栅分光计
grating spectroscope 光栅分光镜
grating spectrum 光栅光谱
gravimeter 重力计
gravimetric baseline 重力基线
gravimetry 重力测量 | 重力測量 [術]〔台〕
gravipause 重力分界
gravisphere 重力范围 又见：*sphere of gravity*
gravitating disk 引力盘
gravitation 引力 又见：*attractive force; gravitational force*
gravitational acceleration 引力加速度，重力加速度
gravitational astronomy 引力天文学 | 重力天文學〔台〕
gravitational attraction 引力吸引

gravitational bend angle　引力偏折角
gravitational binary　引力双星 | 重力雙星〔台〕
　　又见：*gravitational double star*
gravitational bremsstrahlung　引力轫致辐射 | 重力制動輻射〔台〕
gravitational capture　引力俘获
gravitational clustering　引力成团 | 重力成團〔台〕
gravitational collapse　引力坍缩 | 重力塌縮〔台〕
gravitational condensation　引力凝聚 | 重力凝聚〔台〕
gravitational constant　引力常数 | 重力常數〔台〕
　　又见：*constant of gravitation; gravitation constant*
gravitational contraction　引力收缩 | 重力收縮〔台〕
gravitational darkening　引力昏暗
gravitational deflection　引力弯曲，引力偏折 | 重力彎曲，重力偏折〔台〕
gravitational differentiation　引力分异 | 重力分化〔台〕
gravitational displacement　引力位移 | 重力位移〔台〕
gravitational distortion　引力畸变
gravitational double star　引力双星 | 重力雙星〔台〕　又见：*gravitational binary*
gravitational effect　引力效应 | 重力效應〔台〕
　　又见：*gravitation effect*
gravitational encounter　引力交会
gravitational energy　引力能 | 重力能〔台〕
gravitational equilibrium　引力平衡 | 重力平衡〔台〕
gravitational field　引力场 | 重力場〔台〕
gravitational focusing　引力聚焦
gravitational force　引力　又见：*attractive force; gravitation*
gravitational freeze-out　引力冻结
gravitational instability　引力不稳定性 | 重力不穩定 [性]〔台〕
gravitational interaction　引力作用
gravitational lens　引力透镜 | 重力透鏡〔台〕
gravitational lens effect　引力透镜效应 | 重力透鏡效應〔台〕　又见：*gravitational lensing*
gravitational lensing　引力透镜效应 | 重力透鏡效應〔台〕　又见：*gravitational lens effect*
gravitational lensing effect　引力透镜效应
gravitational mass　引力质量 | 重力質量〔台〕
gravitational micro-lens　微引力透镜 | 微重力透鏡〔台〕　又见：*microgravitational lens; microlens*
gravitational micro-lensing　微引力透镜效应 | 微重力透鏡效應〔台〕

gravitational paradox　引力佯谬 | 重力佯謬〔台〕
　　即：*Seeliger paradox*
gravitational potential　引力势 | 重力勢〔台〕
gravitational potential well　引力势阱 | 重力勢阱〔台〕
gravitational radiation　引力辐射 | 重力輻射〔台〕
gravitational radius　引力半径 | 重力半徑〔台〕
gravitational redshift　引力红移 | 重力紅 [位] 移〔台〕
gravitational relaxation　引力弛豫
gravitational scattering　引力散射
gravitational shift　引力偏移
gravitational strain　引力应变
gravitational synchrotron radiation　引力同步加速辐射 | 重力同步 [加速] 輻射〔台〕
gravitational thermodynamics　引力热力学
gravitational tide　引力潮 | 重力潮〔台〕
gravitational time dilation　引力时间延缓
gravitational time scale　引力时标
gravitational wave　（＝GW）引力波 | 重力波〔台〕
gravitational wave astronomy　引力波天文学 | 重力波天文學〔台〕
gravitational wave source　引力波源
gravitational wave telescope　引力波望远镜 | 重力波望遠鏡〔台〕
gravitational wobble　引力摆动
gravitation constant　引力常数 | 重力常數〔台〕
　　又见：*gravitational constant; constant of gravitation*
gravitation effect　引力效应 | 重力效應〔台〕
　　又见：*gravitational effect*
gravitation law　引力定律
gravitino　引力微子 | 重力微子〔台〕
graviton　引力子 | 重力子〔台〕
gravity　重力
gravity anomaly　重力异常
gravity assist　重力助推
gravity center　重心　又见：*center of gravity*
gravity darkening　重力昏暗
gravity drift　重力漂移
gravity driving clock　重力转仪钟
gravity field　重力场
gravity gradient　重力梯度
Gravity Gradient Satellite Experiment　（＝GGSE）重力梯度稳定实验卫星
Gravity Gradient Test Satellite　（＝GGTS）重力梯度实验卫星
gravity gradiometer　重力梯度仪

Gravity Probe B （=GP-B）引力探测器 B（美国科学卫星）

Gravity Recovery and Climate Experiment （=GRACE）重力测量和气候实验（美国科学卫星）

Gravity Recovery and Interior Laboratory （=GRAIL）圣杯号（美国月球探测器）

gravity wave 重力波

gravothermal catastrophe 引力热灾变

grazing eclipse 掠食

grazing incidence 掠射

grazing incidence optics 掠射光学

grazing incidence spectrograph 掠射摄谱仪

grazing incidence system 掠射系统

grazing incidence telescope 掠射望远镜 | 掠入射望遠鏡〔台〕

grazing occultation 掠掩

GRB （=γ-ray burst）γ 射线暴 | γ 射線爆發〔台〕

Great Attractor 巨引源 | 巨重力源〔台〕

great circle 大圆 又见: orthodrome

Great cluster of Hercules 武仙大星团（NGC 6205）

Great Cold Spot 大冷斑

Great Dark Spot （=GDS）大暗斑

Greater Cold 大寒（节气） 又见: Severe Cold

Greater Heat 大暑（节气）

Greater Snow 大雪（节气） 又见: Heavy Snow

greatest brilliancy 最大亮度 | 最亮〔台〕

greatest eastern elongation 东大距 又见: eastern greatest elongation; eastern elongation

greatest elongation （=GE）大距 又见: elongation

greatest north latitude 最大北黄纬

greatest south latitude 最大南黄纬

greatest western elongation 西大距 又见: western elongation

great inequality 中心差 又见: equation of the center

Great Looped Nebula 大圈星云（NGC 2070）

Great Nebula in Orion 猎户[大]星云 | 獵戶[座]大星雲〔台〕（M42） 又见: Orion Nebula; Great Orion Nebula

Great Orion Nebula 猎户[大]星云 | 獵戶[座]大星雲〔台〕（M42） 又见: Orion Nebula; Great Nebula in Orion

Great Red Spot （=GRS）大红斑

Great Rift 大暗隙 | 大裂縫[銀河]〔台〕

Great September Comet 九月大彗星

Great sequence 大序 | 大星序〔台〕

Great square of Pegasus 飞马大四边形 | 飛馬四邊形〔台〕（星组）

great star 景星 又见: splendid star

Great Wall 巨壁 | 長城〔台〕

Great White Spot （=GWS）大白斑

great year 大年（岁差周期，约 25765 年）

Grecian era 希腊纪元

Greek alphabet 希腊字母命名

Greek group 希腊群

Green Bank Telescope （=GBT）绿堤射电望远镜 | 格林班克電波望遠鏡〔台〕

Green crater 格林环形山（月球）

green flash 绿闪 | 綠閃光〔台〕

greenhouse effect 温室效应

green line 绿谱线

green segment 绿闪瞬间

Green's function 格林函数

green sun 绿闪太阳

Green theorem 格林定理

Greenwich apparent noon 格林尼治视午 | 格林[威治]視午〔台〕

Greenwich apparent sidereal time 格林尼治视恒星时 | 格林[威治]視恆星時〔台〕

Greenwich apparent time 格林尼治视时 | 格林[威治]視時〔台〕

Greenwich civil time （=GCT）格林尼治民用时 | 格林[威治]民用時〔台〕

Greenwich hour angle （=GHA）格林尼治时角

Greenwich interval 格林尼治时间间隔

Greenwich lunar time 格林尼治太阴时

Greenwich mean astronomical time （=GMAT）格林尼治平天文时

Greenwich mean noon （=GMN）格林尼治平午 | 格林[威治]平午〔台〕

Greenwich mean sidereal time （=GMST）格林尼治平恒星时 | 格林[威治]平恆星時〔台〕

Greenwich mean time （=GMT）格林尼治平时 | 格林[威治]平時〔台〕

Greenwich meridian 格林尼治子午线，格林尼治子午圈 | 格林[威治]子午線〔台〕

Greenwich Observatory 格林尼治天文台

Greenwich sidereal date （=GSD）格林尼治恒星日期 | 格林[威治]恆星日期〔台〕

Greenwich sidereal day number 格林尼治恒星日数

Greenwich sidereal time （=GST）格林尼治恒星时 | 格林[威治]恆星時〔台〕

Gregorian antenna 格里式天线

Gregorian calendar 格里历
Gregorian telescope 格里望远镜
Gregorian year 格里年
gregorite 钛铁矿
Greip 土卫五十一
Greisen-Zatsepin-Kuzmin limit （＝GZK limit）格莱森－查泽品－库兹敏极限
grens 透镜棱栅
grey atmosphere 灰大气 | 灰[色]大氣〔台〕
grey body 灰体
grey-body radiation 灰体辐射
grey hole 灰洞
grey-scale map 灰度图
grid 栅格 | 格栅，视栅〔台〕
grid collimator 栅格准直器
gridding 栅格化
Griffith Observatory 格里菲斯天文台
Grigg-Skjellerup Comet 格里格－斯克叶勒鲁普彗星
Grigg-Skjellerup comet 格里格－斯基勒鲁普彗星
Grimaldi crater 格里马尔迪环形山（月球）
grism 棱栅
grism spectrograph 棱栅摄谱仪
Grissom crater 格里索姆环形山（月球）
Groombridge's Catalogue of Circumpolar Stars （＝Gr.）格鲁姆布里奇拱极星表
gross shutter 总快门
Grotrian diagram 格罗特利安图
ground absorption 地面吸收
ground-based astronomy 地面天文学
ground-based observation 地基观测
  又见：Earth-based observation
ground-based observatory 地基天文台
ground-based telescope 地基望远镜
  又见：Earth-based telescope
ground cosmic rays enhancement （＝GCE）地面宇宙线增强
ground glass 毛玻璃
ground-layer adaptive optics （＝GLAO）地面层自适应光学
ground level 基级 | 基级阶，地面高〔台〕
ground level event （＝GLE）地面粒子事件
ground plane 地平面 又见：horizontal plane; horizon plain
ground radiation 地面辐射
ground screen 地面屏蔽
ground state 基态
ground station 地面站

ground timing system 地面定时系统
ground wave propagation 地波传播
ground wave radio signal 地波无线电信号
group correction 组间改正
group delay 群延迟
group frequency 群频率
group mean latitude 组平均纬度
group motion 群动
group of asteroids 小行星群 又见：asteroid group
group of comets 彗星群 又见：comet group
group of galaxies 星系群
group of stars 恒星群 又见：star group
group parallax 星群视差 | 星群视差，星團視差〔台〕 又见：cluster parallax
group phenomena 群现象
group representation 群表象
group velocity 群速度 | [波]群速[度]〔台〕
growth curve 生长曲线 又见：curve of growth
growth defect 生长缺陷
growth factor 增长因子
growth function 增长函数
growth rate 生长速率
growth time 生长时标
GRS ①（＝Geophysics Research Satellite）地球物理研究卫星 ②（＝Great Red Spot）大红斑
Gru （＝Grus）天鹤座
Grumium 天棓一（天龙座ξ） 又见：Genam; Genan
Grus （＝Gru）天鹤座
Grus Quartet 天鹤四重星系
GSC （＝Guide Star Catalogue）GSC 导星星表
GSD （＝Greenwich sidereal date）格林尼治恒星日期 | 格林[威治]恆星日期〔台〕
GSFC （＝Goddard Space Flight Center）戈达德航天中心 | 哥達德太空飛行中心〔台〕（美国）
GSMT （＝Giant Segmented Mirror Telescope）巨型拼合镜面望远镜
GSO scintillator 硅酸钆闪烁体
GST （＝Greenwich sidereal time）格林尼治恒星时 | 格林[威治]恆星時〔台〕
G star G型星 又见：G-type star
G subdwarf G型亚矮星 | G型次矮星〔台〕
G subgiant G型亚巨星 | G型次巨星〔台〕
G supergiant G型超巨星
GTC （＝Gran Telescopio CANARIAS）【西】加那利大型望远镜
GTRS （＝Geocentric Terrestrial Reference System）地心地球参考系
G-type asteroid G型小行星

G-type star G 型星　又见：*G star*
Guangcaishiye 光彩事业（小行星 7497 号）
Guangdong 广东（小行星 2185 号）
Guangxi 广西（小行星 2655 号）
Guangzhou 广州（小行星 3048 号）
Guchaohao 谷超豪（小行星 171448 号）
Gueiren 归仁（小行星 185216 号）
Guericke crater 居里克环形山（月球）
guest star 客星（彗星、新星或超新星）
guidance 导星，制导 | 導引〔台〕
guide meridian 参考子午线
guider ① 导星装置　又见：*guiding device* ② 导星镜　又见：*guiding telescope; guide telescope; guidescope*
guidescope 导星镜　又见：*guiding telescope; guide telescope; guider*
guide star 引导星　又见：*guiding star*
Guide Star Catalogue （=GSC）GSC 导星星表
guide system 导星系统
guide telescope 导星镜　又见：*guiding telescope; guider; guidescope*
guiding 导星
guiding center 导中心 | 導向中心〔台〕
guiding device 导星装置　又见：*guider*
guiding error 导星误差
guiding microscope 导星测微镜
guiding star 引导星　又见：*guide star*
guiding system 导星系统
guiding telescope 导星镜　又见：*guide telescope; guider; guidescope*
Guillaume crater 纪尧姆环形山（月球）
guillotine factor 截断因子
Guizhou 贵州（小行星 2632 号）
GuL （=Geschichte und Literatur des Lichtwechsels der Veränderlichen Sterne）【德】《变星光变史和文献》（变星资料汇编）
Gula Mons 古拉山（金星）
Gum crater 古姆环形山（月球）
Gum Nebula 古姆星云 | 甘姆星雲〔台〕（Gum 12）
Gunn effect 冈恩效应
Gunn-Peterson effect 冈恩－彼得森效应
Gunn-Peterson test 冈恩－彼得森检验
Gunn-Peterson trough 冈恩－彼得森槽（类星体）
Guo Shou-Jing 郭守敬（小行星 2012 号）
Guo Shoujing Telescope 大天区面积多目标光纤光谱天文望远镜，郭守敬望远镜 | 大天區面積多目標光纖光譜望遠鏡〔台〕　又见：*Large Sky Area Multi-Object Fiber Spectroscopic Telescope*
GUT （=grand unified theory）大统一理论
Gutenberg crater 谷登堡环形山（月球）
Guyot crater 盖奥特环形山（月球）
Guyuzhou 顾宇洲（小行星 23758 号）
GW （=gravitational wave）引力波 | 重力波〔台〕
GWS （=Great White Spot）大白斑
GW Virginis instability strip 室女 GW 不稳定带
GW Virginis star 室女 GW 型星 | 室女[座]GW 型星〔台〕　又见：*GW Vir star*
GW Virginis variable （=GW Vir variable）室女 GW 型变星　即：*GW Virginis star; GW Vir star*
GW Vir star 室女 GW 型星 | 室女[座]GW 型星〔台〕　又见：*GW Virginis star*
GW Vir variable （=GW Virginis variable）室女 GW 型变星　即：*GW Virginis star; GW Vir star*
GX （=Galactic X-ray source）银河 X 射线源
Gylden method 吉尔当法
gyration period 回转周期
gyration time 回转时间
gyrocompass 陀螺罗盘
gyroemission 回旋加速发射　又见：*cyclotron emission*
gyrofrequency 回旋加速频率 | 迴旋頻率〔台〕　又见：*cyclotron frequency*
gyrofrequency plasmon 回转频率等离激元
gyromagnetic radiation 磁回旋辐射 | 磁迴轉輻射〔台〕
gyroradius 回旋半径　又见：*radius of gyration*
gyrorelaxation 回转弛豫 | 迴轉張弛〔台〕
gyroresonance 回旋共振　又见：*cyclotron resonance*
gyroresonance radiation 回旋共振辐射　又见：*gyroresonant radiation*
gyroresonant absorption 回旋共振吸收
gyroresonant radiation 回旋共振辐射　又见：*gyroresonance radiation*
gyrorotor 陀螺转子
gyroscope 陀螺仪　又见：*gyrostat*
gyrosextant 陀螺六分仪 | 迴轉六分儀，陀螺六分儀〔台〕
gyrostat 陀螺仪　又见：*gyroscope*
gyrosynchrotron radiation 回旋同步加速辐射
GZK limit （=Greisen-Zatsepin-Kuzmin limit）格莱森－查泽品－库兹敏极限

H

H （=hydrogen）氢（1号元素）
HA ①（=Harvard Annual）《哈佛天文台纪事》② （=hour angle）时角
haab 哈布年（玛雅历）
Haber crater 哈伯环形山（月球）
habitability 宜居性 | 適居性，可居住性〔台〕
habitable planet 宜居行星 | 適居行星〔台〕
habitable zone 宜居带 | 適居區〔台〕
H abundance 氢丰度 又见：hydrogen abundance
Hadar 马腹一（半人马座 β） 又见：Agena
Hades 木卫九（曾用名） 即：Sinope
Hadley circulation 哈德利环流
Hadley Rille 哈德利沟纹（月球）
Hadriaca Patera 哈德良托边火山（火星）
hadron 强子
hadron barrier 强子势垒
hadron era 强子期 | 強子時代〔台〕
Haedus 柱二（御夫座 ζ） 又见：Hoedus I
hafnium （=Hf）铪（72号元素）
Hagedorn equation of state 哈格多恩状态方程 | 赫格登狀態方程〔台〕
Hagen crater 哈根环形山（月球）
Hagongda 哈工大（小行星55838号）
Hahn crater 哈恩环形山（月球）
Haig mount 黑格基架
Hainan 海南（小行星3024号）
Hainzel crater 海因泽尔环形山（月球）
Hakucho 天鹅号（日本X射线天文卫星） 又见：Corsa-B
halation 光晕 | 暈光作用〔台〕
Halawe 芝麻片糖（小行星518号）
HALCA （=Highly Advanced Laboratory for Communications and Astronomy）哈尔卡实验室（日本射电卫星）
Hale crater ①海尔环形山（月球）②海尔环形山（火星）
Hale Observatories 海尔天文台

Hale period 海尔周期
Hale sector boundary 海尔扇形边界
Hale's law 海尔定律
Hale Telescope 海尔望远镜
half-coded field of view 半编码视场
half-life 半衰期
half-light radius 半光半径
half-maximum line breadth 半峰线宽
half moon 半月
half-peak width 半峰宽度
half-power beam width （=HPBW）半功率束宽
half-wave dipole 半波偶极子 | 半波雙極〔台〕
half-wave plate 半波晶片 | 半波[晶]片〔台〕
half width 半宽
half width at half maximum （=HWHM）半峰半宽
Halimede 海卫九
Halley crater 哈雷环形山（月球）
Halley's comet 哈雷彗星
Halley-type comet 哈雷型彗星
Hallstatt circle 哈尔史塔特周期（太阳活动）
halo 晕
halo assembly bias 晕组装偏袒
halo bias 晕偏袒
halo concentration 晕聚集度
halo-core structure 晕核结构
halo creation rate 晕产生率
halo density profile 晕密度轮廓
halo destruction rate 晕破坏率
halo dwarf 晕族矮星
halo globular cluster 晕族球状星团
halo mass function 晕质量函数
halo merger rate 晕并合率
halo model 晕模型
halo object 晕族天体

halo occupation distribution （=HOD）暗晕占居数分布
halo of galaxy 星系晕　又见：*galactic halo*
halo of radio-burst region 射电暴晕
halo orbit 晕轨道
halo population 晕族，极端星族 II | [银] 晕星族，極端第二星族〔台〕　又见：*extreme population II*
halo radiation 晕辐射
halo radius 晕半径
Halo ring 哈洛环
halo star 晕族星 | [银] 晕族星〔台〕
halo-tail structure 晕－尾结构
Hamal 娄宿三（白羊座α）　又见：*El Nath*
Hamburg Observatory 汉堡天文台
Hamburg variable （=HBV）汉堡天文台变星
Hamiltonian function 哈密顿函数
Hamiltonian operator 哈密顿算符
Hamiltonian system 哈密顿系统
Hamilton-Jacobi method 哈密顿－雅可比方法
Hamilton principle 哈密顿原理
Hamilton's equation 哈密顿方程
Hamlet Crater 哈姆雷特陨击坑（天卫四）
Han 韩，天市右垣十一（蛇夫座ζ）
Handbuch der Astrophysik （=HdAp）【德】《天体物理手册》
H and K lines H－K 谱线
h and χ Persei 英仙双星团（NGC 869/884）
Hanggao 杭高（小行星 48700 号）
hanging level 悬水准　又见：*suspended level*
Hangtianyuan 航天员（小行星 35313 号）
Hankel function 汉克函数
Hankel transform 汉克变换
Hanle effect 汉勒效应
Hanning method 汉宁方法
Hanno crater 汉诺环形山（月球）
Hansen coefficient 汉森系数
Hansen theory 汉森理论
HAO （=High Altitude Observatory）高山天文台（美国）
Harbin 哈尔滨（小行星 2851 号）
harbinger 前兆
hard binary 硬双星
hard-clipped auto-correlation 强削波自相关
hard-clipped auto-correlator 强削波自相关器
hard crown-glass 硬冕玻璃
hard electron 高能电子
hardening rate 硬化率

hard image 高反差图像
hardware 硬件
Hard X-ray Modulation Telescope （=HXMT）硬 X 射线调制望远镜 | 硬 X 光調制望遠鏡〔台〕（中国天文卫星）
hard X-rays 硬 X 射线
hard X-ray source 硬 X 射线源
hard γ-rays 硬 γ 射线
Haris 招摇（牧夫座γ）　又见：*Seginus*
Harkins rule 哈金斯定则
Harlan crater 哈伦环形山（月球）
Harman-Seaton sequence 哈曼－西顿序列 | 哈曼－西頓序〔台〕
Harmonia 谐神星（小行星 40 号）
harmonic analysis 谐波分析
harmonic component 谐分量
harmonic coordinate 调和坐标
harmonic frequency 谐频
harmonic gauge 谐和规范
harmonic law 调和定律 | 諧和定律〔台〕
harmonic motion 谐运动
harmonic oscillator 谐振子，谐振荡器
harmonic overtone 谐波　又见：*overtone*
harmonic pair 谐波对
Haro classification 阿罗分类
Haro galaxy 阿罗星系 | 哈羅星系〔台〕
Harpalyke 木卫二十二
Harriot crater 哈里奥特环形山（月球）
Harrison-Zel'dovich-Peebles spectrum 哈里森－泽尔多维奇－皮布尔斯谱
Harrison-Zel'dovich spectrum 哈里森－泽尔多维奇谱
Hartebeesthoek Radio Astronomy Observatory （=HartRAO）哈特射电天文台
Hartley 2 Comet 哈特利 2 号彗星
Hartmann-Cornu formula 哈特曼－考纽公式 | 哈特曼－科紐公式〔台〕
Hartmann crater 哈特曼环形山（月球）
Hartmann diaphragm 哈特曼光阑
Hartmann dispersion formula 哈特曼色散公式
Hartmann test 哈特曼检验
HartRAO （=Hartebeesthoek Radio Astronomy Observatory）哈特射电天文台
Hartwig crater 哈特维希环形山（月球）
Harvard Annual （=HA）《哈佛天文台纪事》
Harvard Bulletin （=HB）《哈佛天文台公报》
Harvard Circular （=HC）《哈佛天文台简报》

Harvard classification　哈佛分类 | 哈佛分類法〔台〕
Harvard College Observatory　（＝HCO）哈佛天文台
Harvard Photometry　（＝HP）哈佛测光星表 | 哈佛恆星測光表〔台〕
Harvard pulsar　（＝HP）哈佛天文台脉冲星
Harvard Region　哈佛选区 | 哈佛天區〔台〕
Harvard Revised Photometry　哈佛测光星表修订版 | 哈佛恆星測光表修訂版〔台〕　又见：*Revised Harvard Photometry*
Harvard-Smithsonian Center for Astrophysics　（＝CfA）哈佛史密松天体物理中心
Harvard-Smithsonian Reference Atmosphere　（＝HSRA）哈佛－史密松参考大气
Harvard Standard Region　（＝HSR）哈佛标准选区
Harvard variable　（＝HV）哈佛天文台变星
harvest moon　获月
Harvey crater　哈维环形山（月球）
Hase crater　哈泽环形山（月球）
Hat Creek Observatory　帽子溪射电天文台（美国）
Hati　土卫四十三
Haughton crater　霍顿陨星坑（地球）
Haumea　妊神星（小行星 136108 号，矮行星）
Hausen crater　豪森环形山（月球）
Haute Provence Observatory　（＝OHP）上普罗旺斯天文台
Haviland crater　哈维兰陨星坑 | 哈威蘭隕石坑〔台〕（地球）
HAWC　（＝High-Altitude Water Cherenkov observatory）高海拔水体切伦科夫天文台
Hawking-Hartle wave function　霍金－哈特尔波函数
Hawking radiation　霍金辐射
Hawking's rule　霍金定则
Hawking temperature　霍金温度
Hayabusa　隼鸟号（日本彗星探测器）
Hayashi limit　林忠四郎极限 | 林[忠四郎]極限〔台〕
Hayashi line　林忠四郎线
Hayashi model　林忠四郎模型
Hayashi phase　林忠四郎阶段
Hayashi theorem　林忠四郎定理
Hayashi track　林忠四郎迹程 | 林[忠四郎]軌跡〔台〕
Hayn crater　海因环形山（月球）
Hay spot　哈伊斑

Haystack Observatory　海斯塔克天文台
HB　（＝Harvard Bulletin）《哈佛天文台公报》
HBV　（＝Hamburg variable）汉堡天文台变星
HC　（＝Harvard Circular）《哈佛天文台简报》
HCG　（＝Hickson Compact Group）希克森致密星系群
HCO　（＝Harvard College Observatory）哈佛天文台
H-component　H 分量
HdAp　（＝Handbuch der Astrophysik）【德】《天体物理手册》
HD Catalogue　（＝Henry Draper catalogue）HD 星表
HD classification　（＝Henry Draper classification）HD 恒星光谱分类
HD curve　（＝Hurter-Driffield curve）赫特－德里菲尔德曲线，HD 曲线
HDE　（＝Henry Draper Extension）HD 星表补编
H deficient star　贫氢星 | 貧氫恆星〔台〕
　又见：*hydrogen-deficient star; hydrogen-poor star*
HDF　（＝Hubble Deep Field）哈勃深场 | 哈柏深空區〔台〕
HDM scenario　（＝hot dark matter scenario）热暗物质图景
HD system　（＝Henry Draper system）HD 分类系统
He　（＝helium）氦（2 号元素）
He abundance　（＝helium abundance）氦丰度
head bands　彗头谱带
head of comet　彗头　又见：*cometary head; comet head*
head-on collision　正面碰撞 | [對] 正碰撞〔台〕
head-on cross-section　正截面
head-on encounter　正交会
head-tail galaxy　头尾星系 | 首尾星系〔台〕
head-tail structure　头尾结构 | 首尾結構〔台〕
HEAO　（＝High Energy Astronomical Observatory）高能天文台 | 高能天文衛星〔台〕
HEAO-2　（＝Einstein Observatory）爱因斯坦天文台
Heart　心宿（二十八宿）
HEAsoft　（＝High Energy Astrophysics Software）高能天体物理软件包
heat-absorbing glass　吸热玻璃
heat barrier　热障
heat capacity　热容
heat conduction　热导
heat content　热函　又见：*total heat*
heat convection　热对流

heat death 热寂
heat diffusion 热扩散 又见：thermal diffusion
heat dissipation 热耗散
heat exchange 热交换
heat flow 热流
heat flux 热流量
heat index （＝H.I.）热指数
heating rate 加热速率
heat instability 热不稳定性 又见：thermal instability
heat loss 热耗
heat-loss function 热耗函数
heat of crystallization 结晶热
heat of desorption 退吸热
heat radiation 热辐射｜總輻射，熱輻射〔台〕
又见：bolometric radiation
heat-resistant glass 耐热玻璃
heat source 热源 又见：thermal source
heat transfer 热转移
heat transfer coefficient 输热系数｜熱轉移係數〔台〕
heavenly stem 天干 又见：celestial stem
Heaviside crater 亥维赛环形山（月球）
Heaviside layer 亥维塞层｜海維賽層〔台〕 即：E layer
Heaviside step function 亥维赛阶梯函数
heavy element 重元素
heavy element star 重元素星
heavy lepton 重轻子
heavy meson 重介子
heavy metal star 重金属星
heavy nucleus 重核
heavy particle 重粒子
Heavy Snow 大雪（节气） 又见：Greater Snow
heavy weight star 特大质量星｜超大質量恆星〔台〕 又见：supermassive star
HEB （＝hot electron bolometer）热电子测辐射热计
Hebe 韶神星（小行星 6 号）
Hebei 河北（小行星 2505 号）
HEB mixer （＝superconducting hot electron bolometer mixer）超导 HEB 混频器
Hecataeus crater 赫卡泰奥斯环形山（月球）
hectometer wave 百米波｜百公尺波〔台〕
Hecuba 犬后星（小行星 108 号）
Hecuba gap 犬后星空隙
Hecuba group 犬后星群
Hedda 赫达（小行星 207 号）

Hédervári crater 海代尔瓦里环形山（月球）
hedgehog solution 刺猬解
hedgerow prominence 篱笆状日珥
Hedin crater 斯文赫定环形山（月球）
HEDLA （＝high energy density laboratory astrophysics）高能量密度实验室天体物理学
HED meteorite （＝howardite-eucrite-diogenite meteorite）HED 陨星，古铜钙长无球粒－钙长辉长－奥长古铜无球粒陨星
HEDP （＝high energy density physics）高能量密度物理学
Heeschen-Poskovski relationship 希申－博斯科夫斯基关系
Hegemone 木卫三十九
Heggie's law 赫吉定律
HEGRA （＝High-Energy-Gamma-Ray Astronomy telescope）高能伽马射线天文望远镜
height 高度，地平纬度
height above sea level 海拔
height equation 高差｜高度方程〔台〕
height finder 高度计，测高仪 又见：height indicator
height indicator 高度计，测高仪 又见：height finder
height of disappearance 消失点高度 又见：end height
heiligenschein 【德】灵光
Heilongjiang 黑龙江（小行星 2380 号）
Heinsius crater 海因修斯环形山（月球）
Heisenberg uncertainties principle 海森伯不确定性原理，海森伯测不准原理
Hekate 权神星（小行星 100 号）
Hektor 赫克托（小行星 624 号）
Helena 拐神星（小行星 101 号）
Helene 土卫十二
heliacal cycle 太阳周 又见：cycle of the sun
heliacal rising 偕日升，晨出 又见：acronical rising; acronychal rising
heliacal setting 偕日落，夕没｜偕日降，夕沒〔台〕 又见：acronical setting; acronychal setting
heliacal year 偕日升年
helical angle 螺旋角 又见：helix angle
helical antenna 螺旋天线
helical field 螺旋场
helical orbit 螺旋轨道 又见：spiral orbit
helical symmetry 螺旋对称｜螺旋對稱[性]〔台〕
helicity 螺旋性，螺度｜螺度〔台〕
helicon 螺旋波
Helike 木卫四十五

heliocentric angle 日心角
heliocentric coordinate 日心坐标
heliocentric coordinate network 日心坐标网 | 日面坐標網〔台〕
heliocentric coordinate system 日心坐标系
heliocentric correction 日心改正
heliocentric distance 日心距 | 日心距離〔台〕
heliocentric ephemeris 日心历表
heliocentric gravitational constant 日心引力常数
heliocentric Julian date （=HJD）日心儒略日
heliocentric latitude 日心纬度
heliocentric longitude 日心经度
heliocentric orbit 日心轨道
heliocentric parallax 日心视差
heliocentric phenomena 日心天象
heliocentric position 日心位置
heliocentric radial velocity 日心视向速度
heliocentric system 日心体系 | 日心[宇宙]體系〔台〕（宇宙体系）
heliocentric theory 日心说 | 日心[學]說〔台〕（宇宙学说）
heliocentric velocity 日心速度
heliogeophysics 太阳地球物理 | 太陽地球物理[學]〔台〕
heliogram 太阳照相图
heliograph 太阳照相仪 又见：*photoheliograph*
heliographic chart 日面图
heliographic coordinate 日面坐标
heliographic coordinate system 日面坐标系
heliographic distribution 日面分布
heliographic latitude 日面纬度 又见：*heliolatitude; solar latitude; sun's latitude*
heliographic longitude 日面经度
又见：*heliolongitude; sun's longitude*
heliographic pole 日面极
heliolatitude 日面纬度 又见：*heliographic latitude; solar latitude; sun's latitude*
heliolongitude 日面经度 又见：*heliographic longitude; sun's longitude*
heliomagnetosphere 日球磁层
heliometer 量日仪
heliometry 量日测量
heliomicrometer 太阳测微计
heliopause 日球层顶
Helios 太阳神（西德-美国太阳探测器）
helioscope 太阳目测镜 | 太陽目視[觀測]鏡〔台〕
helioseismology 日震学
heliosheath 日球层鞘

heliosphere 日球层 | 太陽圈，日光層〔台〕
heliostat 定日镜
heliotail 日球层尾
heliotrope 日光回照器，太阳反射仪
helium （=He）氦（2号元素）
helium abundance （=He abundance）氦丰度
helium burning 氦燃烧
helium content 氦含量
helium core 氦核
helium detonation 氦起爆
helium era 氦时期
helium flash 氦闪
helium ionization zone 氦电离区
helium Lyman α forest 氦莱曼α森林
helium main-sequence 氦主序
helium method 氦测法
helium-poor star 贫氦星
helium production 氦合成
helium-rich core 富氦核
helium-rich star 富氦星
helium shell 氦壳
helium shell flash 氦壳闪
helium star 氦星
helium-strong star 强氦星
helium variable 氦变星
helium-weak star 弱氦星
helium white dwarf 氦白矮星
helix angle 螺旋角 又见：*helical angle*
helix array 螺旋天线阵
helix feed 螺旋馈源
Helix Galaxy 螺旋星系（NGC 2685）
Helix Nebula 螺旋星云（NGC 7293）
Hellas Basin 希腊盆地（火星）
Hellas Planitia 希腊平原（火星）
helmet streamer 盔状流
Helmholtz contraction 亥姆霍兹收缩
Helmholtz contraction time 亥姆霍兹收缩时间
Helmholtz crater 亥姆霍兹环形山（月球）
Helmholtz free energy 亥姆霍兹自由能
Helmholtz-Kelvin contraction 亥姆霍兹－开尔文收缩
Helmholtz time scale 亥姆霍兹时标
hemispherical albedo 半球反照率
Hemmungspunkt 【德】爆炸点
HEMT amplifier （=high-electron-mobility transistor amplifier）高电子迁移率晶体管放大器
Henan 河南（小行星2085号）

Henbury crater 亨布里陨星坑 | 亨布里隕石坑〔台〕（地球）
Henon-Helies model 埃农－海利斯模型
Henry Draper catalogue （＝HD Catalogue）HD 星表
Henry Draper classification （＝HD classification）HD 恒星光谱分类　又见：*Henry Draper stellar classification*
Henry Draper Extension （＝HDE）HD 星表补编
Henry Draper stellar classification HD 恒星光谱分类　又见：*Henry Draper classification*
Henry Draper system （＝HD system）HD 分类系统
Henyey track 亨耶迹
HEOS （＝highly eccentric orbit satellite）大偏心轨道卫星
Hephaistos 冶神星（小行星 2212 号）
heptet 七重线　又见：*septet*
Her （＝Hercules）武仙座
Hera ① 后神星（小行星 103 号）② 木卫七（曾用名）　即：*Elara*
Heraclitus crater 赫拉克利特环形山（月球）
Herbig Ae/Be star 赫比格 Ae/Be 型星
Herbig Ae star 赫比格 Ae 型星
Herbig Be star 赫比格 Be 型星
Herbig emission-line star 赫比格发射线星
Herbig-Haro flow （＝HH flow）赫比格－阿罗流
Herbig-Haro nebula （＝HH nebula）赫比格－阿罗星云，HH 星云 | 赫比格－哈羅星雲，HH 星雲〔台〕
Herbig-Haro object （＝HH object）赫比格－阿罗天体，HH 天体 | 赫比格－哈羅天體，HH 天體〔台〕
Herbig-Haro shock wave 赫比格－阿罗激波
Herbig star 赫比格变星
Hercules ①（＝Her）武仙座 ② 北河三（双子座 β）　又见：*Pollux*
Hercules Cluster 武仙星系团
Hercules Ridge 武仙脊
Herculina 大力神星（小行星 532 号）
Hermes 使神星（小行星 69230 号）
Hermippe 木卫三十
Hermite crater 埃尔米特环形山（月球）
Hermitian matrix 厄密矩阵
Hernquist profile 赫恩奎斯特轮廓
herpolhode 空间极迹
herringbone structure 人字形结构

Herschel crater ① 赫歇尔环形山（月球）② 赫歇尔环形山（火星）③ 赫歇尔环形山（土卫一）
Herschel effect 赫歇尔效应
Herschel Gap 赫歇尔环缝（土星）
Herschelian telescope 赫歇尔望远镜 | 赫歇耳式望遠鏡〔台〕
Herschel infrared space telescope 赫歇尔红外空间望远镜 | 赫歇耳紅外[線]太空望遠鏡〔台〕　即：*HSO*
Herschel-Rigollet comet 赫歇尔－里戈莱彗星 | 赫歇耳－里格雷彗星〔台〕
Herschel Space Observatory （＝HSO）赫歇尔空间天文台
Herschel Telescope 赫歇尔望远镜
Herschel wedge 赫歇尔光劈
Herse 木卫五十
Hersilia 埃西莉亚（小行星 206 号）
Hertha 沃神星（小行星 135 号）
Hertha family 沃神星族
Hertz crater 赫兹环形山（月球）
Hertzsprung crater 赫茨普龙环形山（月球）
Hertzsprung gap 赫氏空隙
Hertzsprung-Russell diagram （＝HR diagram）赫罗图 | 赫羅圖，HR 圖〔台〕
Hesperia 夕神星（小行星 69 号）
Hesperus 长庚，昏星 | 長庚星〔台〕
　又见：*Vesper; Vesperus*
HESS （＝High Energy Stereoscopic System）高能立体视野望远镜
Hess crater 赫斯环形山（月球）
Hess diagram 赫斯[频数]图
HESSI （＝High Energy Solar Spectroscopy Imager）高能太阳光谱成像探测器
Hessian matrix 海森矩阵
Hestia ① 司祭星（小行星 46 号）② 木卫六（曾用名）　即：*Himalia*
Hestia gap 司祭星空隙
HET （＝Hobby-Eberly Telescope）霍比－埃伯利望远镜 | 哈比－艾柏利望遠鏡〔台〕（美国）
HETE （＝High Energy Transient Explorer）高能暂现源探测器 | 高能瞬變源探測器〔台〕
HETE 2 （＝High Energy Transient Explorer 2）高能暂现源探测器 2 号
hetegony 伴星起源说
heterochromatic magnitude 混色星等
heterochromatic photometry 混色测光 | 多色測光術〔台〕
heterodyne detector 外差检测器

heterodyne infrared interferometer 外差式红外干涉仪

heterodyne optical interferometer 外差式光学干涉仪

heterodyne oscillator 外差振荡器

heterodyne receiver 外差式接收机

heterodyne spectrometer 外差式频谱仪

heterogeneity 非均匀性

heterogeneous body 非均匀体

heterogeneous spherical earth 非均匀球状地球

heteronuclear molecule 异核分子

heteronucleus 异态核

heteroscian region 日影异向区

heterosphere 非均匀层

heterotic string 杂化弦

Hevelius crater 赫维留环形山（月球）

Hevelius formation 赫维留结构

hexahedrite 六面体陨铁 | 六面體 [式] 隕鐵〔台〕

hexapod mount 六杆支撑

hexapod mounting 六杆式装置

Heymans crater 海曼斯环形山（月球）

Heyuan 河源（小行星 3746 号）

Heze 角宿二（室女座 ζ）

Hf （＝hafnium）铪（72 号元素）

Hg （＝mercury）汞（80 号元素）

HgCdTe infrared array 碲镉汞红外面阵

H. G. Wells crater 赫·乔·威尔斯环形山（月球）

HH flow （＝Herbig-Haro flow）赫比格－阿罗流

HH nebula （＝Herbig-Haro nebula）赫比格－阿罗星云，HH 星云 | 赫比格－哈羅星雲，HH 星雲〔台〕

HH object （＝Herbig-Haro object）赫比格－阿罗天体，HH 天体 | 赫比格－哈羅天體，HH 天體〔台〕

H.I. （＝heat index）热指数

HI absorption 中性氢吸收

Hickson Compact Group （＝HCG）希克森致密星系群

Hickson compact group 希克森致密群

HI cloud 中性氢云

HI complex 中性氢复合体

HI content 中性氢含量

Hidalgo 希达尔戈 | 希達戈〔台〕（小行星 944 号）

hidden companion 隐伴星　又见：*invisible companion*

hidden magnetic flux 隐磁流

hidden mass 隐质量

hidden matter 隐物质

hidden object 隐天体

HI emission 中性氢发射

hierarchical clustering 等级式成团 | 階式成團〔台〕

hierarchical cosmology 等级式宇宙学 | 階式宇宙論〔台〕

hierarchical merging 等级式并合

hierarchical model 等级式模型 | 階式模型〔台〕

hierarchical structure 等级式结构 | 階式結構〔台〕

hierarchical structure formation 等级式结构形成

hierarchical universe 等级式宇宙 | 階式宇宙〔台〕

Higgs boson 希格斯玻色子

Higgs field 希格斯场

Higgs mechanism 希格斯机制

high-accuracy timing 高精度计时

high altitude cloud 高空云

High Altitude Observatory （＝HAO）高山天文台（美国）

high altitude orbit 高轨道

high altitude station 高山观测站

High-Altitude Water Cherenkov observatory （＝HAWC）高海拔水体切伦科夫天文台

high band filter 高通道滤波器

high contrast plate 高衬度底片 | 高對比底片〔台〕

high corona 高层日冕

high dispersion 高色散

high dispersion spectroscopy 高色散分光

high dispersion spectrum 高色散光谱
　又见：*highly dispersed spectrum*

high-earth orbit 远地轨道　又见：*high-orbit*

high-electron-mobility transistor amplifier （＝HEMT amplifier）高电子迁移率晶体管放大器

High Energy Astronomical Observatory （＝HEAO）高能天文台 | 高能天文衛星〔台〕

high energy astronomy 高能天文学

high energy astrophysics 高能天体物理学 | 高能天文物理學〔台〕

High Energy Astrophysics Software （＝HEAsoft）高能天体物理软件包

high-energy component 高能成分

high energy density laboratory astrophysics （＝HEDLA）高能量密度实验室天体物理学

high energy density physics （＝HEDP）高能量密度物理学

high-energy galaxy 高能星系

High-Energy-Gamma-Ray Astronomy telescope （＝HEGRA）高能伽马射线天文望远镜

high-energy particle 高能粒子 又见：*energetic particle*
high-energy process 高能过程
high-energy radiation 高能辐射
High Energy Solar Spectroscopy Imager （＝HESSI）高能太阳光谱成像探测器
High Energy Stereoscopic System （＝HESS）高能立体视野望远镜
high-energy tail 高能尾
High Energy Transient Explorer （＝HETE）高能暂现源探测器 | 高能瞬變源探測器〔台〕
High Energy Transient Explorer 2 （＝HETE 2）高能暂现源探测器 2 号
high-field pulsar 强磁场脉冲星
high-frequency radiation 高频辐射
high-intensity emission 高强度发射
high ion 高电离离子
highland 高地，高原
highland light plain 高地亮平原
high latitude 高纬度
high-latitude flare 高纬度耀斑 | 高緯 [度] 閃焰〔台〕
high-latitude prominence 高纬度日珥
high-latitude spot 高纬度黑子 | 高緯黑子〔台〕
high-luminosity star 高光度星 | 高光度 [恆] 星〔台〕 又见：*luminous star*
Highly Advanced Laboratory for Communications and Astronomy （＝HALCA）哈尔卡实验室（日本射电卫星）
highly dispersed remnant 高弥散遗迹
highly dispersed spectrum 高色散光谱 又见：*high dispersion spectrum*
highly eccentric orbit satellite （＝HEOS）大偏心轨道卫星
highly evolved object 演化晚期天体
highly evolved star 演化晚期星
highly ionized matter 高电离物质
highly polarized quasar （＝HPQ）高偏振类星体
high magnetic arcade （＝HMA）高磁拱
high-mass binary 大质量双星
high-mass planet 大质量行星
high-mass star 大质量星 又见：*massive star*
high-mass X-ray binary （＝HMXB）大质量 X 射线双星
high-metallicity cluster ①高金属度星团 | 高金屬豐度星團〔台〕②高金属度星系团 | 高金屬豐度星系團〔台〕
high-orbit 远地轨道 又见：*high-earth orbit*

high-orbit space telescope 远地轨道望远镜
high-order correlation function 高阶相关函数
high-order perturbation theory 高阶扰动理论
high-peak bias 高峰偏袒
high-power telescope 高倍率望远镜
High Precision Parallax Collecting Satellite （＝Hipparcos）依巴谷天文卫星 | 依巴谷衛星〔台〕
high redshift （＝high z）高红移
high-redshift galaxy 高红移星系
high resolution 高分辨率
high-resolution detector 高分辨检测器
High Resolution Fly's Eye cosmic ray detector （＝HiRes）高分辨率蝇眼宇宙线探测器
high-resolution spectrograph 高分辨摄谱仪
high-resolution spectroscopy 高分辨分光
high-resolution spectrum 高分辨光谱
high-resolution telescope 高分辨望远镜
high seismic-wave velocity 高地震波速度
high-speed camera 高速照相机
high-speed development 高速显影
high-speed encounter 高速交会
high-speed photometer 高速光度计
high-speed photometry 高速测光
high-speed solar wind 高速太阳风
high-speed spectroscopy 高速分光
high-speed telescope 强光力望远镜 又见：*fast telescope*
high-temperature flare 高温耀斑
high vacuum 高真空
high-velocity cloud （＝HVC）高速云
high-velocity dwarf 高速矮星
high-velocity object 高速天体
high-velocity star 高速星
high water 大潮 又见：*spring tide*
high z （＝high redshift）高红移
high-z object 高红移天体
Hi'iaka 妊卫一（矮行星 Haumea 卫星）
HII cloud 电离氢云 | 氫離子雲〔台〕
HII condensation 电离氢凝聚体
HII galaxy 电离氢星系 | 氫離子星系〔台〕
HII region 电离氢区 | HII 區，氫離子區〔台〕 又见：*ionized hydrogen region*
Hilbert crater 希尔伯特环形山（月球）
Hilbert space 希尔伯特空间
Hilbert transformation 希尔伯特变换
Hilda 希尔达 | 希耳達〔台〕（小行星 153 号）

Hilda asteroid 希尔达型小行星
Hilda group 希尔达群 | 希耳達群〔台〕
Hill-Brown theory 希尔－布朗理论 | 希耳－布朗理論〔台〕
Hill element 希尔根数 | 希耳根數〔台〕
Hill equation 希尔方程
Hill problem 希尔问题 | 希耳問題〔台〕
Hill's approximation 希尔近似
Hill sphere 希尔球
Hill stability 希尔稳定性 | 希耳穩定性〔台〕
hilltop inflation 坡顶暴胀
Hiltner-Hall effect 希尔特纳－霍尔效应
Himalia 木卫六 | 木衛六，希默利亞〔台〕
H index 氢指数
Hind's Crimson star （＝R Lep）欣德深红星
Hind's Nebula 欣德星云（NGC 1554）
Hind's Variable Nebula 欣德变星云（NGC 1554）
Hinode 日出号（日本太阳卫星） 又见: Solar-B
Hinotori 火鸟号（日本太阳探测器）
又见: Astro-A
H-ion 负氢离子 又见: negative hydrogen ion
HI Parkes All Sky Survey （＝HIPASS）帕克斯中性氢巡天
HIPASS （＝HI Parkes All Sky Survey）帕克斯中性氢巡天
Hipparchus crater 依巴谷环形山（月球）
Hipparcos （＝High Precision Parallax Collecting Satellite）依巴谷天文卫星 | 依巴谷衛星〔台〕
Hipparcos Catalogue 依巴谷星表
Hipparcos-Tycho Catalogue 依巴谷－第谷星表
Hippocrates crater 希波克拉底环形山（月球）
Hirayama crater 平山环形山（月球）
Hirayama family 平山族 | 平山 [家族] 分類〔台〕
HI region 中性氢区 | HI 區，氢原子區〔台〕
又见: neutral hydrogen zone; neutral hydrogen region
HiRes （＝High Resolution Fly's Eye cosmic ray detector）高分辨率蝇眼宇宙线探测器
Hisaki 火崎号（日本极紫外天文卫星）
又见: SPRINT-A
hiss 嘶声
histogram 直方图
history of astronomy 天文学史
HI stream 中性氢流
Hiten 飞天号（日本空间探测器）
HJD （＝heliocentric Julian date）日心儒略日
HLA （＝Hubble Legacy Archive）哈勃遗珍档案
H line 氢线，电离钙 H 线 | H 線〔台〕
HMA （＝high magnetic arcade）高磁拱

H magnitude H 星等
HMC （＝horizontal transit circle）水平子午环 | 地平 [式] 子午環〔台〕
HMXB （＝high-mass X-ray binary）大质量 X 射线双星
HN （＝hypernova）极超新星 | 特超新星〔台〕
Ho （＝holmium）钬（67 号元素）
Hoag's Object 霍格天体（PGC 54559）
Hoba meteorite 霍巴陨星 | 霍巴隕鐵〔台〕
Hoba West meteorite 西霍巴陨星
Hobby-Eberly Telescope （＝HET）霍比－埃伯利望远镜 | 哈比－艾柏利望遠鏡〔台〕（美国）
HOD （＝halo occupation distribution）暗晕占居数分布
hodoscope 描迹仪
Hoedus I 柱二（御夫座 ζ） 又见: Haedus
Hoedus II 柱三（御夫座 η）
Hoffmeister's Cloud 霍夫麦斯特星云
Hohhot 呼和浩特（小行星 72060 号）
Hohmann orbit 霍曼轨道
Hohmann transfer 霍曼转移
Hohmann transfer orbit 霍曼转移轨道
Holeungholee 何梁何利（小行星 4431 号）
Holmberg classification 霍姆伯格分类
Holmberg criterion 霍姆伯格判据
Holmberg radius 霍姆伯格半径 | 洪伯半徑〔台〕
holmium （＝Ho）钬（67 号元素）
holmium star 钬星
hologram 全息图
holograph 全息照相
holographic dark energy 全息暗能量
holographic grating 全息光栅
holographic principle 全息原理
holography 全息照相法 | 全像術〔台〕
holomorphic function 全纯函数
holonomic constraint 完整约束
Holtsmark approximation 霍茨马克近似
Holtsmark broadening 霍茨马克致宽
Homam 雷电一（飞马座 ζ）
Homestake experiment 霍姆斯特克实验
Hommel crater 霍梅尔环形山（月球）
homocentric sphere model 同心球模型
homochromatic photometry 同色测光
homodyne 零拍，零差
homoeoid 同形体
homoeoid theorem 同形体定理
homogeneity 均匀性

homogeneity of universe 宇宙均匀性
homogeneity problem 均匀性问题（宇宙学）
homogeneous atmosphere 均匀大气 | 均質大氣〔台〕
homogeneous coefficient 齐次系数
homogeneous field 均匀场
homogeneous light 单色光　又见：*monochromatic light*
homogeneous medium 均匀介质
homogeneous radiation 均匀辐射
homogeneous series 同质序列 | 均質序列〔台〕
homogeneous sphere 均匀球
homogeneous star 均匀星
homogeneous universe 均匀宇宙 | 均質宇宙〔台〕
homogenization 匀化
homographic solution 同形解
homologous deformation 保形变形 | 同調變形〔台〕
homologous design 保形设计
homologous flare 相似耀斑 | 相似閃焰〔台〕
homologous gaseous sphere 同模气体球
homologous radio burst 同系射电暴 | 同調電波爆發〔台〕
homologous star 同系星
homologous transformation 同调变换
　　又见：*homology transformation*
homology transformation 同调变换
　　又见：*homologous transformation*
homonuclear molecule 共核分子
homopause 同质层顶
homosphere 匀质大气
homothetic solution 位似解
homotopy group 同伦群
Homunculus Nebula 侏儒星云
honeycomb mirror 蜂巢式反射镜
Hong Kong 香港（小行星3297号）
Hooke crater 胡克环形山（月球）
Hooker Telescope 胡克望远镜
Hopf bifurcation 霍普夫分岔
Hopf-Cole substitution 霍普夫－科尔替换
Hopkins Observatory 霍普金斯天文台
Hopmann crater 霍普曼环形山（月球）
hop sky-wave 反射天波
Hor （＝Horologium）时钟座
horizon 视界
horizon circle 地平经仪

horizon coordinate system 地平坐标系
　　又见：*horizontal coordinate system*
horizon crossing 视界穿越
horizon dip 地平俯角　又见：*dip of the horizon*
horizon distance 视界距
horizon entry 视界进入点
horizon glass 水平镜　又见：*horizontal mirror*
horizon mass 视界质量
horizon of the universe 宇宙视界　又见：*universal horizon*
horizon plain 地平面　又见：*horizontal plane; ground plane*
horizon problem 视界问题
horizon ring 地平环，阴纬环，地浑
horizontal acceleration 水平加速度
horizontal angle 水平角
horizontal axis 水平轴
horizontal branch 水平支 | 水平分支〔台〕
horizontal-branch star 水平支恒星
horizontal circle ① 地平圈 ② 水平度盘
horizontal component 水平分量
horizontal coordinate 地平坐标
horizontal coordinate system 地平坐标系
　　又见：*horizon coordinate system*
horizontal meridian circle 水平子午环 | 地平[式]子午環〔台〕　又见：*horizontal transit circle*
horizontal mirror 水平镜　又见：*horizon glass*
horizontal mounting 地平式装置 | 地平[式]裝置〔台〕
horizontal parallax （＝HP）地平视差
horizontal pendulum 水平摆
horizontal plane ① 水平面 | 地平面，水平面〔台〕② 地平面　又见：*ground plane; horizon plain*
horizontal refraction 地平大气折射　又见：*geodetic refraction*
horizontal resonance 水平向共振
horizontal solar telescope 水平式太阳望远镜 | 水平[式]太陽望遠鏡〔台〕
horizontal spectrograph 水平式摄谱仪
horizontal sundial 水平式日晷 | 水平[式]日晷〔台〕
horizontal telescope 水平式望远镜 | 水平[式]望遠鏡〔台〕
horizontal thread 横丝
horizontal transit circle （＝HMC）水平子午环 | 地平[式]子午環〔台〕　又见：*horizontal meridian circle*

horizontal transit instrument 水平式中星仪 | 地平[式]中星儀〔台〕
horizontal zenith telescope 水平式天顶仪 | 水平[式]天頂儀〔台〕
Horn 角宿（二十八宿）
horn antenna 喇叭天线 | 喇叭[形]天線〔台〕
horn feed 喇叭馈源
horn-reflector antenna 喇叭式反射面天线
horologe ① 时计 又见: *chronometer; time piece* ② 钟表，日晷
Horologium (=Hor) 时钟座
horology 钟表学
horoscope 天宫图（占星）
Horrebow-Talcott method 赫瑞堡－太尔各特法 | 赫瑞鮑－太爾各特法〔台〕
Horsehead Nebula 马头星云（Barnard 33）
Horseshoe Nebula 马蹄星云（M17） 即: *Omega Nebula; ω Nebula*
horseshoe orbit 马蹄形轨道
horseshoe-shaped curve 马蹄形曲线
host galaxy 寄主星系 | 宿主星系〔台〕
host star 寄主星
host system 主系统
hot Big Bang 热大爆炸
hot Big Bang model 热大爆炸模型
hot component 热成分
hot component star 热子星
hot coronal gas 热冕气体
hot dark matter 热暗物质
hot dark matter scenario (=HDM scenario) 热暗物质图景
hot dwarf 热矮星
hot electron bolometer (=HEB) 热电子测辐射热计
hot Jupiter exoplanet 热类木星
hot pixel 热像元
hot R Coronae Borealis star 高温北冕 R 型星
hot relic particle 热遗迹粒子
hot spot 热斑 又见: *warm spot*
hot star 热星
hot subdwarf 热亚矮星
hot universe 热宇宙
hour 时 | 小時〔台〕
hour angle (=HA) 时角
hour-angle axis 时角轴，赤经轴 | 赤經軸〔台〕
hour-angle difference 时角差

hour circle ① 时圈 ② 时角度盘 ③ 赤经度盘
hour glass 沙漏 | 沙漏 [鐘]〔台〕 又见: *sand clock; sand glass; sandy clock*
Hourglass Nebula 沙漏星云
hour index 时标 又见: *time scale; time mark*
hourly motion 每时运动
hourly variation 每时变化
hour mark 时号 又见: *time signal*
Hour star 辰星
hour system 时法
House 营室，室宿 | 室宿〔台〕（二十八宿）又见: *Encampment*
house 宫 又见: *sign*
Houzeau crater 乌佐环形山（月球）
Howardite 古铜钙长无球粒陨石
howardite-eucrite-diogenite meteorite (=HED meteorite) HED 陨石，古铜钙长无球粒－钙长辉长－奥长古铜无球粒陨星
Hoyin 何贤（小行星 5045 号）
Hoyle-Narlikar cosmology 霍伊尔－纳里卡宇宙学 | 霍伊耳－納里卡宇宙學〔台〕
Hoyle-Narlikar theory 霍伊尔－纳里卡理论
HP ①（=Harvard Photometry）哈佛测光星表 | 哈佛恆星測光表〔台〕②（=Harvard pulsar）哈佛天文台脉冲星 ③（=horizontal parallax）地平视差
HPBW （=half-power beam width）半功率束宽
h Persei cluster 英仙 h 星团（NGC 869）
HPQ （=highly polarized quasar）高偏振类星体
HR diagram （=Hertzsprung-Russell diagram）赫罗图 | 赫羅圖，HR 圖〔台〕
HR number HR 星表序号
HSE （=hydrostatic equilibrium）流体静力平衡
HSO （=Herschel Space Observatory）赫歇尔空间天文台
HSR （=Harvard Standard Region）哈佛标准选区
HSRA （=Harvard-Smithsonian Reference Atmosphere）哈佛－史密松参考大气
HST （=Hubble Space Telescope）哈勃空间望远镜 | 哈柏太空遠鏡〔台〕
HS variable star （=Hubble-Sandage variable star）哈勃－桑德奇型变星，HS 型变星 | 哈柏－桑德奇型變星，HS 型變星〔台〕
Huangpu 黄浦（小行星 3502 号）
Huangrunqian 黄润乾（小行星 120569 号）
Huangshan 黄山（小行星 79316 号）
Huangsushu 黄授书（小行星 3014 号）
Hubble age 哈勃年龄 | 哈柏年齡〔台〕

Hubble Atlas of Galaxies 《哈勃星系图册》
Hubble classification 哈勃分类 | 哈柏分類〔台〕
Hubble constant 哈勃常数 | 哈柏常數〔台〕
Hubble crater 哈勃环形山（月球）
Hubble Deep Field （＝HDF）哈勃深场 | 哈柏深空區〔台〕
Hubble diagram 哈勃图 | 哈柏圖〔台〕
Hubble distance 哈勃距离 | 哈柏距離〔台〕
Hubble drag 哈勃拖曳
Hubble effect 哈勃效应
Hubble expansion 哈勃膨胀
Hubble flow 哈勃流 | 哈柏流〔台〕
Hubble Heritage Project 哈勃传承计划
Hubble horizon 哈勃视界
Hubble law 哈勃定律 | 哈柏定律〔台〕
又见：Hubble's law
Hubble Legacy Archive （＝HLA）哈勃遗珍档案
Hubble length 哈勃长度
Hubble Nebula 哈勃星云（NGC 2261）
又见：Hubble's Nebula
Hubble-Oemler law 哈勃－欧姆勒律
Hubble parameter 哈勃参数 | 哈柏參數〔台〕
Hubble radius 哈勃半径 | 哈柏半徑〔台〕
Hubble rate 哈勃率
Hubble relation 哈勃关系 | 哈柏關係〔台〕
Hubble-Reynolds law 哈勃－雷诺规律
Hubble-Sandage variable 哈勃－桑德奇型变星
Hubble-Sandage variable star （＝HS variable star）哈勃－桑德奇型变星，HS 型变星 | 哈柏－桑德奇型變星，HS 型變星〔台〕
Hubble's classification system 哈勃分类系统
Hubble sequence 哈勃序列 | 哈柏序列〔台〕
Hubble's law 哈勃定律 | 哈柏定律〔台〕
又见：Hubble law
Hubble's Nebula 哈勃星云（NGC 2261）
又见：Hubble Nebula
Hubble Space Telescope （＝HST）哈勃空间望远镜 | 哈柏太空望遠鏡〔台〕
Hubble sphere 哈勃球
Hubble stage 哈勃分类参数 | 哈柏分類參數〔台〕
Hubble's Variable Nebula 哈勃变光星云（NGC 2261）
Hubble test 哈勃检验
Hubble time 哈勃时间 | 哈柏時間〔台〕
Hubble Ultra Deep Field （＝HUDF）哈勃极深场 | 哈柏極深空區〔台〕
Hubble velocity 哈勃速度
Hubei 湖北（小行星 2547 号）

HUDF （＝Hubble Ultra Deep Field）哈勃极深场 | 哈柏極深空區〔台〕
Huggins crater 哈金斯环形山（月球）
Hugoniot curve 于戈尼奥曲线
Hugoniot equation 于戈尼奥方程
Hugoniot relation 于戈尼奥关系
Hugoniot shock 于戈尼奥激波
Huhunglick 胡鸿烈（小行星 34778 号）
Huichiming 许智明（小行星 5390 号）
Hulse-Taylor pulsar 赫尔斯－泰勒脉冲星
human eye 肉眼 又见：naked eye
human space flight 载人空间飞行 | 載人太空飛行〔台〕
Humason-Zwicky star （＝HZ star）HZ 型星，哈马森－兹威基型星 | HZ 星，哈馬遜－茲威基星〔台〕
Humboldt crater 洪堡环形山（月球）
Hume crater 休姆环形山（月球）
humidity 湿度 又见：degree of wetness
hump 驼峰
hump Cepheid 驼峰造父变星 | 拱峰造父變星〔台〕
Hunan 湖南（小行星 2592 号）
Hund's rule 洪德定则
Hungaria 匈牙利（小行星 434 号）
Hungaria group 匈牙利群
Hungarian Virtual Observatory （＝HVO）匈牙利虚拟天文台
Hun Kal crater 珲卡尔环形山（水星）
hunter's moon 狩月
Hurter-Driffield curve （＝HD curve）赫特－德里菲尔德曲线，HD 曲线
Husband Hill 赫斯本德山（火星）
Hutton crater 赫顿环形山（月球）
Huxley crater 赫胥黎环形山（月球）
Huya 雨神星（小行星 38628 号）
Huygenian eyepiece 惠更斯目镜 又见：Huygens eyepiece
Huygenian region 惠更斯区 | 惠岡區〔台〕
Huygenian telescope 惠更斯望远镜 又见：Huygens telescope
Huygens crater 惠更斯环形山（火星）
Huygens eyepiece 惠更斯目镜 又见：Huygenian eyepiece
Huygens Gap 惠更斯环缝（土星）
Huygens lander 惠更斯号着陆器（土卫六）
Huygens' principle 惠更斯原理
Huygens probe 惠更斯空间探测器 | 惠更斯號[探測器]〔台〕

Huygens telescope　惠更斯望远镜　又见：*Huygenian telescope*
HV　（＝Harvard variable）哈佛天文台变星
HVC　（＝high-velocity cloud）高速云
HVO　（＝Hungarian Virtual Observatory）匈牙利虚拟天文台
HWHM　（＝half width at half maximum）半峰半宽
HXMT　（＝Hard X-ray Modulation Telescope）硬 X 射线调制望远镜 | 硬 X 光調制望遠鏡〔台〕（中国天文卫星）
Hya　（＝Hydra）长蛇座
Hyad　毕团星
Hyades　毕星团 | 畢宿星團〔台〕
Hyades group　毕宿星群
Hyades supercluster　毕宿超级星团 | 畢宿超星團〔台〕
Hyadum I　毕宿四（金牛座 γ）　又见：*Prima Hyadum*
Hyadum II　毕宿三（金牛座 δ^1）
hybrid　混频环　又见：*hybrid ring*
hybrid beam optics　混合束光学
hybrid-chromosphere star　混合色球星
hybrid CMOS　混合型 CMOS
hybrid infrared array　混合型红外面阵
hybrid mapping　混合成图, 混合成像
hybrid ring　混频环　又见：*hybrid*
hybrid simulation　混合模拟
hybrid star　混合大气星
hybrid system　混杂系统
hybrid telescope　混合式望远镜
Hydra　① 冥卫三 ②（＝Hya）长蛇座
Hydra-Centaurus supercluster　长蛇－半人马超星系团
Hydra Cluster　长蛇星系团（Abell 1060）
Hydra Ridge　长蛇脊
hydrocarbon　烃
hydrodynamical simulation　流体动力学模拟
hydrodynamic instability　流体动力不稳定性
hydrodynamics　流体动力学　又见：*fluid dynamics*
hydrodynamic time scale　流体动力时标
hydrodynamic turbulence　流体动力湍动
hydrogen　（＝H）氢（1 号元素）
hydrogen abundance　氢丰度　又见：*H abundance*
hydrogen beam standard　氢束标准
hydrogen burning　氢燃烧
hydrogen clock　氢钟 | 氢 [原子] 鐘〔台〕
hydrogen cloud　氢云
hydrogen content　氢含量
hydrogen convection layer　氢对流层
hydrogen corona　氢冕
hydrogen cyanide　氰化氢（HCN）
hydrogen cycle　氢循环
hydrogen-deficient carbon star　贫氢富碳星
hydrogen-deficient donor　缺氢输质星
hydrogen-deficient star　贫氢星 | 貧氫恆星〔台〕
　　又见：*H deficient star; hydrogen-poor star*
hydrogen distribution　氢分布
hydrogen emission region　氢发射区
hydrogen flocculus　氢谱斑
hydrogen geocorona　氢地冕
hydrogen halo　氢晕
hydrogen ion　氢离子
hydrogen isocyanide　异氰化氢（HNC）
hydrogen-like atom　类氢原子
hydrogen line　氢线
hydrogen main sequence　氢主序
hydrogen maser　氢微波激射 | 氫邁射〔台〕
hydrogen-metal ratio　氢－金属比
hydrogen nebula　氢星云
hydrogenous atmosphere　氢型大气
hydrogen planet　氢行星
hydrogen-poor star　贫氢星 | 貧氫恆星〔台〕
　　又见：*hydrogen-deficient star; H deficient star*
hydrogen prominence　氢日珥
hydrogen recombination　氢复合
hydrogen recombination line　氢复合谱线
hydrogen region　氢区
hydrogen-rich donor　富氢输质星
hydrogen spectrum　氢光谱
hydrogen star　氢星
hydrogen sulfide　硫化氢（H_2S）
hydrokinetics　流体动理学
hydromagnetic bubble　磁流泡
hydromagnetic cavity　磁流穴
hydromagnetic dynamo　磁流发电机
　　又见：*hydromagnetodynamo*
hydromagnetic dynamo mechanism　磁流发电机机制　又见：*hydromagnetodynamo mechanism*
hydromagnetic instability　磁流不稳定性
hydromagnetics　磁流力学 | 磁流 [體] 力學〔台〕
hydromagnetic shock-wave　磁流体激波
hydromagnetic wave　磁流体波
hydromagnetodynamo　磁流发电机
　　又见：*hydromagnetic dynamo*

hydromagnetodynamo mechanism　磁流发电机机制　又见：*hydromagnetic dynamo mechanism*
hydrosphere　水圈，水界 | 水圈〔台〕
hydrostatic equilibrium　（＝HSE）流体静力平衡
hydrostatics　流体静力学
hydroxyl　羟基（OH）
hydroxyl maser　（＝OH maser）羟基微波激射 | 羟基迈射〔台〕
hydroxyl radical　（＝OH radical）羟基
Hydrus　（＝Hyi）水蛇座
Hygiea　健神星（小行星 10 号）
Hyginus crater　希吉努斯环形山（月球）
Hyi　（＝Hydrus）水蛇座
hyperactivity　超强活动
hyperbola　双曲线
hyperbolic comet　双曲线轨道彗星 | 雙曲線[軌道]彗星〔台〕
hyperbolic meteor　双曲线轨道流星 | 雙曲線流星〔台〕
hyperbolic orbit　双曲线轨道
hyperbolic space　双曲空间
hyperbolic velocity　双曲线速度
hyperboloid　双曲面 | 雙曲體〔台〕
hyperboloidal mirror　双曲面反射镜　又见：*hyperboloid mirror*
hyperboloidal reflector　双曲面反射望远镜　又见：*hyperboloid reflector*
hyperboloid mirror　双曲面反射镜　又见：*hyperboloidal mirror*
hyperboloid reflector　双曲面反射望远镜　又见：*hyperboloidal reflector*
hypercharge　超荷
hyperfine interaction　超精细相互作用
hyperfine line　超精细谱线
hyperfine resonance　超精细共振
hyperfine splitting　超精细分裂
hyperfine structure　超精细结构
hyperfine transition　超精细跃迁
hypergalaxy　超星系　又见：*supergalaxy*
hypergiant [star]　特超巨星
hypergolic propellant　自燃推进器
hypergranulation　超米粒组织　又见：*supergranulation*
hypergranule　超米粒 | 超米粒組織〔台〕　又见：*supergranule*
hypering　超化
Hyperion　土卫七
hypermetagalaxy　超总星系
hypernova　（＝HN）极超新星 | 特超新星〔台〕
hyperon　超子
hyperon star　超子星
hyperplane　超平面
hypersensitization　敏化 | 增感，敏化〔台〕　又见：*sensitization; hypersensitizing*
hypersensitizing　敏化 | 增感，敏化〔台〕　又见：*sensitization; hypersensitization*
hyperspace　多维空间　又见：*multidimensional space*
hypersphere　多维球 | 超球面〔台〕
hypersthene　紫苏辉石
hypersthene achondrite　紫苏无球粒陨石
hypersurface　超曲面
hypersurface of simultaneity　同时性超曲面
hypertelescope　超望远镜
hypervelocity accelerator　超高速加速器
hypervelocity impact　超高速碰撞
hypothesis　假说 | 假說，假設〔台〕
hypothetical parallax　理想视差
hypothetical planet　假想行星
Hyrrokkin　土卫四十四
HZ Herculis star　武仙 HZ 型星
HZ star　（＝Humason-Zwicky star）HZ 型星，哈马森－兹威基型星 | HZ 星，哈馬遜－茲威基星〔台〕
Hα emission-line star　Hα 发射线星
Hα photometry　Hα 测光
Hα spectroscopy　Hα 分光
Hα survey　Hα 巡天
Hβ photometry　Hβ 测光

I

I （=iodine）碘（53号元素）
IAF （=International Astronautical Federation）国际航天联合会
IAG （=International Association of Geodesy）国际测地协会
IAGA （=International Association of Geomagnetism and Aeronomy）国际地磁和高层大气物理协会
IAGC （=International Association of Geochemistry and Cosmochemistry）国际地球化学和宇宙化学协会
IAMAP （=International Association of Meteorology and Atmospheric Physics）国际气象和大气物理协会
Ianthe 佳女星（小行星98号）
Iapetus 土卫八
IASPEI （=International Association of Seismology and Physics of the Earth's Interior）国际地震和地球内部物理协会
IASY （=International Active Sun Year）国际活动太阳年
IATME （=International Association of Terrestrial Magnetism and Electricity）国际地磁和地电协会
IAU （=International Astronomical Union）国际天文学联合会
IAUC （=IAU Circular）IAU快报
IAU Circular （=IAUC）IAU快报
IAU galactic coordinate system IAU银道坐标系
IAU galactocentric distance IAU银心距
IAU Oort constants IAU奥尔特常数
IAU Planetary Photograph Center IAU行星照相中心
IAU System of Astronomical constants IAU天文常数系统
IAU Working Group for Future large scale facilities IAU今后大型设施工作组
IAU Working Group for Planetary system nomenclature IAU行星系命名工作组
IAU Working Group in History of astronomy IAU天文史工作组
IAU Working Group on Astronomiacl data IAU天文数据工作组
IAU Working Group on Astronomical standards IAU天文标准工作组
IAU Working Group on Comets IAU彗星工作组
IAU Working Group on Designations IAU命名工作组
IAU Working Group on Earth rotation in the HIPPARCOS reference frame IAU依巴谷参考架内地球自转工作组
IAU Working Group on Lunar based astronomy IAU月基天文工作组
IAU Working Group on Minor planets and Meteorites IAU小行星和陨星工作组
IAU Working Group on Near Earth objects IAU近地天体工作组
IAU Working Group on Parallax standards IAU视差标准星工作组
IAU Working Group on Peculiar stars IAU特殊星工作组
IAU Working Group on Planetary nebulae IAU行星状星云工作组
IAU Working Group on Reference frames IAU参考架工作组
IAU Working Group on Satellites IAU卫星工作组
IAU Working Group on Spectroscopic data archives IAU分光资料档案工作组
IAU Working Group on Supernovae IAU超新星工作组
IAU Working Group on Wide-field imaging IAU大视场成像工作组
IBC detector （=impurity band conduction detector）杂带导通探测器
IBEX （=Interstellar Boundary Explorer）星际边界探测者
Ibn Yunus crater 伊本·尤努斯环形山（月球）

IBVS （=Information Bulletin on Variable Stars）《变星快报》（匈牙利期刊）

IC （=Index Catalogue of Nebulae and Clusters of Stars）星云星团新总表续编 | 索引星表，IC 星表〔台〕

Icarus ①《伊卡洛斯》|《伊卡若斯》〔台〕（美国期刊）② 伊卡洛斯 | 伊卡若斯〔台〕（小行星 1566 号）

Icarus crater 伊卡洛斯环形山（月球）

ICE ①（=International Cometary Explorer）国际彗星探测器 ②（=inverse Compton effect）逆康普顿效应 | 逆康卜吞效應〔台〕

ice age 冰期 又见：*glacial period*

ice Cherenkov neutrino telescope 冰切伦科夫式中微子望远镜

IceCube Neutrino Observatory 冰立方中微子天文台

ice dwarf 冰矮天体

ice halo 冰晕

Iceland spar 冰洲石

ice-rich asteroid 富冰小行星

ice-rich mantle 富冰幔

ICET （=International Center of Earth Tide）国际固体潮中心

IC II （=Second Index Catalogue of Nebulae and Clusters of Stars）星云星团新总表续编 II

ICM ①（=intercloud matter）云际物质 ②（=intercloud medium）云际介质 | [星] 雲際介質〔台〕③（=intercluster matter）星团际物质 ④（=intercluster matter）星系团际物质 ⑤（=intracluster matter）星团内物质 ⑥（=intracluster matter）星系团内物质 ⑦（=intracluster medium）星团内介质 ⑧（=intracluster medium）星系团内介质

iconoscope 光电摄像管

ICRF （=International Celestial Reference Frame）国际天球参考架 | 國際天球參考坐標〔台〕

ICRS （=International Celestial Reference System）国际天球参考系

ICRS place ICRS 位置

icy comet 冰质彗星

icy conglomerate model 冰质团块模型 | 冰凍團塊模型〔台〕

icy-dwarf planet 冰质矮行星

icy-giant planet 冰质巨行星

icy planet 冰质行星

icy satellite 冰质卫星

IDA ①（=integrated diode array）集成二极管阵 ②（=International Dark-Sky Association）国际夜空保护协会

Ida 艾达（小行星 243 号）

ideal black-body 理想黑体

ideal coordinate 理想坐标

ideal gas 理想气体 又见：*perfect gas*

ideal plasma 理想等离子体

ideal receiver 理想接收机

ideal resonance 理想共振

identification 证认 | 識別〔台〕

identification chart 证认图 | 識別圖〔台〕

identified asteroid 已证认小行星

identified flying object （=IFO）已证认飞行物 | 已監定飛行物〔台〕

identified infrared source 已证认红外源

identified moving object 已证认移动天体

identified radio source 已证认射电源

identified satellite 已证认卫星

identified ultraviolet source 已证认紫外源

identified X-ray source 已证认 X 射线源

identified γ-ray source 已证认 γ 射线源

idiometer 人差仪

IDP （=interplanetary dust particle）行星际尘粒

IDS （=image-dissector scanner）析像扫描器

IEH （=International Extreme-UV Hitchhiker）国际极紫外飞行器（美国航天飞机载荷）

IERS （=International Earth Rotation and Reference Systems Service）国际地球自转服务

IFO （=identified flying object）已证认飞行物 | 已監定飛行物〔台〕

IFRB （=International Frequency Registration Board）国际频率登记委员会

IFU （=integral field unit）积分视场单元

IGC （=International Geophysical Cooperation）国际地球物理协作

IGM ①（=intergalactic matter）星系际物质 ②（=intergalactic medium）星系际介质

igneous rock 火成岩 又见：*pyrogenetic rock*

ignition temperature 点火温度

ignorable coordinate 可遗坐标

IGY （=International Geophysical Year）国际地球物理年

IHY （=International Heliophysical Year）国际太阳物理年

IIBAE （=International Information Bureau on Astronomical Ephemerides）国际天文历表信息局

Ijiraq 土卫二十二

Ikeya-Seki comet 池谷－关彗星 | 池谷－關[氏]彗星〔台〕

Ikeya-Zhang Comet 池谷－张彗星

ILE (＝Improved Lunar Ephemeris）改进月球历表
illuminance [光] 照度 | 照明度〔台〕
　　又见：*illumination; luminous flux density*
illuminated area 照亮面积
illuminated disk 照亮圆面
illuminated hemisphere 照亮半球 | 受亮半球〔台〕
illuminated zone 照亮区
illuminating source 施照源
illuminating star 施照星
illumination 光 照度 | 照明度〔台〕
　　又见：*illuminance; luminous flux density* ① 照明
illuminator 施照体，施照器
illuminometer 照度计
illustration 图例
ilmenite 钛铁
ILO (＝International Latitude Observatory）国际纬度台
ILS (＝International Latitude Service）国际纬度服务 | 國際緯度服務處〔台〕
image amplifier 像增强器 | 像加强器〔台〕
　　又见：*image intensifier*
image convertor 变像管
image deconvolution 图像去卷积，图像反卷积，图像解卷积
image degradation 图像劣化
image derotator 像消旋器 | 像消轉器〔台〕
image diameter 星像直径
image dissector 析像管　又见：*dissector*
image-dissector scanner (＝IDS）析像扫描器
image distortion 图像畸变 | 影像扭曲〔台〕
　　又见：*map distortion*
image field 像场
image formation 成像　又见：*formation of image; imagery; imaging*
image-forming device 成像器件
image frequency 像频，帧频
image intensifier 像增强器 | 像加强器〔台〕
　　又见：*image amplifier*
image intensifier-dissector 像增强-析像管
image isocon 分流直像管　又见：*isocon*
image orthicon 正析像管
image photometry 成像测光　即：*2-dimensional photometry*
image photon-counting system (＝IPCS）图像光子计数器，图像光子计数系统
image plane 像平面
image-plane beam combiner 像面合束器

image-plane interferometer 像面干涉仪
image plane scanning 像面扫描
image processing 图像处理 | 影像處理〔台〕
image reconstruction 图像重建 | 影像重建〔台〕
Image Reduction and Analysis Facility (＝IRAF）图像复原和分析软件 | 圖像復原和分析套件〔台〕
image restoration 图像复原 | 影像復原〔台〕
image rotator 旋像器
imagery 成像　又见：*image formation; formation of image; imaging*
image sensing 图像传感
image sharping 星像增锐
image-sharping telescope 星像增锐望远镜
image slicer 星像切分器 | 像切分儀〔台〕
image space 像方，像空间 | 像空間，像宇〔台〕
image spread 星像扩散度
image surface 像面
image synthesis 图像综合 | 影像合成〔台〕
image synthesis array 图像综合望远镜阵
image trailer 像迹仪
image tube 像管 | 電子[像]管〔台〕
image tube spectrograph 像管摄谱仪
imaginary axis 虚轴
imaginary quantity 虚量
imaginary source 假想源
imaging 成像　又见：*image formation; formation of image; imagery*
imaging Cherenkov telescope 成像切伦科夫望远镜
imaging-chip technology 星像切割技术
Imaging Fourier transform spectrometer 成像傅里叶变换光谱仪
imaging instrument 成像仪器
imaging photometer 成像光度计
imaging polarimetry 成像偏振测量
imaging polarimter 成像偏振计
imaging proportional counter (＝IPC）成像正比计数器
imaging spectrograph 成像摄谱仪
imaging spectrophotometry 成像分光光度测量
imaging spectroscopy 成像分光
I-magnitude Ⅰ星等
Imbrian System 雨海地层系统（月球）
Imbrium Basin 雨海盆地（月球）
IME (＝International Magnetospheric Explorer）国际磁层探测器

IMEX (=Inner Magnetosphere Explorer) 内磁层探测器（美国）
IMF ① (=initial mass function) 初始质量函数 | 初始質量[分佈]函數〔台〕② (=interplanetary magnetic field) 行星际磁场 ③ (=interstellar magnetic field) 星际磁场
imitator 模拟器
immersed echelle 浸渍阶梯光栅
immersed grating 浸没光栅 又见: *immersion grating*
immersion 掩始 又见: *disappearance*
immersion grating 浸没光栅 又见: *immersed grating*
immersion time 掩始时刻
IMP (=Interplanetary Monitoring Platform) 行星际监测站
impact basin 陨击盆地
impact broadening 碰撞致宽 又见: *collisional broadening; collision broadening*
impact crater 陨击坑（行星地貌）又见: *crater*
impact excitation 碰撞激发 又见: *collisional excitation*
impact fluorescence 碰撞致荧
impact hypothesis 碰撞假说 又见: *collision hypothesis*
impact ionization 碰撞电离 | 碰撞游離〔台〕 又见: *collisional ionization; collision ionization; ionization by collision*
impactite 陨击岩
impact lightflash 碰撞致闪
impact parameter 碰撞参数 又见: *collision parameter*
impact strength 碰撞力度
impact zone 碰撞带
impedance matching 阻抗匹配
imperfact absorption 非理想吸收
imperfect fluid 非理想流体
imperfect gas 非理想气体
imperfect scattering 非理想散射
imperial astronomer ① 司天监（宋元官职）② 太史令（秦汉官职）又见: *Astronomer Royal* ③ 钦天监监正（明清官职）又见: *director of imperial observatory*
imperial observatory 司天监（古代官署）
impersonal astrolabe 超人差等高仪
impersonal micrometer 超人差测微计
implicit function 隐函数
implosion 爆缩
importance of a flare 耀斑级别 | 閃焰級別〔台〕
imprisoned radiation 束缚辐射

Improved Lunar Ephemeris (=ILE) 改进月球历表
improvement of orbit 轨道改进 又见: *orbit improvement*
impulse 脉冲 又见: *pulse*
impulse approximation 脉冲近似
impulse counter 脉冲计数器 又见: *pulse counter*
impulse response 脉冲响应
impulsive burst 脉冲暴 | 脈衝爆發〔台〕
impulsive hard phase 脉冲急速相
impulsive phase 脉冲相
impulsive solar flare 脉冲太阳耀斑
impurity 杂质
impurity band conduction detector (=IBC detector) 杂带导通探测器
IMS (=International Magnetosphere Study) 国际磁层研究
In (=indium) 铟（49号元素）
in-between object 中介天体
incidence angle 入射角 又见: *incident angle; angle of incidence*
incidental prominence 偶现日珥
incident angle 入射角 又见: *incidence angle; angle of incidence*
incident beam 入射波束
incident direction 入射方向
incident intensity 入射强度
incident power 入射功率
incident wave 入射波
incipient star 早期恒星
inclination 倾角，交角，水平差 | 交角，傾角〔台〕 又见: *angle of inclination*
inclination of orbit 轨道倾角，轨道交角
inclination resonance 倾斜共振
inclinometer 磁倾计，倾斜仪
incoherent averager 非相干平均器
incoherent averaging 非相干平均
incoherent emission 非相干发射 又见: *non-coherent emission*
incoherent scattering 非相干散射 | 非同調散射〔台〕 又见: *non-coherent scattering*
incoming signal 入射信号
incoming trajectory 进入轨道
incommensurability 不可通约性 | 不可通約[性]〔台〕
incompressibility 不可压缩性
incompressible liquid 不可压缩流体
increasing wave 增长波

increment 增量
increscent moon 盈月　又见：*waxing moon*
Ind （＝Indus）印第安座
independent catalogue 独立星表
independent day number 独立日数
independent equatorial coordinate 独立赤道坐标
independent variable 独立变量
indeterminate principle 测不准原理
　　又见：*uncertainty principle*
index arm 指标臂
Index Catalogue of Nebulae and Clusters of Stars （＝IC）星云星团新总表续编｜索引星表，IC星表〔台〕
index correction 指标改正
index error 指标误差｜指標[誤]差〔台〕
index mirror 指标镜
index of inertia 惯性指数
index of refraction 折射率　又见：*refractive index*
index of rotation 旋转指数
Indian era 印度纪元
indicatometer 测光指示计
indicator 指示器
indicator-rod 漏箭
indicatrix 指示量
indicatrix of scattering 散射指示量
indiction 小纪（15年）　又见：*small cycle*
indifferent equilibrium 随遇平衡　又见：*neutral equilibrium*
indirect observation 间接观测
indium （＝In）铟（49号元素）
indium antimonide 锑化铟
individual error 人差｜人[為]差，個人誤差〔台〕
　　又见：*personal difference; personal equation; personal error*
individual-particle model 单粒子模型
individual plasma 孤立等离子体
indochinite 印支陨体
induced absorption 受迫吸收｜強迫吸收〔台〕
　　又见：*forced absorption*
induced combination 受迫组合
induced Compton scattering 受迫康普顿散射
induced dipole radiation 受迫偶极辐射
induced emission 受迫发射｜強迫發射，誘發發射〔台〕　又见：*forced emission*
induced gravity 诱导引力
induced metric 诱导度规
induced rate 诱发率
induced recombination 受迫复合｜誘發復合〔台〕
induced scattering 受迫散射
induced star formation 诱发恒星形成
induced symmetry-breaking 诱发对称性破缺
induced transition 受迫跃迁｜強迫躍遷，誘發躍遷〔台〕　又见：*forced transition*
induced γ-ray background 感生γ射线背景
induction acceleration 感应加速度
induction drag 感生阻力
induction-type acceleration mechanism 感应型加速机制
inductosyn 感应式传感器
Indus （＝Ind）印第安座
inelastic collision 非弹性碰撞　又见：*inelastic impact; non-elastic collision*
inelastic encounter 非弹性交会
inelastic impact 非弹性碰撞　又见：*inelastic collision; non-elastic collision*
inelastic scattering 非弹性散射
inequality 不等性
inertia force 惯性力
inertial confinement 惯性约束
inertial coordinate system 惯性坐标系
inertial field 惯性场
inertial flight 惯性飞行
inertial force of rotation 转动惯性力
inertial frame 惯性架
inertial guidance 惯性导航
inertial mass 惯性质量
inertial positioning system （＝IPS）惯性定位系统
inertial radio source coordinate system （＝IRCS）射电源惯性坐标系
inertial reference frame 惯性参考架
inertial reference system 惯性参考系
inertial surveying system （＝ISS）惯性勘测系统
inertial system 惯性系
inertial time 惯性时
infall velocity 降落速度
inference 推论
inferior conjunction 下合
inferior ecliptic limit 下食限
inferior planet 内行星｜地内行星〔台〕
　　又见：*interior planet*
inferior tide 下高潮
infinite conductivity 无限传导率｜無限導率〔台〕
Infinite Empty Space 宣夜说　又见：*theory of expounding appearance; theory of infinite heavens; theory of expounding appearance in the night sky*

infinite hierarchy model 无限层次模型
infinite hierarchy structure 无限层次结构
infinite hierarchy universe 无限层次宇宙
infinitesimal disturbance 微扰
infinitesimal increment 无限小增量
infinitesimal transformation 无限小变换
infinite space 无限空间
infinite universe 无限宇宙
infinite wave train 无限波列
inflation 暴胀
inflationary cosmological model 暴胀宇宙模型
inflationary epoch 暴胀期　又见：*inflationary era*
inflationary era 暴胀期　又见：*inflationary epoch*
inflationary phase 暴胀阶段
inflationary phase transition 暴胀相变
inflationary scenario 暴胀演化图像
inflationary universe 暴胀宇宙
inflation model 暴胀模型
inflation preheating 暴胀预热
inflation reheating 暴胀再热
inflaton 暴胀子
INFN （＝Istituto Nazionale di Fisica Nucleare）【意】国家核物理研究院（意大利）
Information Bulletin on Variable Stars （＝IBVS）《变星快报》（匈牙利期刊）
information capacity 信息容量　又见：*data capacity*
information processing 信息处理
information storage 信息存储
information transform 信息传输
infralateral arc 外侧晕弧
infrared （＝IR）红外
infrared albedo 红外反照率
infrared array 红外面阵
Infrared Astronomical Satellite （＝IRAS）红外天文卫星｜紅外 [線] 天文衛星〔台〕
infrared astronomy 红外天文学｜紅外 [線] 天文學〔台〕
infrared astrophysics 红外天体物理
infrared background 红外背景
infrared background radiation 红外背景辐射
infrared bolometer 红外测辐射热计
infrared camera 红外照相机
infrared CCD 红外 CCD
infrared cirrus 红外卷云
infrared color index 红外色指数｜紅外 [線] 色指數〔台〕
infrared corona 红外冕｜紅外 [線] 冕〔台〕

infrared counterpart 红外对应体
infrared detector 红外探测器
infrared divergence 红外发散
infrared emanation 红外辐射｜紅外 [線] 輻射〔台〕　又见：*infrared radiation*
infrared emanator 红外辐射源
infrared emission 红外发射
infrared excess 红外超｜紅外超量〔台〕
infrared-excess object 红外超天体｜紅外超量天體〔台〕
infrared flux 红外流量
infrared-flux method 红外流量法
infrared galaxy 红外星系｜紅外 [線] 星系〔台〕
infrared glow 红外天光
infrared helioseismology 红外日震学｜紅外 [線] 日震學〔台〕
infrared horizon sensor 红外地平仪
Infrared Imager/Spectrograph （＝IRIS）红外成像器/摄谱仪
infrared imaging 红外成像
infrared imaging spectroscopy 红外成像分光
infrared index 红外指数
infrared interference filter 红外干涉滤光片
infrared interferometer 红外干涉仪
infrared interferometry 红外干涉测量
infrared luminosity 红外光度
infrared magnitude 红外星等｜紅外 [線] 星等〔台〕
infrared modulation 红外调制
infrared object 红外天体｜紅外 [線] 天體〔台〕
infrared observatory 红外天文台
infrared photography 红外照相｜紅外 [線] 照相〔台〕
infrared photometer 红外光度计
Infrared Photometer/Spectrometer （＝IRPS）红外光度计/分光仪
infrared photometry 红外测光
infrared polarimetry 红外偏振测量
infrared radiation 红外辐射｜紅外 [線] 輻射〔台〕　又见：*infrared emanation*
infrared radiometer 红外辐射计
infrared radiometry 红外辐射测量
infrared solar radiation 太阳红外辐射　又见：*solar infrared radiation*
infrared source 红外源｜紅外 [線] 源〔台〕
infrared space astronomy 红外空间天文学
Infrared Space Observatory （＝ISO）红外空间天文台｜紅外 [線] 太空天文臺〔台〕

Infrared Spatial Interferometer （＝ISI）红外空间干涉仪
infrared spectrometer 红外分光仪
infrared spectrophotometry 红外分光光度测量
infrared spectroscopy 红外分光
infrared spectrum 红外光谱
infrared star 红外星 | 紅外 [線] 星〔台〕
infrared stellar radiation 恒星红外辐射
又见: *stellar infrared radiation*
infrared-submillimeter background 红外－亚毫米波背景
infrared sun 红外太阳 | 紅外 [線] 太陽〔台〕
infrared telescope 红外望远镜 | 紅外 [線] 望遠鏡〔台〕
Infrared Telescope in Space （＝IRTS）IRTS空间红外望远镜 | 紅外 [線] 太空望遠鏡〔台〕
infrared temperature 红外温度
infrared window 红外窗口
ING （＝Isaac Newton Group）牛顿望远镜群
InGaAs infrared array 铟镓砷红外面阵
ingenious armillary sphere 玲珑仪 又见: *ingenious planetarium*
ingenious planetarium 玲珑仪 又见: *ingenious armillary sphere*
Inghirami crater 因吉拉米环形山（月球）
Inglis-Teller limit 英格利斯－特勒极限
ingress 入凌，进食 | 初切〔台〕
inhomogeneity 不均匀性 又见: *non-uniformity; unevenness*
inhomogeneity scale 不均匀性尺度
inhomogeneous medium 不均匀介质
inhomogeneous nucleosynthesis 非均匀核合成
initial condition 初始条件
initial data 初始数据
initial earth 初始地球
initial luminosity function 初始光度函数
initial magnetic field 初始磁场
initial main sequence 初始主序
initial main-sequence star 初始主序星
initial mass distribution 初始质量分布
initial mass function （＝IMF）初始质量函数 | 初始質量 [分佈] 函數〔台〕
initial orbit 初始轨道，初轨 | 初 [始] 軌 [道]〔台〕
又见: *preliminary orbit; primitive orbit*
initial perturbation power spectrum 初始扰动功率谱
initial phase 初始阶段，初相
initial planet 初始行星

initial rise 初升
initial singularity 初始奇点
initial star 初始恒星
initial state 初态
initial sun 初始太阳
initial value 初值 又见: *starting value*
initial value problem 初值问题
initial zone 零时区 又见: *zero zone*
injected plasma 注入等离子体
injection source 注入源
inner bremsstrahlung 内轫致辐射
inner coma 内彗发
inner core 内核
inner corona 内冕 | 內日冕〔台〕
inner edge 内缘
inner electron 内层电子
inner halo cluster 内晕族星团
inner inner Lindblad resonance 内内林德布拉德共振
inner Lagrangian point 内拉格朗日点
inner Lindblad radius 内林德布拉德半径
inner Lindblad resonance 内林德布拉德共振
Inner Magnetosphere Explorer （＝IMEX）内磁层探测器（美国）
innermost electron 最内层电子
innermost planet 最内行星
inner planet 带内行星 | 內行星〔台〕
inner radiation zone 内辐射带
inner shell 内壳层
inner solar system 内太阳系
inner structure 内部结构 又见: *internal structure; internal constitution*
input signal 输入信号
input terminal 输入端
insertion loss 介入损耗
InSight （＝Interior Exploration using Seismic Investigations, Geodesy and Heat Transport）洞察火星任务
in situ acceleration 原位加速
in situ brightening 原位增亮
in situ observation 实地观测
in situ survey 实地巡测
insolation 日射 | 入日射〔台〕
instability 不稳定性 | 不穩定〔台〕
instability region 不稳定区
instability strip 不稳定带 又见: *instability zone*
instability zone 不稳定带 又见: *instability strip*

instantaneous elements	瞬时根数
instantaneous latitude	瞬时纬度
instantaneous longitude	瞬时经度
instantaneous pole	瞬时极
instantaneous power	瞬时功率
instantaneous recombination	瞬时复合
instantaneous recycling approximation	瞬时循环近似
instantaneous spectrum	瞬时光谱，瞬时谱
instantaneous value	瞬时值
instantaneous velocity	瞬时速度
instanton	瞬子
Institut de Radioastronomie Millimétrique	（＝IRAM）【法】毫米波射电天文所（法国）
instrumental azimuth	仪器方位 | 儀器方位角〔台〕
instrumental broadening	仪器致宽
instrumental constant	仪器常数
instrumental contour	仪器轮廓　又见：*instrumental profile*
instrumental correction	仪器改正
instrumental effect	仪器效应
instrumental error	仪器误差 | 儀器[誤]差〔台〕
instrumental latitude	仪器纬度
instrumental longitude	仪器经度
instrumental magnitude	仪器星等
instrumental polarization	仪器偏振
instrumental profile	仪器轮廓　又见：*instrumental contour*
instrumental refraction	仪器内大气折射 | 儀器内[大氣]折射〔台〕
instrumental response	仪器响应
instrumentation	附属仪器，仪器
instrument for solar and lunar eclipses	日月食仪
insulation	绝缘
insulator	绝缘体
INT	（＝Isaac Newton Telescope）INT 牛顿望远镜 | 牛頓望遠鏡〔台〕
integrability	可积性
integrable system	可积系统
INTEGRAL	（＝International Gamma-Ray Astrophysics Laboratory）国际 γ 射线天体物理实验室 | 國際 γ 射線天文物理實驗室〔台〕
integral atomic time	累积原子时
integral curve	积分曲线
integral field spectrograph	集成视场摄谱仪
integral field unit	（＝IFU）积分视场单元
integral linear polarization	累积线偏振
integral observatory	集成天文台
integral of motion	运动积分
integral scattering	积分散射
Integral Sign galaxy	积分号星系（UGC 3697）
integral sign warp	积分号式翘曲
integrated absorption coefficient	累积吸收系数
integrated blue magnitude	累积蓝星等
integrated brightness	累积亮度
integrated color index	累积色指数
integrated cross-section	积分截面
integrated diode array	（＝IDA）集成二极管阵
integrated Doppler	积分多普勒
integrated emission	累积发射
integrated flux	累积流量 | 累積通量〔台〕
integrated intensity	累积强度
integrated magnitude	累积星等
integrated noise	累积噪声
integrated noise temperature	累积噪声温度
integrated photoelectric magnitude	累积光电星等
integrated photographic magnitude	累积照相星等
integrated photonic spectrograph	集成光子学摄谱仪
integrated photovisual magnitude	累积仿视星等
integrated pulse	累积脉冲
integrated radiation	累积辐射
integrated red magnitude	累积红星等
integrated Sachs-Wolfe effect	（＝ISW effect）累积萨克斯－沃尔夫效应，ISW 效应
integrated sky brightness	累积天空亮度
integrated spectrum	累积光谱
integrating ionization chamber	累积电离室
integrating photometer	积分光度计
integration time	积分时间
intense burst	强暴　又见：*strong burst*
intense shock-wave	强激波
intense source	强源　又见：*strong source*
intensified CCD	增强 CCD
intensifier	增强器
intensity	强度　又见：*strength*
intensity distribution	强度分布
intensity interferometer	强度干涉仪
intensity interferometry	强度干涉测量
intensity mapping	强度映射
interacting binary	相互作用双星，相互作用双重星系 | 互作用雙星〔台〕
interacting close binary	相互作用密近双星
interacting galaxy	相互作用星系 | 互作用星系〔台〕

interacting prominence	互扰日珥
interaction	相互作用
interaction constant	相互作用常数
interarm object	臂际天体
interarm star	臂际星
interastral object	星际天体
intercalary cycle	闰周
intercalary day	闰日　又见：*leap day; bissextile day; epagomenal day*
intercalary month	闰月　又见：*leap month; intercalated month*
intercalary year	闰年　又见：*leap year; bissextile year; embolismic year*
intercalated month	闰月　又见：*leap month; intercalary month*
intercalation	置闰
intercept age	截距法年龄
interchange reaction	交换反应
intercloud extinction	云际消光
intercloud gas	云际气体｜[星] 雲際氣體〔台〕
intercloud matter	（＝ICM）云际物质
intercloud medium	（＝ICM）云际介质｜[星] 雲際介質〔台〕
intercloud object	云际天体
intercloud star	云际星
intercluster matter	①（＝ICM）星团际物质 ②（＝ICM）星系团际物质
intercombination	相互组合
intercombination line	互组谱线
intercombination transition	互组跃迁
intercomparison	相互比较
intercontinental synchronization	洲际同步
Intercosmos program	国际宇宙计划（苏联）　又见：*Interkosmos program*
Interface Region Imaging Spectrograph	（＝IRIS）过渡区成像摄谱仪（美国）
interference filter	①干涉滤光片｜干涉濾[光]鏡〔台〕②干涉滤波器
interference fringe	干涉条纹
interference pattern	干涉图样
interference spectroscopy	干涉分光
interference system	干涉系统
interferogram	干涉图
interferometer	干涉仪
interferometer pattern	干涉仪图样
interferometer phase	干涉仪相位
interferometer polar diagram	干涉极坐标方向图
interferometer radar	干涉雷达
interferometric astrometry	干涉天体测量
interferometric binary	干涉双星
interferometric nulling	干涉相消
interferometry	干涉测量｜干涉法〔台〕
interflare matter	耀斑际物质
intergalactic absorption	星系际吸收
intergalactic bridge	星系际桥
intergalactic cloud	星系际云
intergalactic dust	星系际尘埃
intergalactic extinction	星系际消光
intergalactic gas	星系际气体
intergalactic magnetic field	星系际磁场
intergalactic matter	（＝IGM）星系际物质
intergalactic medium	（＝IGM）星系际介质
intergalactic nebula	星系际星云
intergalactic object	星系际天体
intergalactic space	星系际空间
intergranular area	米粒间区
intergranular material	米粒际物质
intergration along the line of sight	视向积分
interior contact	内切
Interior Exploration using Seismic Investigations, Geodesy and Heat Transport	（＝InSight）洞察火星任务
interior ingress	内初凌
interior planet	内行星｜地内行星〔台〕　又见：*inferior planet*
interior solution	内解
Interkosmos program	国际宇宙计划（苏联）　又见：*Intercosmos program*
interline transfer	行间转移
interlocking	联锁作用｜連鎖[效應]〔台〕
interloper	窜入星
interlunation	无月期间
intermediary solution	过渡解
intermediate band	中波带
intermediate band photometry	中带测光　又见：*medium-band photometry*
intermediate boson	中间玻色子
intermediate component	中介子系
intermediate coordinate	中介坐标
intermediate corona	中介日冕｜居間日冕〔台〕
intermediate coupling	中介耦合｜居間耦合〔台〕
intermediate declination	中间赤纬
intermediate drift burst	中介漂暴
intermediate equator	中间赤道
intermediate group	子夜星组｜子夜[星]組〔台〕

intermediate ionization　中度电离
intermediate-mass black hole　中等质量黑洞
intermediate-mass star　中等质量恒星
intermediate orbit　中间轨道
intermediate place　中间位置
intermediate polar　中介偏振星　即：*DQ Her star*；又见：*intermediate polar system*
intermediate polar system　中介偏振星　即：*DQ Her star*；又见：*intermediate polar*
intermediate population　中介星族
intermediate-population star　中介星族恒星
intermediate right ascension　中间赤经
intermediate spectrum type　中介光谱型
intermediate subsystem　中介次系
intermediate-type star　中介光谱型恒星
intermediate vector boson　中间矢量玻色子
intermediate-velocity cloud　中速云
intermittency effect　间歇效应
intermittent region　断续区
intermolecular Stark effect　分子际斯塔克效应 | 分子際史塔克效應〔台〕
internal adjustment　内部平差
internal agreement　内部符合
internal constitution　内部结构　又见：*inner structure; internal structure*
internal energy　内能
internal Faraday rotation　内禀法拉第旋转
internal field　内场
internal flat field　内部平场
internal force　内力
internal memory　内存储器
internal metrology system　内计量系统
internal motion　内部运动
internal shock wave　内激波
internal shock wave model　内激波模型
internal structure　内部结构　又见：*inner structure; internal constitution*
internal temperature　内部温度
International Active Sun Year　（=IASY）国际活动太阳年
international angstrom　国际埃
International Association of Geochemistry and Cosmochemistry　（=IAGC）国际地球化学和宇宙化学协会
International Association of Geodesy　（=IAG）国际测地协会

International Association of Geomagnetism and Aeronomy　（=IAGA）国际地磁和高层大气物理协会
International Association of Meteorology and Atmospheric Physics　（=IAMAP）国际气象和大气物理协会
International Association of Seismology and Physics of the Earth's Interior　（=IASPEI）国际地震和地球内部物理协会
International Association of Terrestrial Magnetism and Electricity　（=IATME）国际地磁和地电协会
International Astronautical Federation　（=IAF）国际航天联合会
International Astronomical Union　（=IAU）国际天文学联合会
International Atomic Time　国际原子时
International Celestial Reference Frame　（=ICRF）国际天球参考架 | 國際天球參考坐標〔台〕
International Celestial Reference System　（=ICRS）国际天球参考系
International Center of Earth Tide　（=ICET）国际固体潮中心
international color index　国际色指数
International Cometary Explorer　（=ICE）国际彗星探测器
International Comet Explorer　国际彗星探测器
International Dark-Sky Association　（=IDA）国际夜空保护协会
international date line　国际变日线 | 國際換日線〔台〕
International Earth Rotation and Reference Systems Service　（=IERS）国际地球自转服务
international ellipsoid　国际椭球体
International Extreme-UV Hitchhiker　（=IEH）国际极紫外飞行器（美国航天飞机载荷）
International Frequency Registration Board　（=IFRB）国际频率登记委员会
International Gamma-Ray Astrophysics Laboratory　（=INTEGRAL）国际γ射线天体物理实验室 | 國際γ射線天文物理實驗室〔台〕
International Geophysical Cooperation　（=IGC）国际地球物理协作
International Geophysical Year　（=IGY）国际地球物理年
International Heliophysical Year　（=IHY）国际太阳物理年

International Information Bureau on Astronomical Ephemerides （=IIBAE）国际天文历表信息局

International Journal of Modern Physics D 《国际现代物理学杂志 D 辑》（新加坡期刊）

International Latitude Observatory （=ILO）国际纬度台

International Latitude Service （=ILS）国际纬度服务 | 國際緯度服務處〔台〕

International Latitude Station 国际纬度站

International Magnetosphere Study （=IMS）国际磁层研究

International Magnetospheric Explorer （=IME）国际磁层探测器

international magnitude system 国际星等系统

International Occultation Timing Association （=IOTA）国际掩食测时协会

International Polar Motion Service （=IPMS）国际极移服务 | 國際極移服務處〔台〕

International Polar Year （=IPY）国际极地年

International Quiet Sun Year （=IQSY）国际宁静太阳年

International Radio Consultative Committee （=CCIR）国际电信咨询委员会

International Rapid Latitude Service （=IRLS）国际纬度快速服务

International Satellite for Ionospheric Studies （=ISIS）国际电离层科学卫星

International Scientific Optical Network （=ISON）国际光学监测网

International Scientific Radio Union （=ISRU）国际无线电科学联合会

International Solar and Terrestrial Service （=ISTS）国际日地服务

International Space Station （=ISS）国际空间站 | 國際太空站〔台〕

International Sun-Earth Explorer （=ISEE）国际日地探测器

international sunspot number 国际太阳黑子数

international system of units （=SI）国际单位制

International Telecommunication Union （=ITU）国际电信联盟

International Terrestrial Reference Frame （=ITRF）国际地球参考架 | 國際地球參考坐標〔台〕

International Terrestrial Reference System （=ITRS）国际地球参考系

International Ultraviolet Explorer （=IUE）国际紫外探测器

International Union of Amateur Astronomers （=IUAA）国际天文爱好者联合会

International Union of Geodesy and Geophysics （=IUGG）国际测地和地球物理联合会

International Union of Radio Science （=URSI）国际无线电科学联合会

International Virtual Observatory Alliance （=IVOA）国际虚拟天文台联盟

International Year of Astronomy 2009 （=IYA2009）国际天文年 2009

International Years of the Quiet Sun （=IQSY）国际太阳宁静年

interparticle distance 粒子间距

interplanetary absorption 行星际吸收

interplanetary blast 行星际风暴

interplanetary dust 行星际尘埃

interplanetary dust particle （=IDP）行星际尘粒 又见: *interplanetary grain*

interplanetary exploration 行星际探索

interplanetary extinction 行星际消光

interplanetary flight 行星际飞行

interplanetary gas 行星际气体

interplanetary grain 行星际尘粒 又见: *interplanetary dust particle*

interplanetary light 行星际光

interplanetary magnetic field （=IMF）行星际磁场

interplanetary magnetic storm 行星际磁暴

interplanetary matter （=IPM）行星际物质

interplanetary medium （=IPM）行星际介质

interplanetary meteor 行星际流星

Interplanetary Monitoring Platform （=IMP）行星际监测站

interplanetary navigation 行星际航行

Interplanetary Network （=IPN）行星际观测网

interplanetary orbit 行星际轨道 又见: *interplanetary trajectory*

interplanetary particle 行星际粒子

interplanetary plasma 行星际等离子体

interplanetary probe 行星际探测器

interplanetary rocket 行星际火箭

interplanetary scattering （=IPS）行星际散射

interplanetary scintillation （=IPS）行星际闪烁

interplanetary sector 行星际扇形结构 | 行星際扇形區〔台〕

interplanetary space 行星际空间

interplanetary trajectory 行星际轨道 又见: *interplanetary orbit*

interplanetary transfer trajectory	行星际转移轨道
interpolating factor	内插因子
interpolation	内插丨内插 [法]〔台〕
interpolation polynomial	内插多项式
interpretation astronomy	通俗天文
interpulse	中介脉冲
interspicular region	针状物际区
interstellar absorption	星际吸收
interstellar absorption band	星际吸收带
interstellar absorption line	星际吸收线
interstellar band	星际谱带
Interstellar Boundary Explorer	(=IBEX) 星际边界探测者
interstellar bubble	星际泡
interstellar calcium	星际钙
interstellar chemistry	星际化学
interstellar cloud	星际云
interstellar communication	星际通信
interstellar complex	星际复合体
interstellar depletion	星际耗损
interstellar diffuse matter	星际弥漫物质
interstellar dust	星际尘埃
interstellar dust cloud	星际尘云
interstellar emission	星际发射
interstellar extinction	星际消光
interstellar field	星际场
interstellar gas	星际气体
interstellar gas cloud	星际气体云
interstellar gas-dust cloud	星际气体尘埃云
interstellar grain	星际尘粒
interstellar hydrogen	星际氢
interstellar hydrogen line	星际氢线
interstellar line	星际谱线丨星際[譜]線〔台〕
interstellar magnetic field	(=IMF) 星际磁场
interstellar maser	星际微波激射
interstellar matter	星际介质，星际物质 又见：*interstellar medium*
interstellar medium	(=ISM) 星际介质，星际物质 又见：*interstellar matter*
interstellar meteor	星际流星丨恆星際流星〔台〕
interstellar molecular cloud	星际分子云
interstellar molecular line	星际分子谱线
interstellar molecule	星际分子
interstellar parallax	星际视差
interstellar particle	星际粒子丨星際質點〔台〕
interstellar plasma	星际等离子体
interstellar polarization	星际偏振
interstellar radiation	星际辐射
interstellar radiation field	星际辐射场
interstellar reddening	星际红化
interstellar scintillation	(=ISS) 星际闪烁
interstellar space	星际空间丨[恆]星際空間〔台〕
interstellar stream	星际股流
interstellar wind	星际风
intervening galaxy	居间星系
intracluster gas	① 星团内气体 ② 星系团内气体
intracluster light	星系团内光
intracluster matter	①（=ICM）星团内物质 ②（=ICM）星系团内物质
intracluster medium	①（=ICM）星团内介质 ②（=ICM）星系团内介质
intraday variation	日内变化
intragalactic communication	银河系内通信
intra-Jovian planet	木内行星
intra-Mercurial planet	水内行星 又见：*intra-Mercurian planet*
intra-Mercurian planet	水内行星 又见：*intra-Mercurian planet*
intranetwork element	网内元
intranetwork field	网内场，网络内场丨網路內場〔台〕
intrapulse time	脉冲间歇时间
intrinsic accuracy	本征精度
intrinsic brightness	内禀亮度丨内稟亮度，本身亮度〔台〕
intrinsic color	本征颜色
intrinsic color index	本征色指数
intrinsic dispersion	内禀弥散度
intrinsic energy	内禀能
intrinsic luminosity	本征光度丨内稟光度，本身光度〔台〕
intrinsic magnetic moment	内禀磁矩
intrinsic magnitude	本征星等丨本身星等〔台〕
intrinsic redshift	内禀红移
intrinsic reproducibility	固有复制性丨固有重现性〔台〕
intrinsic temperature fluctuation	内禀温度涨落
intrinsic variable	本征变星
intrinsic variable star	内因变星
invariable plane	不变平面 又见：*invariant plane*
invariant curve	不变曲线
invariant distance	不变距离
invariant pendulum	定长摆
invariant plane	不变平面 又见：*invariable plane*

invariant point 不变点
invariant relation 不变关系
invariant surface 不变曲面
invasion 袭
inverse azimuth 反方位角 又见：*reverse azimuth*
inverse bremsstrahlung 逆韧致辐射 | 逆制動輻射〔台〕
inverse Compton effect （＝ICE）逆康普顿效应 | 逆康卜吞效應〔台〕
inverse Compton limit 逆康普顿极限
inverse Compton scattering 逆康普顿散射
inverse covariance matrix 逆协方差矩阵
inverse maser 逆微波激射
inverse maser effect 逆微波激射效应
inverse P Cygni profile 逆天鹅P型星谱线轮廓
inverse plasmon scattering 逆等离子体激元散射
inverse problem 逆问题
inverse square law 平方反比律
inverse Zeeman effect 逆塞曼效应
inversion 反演
inversion layer 反变层 又见：*reversing layer*
inversion transition 反演跃迁
inverted image 倒像
inverted pendulum 倒立摆
inverting eyepiece 倒像目镜
inverting prism 倒像棱镜 又见：*reversing prism*
inverting telescope 倒像望远镜
Investigator 参宿（二十八宿） 又见：*Triads; Three-stars*
invisible axion 隐轴子
invisible companion 隐伴星 又见：*hidden companion*
invisible matter 隐物质，不可见物质 | 不可見物質〔台〕
involution 对合
involutory transformation 对合变换
Io ①木卫一 ②犉神星（小行星85号）
Iocaste 木卫二十四
iodine （＝I）碘（53号元素）
iodine absorption cell 碘吸收池
Ioffe crater 约费环形山（月球）
ion 离子
ion-acoustic current 离子声流
ion-acoustic velocity 离子声速 又见：*iono-acoustic velocity; ion-sound speed*
ion-acoustic wave 离子声波，离子波 又见：*ion-sound wave*
ion composition 离子成分

ion-cyclotron cut-off 离子回旋截止
ion-cyclotron resonance 离子回旋共振
ion-cyclotron wave 离子回旋波
ion density 离子密度
ion-gyro-frequency 离子回旋频率
ionicity 电离度 | 游離度〔台〕 又见：*degree of ionization; ionizability; ionization degree*
ionic state 电离态 又见：*state of ionization*
ionizability 电离度 | 游離度〔台〕 又见：*degree of ionization; ionicity; ionization degree*
ionization 电离 | 游離〔台〕
ionization by collision 碰撞电离 | 碰撞游離〔台〕 又见：*collisional ionization; collision ionization; impact ionization*
ionization cross-section 电离截面
ionization degree 电离度 | 游離度〔台〕 又见：*degree of ionization; ionicity; ionizability*
ionization equilibrium 电离平衡
ionization fraction 电离度
ionization front 电离波前 | 游離界面〔台〕
ionization level 电离级 又见：*stage of ionization*
ionization loss 电离耗损
ionization parameter 电离参数
ionization potential 电离电势 | 游離電位，游離電勢〔台〕
ionization temperature 电离温度 | 游離溫度〔台〕
ionization zone 电离带
ionized hydrogen region 电离氢区 | HII區，氫離子區〔台〕 又见：*HII region*
ionizing background 电离背景
ionizing efficiency 电离效率 | 游離效應〔台〕
ionizing radiation 电离辐射
ionizing shock wave 电离激波
iono-acoustic velocity 离子声速 又见：*ion-acoustic velocity; ion-sound speed*
ionopause 电离层顶
ionosonde 电离层探测器 | 游離層探測器〔台〕
ionosphere 电离层 | 游離層〔台〕
ionosphere disturbance 电离层扰动
ionospheric eclipse 电离层食 | 游離層食〔台〕
ionospheric effect 电离层效应
ionospheric refraction 电离层折射
ionospheric refraction correction 电离层折射改正
ionospheric satellite 电离层探测卫星
ionospheric scintillation 电离层闪烁
ionospheric storm 电离层暴 | 游離層暴〔台〕
ion plasma frequency 离子等离子体频率
ion plasma oscillation 离子等离子体振荡

ion plasma wave	离子等离子体波
ion rocket	离子火箭
ion-sound solitary wave	离子声孤波
ion-sound speed	离子声速　又见: *ion-acoustic velocity; iono-acoustic velocity*
ion-sound turbulence	离子声湍动
ion-sound wave	离子声波，离子波　又见: *ion-acoustic wave*
ion spot	离子斑
ion tail	离子彗尾　又见: *plasma tail*
ion temperature	离子温度
ion trap	离子[陷]阱
ion wave	离子波
IOTA	(＝International Occultation Timing Association) 国际掩食测时协会
Iota Aquarid meteor shower	宝瓶ι流星雨
IPC	(＝imaging proportional counter) 成像正比计数器
IPCS	(＝image photon-counting system) 图像光子计数器，图像光子计数系统
IPHAS	(＝Isaac Newton Telescope Photometric H-alpha Survey) 牛顿望远镜Hα测光巡天
Iphigenia	祭神星（小行星112号）
IPM	① (＝interplanetary matter) 行星际物质 ② (＝interplanetary medium) 行星际介质
IPMS	(＝International Polar Motion Service) 国际极移服务	國際極移服務處〔台〕
IPN	(＝Interplanetary Network) 行星际观测网
IPS	① (＝inertial positioning system) 惯性定位系统 ② (＝interplanetary scattering) 行星际散射 ③ (＝interplanetary scintillation) 行星际闪烁
IPY	(＝International Polar Year) 国际极地年
IQSY	① (＝International Quiet Sun Year) 国际宁静太阳年 ② (＝International Years of the Quiet Sun) 国际太阳宁静年
IR	(＝infrared) 红外
Ir	(＝iridium) 铱（77号元素）
IRAF	(＝Image Reduction and Analysis Facility) 图像复原和分析软件	圖像復原和分析套件〔台〕
IRAM	(＝Institut de Radioastronomie Millimétrique)【法】毫米波射电天文所（法国）
IRAS	(＝Infrared Astronomical Satellite) 红外天文卫星	紅外[線]天文衛星〔台〕
iraser	红外激射
IRAS-Iraki-Alcock comet	IRAS－荒木－阿尔科克彗星
IRC	(＝Caltech Infrared Catalogue) 加州理工学院红外源表
IRCS	(＝inertial radio source coordinate system) 射电源惯性坐标系
IRC source	IRC红外源	IRC源〔台〕
Irene	司宁星（小行星14号）
iridium	(＝Ir) 铱（77号元素）
IRIS	① (＝Infrared Imager/Spectrograph) 红外成像器/摄谱仪 ② (＝Interface Region Imaging Spectrograph) 过渡区成像摄谱仪（美国）
Iris	① 虹神号	虹神星〔台〕（欧洲科学卫星） ② 虹神星（小行星7号）
iris aperture	可变孔径
iris diaphragm	可变光阑
iris diaphragm photometer	光瞳光度计　又见: *iris photometer*
iris microphotometer	光瞳测微光度计
Iris Nebula	彩虹星云（NGC 7023）
iris photometer	光瞳光度计　又见: *iris diaphragm photometer*
iris shutter	虹膜快门
IRLS	(＝International Rapid Latitude Service) 国际纬度快速服务
iron	(＝Fe) 铁（26号元素）
iron meteorite	铁陨星，铁陨石，陨铁	隕鐵，鐵[質]隕石〔台〕　又见: *aerosiderite; siderite; meteoric iron*
iron peak	铁峰
iron-peak element	铁峰元素
iron-poor star	贫铁星
iron-rich star	富铁星
iron star	铁星
IRPS	(＝Infrared Photometer/Spectrometer) 红外光度计/分光仪
irradiance	辐照度
irradiation	辐照
irreducible mass	不可约质量
irregular anagalactic nebula	不规则河外星云
irregular cluster	① 不规则星团 ② 不规则星系团　又见: *irregular cluster of galaxies*
irregular cluster of galaxies	不规则星系团　又见: *irregular cluster*
irregular galaxy	不规则星系　又见: *abnormal galaxy*
irregularity	不规则性
irregular nebula	不规则星云
irregular orbit	混沌轨道，不规则轨道	混沌軌道〔台〕　又见: *chaotic orbit*
irregular rotation	不规则自转
irregular satellite	不规则卫星
irregular variable	不规则变星

irreversibility 不可逆性 又见：non-reversibility
irreversible process 不可逆过程
Irr I galaxy I 型不规则星系
Irr II galaxy II 不规则型星系
irrotational motion 无旋运动
IRTF （＝NASA Infrared Telescope Facility）[美国] 航天局红外望远镜 | NASA 紅外望遠鏡〔台〕
irtron 红外光电管
IRTS （＝Infrared Telescope in Space）IRTS 空间红外望远镜 | 紅外 [線] 太空望遠鏡〔台〕
Isaac Newton Group （＝ING）牛顿望远镜群
Isaac Newton Telescope （＝INT）INT 牛顿望远镜 | 牛頓望遠鏡〔台〕
Isaac Newton Telescope Photometric H-alpha Survey （＝IPHAS）牛顿望远镜 Hα 测光巡天
Isaev crater 伊萨耶夫环形山（月球）
ISEE （＝International Sun-Earth Explorer）国际日地探测器
isenthalpic curve 等焓线
isentropic change 等熵变化
isentropic curve 等熵线
isentropic gas 等熵气体
isentropic initial condition 等熵初始条件
isentropic perturbation 等熵扰动
Ishtar Terra 伊斯塔高地（金星）
ISI （＝Infrared Spatial Interferometer）红外空间干涉仪
Isidis Planitia 伊希地平原（火星）
ISIS （＝International Satellite for Ionospheric Studies）国际电离层科学卫星
Isis 育神星（小行星 42 号）
islamic calendar 伊斯兰历 | 回曆〔台〕
 又见：Mohammedan calendar
island universe 岛宇宙
ISM （＝interstellar medium）星际介质，星际物质
ISO （＝Infrared Space Observatory）红外空间天文台 | 紅外 [線] 太空天文台〔台〕
isobar 等压线 又见：isopiestic; isostatics
isochoric change 恒容变化
isochron 等龄线 | 等時線〔台〕 又见：isochrone
isochron age 等时线法年龄
isochronal line 等时线 又见：synchrone
isochrone 等龄线 | 等時線〔台〕 又见：isochron
isochrone method 等龄法
isochrone potential 等时势
isochronism 等龄性，等时性
isochronous correspondence 等时对应

isoclinal line 等倾线 又见：isocline
isocline 等倾线 又见：isoclinal line
isocon 分流直像管 又见：image isocon
isocurvature 等曲率
isocurvature CDM model 等曲率冷暗物质模型
isocurvature fluctuation 等曲率涨落
isocurvature initial condition 等曲率初条件
isocurvature model 等曲率模型
isocurvature perturbation 等曲率扰动
isocyanic acid 异氰酸（HNCO）
isodynamic line 等力线 又见：syndyname
isoelectronic sequence 等电子序 | 等電子序列〔台〕
isoelectronic spectrum 等电子谱
isoenergetic displacement 等能位移
isoenergetic stability 等能稳定性
isogauss contour 等高斯轮廓
isogonic line 等偏线 | 等 [磁] 偏線〔台〕
isolated burst 孤立暴
isolated galaxy 孤立星系
isolated intergalactic cloud 星系际孤立云
isolated star 孤立星
isolated system 孤立系
isolating integral 孤立积分
isolator 隔离器，单向器
isomer 同分异构体 | 異構物〔台〕
isometric latitude 等量纬度
isometry group 等长群
ISON （＝International Scientific Optical Network）国际光学监测网
Isonoe 木卫二十六
isoperiodic orbit 等周期轨道
isophase 等相线
isophotal radius 等光度半径
isophotal wavelength 等光波长
isophote 等照度线 | 等光強線〔台〕
 又见：isophotiur
isophote ellipticity 等照度线椭率
isophote twisting 等照度线扭曲
isophotic contour 等照度轮廓
isophotiur 等照度线 | 等光強線〔台〕
 又见：isophote
isophotometer 等光度计
isophotometric atlas 等光度图
isophotometry 等光度测量
isopiestic 等压线 又见：isostatics; isobar
isoplanatic angle 等晕角

isoplanatic path angle 等晕斑角
isopleth 等值线
isopycnal 等密线
isopycnic 等密面
isorotation 共转
isosceles triangular solution 等腰三角形解
isospin 同位旋 | 同重旋〔台〕 又见：*isotopic spin*
isostasy 均衡 | 地壳均衡〔台〕
isostatics 等压线 又见：*isopiestic; isobar*
isotach 等速线
isotherm 等温线
isothermal atmosphere 等温大气
isothermal core 等温核 | 等温核心〔台〕
isothermal equilibrium 等温平衡
isothermal gas 等温气体
isothermal gas sphere 等温气体球
isothermal jump 等温跳变
isothermal perturbations 等温扰动
isothermal plasma 等温等离子体
isothermal process 等温过程
isothermal region 等温区
isothermal sheet 等温片
isothermal sound speed 等温声速
isothermal sphere 等温球
isothermal theory 等温理论
isothiocyanic acid 硫代异氰酸（HNCS）
isotone 同中子异核素
isotope 同位素
isotope age 同位素年龄
isotope assay 同位素验定
isotope effect 同位素效应
isotope frequency 同位素频率
isotopic abundance 同位素丰度
isotopic composition 同位素成分
isotopic content 同位素含量
isotopic dating 同位素纪年 | 同位素定年〔台〕
isotopic spin 同位旋 | 同重旋〔台〕 又见：*isospin*
isotropic antenna 各向同性天线
isotropic body 各向同性体
isotropic conductivity 各向同性传导率
isotropic correlation function 各向同性相关函数
isotropic distribution 各向同性分布
isotropic medium 各向同性介质
isotropic power spectrum 各向同性功率谱
isotropic propagation 各向同性传播
isotropic scattering 各向同性散射
isotropic tensor 各向同性张量
isotropic universe 各向同性宇宙
isotropic velocity dispersion 各向同性速度弥散度
isotropization 各向同性化
isotropy 各向同性
Israel-Robinson theorem 伊斯雷尔－鲁宾逊定理 | 伊色列－羅賓森定理〔台〕
ISRU （=International Scientific Radio Union）国际无线电科学联合会
ISS ①（=International Space Station）国际空间站 | 國際太空站〔台〕②（=inertial surveying system）惯性勘测系统 ③（=interstellar scintillation）星际闪烁 ④（=scattering by interstellar media）星际散射
Istituto Nazionale di Fisica Nucleare （=INFN）【意】国家核物理研究院（意大利）
ISTS （=International Solar and Terrestrial Service）国际日地服务
ISW effect （=integrated Sachs-Wolfe effect）累积萨克斯－沃尔夫效应，ISW 效应
Italian Virtual Observatory （=Vobs.it）意大利虚拟天文台
iteration 迭代
iterative method 迭代法 | 叠代渐近法〔台〕
Ithaca Chasma 伊萨卡深谷（土卫三）
Itokawa 丝川，糸川（小行星 25143 号）
ITRF （=International Terrestrial Reference Frame）国际地球参考架 | 國際地球參考坐標〔台〕
ITRS （=International Terrestrial Reference System）国际地球参考系
ITU （=International Telecommunication Union）国际电信联盟
IUAA （=International Union of Amateur Astronomers）国际天文爱好者联合会
IUE （=International Ultraviolet Explorer）国际紫外探测器
IUGG （=International Union of Geodesy and Geophysics）国际测地和地球物理联合会
IVOA （=International Virtual Observatory Alliance）国际虚拟天文台联盟
Ixion 伊克西翁（小行星 28978 号）
IYA2009 （=International Year of Astronomy 2009）国际天文年 2009
Izar 梗河一（牧夫座 ε） 又见：*Mirach; Mirak; Mirrak; Pulcherrima*

J

Jabbah 键闭（天蝎座 ν）
Jabhat al Akrab 钩铃一和钩铃二（天蝎座 ω）
JAC （＝Joint Astronomy Centre）联合天文中心（夏威夷）
jackknife resampling 刀切法复采样
Jackson crater 杰克逊环形山（月球）
Jacobi constant 雅可比常数
Jacobi crater 雅可比环形山（月球）
Jacobi ellipsoid 雅可比椭球 | 亞可比橢圓體〔台〕
Jacobi identity 雅可比恒等式
Jacobi limit 雅可比极限
Jacobi's integral 雅可比积分 | 亞可比積分〔台〕
Jacobi-type bar 雅可比型棒
Jacobus Kapteyn Telescope （＝JKT）卡普坦望远镜
Jaffe profile 贾菲轮廓
Jain calendar 耆那历（印度古历）
James Clerk Maxwell Telescope （＝JCMT）麦克斯韦望远镜 | 馬克斯威 [次毫米波] 望遠鏡〔台〕
James Webb Space Telescope （＝JWST）韦布空间望远镜 | 韋柏太空望遠鏡〔台〕
jansky （＝Jy）央 | 顏 [斯基]〔台〕（天体射电流量密度单位）
Jansky crater 央斯基环形山（月球）
Jansky noise 央斯基噪声
Janssen crater 让桑环形山（月球）
Janus 土卫十
Japan Aerospace Exploration Agency （＝JAXA）日本航天局（日本宇宙航空研究开发机构）
Japanese era 日本纪元
Japanese Virtual Observatory （＝JVO）日本虚拟天文台
Japan Standard Time （＝JST）日本标准时
Jarnsaxa 土卫五十
Jason crater 伊阿宋环形山（土卫九）
JAXA （＝Japan Aerospace Exploration Agency）日本航天局（日本宇宙航空研究开发机构）

JBIS （＝Journal of The British Interplanetary Society）《英国行星际学会志》（英国期刊）
JCMT （＝James Clerk Maxwell Telescope）麦克斯韦望远镜 | 馬克斯威 [次毫米波] 望遠鏡〔台〕
JD （＝Julian date）儒略日期
JDEM （＝Joint Dark Energy Mission）联合暗能量任务
Jeanne crater 珍妮环形山（金星）
Jeans crater 金斯环形山（月球）
Jeans criterion 金斯判据
Jeans equation 金斯方程
Jeans escape formula 金斯逃逸公式
Jeans instability 金斯不稳定性
Jeans length 金斯长度
Jeans mass 金斯质量
Jeans model 金斯模型
Jeans rate 金斯速率
Jeans spheroid 金斯球体
Jeans swindle 金斯假说
Jeans theorem 金斯定理
Jeans velocity 金斯速度
Jeans wavelength 金斯波长
Jeans wavenumber 金斯波数
JED （＝Julian ephemeris date）儒略历书日期
Jed Prior 梁，天市右垣九（蛇夫座 δ） 又见：Yed Prior
Jeffreys Gap 杰弗里斯环缝（土星）
Jenner crater 詹纳环形山（月球）
jet 喷流
jet galaxy 喷流星系
jet-like feature 喷流形特征
jet-like structure 喷流形结构 又见：jetlike structure
jetlike structure 喷流形结构 又见：jet-like structure
Jet Propulsion Laboratory （＝JPL）喷气推进实验室 | 噴射推進實驗室〔台〕
jet stream 急流（大气）
jet terminal shock 喷流终点激波

Jewel Box　宝盒星团（NGC 4755）
Jewish calendar　犹太历
Jewish era　犹太纪元
J. Herschel crater　约·赫歇尔环形山（月球）
Jhongda　中大（小行星 145534 号）
Jiangxi　江西（小行星 2617 号）
Jia-xiang　张家祥（小行星 4760 号）
Jilin　吉林（小行星 2398 号）
Jiling meteorite　吉林陨星｜吉林隕石〔台〕
　又见：*Jilin meteorite; Kirin meteorite*
Jilin meteorite　吉林陨星｜吉林隕石〔台〕
　又见：*Jiling meteorite; Kirin meteorite*
JIMO　（＝Jupiter Icy Moons Orbiter）冰质木卫轨道飞行器
Jinxiuzhonghua　锦绣中华（小行星 3088 号）
Jinyilian　金怡濂（小行星 100434 号）
Jinyong　金庸（小行星 10930 号）
jitter　颤动
j-j coupling　j－j 耦合
JKT　（＝Jacobus Kapteyn Telescope）卡普坦望远镜
J-magnitude　J 星等
Job's Coffin　乔布之棺（海豚座星组）
Jodrell Bank Pulsar　（＝JP）焦德雷班克脉冲星
Johanna　约翰娜（小行星 127 号）
Johnson color system　约翰逊颜色系统
Johnson-Morgan photometric system　约翰逊－摩根测光系统
Johnson-Morgan photometry　约翰逊－摩根测光｜強生－摩根測光〔台〕
Johnson-Morgan system　约翰逊－摩根系统｜強生－摩根系統〔台〕
Johnson noise　约翰逊噪声｜強生雜訊〔台〕
Johnson photometry　约翰逊测光
Johnson Space Center　（＝JSC）约翰逊空间中心｜詹森太空中心〔台〕
Joint Astronomy Centre　（＝JAC）联合天文中心（夏威夷）
Joint Dark Energy Mission　（＝JDEM）联合暗能量任务
Joint Organization of Solar Observation　（＝JOSO）太阳联合观测组织｜JOSO[太陽聯合觀測組織]〔台〕
Joliot crater　约里奥环形山（月球）
Jordan elementary time　约旦单元时间
Josephson effect　约瑟夫森效应
Josephson junction mixer　约瑟夫森结混频器

JOSO　（＝Joint Organization of Solar Observation）太阳联合观测组织｜JOSO[太陽聯合觀測組織]〔台〕
Joule crater　焦耳环形山（月球）
Joule dissipation　焦耳耗散
Joule energy　焦耳能
Joule heat　焦耳热
Joule heating　焦耳致热
Journal of Astrophysics and Astronomy　《天体物理学与天文学杂志》（美国期刊）
Journal of Cosmology and Astroparticle Physics　《宇宙学与天体粒子物理学学报》（英国期刊）
Journal of Geophysical Research　《地球物理学研究杂志》（美国期刊）
Journal of the BIS　（＝Journal of The British Interplanetary Society）《英国行星际学会志》（英国期刊）
Journal of The British Interplanetary Society　（＝JBIS; Journal of the BIS）《英国行星际学会志》（英国期刊）
Journal of The Korean Astronomical Society　《韩国天文学会志》（韩国期刊）
Jovian atmosphere　木星大气　又见：*Jupiter atmosphere*
Jovian burst　木星暴｜木星爆〔台〕　又见：*Jupiter burst*
Jovian exoplanet　系外类木行星｜[太陽] 系外類木行星〔台〕　又见：*exo-Jupiter; Jupiter-like exoplanet*
Jovian family　木族　又见：*Jupiter's family*
Jovian ionosphere　木星电离层　又见：*Jupiter ionosphere*
Jovian magnetosphere　木星磁层　又见：*Jupiter magnetosphere*
Jovian planet　类木行星
Jovian plasmasphere　木星等离子层　又见：*Jupiter plasmasphere*
Jovian radiation belt　木星辐射带　又见：*Jupiter radiation belt*
Jovian radio burst　木星射电暴　又见：*Jupiter radio burst*
Jovian radio radiation　木星射电　又见：*Jupiter radio radiation*
Jovian ring　木星环　又见：*Jupiter's ring; ring of Jupiter*
Jovian ringlet　木星细环　又见：*Jupiter's ringlet*
Jovian satellite　木卫｜木[星]衛[星]〔台〕
　又见：*Jupiter's satellite; satellite of Jupiter*
Jovian seismology　木震学
jovicentric coordinate　木心坐标｜木星[中心]坐標〔台〕　又见：*zenocentric coordinate*
jovicentric latitude　木心纬度

jovicentric longitude　木心经度
jovicentric orbit　木心轨道
jovigraphic coordinate　木面坐标 | 木 [星表] 面坐標〔台〕　又见：zenographic coordinate
jovigraphic latitudem　木面纬度　又见：zenographic latitude
jovigraphic longitude　木面经度　又见：zenographic longitude
JP　① （＝Jodrell Bank Pulsar）焦德雷班克脉冲星　② （＝Julian period）儒略周期
JPL　（＝Jet Propulsion Laboratory）喷气推进实验室 | 噴射推進實驗室〔台〕
JSC　（＝Johnson Space Center）约翰逊空间中心 | 詹森太空中心〔台〕
JST　（＝Japan Standard Time）日本标准时
J-type star　J 型星
Juewa　九华星（小行星 139 号）
JUICE　（＝Jupiter Icy moons Explorer）冰质木卫探测器
Jules Verne crater　儒勒·凡尔纳环形山（月球）
Julia　姝女星（小行星 89 号）
Julian calendar　儒略历
Julian century　儒略世纪
Julian date　（＝JD）儒略日期
Julian day　儒略日
Julian day calendar　儒略日历
Julian day number　儒略日数　又见：Julian number
Julian ephemeris century　儒略历书世纪
Julian ephemeris date　（＝JED）儒略历书日期
Julian epoch　儒略历元
Julian era　儒略纪元
Julian number　儒略日数　又见：Julian day number
Julian period　（＝JP）儒略周期
Julian year　儒略年
Juliet　天卫十一 | 天 [王] 衛十一〔台〕
Julius Caesar crater　儒略·凯撒环形山（月球）
jump condition　跳变条件 | 跳渡條件〔台〕
jump relation　跳变关系
June Lyrid meteor shower　六月天琴流星雨 | 六月天琴 [座] 流星雨〔台〕
June Solstice　夏至（节气）　又见：Summer Solstice
June solstice　夏至点　又见：summer solstice; first point of Cancer
Juno　① 婚神星（小行星 3 号）② 朱诺号（美国木星探测器）
Jupiter　① 岁星 ② 木星
Jupiter atmosphere　木星大气　又见：Jovian atmosphere

Jupiter burst　木星暴 | 木星爆〔台〕　又见：Jovian burst
Jupiter comet family　木族彗星
Jupiter-crosser　越木天体 | 越木小行星〔台〕
Jupiter-crossing asteroid　越木小行星
Jupiter cycle　岁星纪年
Jupiter Icy moons Explorer　（＝JUICE）冰质木卫探测器
Jupiter Icy Moons Orbiter　（＝JIMO）冰质木卫轨道飞行器
Jupiter ionosphere　木星电离层　又见：Jovian ionosphere
Jupiter-like exoplanet　系外类木行星 | [太陽] 系外類木行星〔台〕　又见：Jovian exoplanet; exo-Jupiter
Jupiter magnetosphere　木星磁层　又见：Jovian magnetosphere
Jupiter plasmasphere　木星等离子层　又见：Jovian plasmasphere
Jupiter radiation belt　木星辐射带　又见：Jovian radiation belt
Jupiter radio burst　木星射电暴　又见：Jovian radio burst
Jupiter radio radiation　木星射电　又见：Jovian radio radiation
Jupiter's asteroid family　木族小行星
Jupiter's comet family　木族彗星 | 木 [星] 族彗星〔台〕　又见：Jupiter's family of comets; comet of Jupiter family; Jupiter-type comet
Jupiter's family　木族　又见：Jovian family
Jupiter's family of comets　木族彗星 | 木 [星] 族彗星〔台〕　又见：Jupiter's comet family; comet of Jupiter family; Jupiter-type comet
Jupiter's Ghost　木魂星云（NGC 3242）　又见：Ghost of Jupiter
Jupiter's ring　木星环　又见：Jovian ring; ring of Jupiter
Jupiter's ringlet　木星细环　又见：Jovian ringlet
Jupiter's satellite　木卫 | 木 [星] 衛 [星]〔台〕　又见：Jovian satellite; satellite of Jupiter
Jupiter-Trojan asteroid　木星－特洛伊族小行星
Jupiter-type comet　木族彗星 | 木 [星] 族彗星〔台〕　又见：Jupiter's comet family; Jupiter's family of comets; comet of Jupiter family
Juza　上辅，紫微右垣三（天龙座 λ）　又见：Gianfar; Giansar
J-value　J 值
JVLA　（＝Karl G. Jansky Very Large Array）央斯基甚大阵
JVO　（＝Japanese Virtual Observatory）日本虚拟天文台

JWST （=James Webb Space Telescope）韦布空间望远镜 | 韋柏太空望遠鏡〔台〕

Jy （=jansky）央 | 顏[斯基]〔台〕（天体射电流量密度单位）

K

K ① （＝kelvin）开（绝对温标）
② （＝potassium）钾（19号元素）
k+a galaxy　k+a 星系
Ka band　Ka 波段
Kachina Chasmata　克奇纳深谷（天卫一）
Kaff　王良一（仙后座 β）　又见：Caph; Chaf
Kaffaljidhmah　天囷八（鲸鱼座 γ）
KAGUYA　月球学与工程探测器，月亮女神号（日本月球探测器）　又见：Selenological and Engineering Explorer
Kähler potential　卡勒势
Kaifeng　开封（小行星 35366 号）
Kaiser effect　凯泽效应
Kaitain　外屏七（双鱼座 α）　又见：Alrescha; Okda
Kakinchan　陈嘉键（小行星 23165 号）
Kale　木卫三十七
Kallichore　木卫四十四
Kalliope　司赋星（小行星 22 号）
Kalman filter　卡尔曼滤波器
Kalman filter theory　卡尔曼滤波理论
Kalnajs disk　卡纳斯盘
Kaluza-Klein particle　卡卢察－克莱因粒子
Kaluza-Klein theory　卡卢察－克莱因理论
Kalyke　木卫二十三
Kalypso　岛神星（小行星 53 号）
kamacite　锥纹石
Kamensk crater　卡缅斯克陨星坑（地球）
Kamerlingh Onnes crater　开默林·昂内斯环形山（月球）
KAM theorem　（＝Kolmogorov-Arnold-Moser theorem）KAM 定理
Kane crater　凯恩环形山（月球）
Kant crater　康德环形山（月球）
Kant-Laplace nebular theory　康德－拉普拉斯星云说｜康得－拉普拉斯星雲說〔台〕
Kant-Laplace theory　康德－拉普拉斯学说
Kantowski-Sachs solution　康托斯基－萨克斯解

Kant theory　康德学说
KAO　（＝Kuiper Airborne Observatory）柯伊伯机载天文台｜古柏機載天文台〔台〕
Kaohsiung　高雄（小行星 215080 号）
Kaokuen　高锟（小行星 3463 号）
Kaon　κ 介子
Kappa Crucis Cluster　南十字 κ 星团
Kappa Cygnid meteor shower　天鹅 κ 流星雨
Kappa mechanism　κ 机制
Kapteyn Astronomical Laboratory　卡普坦天文实验室
Kapteyn Region　卡普坦天区
Kapteyn Selected Area　卡普坦选区
Kapteyn's star　卡普坦星
Kapteyn Telescope　卡普坦望远镜
Kapteyn Universe　卡普坦宇宙
Kara crater　卡拉河陨星坑（地球）
Karakul crater　卡拉库尔湖陨星坑（地球）
Karatepe　卡拉泰佩地区（火星）
Karhunen-Loeve method　（＝KL method）卡尔胡宁－勒夫方法，KL 方法
Kari　土卫四十五
Karl G. Jansky Very Large Array　（＝JVLA）央斯基甚大阵
Karl-Schwarzschild Observatory　施瓦西天文台
karma month　世间月
Karpinskiy crater　卡尔平斯基环形山（月球）
Karrer crater　卡勒环形山（月球）
Kasei Valles　卡塞谷（火星）
Kashimamachi Station　鹿岛射电天文站（日本）
Kasner solution　卡斯纳解
Kassandra　见神星（小行星 114 号）
Kästner crater　克斯特纳环形山（月球）
kation　正离子　又见：positive ion
K'atun　卡顿（玛雅历，7200 天）
Kaus Australis　箕宿三（人马座 ε）
Kaus Borealis　斗宿二（人马座 λ）

Kaus Meridionalis　箕宿二（人马座 δ）
　又见：*Media*
kayser　凯塞（波数单位）
Kazan University Observatory　喀山大学天文台
K band　K 波段
KBO　（＝Kuiper belt object）柯伊伯带天体｜古柏帶天體〔台〕
K-capture　K 俘获
K component　K 成分｜K 分量〔台〕
K corona　K 冕｜K 日冕，連續 [光譜] 日冕〔台〕
K coronameter　K 冕光度计｜K 冕光度計〔台〕
K-correction　K 改正｜K 修正〔台〕
K doublet　K 双重线
K-doubling　K 双重性
KDP crystal　（＝Potassium Dihydrogen Phosphate crystal）KDP 晶体
K dwarf　K 型矮星
Keck II Telescope　凯克望远镜 II｜凱克 II 望遠鏡〔台〕
Keck Interferometer　凯克干涉仪
Keck I Telescope　凯克望远镜 I｜凱克 I 望遠鏡〔台〕
K-edge　K 限
Keeler crater　基勒环形山（月球）
Keeler Gap　基勒环缝（土星）
K-effect　K 效应
Keid　九州殊口增七，九州殊口增十一（波江座 o）
　又见：*Kied*
Ke Kouan　骑官四（豺狼座 β）　又见：*Ke Kwan*
Kekulé crater　凯库勒环形山（月球）
Ke Kwan　骑官四（豺狼座 β）　又见：*Ke Kouan*
Keller-Meyerott opacity　凯勒－麦耶洛特不透明度
Kellner eyepiece　凯尔纳目镜　又见：*Kellner ocular*
Kellner ocular　凯尔纳目镜　又见：*Kellner eyepiece*
kelvin　（＝K）开（绝对温标）
Kelvin contraction　开尔文收缩｜卡耳文收縮〔台〕
Kelvin-Helmholtz contraction　开尔文－亥姆霍兹收缩｜克耳文－亥姆霍茲收縮〔台〕
Kelvin-Helmholtz instability　开尔文－亥姆霍兹不稳定性｜克耳文－亥姆霍茲不穩定性〔台〕
Kelvin-Helmholtz mechanism　开尔文－亥姆霍兹机制｜克赫歷程〔台〕
Kelvin-Helmholtz time scale　开尔文－亥姆霍兹时标｜克耳文－亥姆霍茲時標〔台〕
Kelvin scale　开尔文温标
Kelvin's circulation theorem　开尔文环流定理
Kelvin temperature scale　开氏温标

Kelvin time scale　开氏时标｜卡耳文時標〔台〕
Kemble's Cascade　甘伯串珠（鹿豹座星组）
Kennedy Space Center　（＝KSC）肯尼迪空间中心｜甘迺迪空間中心〔台〕
Kennelly-Heaviside layer　肯内利－亥维塞层
　即：*E layer*
Kennicutt-Schmidt law　肯尼科特－施密特定律
Kepler　开普勒（小行星 1134 号）
Kepler crater　开普勒环形山（月球）
Kepler element　开普勒根数
Kepler ellipse　开普勒椭圆
Kepler equation　开普勒方程｜克卜勒方程〔台〕
　又见：*Kepler's equation*
Keplerian disk　开普勒盘｜克卜勒盤〔台〕
　又见：*Kepler's disk*
Keplerian motion　开普勒运动｜克卜勒運動〔台〕
　又见：*Kepler motion*
Keplerian orbit　开普勒轨道｜克卜勒軌道〔台〕
　又见：*Kepler orbit*
Keplerian region　开普勒区
Keplerian rotation　开普勒自转｜克卜勒自轉〔台〕
Keplerian shear　开普勒剪切
Keplerian speed　开普勒速度　又见：*Keplerian velocity*
Keplerian telescope　开普勒望远镜｜克卜勒 [式折射] 望遠鏡〔台〕
Keplerian velocity　开普勒速度　又见：*Keplerian speed*
Kepler mission　开普勒计划
Kepler motion　开普勒运动｜克卜勒運動〔台〕
　又见：*Keplerian motion*
Kepler orbit　开普勒轨道｜克卜勒軌道〔台〕
　又见：*Keplerian orbit*
Kepler potential　开普勒势
Kepler problem　开普勒问题
Kepler's disk　开普勒盘｜克卜勒盤〔台〕
　又见：*Keplerian disk*
Kepler's equation　开普勒方程｜克卜勒方程〔台〕
　又见：*Kepler equation*
Kepler's law　开普勒定律｜克卜勒定律〔台〕
Kepler's nova　开普勒新星（SN 1604）
Kepler's star　开普勒星（SN 1604）
Kepler's supernova　开普勒超新星｜克卜勒超新星〔台〕（SN 1604）
Kerberos　冥卫四
Kerr black hole　克尔黑洞｜克而黑洞〔台〕
Kerr metric　克尔度规
Kerr-Newman black hole　克尔－纽曼黑洞

Kerr-Newman metric　克尔－纽曼度规
k-essence　k 质
ketene　乙烯酮（CH_2CO）
Ketu　计都（黄白道降交点）
Keurusselkä crater　凯乌鲁塞尔凯湖陨星坑（地球）
keV　（＝kiloelectron-volt）千电子伏
key-controlled time signal　键控时号
Keyhole Nebula　钥匙孔星云（NGC 3372）
K giant　K 型巨星
K giant variable　（＝KGV）K 型变光巨星
KGV　（＝K giant variable）K 型变光巨星
Kharkov University Observatory　哈尔科夫大学天文台
Khvol'son crater　赫沃尔松环形山（月球）
Kiang　江涛（小行星 3751 号）
Kiangsu　江苏（小行星 2077 号）
Kibal'chich crater　基巴利契奇环形山（月球）
Kibble mechanism　基布尔机制
Kidinnu crater　西丹努斯环形山（月球）
Kied　九州殊口增七，九州殊口增十一（波江座 o）
　又见：*Keid*
Kiel University Radio Observatory　基尔大学射电天文台
Kiess crater　基斯环形山（月球）
Kiev University Observatory　基辅大学天文台
Kiffa Australis　氐宿增七（天秤座 α）
　又见：*Zubenelgenubi*
Kiffa Borealis　氐宿四（天秤座 β）
　又见：*Zubeneschamali*
Killing vector　基林矢量 | 基林向量〔台〕
kiloelectron-volt　（＝keV）千电子伏
kiloparsec　（＝kpc）千秒差距
Kimura term　木村项 | 木村相〔台〕
K'in　金（玛雅历，1 天）
kinchiltun　金切尔顿（玛雅历，1.152×10^9 天）
kinematical reference frame　运动学参考架
kinematical reference system　运动学参考系
kinematic astronomy　运动学天文学
kinematic cosmology　运动学宇宙学 | 運動宇宙論〔台〕
kinematic density wave　运动学密度波
kinematic distance　运动学距离
kinematic parallax　运动学视差
kinematics　运动学
kinematic viscosity　运动黏度 | 動粘滯率〔台〕
kinetic decoupling　动理学退耦
kinetic energy　动能
kinetic energy tensor　动能张量
kinetic equilibrium　动理学平衡
kinetic instability　动理学不稳定性
kinetic pressure　动压
kinetics　动理学
kinetic temperature　运动温度
kinetic theory　分子运动论，动理 [学理] 论
kinetic theory of gases　气体分子运动论，气体动理 [学理] 论　又见：*gas-kinetic theory*
kinetic viscosity　动理黏度
k-inflation　k 暴胀
King crater　金环形山（月球）
King law　金氏定律
King model　金氏模型
king post mount　主轴式装置
King profile　金氏轮廓
King radius　金氏半径
king-size flare star　巨型耀星
kink instability　扭折不稳定性 | 折扭不稳定性〔台〕
kinotheodolite　电影经纬仪　又见：*cine theodolite*
Kircher crater　基歇尔环形山（月球）
Kirchhoff law　基尔霍夫定律 | 柯希何夫定律〔台〕
　又见：*Kirchhoff's law*
Kirchhoff's law　基尔霍夫定律 | 柯希何夫定律〔台〕　又见：*Kirchhoff law*
Kirin meteorite　吉林陨星 | 吉林陨石〔台〕
　又见：*Jilin meteorite*; *Jiling meteorite*
Kirkpatrick-Baez type X-ray telescope　K－B 型 X 射线望远镜
Kirkwood crater　柯克伍德环形山（月球）
Kirkwood gap　柯克伍德空隙
Kirkwood Observatory　柯克伍德天文台
Kitabpha　虚宿二（小马座 α）　又见：*Kitalpha*
Kitalpha　虚宿二（小马座 α）　又见：*Kitabpha*
Kitt Peak National Observatory　（＝KPNO）基特峰国家天文台（美国）
Kitt Peak Station of Steward Observatory　斯图尔德天文台基特峰站
Kiviuq　土卫二十四
Klaproth crater　克拉普罗特环形山（月球）
Klein-Alfvén cosmology　克莱因－阿尔文宇宙学
Klein-Alfvén hypothesis　克莱因－阿尔文假说
Klein hypothesis　克莱因假说
Kleinmann-Low Nebula　（＝KL Nebula）克莱因曼－洛星云，KL 星云 | 克來曼－樓星雲，KL 星雲〔台〕

Kleinmann-Low object （=KL object）克莱因曼－洛天体，KL 天体 | 克來曼－樓天體，KL 天體〔台〕

Kleinmann-Low source （=KL source）克莱因曼－洛源，KL 源

Klein-Nishina cross-section 克莱因－仁科截面

Klein-Nishina formula 克莱因－仁科公式

K line K 线

Klio 史神星（小行星 84 号）

KL method （=Karhunen-Loeve method）卡尔胡宁－勒夫方法，KL 方法

KL Nebula （=Kleinmann-Low Nebula）克莱因曼－洛星云，KL 星云 | 克來曼－樓星雲，KL 星雲〔台〕

KL object （=Kleinmann-Low object）克莱因曼－洛天体，KL 天体 | 克來曼－樓天體，KL 天體〔台〕

Klotho 纺神星（小行星 97 号）

KL source （=Kleinmann-Low source）克莱因曼－洛源，KL 源

Klute crater 克卢特环形山（月球）

Klymene 伴女星（小行星 104 号）

klystron 速调管

Klytia 芥神星（小行星 73 号）

KM3NeT （=Cubic Kilometre Neutrino Telescope）立方千米中微子望远镜

K-magnitude K 星等

KM matrix （=Kobayashi-Maskawa matrix）小林－益川矩阵

knee whistler 膝啸

Knife Edge Galaxy 刀锋星系（NGC 5907）
  即：*Splinter Galaxy*

knife-edge test 刀口检验 | 刀口測試〔台〕

knock-on spectrum 击出粒子谱

Knudsen number 克努森数

Kobayashi-Maskawa matrix （=KM matrix）小林－益川矩阵

Kocab 帝，北极二（小熊座 β）

Koch crater 科赫环形山（月球）

Kodaikanal Astrophysical Observatory 科代卡纳天体物理台

Kodak astronomical plate 柯达天文底片

Kodak plate 柯达底片

Kodak spectroscopic plate 柯达光谱片

Kohlschütter crater 科尔许特环形山（月球）

Kohoutek comet 科胡特克彗星（C/1973 E1）

koinomatter 正常物质

Kolmogorov-Arnold-Moser theorem （=KAM theorem）KAM 定理

Kolmogorov-Smirnov test （=K-S test）科尔莫戈罗夫－斯米尔诺夫检验

Kolmogorov spectrum 科莫戈洛夫频谱

Kolmogorov turbulence 科莫戈洛夫湍流

Komarov crater 科马罗夫环形山（月球）

Kompaneets equation 科姆帕尼茨方程

Kondratyuk crater 孔德拉秋克环形山（月球）

König eyepiece 科尼希目镜

Konkoly Observatory 康科利天文台

Kore 木卫四十九

Korean Virtual Observatory （=KVO）韩国虚拟天文台

Kormendy relation 科曼蒂关系

Korneforos 河中，天市右垣一（武仙座 β）
  又见：*Kornephoros; Rutilicus*

Kornephoros 河中，天市右垣一（武仙座 β）
  又见：*Korneforos; Rutilicus*

Korolev crater 科罗廖夫环形山（月球）

Koronas （=Complex Orbital Observations Near-Earth of Activity of the Sun）日冕系列卫星（俄罗斯）

Koronas-Foton 日冕质子号卫星（俄罗斯）
  又见：*Coronas-Photon*

Koronis family 科朗尼斯族

Koronis family, asteroids 科朗尼斯族, 小行星

Korsch system 柯尔施光学系统

Kostinskiy crater 科斯京斯基环形山（月球）

Kostinsky effect 科斯京斯基效应

Kovalevskaya crater 柯瓦列夫斯卡娅环形山（月球）

Kozai resonance 古在共振

kpc （=kiloparsec）千秒差距

Kp index Kp 指数

KPNO （=Kitt Peak National Observatory）基特峰国家天文台（美国）

Kr （=krypton）氪（36 号元素）

Krafft crater 克拉夫特环形山（月球）

Kramers-Kronig dispersion relation 克拉默斯－克勒尼希色散关系

Kramers" law 克拉默斯定律 | 克瑞馬定律〔台〕

Kramers opacity 克拉默斯不透明度 | 克瑞馬不透明度〔台〕

Krammers absorption cross-section 克拉默斯吸收截面

Krasovskiy crater 克拉索夫斯基环形山（月球）

Kraus-type radio telescope 克劳斯型射电望远镜

Kraz 轸宿四（乌鸦座 β）
KREEP norite 克里普岩
Kreutz group 克罗伊策群 | 克羅伊茲群〔台〕
  又见：*Kreutz group of comets*
Kreutz group of comets 克罗伊策群 | 克羅伊茲群〔台〕 又见：*Kreutz group*
Kreutz sungrazer 克罗伊策掠日彗星 | 克羅伊茲掠日彗星〔台〕
Kron electrographic camera 克朗电子照相机
Kruskal diagram 克鲁斯卡图 | 克氏圖〔台〕
Kruskal metric 克鲁斯卡度规
Kruskal-Szekeres coordinates 克鲁斯卡－塞凯赖什坐标
krypton （＝Kr）氪（36 号元素）
Krzeminski's star 库兹明斯基星
KSC （＝Kennedy Space Center）肯尼迪空间中心 | 甘迺迪空間中心〔台〕
K-shell K 层
K-S regularization （＝Kustaanheimo-Stiefel regularization）K－S 正规化
K star K 型星 又见：*K-type star*
K-S test （＝Kolmogorov-Smirnov test）科尔莫戈罗夫－斯米尔诺夫检验
K subdwarf K 型亚矮星
K subgiant K 型亚巨星
K supergiant K 型超巨星
KSZ （＝Catalogue of Faint Stars）暗星星表，暗星表 | 暗星表〔台〕
K-term K 项 即：*K-effect*
K-type star K 型星 又见：*K star*
Ku band Ku 波段
Kugler crater 库格勒环形山（月球）

Kuiper Airborne Observatory （＝KAO）柯伊伯机载天文台 | 古柏機載天文台〔台〕
Kuiper belt 柯伊伯带 | 古柏帶〔台〕
Kuiper belt object （＝KBO）柯伊伯带天体 | 古柏帶天體〔台〕
Kuiper disk 柯伊伯盘
Kuiper-Edgeworth belt 柯伊伯－埃奇沃思带 | 古柏－埃奇沃思帶〔台〕
Kuiper Gap 柯伊伯环缝（土星）
Kuiper's star 柯伊伯星
Kulik crater 库利克环形山（月球）
Kuma 天棓二（天龙座 ν）
Kunlun 昆仑（小行星 3613 号）
Kunming 昆明（小行星 3650 号）
Kuo Shou Ching crater 郭守敬环形山（月球）
Kuotzuhao 郭子豪（小行星 20843 号）
Kurchatov crater 库尔恰托夫环形山（月球）
Kurhah 天钩六（仙王座 ξ）
kurtosis 峭度 | 峰態，峭度〔台〕
Kustaanheimo-Stiefel regularization （＝K-S regularization）K－S 正规化
Kutne's Nebula 库特勒星云（LDN 1524）
Kuzmin disk 库兹敏盘
Kvant 量子号（苏联和平号空间站模块）
  又见：*Kwant*
KVO （＝Korean Virtual Observatory）韩国虚拟天文台
Kwant 量子号（苏联和平号空间站模块）
  又见：*Kvant*
Kwasan Observatory 花山天文台（日本）
Kyokko 北极光（日本科学卫星）

L

L3CCD (=low light level CCD) 微光 CCD
La (=lanthanum) 镧（57 号元素）
LAB (=Lyman-α blob) 莱曼 α 团块
laboratory astrophysics 实验室天体物理
laboratory frequency standard 实验室频标
laboratory reference frame 实验室参考架 | 實驗室參考系〔台〕
laboratory reference system 实验室参考系
laboratory spectrum 实验室光谱
Labyrinthus 迷径沟网
Lac (=Lacerta) 蝎虎座
La Caille crater 拉卡耶环形山（月球）
Lacchini crater 拉基尼环形山（月球）
Lace Nebula 花边星云（NGC 6960）
  又见: *Lacework Nebula*
Lacerta (=Lac) 蝎虎座
Lacertid 蝎虎天体 又见: *BL Lacertid; BL Lacertae object*
Lacework Nebula 花边星云（NGC 6960）
  又见: *Lace Nebula*
Lacey-Cole merger rate 莱西—科尔并合率
LaCoste pendulum 拉考斯特摆
lacus 【拉】[月] 湖
Lacus Aestatis 【拉】夏湖（月球）
Lacus Autumni 【拉】秋湖（月球）
Lacus Bonitatis 【拉】仁慈湖（月球）
Lacus Doloris 【拉】忧伤湖（月球）
Lacus Excellentiae 【拉】秀丽湖（月球）
Lacus Felicitatis 【拉】幸福湖（月球）
Lacus Gaudii 【拉】欢乐湖（月球）
Lacus Hiemalis 【拉】冬湖（月球）
Lacus Lenitatis 【拉】温柔湖（月球）
Lacus Luxuriae 【拉】华贵湖（月球）
Lacus Mortis 【拉】死湖（月球）
Lacus Odii 【拉】怨恨湖（月球）
Lacus Perseverantiae 【拉】长存湖（月球）
Lacus Solitudinis 【拉】孤独湖（月球）
Lacus Somniorum 【拉】梦湖（月球）
Lacus Spei 【拉】希望湖（月球）
Lacus Temporis 【拉】时令湖（月球）
Lacus Timoris 【拉】恐怖湖（月球）
Lacus Veris 【拉】春湖（月球）
Lada Terra 拉达高地（金星）
Lade crater 拉德环形山（月球）
LADEE (=Lunar Atmosphere and Dust Environment Explorer) 月球大气及尘埃环境探测器（美国）
LAE (=Lyman alpha emitter) 莱曼 α 发射体
Laetitia 喜神星（小行星 39 号）
laevorotation 左旋
lag 迟滞
Lagalla crater 拉加拉环形山（月球）
LAGEOS (=LAser GEOdynamics Satellite) 激光地球动力学卫星
lag error 迟滞差
lagging of tides 潮汐迟滞
Lagoon Nebula 礁湖星云（NGC 6523）
Lagrange bracket 拉格朗日括号 又见: *Lagrange parenthesis*
Lagrange crater 拉格朗日环形山（月球）
Lagrange invariant 拉格朗日不变量
Lagrange parenthesis 拉格朗日括号
  又见: *Lagrange bracket*
Lagrange particular solution 拉格朗日特解
Lagrange planetary equation 拉格朗日行星运动方程 又见: *Lagrange's planetary equation*
Lagrange point 拉格朗日点 又见: *Lagrangian point*
Lagrange's equation 拉格朗日方程
Lagrange's planetary equation 拉格朗日行星运动方程 又见: *Lagrange planetary equation*
Lagrangian 拉格朗日算子
Lagrangian density 拉格朗日密度
Lagrangian derivative 拉格朗日导数
Lagrangian formulation 拉格朗日表达

Lagrangian function　拉格朗日函数
Lagrangian perturbation theory　拉格朗日扰动理论
Lagrangian point　拉格朗日点　又见：*Lagrange point*
lag time　（＝LT）滞后时间
L'Aigle meteorite shower　莱格勒陨石雨
Laitsaita　赖才达（小行星 23280 号）
Lakshmi Planum　吉祥天女高原（金星）
Lallemand electronic camera　拉勒芒电子照相机
LAMA　（＝Large Liquid-Aperture Mirror Array）大口径液[态]镜[面]阵
Lamarck crater　拉马克环形山（月球）
Lamb crater　兰姆环形山（月球）
lambdameter　波长计
lambert　朗伯（亮度单位）
Lambert albedo　朗伯反照率
Lambert equation　朗伯方程
Lambert series　朗伯级数
Lambert's law　朗伯定律
Lamb shift　兰姆位移
Lamchiuying　林超英（小行星 64288 号）
Lame constant　拉梅常数
Lamé crater　拉梅环形山（月球）
Lame elastic parameter　拉梅弹性参数
lamellar grating　层状光栅
lamella（复数：lamellae）　壳层
Lame parameter　拉梅参数
laminar flow　片流 | 層流，片流〔台〕
Lamont crater　拉蒙特环形山（月球）
LAMOST　（＝Large Sky Area Multi-Object Fiber Spectroscopic Telescope）大天区面积多目标光纤光谱天文望远镜，郭守敬望远镜 | 大天區面積多目標光纖光譜望遠鏡〔台〕
Landau crater　朗道环形山（月球）
Landau damping　朗道阻尼 | 蘭道阻尼〔台〕
Landau energy level　朗道能级
Landau length　朗道长度
Lande factor　朗德因子
Lande g value　朗德 g 值
lander　着陆器
Lande splitting factor　朗德劈裂因子 | 蘭德分裂因子〔台〕
landing rocket　着陆火箭 | 登陸火箭〔台〕
Lane crater　莱恩环形山（月球）
Lane-Emden equation　莱恩－埃姆登方程 | 藍艾方程〔台〕
Langemak crater　朗格马克环形山（月球）

Langevin crater　朗之万环形山（月球）
Langevin theory　朗之万定理
Langley crater　兰利环形山（月球）
Langmuir crater　朗缪尔环形山（月球）
Langmuir frequency　朗缪尔频率
Langmuir oscillation　朗缪尔振荡
Langmuir probe　朗缪尔测量仪
Langmuir wave　朗缪尔波 | 蘭牟而波〔台〕
Langmuir wave instability　朗缪尔波不稳定性
Langmuir wave turbulence　朗缪尔波湍动
Langrenus crater　朗格伦环形山（月球）
lanthanum　（＝La）镧（57 号元素）
Laomedeia　海卫十二
Laotse　老子（小行星 7854 号）
La Pérouse crater　拉彼鲁兹环形山（月球）
Laplace azimuth　拉普拉斯方位角
Laplace coefficient　拉普拉斯系数
Laplace Gap　拉普拉斯环缝（土星）
Laplace plane　拉普拉斯平面　又见：*Laplacian plane*
Laplace point　拉普拉斯点
Laplace resonance　拉普拉斯共振
Laplace's equation　拉普拉斯方程
Laplace's nebular hypothesis　拉普拉斯星云假说
Laplace theory　拉普拉斯学说
Laplace transform　拉普拉斯变换
Laplace vector　拉普拉斯矢量 | 拉普拉斯向量〔台〕
Laplacian　拉普拉斯算子
Laplacian plane　拉普拉斯平面　又见：*Laplace plane*
Laplacian surface　拉普拉斯面
La Plata Observatory　拉普拉塔天文台
Lappajärvi crater　拉帕湖陨星坑（地球）
lapse function　时移函数
Large Altazimuth Telescope　（＝BTA）大型地平装置望远镜（俄罗斯 6 米望远镜）
large-angle anisotropy　大角度各向异性
Large Binocular Telescope　（＝LBT）LBT 大型双筒望远镜 | 大雙筒望遠鏡〔台〕
Large Binocular Telescope Interferometer　（＝LBTI）大型双筒望远镜干涉仪（美国）
large diameter source　大角径源
Large Earth-based Solar Telescope　（＝LEST）大型地基太阳望远镜
Large Electron Positron Collider　（＝LEP）大型正负电子对撞机
Large Field Multi-Color Sky Survey　大视场多色巡天

Large Hadron Collider （=LHC）大型强子对撞机

Large High Altitude Air Shower Observatory （=LHAASO）高海拔宇宙线观测站（中国）

Large Liquid-Aperture Mirror Array （=LAMA）大口径液[态]镜[面]阵

Large Magellanic Cloud （=LMC）大麦云，大麦哲伦云

Large Millimeter Telescope （=LMT）LMT 大型毫米波望远镜（墨西哥）

large number hypothesis 大数假说

large outburst 巨大爆发

large scale beam 大尺度射束

large scale characteristic 大尺度特征

large-scale clustering 大尺度成团性

large scale jet 大尺度喷流

large scale sky map 大尺度天图

large scale structure （=LSS）大尺度结构

large scale structure of the universe 宇宙大尺度结构

large scale turbulence 大尺度湍动

large scale velocity field 大尺度速度场

Large Sky Area Multi-Object Fiber Spectroscopic Telescope （=LAMOST）大天区面积多目标光纤光谱天文望远镜，郭守敬望远镜 | 大天區面積多目標光纖光譜望遠鏡〔台〕 又见：*Guo Shoujing Telescope*

Large Synoptic Survey Telescope （=LSST）大口径全天巡视望远镜

large telescope 大型望远镜

Large Zenith Telescope （=LZT）大型天顶望远镜（加拿大）

Larissa 海卫七

Larmor crater 拉莫尔环形山（月球）

Larmor frequency 拉莫尔频率 | 拉莫爾頻率〔台〕

Larmor precession 拉莫尔进动 | 拉莫爾進動〔台〕

Larmor radius 拉莫尔半径 | 拉莫爾半徑〔台〕

Las Campanas Observatory （=LCO）拉斯坎帕纳斯天文台

Las Campanas Redshift Survey （=LCRS）拉斯坎帕纳斯红移巡天

laser 激光 又见：*light amplification by stimulated emission of radiation*

laser frequency comb （=LFC）激光频率梳

laser geodimeter 激光测距仪 | 雷射測距儀〔台〕 又见：*laser rangefinder*

LAser GEOdynamics Satellite （=LAGEOS）激光地球动力学卫星

laser guided adaptive optics 激光引导自适应光学 | 雷射引導自適應光學〔台〕

laser guide star 激光引导星 | 雷射引導星〔台〕

laser interferometer 激光干涉仪

laser interferometer gravitational wave detector 激光干涉仪型引力波探测器

Laser Interferometer Gravitational-wave Observatory （=LIGO）激光干涉引力波观测台

Laser Interferometer Space Antenna （=LISA）太空激光干涉仪

laser metrology system 激光计量系统

laser radar 激光雷达 又见：*optical laser radar*

laser rangefinder 激光测距仪 | 雷射測距儀〔台〕 又见：*laser geodimeter*

laser ranging 激光测距 | 雷射測距〔台〕

laser reflector 激光反射器

laser satellite 激光卫星

laser tracking 激光跟踪

laser tube 激光管

La Silla Observatory 拉西亚天文台

Lassell ring 拉塞尔环（海王星）

Lasso 娄宿（二十八宿） 又见：*Bond*

last contact 复圆 又见：*last contact of umbra; fourth contact*

last contact of umbra 复圆 又见：*last contact; fourth contact*

last quarter 下弦 又见：*third quarter*

last scattering surface 最后散射面

last significant figure 末位有效数字

La Superba 猎犬 Y

late cluster 晚型星系团

late heavy bombardment 晚期重轰击（月球）

late integrated Sachs-Wolfe effect 晚期累积萨克斯－沃尔夫效应

latent heat 潜热

latent image 潜像

lateral aberration 横向像差

lateral chromatic aberration 横向色差 | 側向色差〔台〕

lateral flexure 横向弯曲 | 側彎曲〔台〕

lateral magnification 横向放大率

lateral refraction 旁折射 | 側折射〔台〕

late-type dwarf 晚型矮星

late-type galaxy 晚型星系

late-type giant 晚型巨星

late-type spiral galaxy 晚型旋涡星系

late-type star 晚型星

late-type subdwarf　晚型亚矮星
late-type subgiant　晚型亚巨星
late-type supergiant　晚型超巨星
late-type variable　晚型变星
latitude　纬度，黄纬 | 緯度〔台〕
latitude circle　① 纬 [度] 圈　又见: *parallel of latitude; parallel circle* ② 黄纬圈　又见: *parallel of latitude*
latitude determination　纬度测定
latitude distribution　纬度分布
latitude effect　纬度效应
latitude line　纬线
latitude of exposure　曝光时限 | 曝光範圍〔台〕
　　又见: *exposure age; exposure latitude*
latitude service　纬度服务
latitude star　测纬星
latitude station　纬度站
latitude variation　纬度变化　又见: *variation of latitude*
latitudinal action　纬向作用量
lattice distortion　晶格畸变
latus rectum　通径 | 正焦弦〔台〕
Laue crater　劳厄环形山（月球）
Laue lens telescope　劳厄透镜望远镜
launch angle　发射角
launcher　发射器　又见: *launch vehicle; launching vehicle*
launching vehicle　发射器　又见: *launch vehicle; launcher*
launch pad　发射台，发射场
launch vehicle　发射器　又见: *launcher; launching vehicle*
launch window　最佳发射时间 | 發射窗口〔台〕
Lauritsen crater　劳里森环形山（月球）
Lautakkin　刘德健（小行星 25065 号）
Lautakshing　刘德诚（小行星 25073 号）
lava　熔岩
Lavoisier crater　拉瓦锡环形山（月球）
law of area　面积定律
law of causation　因果律　又见: *causality*
law of chance　机遇律
law of conservation of angular momentum　角动量守恒定律
law of conservation of energy　能量守恒定律
law of conservation of matter　物质守恒定律
law of conservation of momentum　动量守恒定律
law of distribution of velocity　速度分布律
law of isochronism　等周期定律 | 等周 [期] 定律〔台〕

law of large numbers　大数定律
law of planetary distances　行星距离定律
law of propagation of errors　误差传播定律
law of universal gravitation　万有引力定律
Lawrence Livermore National Laboratory　（＝LLNL）劳伦斯利物莫国家实验室（美国）
lawrencium　（＝Lr）铹（103 号元素）
laws of planetary motion　行星运动定律
Layzer-Irvine equation　雷泽－欧文方程
LBA　（＝Long Baseline Array）长基线望远镜阵 | 長基線 [望遠鏡] 陣〔台〕
L band　L 波段
LBG　（＝Lyman break galaxy）莱曼断裂星系 | 來曼斷裂星系〔台〕
LBI　①（＝long baseline interferometer）长基线干涉仪 ②（＝long baseline interferometry）长基线干涉测量
LBT　（＝Large Binocular Telescope）LBT 大型双筒望远镜 | 大雙筒遠鏡〔台〕
LBTI　（＝Large Binocular Telescope Interferometer）大型双筒望远镜干涉仪（美国）
LBV　（＝luminous blue variable）高光度蓝变星 | 亮藍變星〔台〕
Lb variable　Lb 型变星
LCN　（＝Lyapunov characteristic number）李雅普诺夫特征数
LCO　（＝Las Campanas Observatory）拉斯坎帕纳斯天文台
L component　L 成分
L corona　L 冕
LCROSS　（＝Lunar Crater Observation and Sensing Satellite）月球环形山观测与遥感卫星（美国）
LCRS　（＝Las Campanas Redshift Survey）拉斯坎帕纳斯红移巡天
Lc variable　Lc 型变星
LCVR　（＝liquid crystal variable retarder）液晶相位可变延迟器
LDEF　（＝Long Duration Exposure Facility）长期曝露飞行器
L dwarf　L 型矮星
lead　（＝Pb）铅（82 号元素）
leading arm　导臂 | 前导旋臂〔台〕
leading edge　西边缘，前导边缘　又见: *leading limb; preceding limb*
leading hemisphere　前导半球
leading limb　西边缘，前导边缘　又见: *leading edge; preceding limb*
leading member　前导成员

leading spot 前导黑子，p 黑子 | 先導黑子，前導黑子〔台〕 又见：leading sunspot; preceding sunspot; preceding spot

leading sunspot （＝p-sunspot; p-spot）前导黑子，p 黑子 | 先導黑子，前導黑子〔台〕
又见：preceding sunspot; preceding spot; leading spot

leading wave 导波
lead oxide vidicon 氧化铅光导摄像管
又见：plumbicon
lead screw 导杆
lead selenide 硒化铅
lead sulphide 硫化铅
lead telluride 碲化铅
leakage 渗漏
leaky box model 漏箱模型
Leander McCormick Observatory 利安德·麦考密克天文台
leap day 闰日 又见：bissextile day; epagomenal day; intercalary day
leap-frog integration scheme 蛙跳积分法
leap month 闰月 又见：intercalary month; intercalated month
leap second 闰秒
leap second adjustment 闰秒调整
leap year 闰年 又见：bissextile year; embolismic year; intercalary year
least action 最小作用量
least square adjustment 最小二乘平差
least square fitting 最小二乘拟合 | 最小平方擬合〔台〕 又见：least squares fitting
least square method 最小二乘法 又见：method of least square; method of minimum square
least squares fitting 最小二乘拟合 | 最小平方擬合〔台〕 又见：least square fitting
least square solution 最小二乘解
Lebedev crater 列别捷夫环形山（月球）
Lebedinskiy crater 列别金斯基环形山（月球）
Lebesgue integral 勒贝格积分
Leda ①卵神星（小行星 38 号）②木卫十三
Leechaohsi 李诏熙（小行星 21554 号）
Lee model 李模型
Leesuikwan 李瑞均（小行星 94228 号）
Leetsungdao 李政道（小行星 3443 号）
Leeuwenhoek crater 列文虎克环形山（月球）
Lee wave 李波
Lee-Weinberg limit 李－温伯格极限
left-handed polarization 左旋偏振
legal time 法定时

Legendre crater 勒让德环形山（月球）
Legendre polynomial 勒让德多项式
Le Gentil crater 勒让蒂环形山（月球）
Legs 奎宿（二十八宿） 又见：Stride
Lehmann crater 勒曼环形山（月球）
Leibnitz crater 莱布尼兹环形山（月球）
Leiden University Observatory 莱顿大学天文台
Leipzig University Observatory 莱比锡大学天文台
LEM （＝lunar excursion module）登月舱（美国阿波罗计划）
Lemaitre cosmological model 勒梅特宇宙学模型 | 勒麥特宇宙模型〔台〕
Lemaitre cosmology 勒梅特宇宙学 | 勒麥特宇宙論〔台〕
Lemaitre model 勒梅特模型
Lemaitre regularization 勒梅特正规化
Lemaitre-Tolman-Bondi model （＝LTB model）勒梅特－托尔曼－邦迪模型
Lemaitre universe 勒梅特宇宙 | 勒麥特宇宙論〔台〕
Lemaître crater 勒梅特环形山（月球）
Le Monnier crater 勒莫尼耶环形山（月球）
length of exposure 曝光时间 | 曝光時長〔台〕
Lennard-Jones potential 伦纳德－琼斯势
lens 透镜
lens antenna 透镜天线
lens design 透镜设计
lensed galaxy 受透镜效应星系
lensed quasar 受透镜效应类星体
Lense-Thirring effect 伦塞－西凌效应
lensing cross-section [引力]透镜截面
lensing effect [引力]透镜效应
lensing galaxy [引力]透镜星系 | 重力透鏡星系〔台〕
lensing object [引力]透镜天体
lensing optical depth [引力]透镜光深
lensing potential 透镜势
lensing quasar [引力]透镜类星体
lensing star [引力]透镜恒星
lenslet array 小透镜阵
lens plane 透镜平面
lenticular cloud 荚状云
lenticular galaxy 透镜状星系
Lents crater 楞次环形山（月球） 又见：Lenz crater
Lenz crater 楞次环形山（月球） 又见：Lents crater
Lenz law 楞次定律

LEO （=low earth orbit）近地轨道

Leo ① （=Leo）狮子座 ② （=Leo）狮子宫，鹑火，午宫

Leo Doublet 狮子双重星系（M65/M66）

Leo I galaxy 狮子座 I 星系

Leo Minor （=LMi）小狮座

Leonid meteor shower 狮子流星雨

Leonids 狮子流星群｜狮子[座]流星雨〔台〕

Leopold-Figl Observatory 利奥波德－菲格天文台

Leo Triplet 狮子三重星系　又见：*Trio in Leo*

LEP （=Large Electron Positron Collider）大型正负电子对撞机

Lep （=Lepus）天兔座

leptogenesis 轻子创生

lepton 轻子

lepton asymmetry 轻子不对称性

lepton era 轻子期｜輕子時代〔台〕

leptonic charge 轻子荷

lepton number 轻子数

lepton number density 轻子数密度

lepton synthesis 轻子合成

Lepus （=Lep）天兔座

Lesath 尾宿九（天蝎座 υ）　又见：*Leschath*

Leschath 尾宿九（天蝎座 υ）　又见：*Lesath*

Lesser Cold 小寒（节气）　又见：*Slight Cold*

Lesser Fullness 小满（节气）　又见：*Grain Fills*

Lesser Heat 小暑（节气）　又见：*Slight Heat*

less luminous supergiant 次亮超巨星

LEST （=Large Earth-based Solar Telescope）大型地基太阳望远镜

Leto 明神星（小行星 68 号）

Leto family 明神星族

Letronne crater 勒特罗纳环形山（月球）

Leucippus crater 留基伯环形山（月球）

Leukothea 沉神星（小行星 35 号）

Leuschner Observatory 劳希纳天文台

level bubble 水准气泡

level chamber 水准气泡室

level constant 水准常数

level error 水准差，地平差，水平差｜水準[误]差〔台〕

level-I civilization I 级文明

level-II civilization II 级文明

level-III civilization III 级文明

leveling 水准测量　又见：*levelling*

leveling instrument 水准仪｜水準儀，水準器〔台〕　又见：*levelling instrument*

levelling 水准测量　又见：*leveling*

levelling instrument 水准仪｜水準儀，水準器〔台〕　又见：*leveling instrument*

level sensitivity 水准灵敏度

level surface 水准面

level tester 水准检验仪　又见：*level testing instrument*

level testing instrument 水准检验仪　又见：*level tester*

Leverrier ring 勒威耶环

Levi-Civita crater 列维－齐维塔环形山（月球）

Levi-Civita pseudotensor 列维－奇维塔赝张量

Levi-Civita regularization 列维－齐维塔正规化

levogyrate component 左旋子线

Lexell crater 莱克塞尔环形山（月球）

Lexell's comet 莱克塞尔彗星（D/1770 L1）

Ley crater 莱伊环形山（月球）

LFC （=laser frequency comb）激光频率梳

LF time and frequency 低频时频

LF time and frequency dissemination 低频时频发播

L galaxy L 星系

L* galaxy 特征亮度星系

LHA （=local hour-angle）地方时角

LHAASO （=Large High Altitude Air Shower Observatory）高海拔宇宙线观测站（中国）

Lhasa 拉萨（小行星 7859 号）

LHC （=Large Hadron Collider）大型强子对撞机

Li （=lithium）锂（3 号元素）

Li abundance （=lithium abundance）锂丰度

Liaoning 辽宁（小行星 2503 号）

Liaoyenting 廖彦婷（小行星 23249 号）

Lib ① （=Libra）天秤座 ② （=Libra）天秤宫，寿星，辰宫

Libai 李白（小行星 110288 号）

Liberté arc 自由环弧（海王星）

Libra ① （=Lib）天秤座 ② （=Lib）天秤宫，寿星，辰宫

libration 天平动

libration deviation 天平动偏异，秤动偏异｜天平動偏異〔台〕

libration effect 天平动效应，秤动效应｜天平動效應〔台〕

libration ellipse 天平动椭圆，秤动椭圆｜天平動椭圓〔台〕

libration in latitude 纬天平动｜緯[度]天平動〔台〕

libration in longitude 经天平动 | 經 [度] 天平動〔台〕
libration orbit 秤动轨道
libration point 秤动点 | 天平動點〔台〕
Librids 天秤流星群 | 天秤 [座] 流星雨〔台〕
Lick catalogue 利克星表
Lick index 利克指数
Lick Observatory 利克天文台
Lidaksum 李达三（小行星3812号）
Liebig crater 李比希环形山（月球）
Lie derivative 李导数
Lienard-Wiechert potential 李纳－维谢尔势
Lie series 李级数
Lie transformation 李变换
life-bearing planet 宜居行星 | 有生命行星〔台〕
life support 生命保障
liftoff 起飞
light amplification by stimulated emission of radiation 激光　又见: laser
light amplifier 光放大器
light bridge 亮桥　又见: bright bridge
light coherence 光相干
light cone 光锥　又见: pencil
light cross 光柱
light crossing time 光穿越时间
light crown 轻冕玻璃
light curve 光变曲线
light cylinder 光速圆柱面 | 光柱〔台〕
light day 光日
light echo 回光
light element 轻元素
light element abundance 轻元素丰度
light emission 光发射
lightening 闪光　又见: flash; flashing light; lightening flash; shimmer
lightening flash 闪光　又见: flash; flashing light; lightening; shimmer
light equation 光行时差 | 光 [行时] 差〔台〕　又见: equation of light
lightest supersymmetric particle （=LSP）最轻超对称粒子
light filter 滤光片，滤光器 | 濾光器，濾光鏡〔台〕
light flint 轻火石玻璃
light-gathering aperture 聚光孔径
light-gathering power 聚光本领 | 聚光率〔台〕
light grasp 聚光
light hour 光时

light interference 光干涉
light interferometer 光干涉仪
light interferometry 光干涉测量
lightlike interval 类光间隔
light mantle 发光月幔
light meson 轻介子
light minute 光分
light modulator 调光器
light month 光月
light nebula 发光星云
light of the moon 月光期
light of the night sky 夜天光 | 夜 [間天] 光〔台〕　又见: night glow; night sky light
light period 光变周期　又见: period of light variation
light pipe 光束管
light polarization 光偏振　又见: polarization of light
light pollution 光污染 | 光害，光污染〔台〕
light pressure 光压
light quantization 光量子化
light quantum 光量子 | 光量子，光子〔台〕
light scattering 光散射
light second 光秒
light sensation 光感
light shield 遮光罩
Light Snow 小雪（节气）　又见: Slight Snow
light source 光源　又见: photosource
light step 光阶
light time 光行时
light wave 光波
light weight star 小质量恒星
light year （=ly）光年
LIGO （=Laser Interferometer Gravitational-wave Observatory）激光干涉引力波观测台
Lijiang 丽江（小行星14656号）
likelihood 似然度
likelihood function 似然函数
Lilith 月孛
Liloketai 李陆大（小行星3609号）
limb brightening 临边增亮　又见: brightening towards the limb
limb darkening 临边昏暗 | 周邊減光，臨邊昏暗〔台〕　又见: darkening towards the limb
limb-darkening coefficient 临边昏暗系数
Limber equation 林伯方程
Limber hypothesis 林伯假说
limb flare 边缘耀斑 | 邊緣閃焰〔台〕
limb polarization 临边偏振

limb spot 边缘黑子
limiting apparent magnitude 极限视星等
limiting dispersion 极限色散度，极限频散度
limiting exposure 极限曝光时间
limiting flux 极限流量
limiting frequency 极限频率
limiting magnitude 极限星等
limiting photoelectric magnitude 极限光电星等
limiting photographic magnitude 极限照相星等
limiting photovisual magnitude 极限仿视星等
limiting radius 极限半径
limiting resolution 极限分辨率
Limit of Heat 处暑（节气） 又见：End of Heat
limonite 褐铁矿
Limporyen 林百欣（小行星5539号）
Linchen 林晨（小行星20828号）
Linchinghsia 林青霞（小行星38821号）
Lincoln Near-Earth Object Research（=LINEAR）林肯近地天体研究
Lindblad crater 林德布拉德环形山（月球）
Lindblad dispersion orbit 林德布拉德弥散轨道
Lindblad radius 林德布拉德半径
Lindblad resonance 林德布拉德共振 | 林達博共振〔台〕
Lindblad's ring 林德布拉德环
Lindenau crater 林德瑙环形山（月球）
Linde-Weinberg bound 林德－温伯格约束
Lindstedt equation 林德斯塔方程
linea 【拉】线状结构（行星地貌）
line absorption 线吸收
line absorption coefficient 线吸收系数 | [谱] 線吸收係數〔台〕
LINEAR （=Lincoln Near-Earth Object Research）林肯近地天体研究
linear amplification 线性放大 又见：linear magnification
linear aperture 线孔径
linear bias 线性偏袒
linear bias model 线性偏袒模型
linear bias parameter 线性偏袒参数
linear birefringence 线性双折射
linear correlation 线性相关 又见：linear dependence
linear dependence 线性相关 又见：linear correlation
linear diameter 线直径
linear dichroism 线二色性
linear dispersion 线色散度
linear dispersion relation 线色散关系

linear distance 线距离 | 線距 [離]〔台〕
linear growth factor 线性增长因子
linear growth rate 线性增长率
linear independence 线性无关
linear infall 线性沉降
linearity 线性
linearization 线性化
linearized collisionless Boltzmann equation 线性化无碰撞玻尔兹曼方程
linearized Einstein equation 线性化爱因斯坦方程
linearized equations of motion 线性化运动方程
linearly polarized radiation 线偏振辐射
linear magnification 线性放大 又见：linear amplification
linear molecule 线型分子
linear momentum 线性矩
linear operator 线性算子
linear optics 线性光学
linear perturbation power spectrum 线性扰动功率谱
linear polarization 线偏振
linear polyatomic molecule 线型多原子分子
linear power spectrum 线性功率谱
line array 线型阵
linear regression 线性回归
linear response theory 线性响应理论
linear size 线大小
linear space 线性空间
linear stability 线性稳定性
linear Stark effect 线性斯塔克效应 | 線性史塔克效應〔台〕
linear system 线性系统
linear transformation 线性变换
linear velocity 线速度
line asymmetry 谱线不对称性
line blanketing 谱线覆盖 | 譜線覆蓋〔台〕 又见：line blocking
line blanketing index 谱线覆盖因子 又见：line blocking factor
line blending 谱线混合
line blocking 谱线覆盖 | 譜線覆蓋〔台〕 又见：line blanketing
line blocking factor 谱线覆盖因子 又见：line blanketing index
line broadening 谱线变宽 | 譜線致寬〔台〕
line broadening by rotation 谱线自转致宽
line broadening by turbulence 谱线湍动致宽

line center　线心 | [谱] 線 [中] 心〔台〕　又见：*line core; core of a line*
line contour　谱线轮廓　又见：*line profile*
line core　线心 | [谱] 線 [中] 心〔台〕　又见：*core of a line; line center*
line displacement　谱线位移　又见：*line shift; spectral line shift*
line emission　线发射
line emission cloud　谱线发射云
line emission nebula　谱线发射星云
line feed　线形馈源
line formation　谱线形成　又见：*spectral line formation*
line identification　谱线证认 | 譜線識別〔台〕
line image sensor　线成像敏感器
line intensity　谱线强度　又见：*line strength*
line locking　线锁
line-locking process　线锁过程
line luminosity　谱线光度
line of apsides　拱线　又见：*apsidal line; apse line*
line of collimation　准直线
line of cusps　月角线
line of declination　赤纬线
line of force　力线
line of inversion　反转线
line of nodes　交点线
line of position　位置线 | 方位線〔台〕
　　又见：*position line*
line of right ascension　赤经线
line of sight　视线　又见：*line of vision; visual line*
line-of-sight component　视向分量
line of sight integral　视向积分
line-of-sight velocity　视向速度　又见：*radial velocity*
line of vision　视线　又见：*line of sight; visual line*
line profile　谱线轮廓　又见：*line contour*
line-profile variable　谱线轮廓变星
Liner　（＝low ionization nuclear emission-line region）低电离星系核
line radiation　线辐射
line ratio　谱线比
line ratio method　谱线比法
line receiver　谱线接收机　又见：*spectral line receiver*
line scattering　线散射
line scattering coefficient　线散射系数
line shift　谱线位移　又见：*line displacement; spectral line shift*
line-shifter　谱线位移器
line spectrum　线光谱

line splitting　谱线分裂　又见：*spectral line splitting*
line spread function　线扩散函数
line strength　谱线强度　又见：*line intensity*
line width　线宽 | [谱] 線寬 [度]〔台〕
line wing　线翼 | [谱] 線翼〔台〕
Lingchen　凌晨（小行星 20638 号）
Lintingnien　林庭年（小行星 20822 号）
Linus　赋卫一（小行星 Sappho 的卫星）
Linzexu　林则徐（小行星 7145 号）
Liouville equation　刘维尔方程
Liouville theorem　刘维尔定理 | 劉維爾定理〔台〕
Li-poor star　（＝lithium poor star）贫锂星
Lippmann crater　李普曼环形山（月球）
Lipschitz condition　利普希茨条件
Lipskiy crater　利普斯基环形山（月球）
liquid core　液态核
liquid crystal phase corrector　液晶相位改正器
liquid crystal variable retarder　（＝LCVR）液晶相位可变延迟器
liquid helium　液氦
liquid hydrogen　液氢
liquid mirror telescope　液 [态] 镜 [面] 望远镜
liquid nitrogen　液氮
liquid oxygen　液氧
liquid state　液态
LIR　（＝low intensity reciprocity）低强度倒易性
LIRF　（＝low intensity reciprocity failure）低强度倒易性失效
LIRG　（＝luminous infrared galaxy）亮红外星系
Li-rich star　（＝lithium rich star）富锂星
LISA　（＝Laser Interferometer Space Antenna）太空激光干涉仪
Lisiguang　李四光（小行星 137039 号）
Lissajous figure　利萨茹图形
Lissajous pattern　利萨茹图
Li star　（＝lithium star）锂星
literal algebra　文字代数
lithium　（＝Li）锂（3 号元素）
lithium abundance　（＝Li abundance）锂丰度
lithium niobate　铌酸锂（$LiNbO_3$）
lithium poor star　（＝Li-poor star）贫锂星
lithium rich star　（＝Li-rich star）富锂星
lithium star　（＝Li star）锂星
lithosiderite　石铁陨星 | 石隕鐵〔台〕
　　又见：*siderolite; stony-iron meteorite*
lithosphere　岩石圈　又见：*oxysphere*
lit side　光照侧边

little bang 小爆炸
Little Dipper 小北斗（小熊座星组）
Little Dumbbell 小哑铃星云
Littrow condition 利特洛条件
Littrow spectrograph 利特洛摄谱仪 | 利特羅攝譜儀〔台〕
Liu 廖庆齐（小行星 6743 号）
Liudongyan 刘栋艳（小行星 19874 号）
Liutingchun 刘亭均（小行星 20823 号）
Liutungsheng 刘东生（小行星 58605 号）
Liuzongli 刘宗礼（小行星 10070 号）
Lixiaohua 李晓华（小行星 3556 号）
Liyuan 李元（小行星 6741 号）
Liyulin 李育霖（小行星 20846 号）
LLNL （＝Lawrence Livermore National Laboratory）劳伦斯利物莫国家实验室（美国）
LLR （＝lunar laser ranging）激光测月 | 月球雷射測距〔台〕
LM （＝lunar module）登月舱
L-magnitude L 星等
LMC （＝Large Magellanic Cloud）大麦云，大麦哲伦云
LMi （＝Leo Minor）小狮座
LMT ①（＝Large Millimeter Telescope）LMT 大型毫米波望远镜（墨西哥）②（＝local mean time）地方平时
LMXB （＝low-mass X-ray binary）小质量 X 射线双星
LNH （＝Dirac's large-number hypothesis）狄拉克大数假说
LO （＝local oscillator）本振
load switching 负载开关
Lobachevskiy crater 罗巴切夫斯基环形山（月球）
Lobachevsky space 罗巴切夫斯基空间
lobate ridge 叶状崖脊
lobate scarp 叶状悬崖
Lob crater 洛布环形山（天卫十五）
lobe rotation 波瓣旋转
lobes of radio galaxy 射电星系瓣
lobe sweeping 波瓣开关
lobe sweeping interferometer 扫瓣干涉仪
又见：*swept-lobe interferometer*
lobe sweeping interferometry 扫瓣干涉测量
lobster-eye X-ray telescope 龙虾眼 X 射线望远镜
Lobster Nebula 龙虾星云（M17） 即：*Omega Nebula; ω Nebula*
local apparent noon 地方视正午
local apparent time 地方视时

local approximation 局部近似
Local arm 近域旋臂 | 本地旋臂〔台〕
local bias model 局域偏袒模型
Local Bubble 本地泡（银河）
local civil time 地方民用时
Local cluster of galaxies 本星系团
local effect 局域效应 | 局部性效應〔台〕
local equilibrium 局域平衡
local F-corona 近域 F 冕
local flow 局域流
local galaxy 局域星系 | 本地星系〔台〕
Local Group 本星系群 又见：*Local group of galaxies*
Local group of galaxies 本星系群 又见：*Local Group*
local helioseismology 局部日震学
local hour-angle （＝LHA）地方时角
local hypothesis of quasars 类星体近域假说
local inertial frame 局域惯性架
local inertial system 局域惯性系 | 本地慣性系〔台〕
local interaction 局部相互作用
Local Interstellar Cloud 本星际云
local ionization equilibrium 局部电离平衡
localized source 近域源 | 局部源〔台〕
local luminosity function 局域光度函数
local lunar time 地方太阴时
local mean noon 地方平午
local mean time （＝LMT）地方平时
local meridian ①地方子午圈 ②地方子午线
local mode 局部模型
local noon 地方正午
local object 局域天体
local oscillator （＝LO）本振
local radio source 局域射电源
local reference frame 局域参考架
local reference system 局域参考系 | 本地參考系〔台〕
local refraction 局域大气折射 | 地方[大氣]折射〔台〕
local regularization 局部正规化
local sidereal time （＝LST）地方恒星时
local standard 本地标准
local standard of rest （＝LSR）本地静止标准，局域静止标准 | 本地靜止標準〔台〕
local star 局域恒星 | 本地恆星〔台〕
local star cloud 局域恒星云
local star stream 局域恒星流 | 本星流〔台〕

local stellar system　局域恒星系统 | 本星團〔台〕
　　又见：*local system*
Local supercluster　本超星系团
Local supergalaxy　本超星系
Local System　本恒星系统
local system　局域恒星系统 | 本星團〔台〕
　　又见：*local stellar system*
local temperature　局域温度 | 局部溫度〔台〕
local thermal equilibrium　（＝LTE）局部热平衡
local thermodynamic equilibrium　（＝LTE）局部热动平衡 | 局部熱力平衡〔台〕
local time　（＝LT）地方时
local true time　地方真时
local uniformizer　局部单值化参数
local universe　近域宇宙
local vertical　地方垂线
lock-in　锁定　又见：*locking*
lock-in amplifier　锁相放大器
locking　锁定　又见：*lock-in*
lock-on　捕获
lock-servo system　锁相伺服系统
locus　【拉】轨迹
locus fictus　【拉】视线黄道面交角
Lodygin crater　洛德金环形山（月球）
LOFAR　（＝Low Frequency Array）LOFAR 低频阵
Logancha crater　洛甘察陨星坑（地球）
logarithmic potential　对数势
logarithmic spiral　对数螺旋线
Loge　土卫四十六
lognormal density field　对数正态分布密度场
lognormal distribution　对数正态分布
lognormal model　对数正态分布模型
Logos　逻各斯（小行星 58534 号）
log-periodic antenna　对数周期天线
log-periodic dipole　对数周期偶极
log-periodic dipole array　对数周期偶极阵
log-periodic feed　对数周期馈源
Lohmann-Bottlinger method　洛曼－玻林格方法
loitering model　游荡模型
Lomia　罗女星（小行星 117 号）
Lommel-Seeliger law　隆梅尔－西利格定律
Lommel-Seeliger surface　隆梅尔－西利格面
Lomonosov crater　罗蒙诺索夫大环形山（月球）
London University Observatory　伦敦大学天文台
LONEOS　（＝Lowell Observatory Near-Earth Object Search）洛厄尔近地天体搜索

long-axis tube orbit　长轴管形轨道
Long Baseline Array　（＝LBA）长基线望远镜阵 | 長基線[望遠鏡]陣〔台〕
long baseline interferometer　（＝LBI）长基线干涉仪
long baseline interferometry　（＝LBI）长基线干涉测量
Long Count　长期积日制（玛雅历）
Long Duration Exposure Facility　（＝LDEF）长期曝露飞行器
long exposure　长时间曝光　又见：*long time exposure*
long-focus photographic astrometry　长焦距照相天体测量
longitude　经度
longitude at the epoch　历元经度
longitude circle　黄经圈　又见：*circle of longitude*
longitude distribution　经度分布
longitude effect　经度效应
longitude of ascending node　升交点经度
longitude of descending node　降交点经度
longitude of node　交点经度
longitude of periastron　近星点经度
longitude of perigee　近地点经度
longitude of perihelion　近日点经度
longitude of the ascending node　升交点经度
longitude of the periapsis　近心点经度
longitudial dispersion corrector　纵向色散改正器
longitudinal aberration　纵向像差
longitudinal chromatic aberration　纵向色差
longitudinal component　纵向分量
longitudinal conductivity　纵向传导率
longitudinal diffusion　纵向扩散
longitudinal diffusion coefficient　纵向扩散系数
longitudinal electrostatic wave　纵静电波
longitudinal field　纵场 | 縱場〔台〕
longitudinal filter　纵向滤波器
longitudinal gauge　纵向规范
longitudinal magnetic field　纵向磁场
　　即：*longitudinal field*
longitudinal magnetic flux　纵向磁通量
longitudinal plasma turbulence　纵向等离子体湍动
longitudinal plasma wave　纵向等离子体波
longitudinal pressure　纵向压力
longitudinal propagation　纵向传播
longitudinal spherical aberration　纵向球差 | 縱向球面像差〔台〕
longitudinal turbulence　纵向湍动

longitudinal viscosity 纵向黏度
longitudinal wave 纵波
longitudinal Zeeman effect 纵向塞曼效应
long-lived spot 长寿黑子，长寿太阳黑子 | 長壽黑子〔台〕 又见：long-lived sunspot
long-lived sunspot 长寿黑子，长寿太阳黑子 | 長壽黑子〔台〕 又见：long-lived spot
Longomontanus crater 隆哥蒙塔努斯环形山（月球）
long period Cepheid 长周期造父变星 即：classical Cepheid
long period comet 长周期彗星
long period perturbation 长周期摄动
long period term 长周期项
long period variable （＝LPV）长周期变星
long-range coupling 长程耦合
long-range force 长程力
long range navigation （＝Loran）远程导航，罗兰
long-slit spectrograph 长缝摄谱仪
long-slit spectroscopy 长缝分光
long-slit spectrum 长缝光谱
long-term fluctuation 长期起伏
long-term forecast 长期预报
long-term perturbation 长期摄动 又见：secular perturbation
long-term stability 长期稳定性 | 長期穩定[度]〔台〕 又见：secular stability
long time exposure 长时间曝光 又见：long exposure
long-wave branch 长波支
longwave photometric system 长波测光系统
longwave photometry 长波测光
longwave radiation 长波辐射
lookback distance 回溯距离
look-back time 回溯时间 | 迴顧時〔台〕
look-up table （＝LUT）对照表
loop feed 环形馈源
Loop I 一号圈（银河）
Loop II 二号圈（银河）
Loop III 三号圈（银河）
loop nebula 圈状星云
loop of retrogression 逆行圈
loop orbit 环轨道
loop prominence 环状日珥 | 圈状日珥〔台〕
loop quantum gravity 圈量子引力
loose group 稀疏星系群
loosely wound arm 松卷旋臂 又见：loosely wound spiral arm

loosely wound spiral arm 松卷旋臂 又见：loosely wound arm
loose phase-locking 弱锁相
loose phase-lock loop 弱锁相回路
Loran （＝long range navigation）远程导航，罗兰
Loran-A chain 罗兰A链
Loran-A system 罗兰A系统
Loran-C chain 罗兰C链
Loran-C system 罗兰C系统
Loran triad 罗兰三角形
lord of the ascendant 首座星
Lorentz contraction 洛伦兹收缩 | 勞侖茲收縮〔台〕
Lorentz crater 洛伦兹环形山（月球）
Lorentz damping 洛伦兹阻尼
Lorentz factor 洛伦兹因子
Lorentz-FitzGerald contraction 洛伦兹－菲茨杰拉德收缩
Lorentz force 洛伦兹力 | 勞侖茲力〔台〕
Lorentz force equation 洛伦兹力方程
Lorentz gas 洛伦兹气体
Lorentz group 洛伦兹群
Lorentzian profile 洛伦兹轮廓
Lorentz invariance 洛伦兹不变性
Lorentz line profile 洛伦兹谱线轮廓
Lorentz metric 洛伦兹度规
Lorentz transformation 洛伦兹变换 | 勞侖茲變換〔台〕
Lorentz unit 洛伦兹单位
Lorentz-Weisskopf broadening 洛伦兹－韦斯科夫致宽
Loschmidt number 洛施密特数
loss cone 损失锥
loss-cone instability 损失锥不稳定性
lossless medium 无损耗介质
Lost City meteorite 洛斯特陨星
lost motion 齿隙差
louver shutter 百叶窗快门
Love crater 勒夫环形山（月球）
Love horn 勒夫喇叭
Lovejoy Comet 洛夫乔伊彗星
Lovelace crater 洛夫莱斯环形山（月球）
Lovell Radio Telescope 洛弗尔射电望远镜 | 洛弗爾電波望遠鏡〔台〕
Lovell Telescope 洛弗尔望远镜
Love number 勒夫数

low-altitude satellite 低高度卫星
low-band filter 低通滤波器 又见: *low pass filter*
low-coefficient glass 低膨胀系数玻璃
low-dispersion spectrograph 低色散摄谱仪
low-dispersion spectrometer 低色散分光计
low-dispersion spectroscopy 低色散分光
low-dispersion spectrum 低色散光谱
low earth orbit （＝LEO）近地轨道
low eccentricity 小偏心率
Lowell crater 洛厄尔环形山（月球）
Lowell Observatory 洛厄尔天文台
Lowell Observatory Near-Earth Object Search （＝LONEOS）洛厄尔近地天体搜索
Lowell's band 洛厄尔带｜羅威爾帶〔台〕（火星）
low-energy component 低能成分
lower atmosphere 低层大气
lower branch 下半圈
lower chromosphere 低层色球，低色球｜低層色球〔台〕
lower circle 恒隐圈，下规｜恆隱圈〔台〕
  又见: *circle of perpetual occultation*
lower corona 低层日冕，低日冕
lower culmination 下中天 又见: *lower transit*
lowered isothermal model 截断等温模型
lower limb 下边缘
lower limit 下限
lower main sequence 下主星序
lower mantle 下地幔
lower transit 下中天 又见: *lower culmination*
low-expansion glass 低膨胀玻璃
Low Frequency Array （＝LOFAR）LOFAR 低频阵
low frequency fluctuation 低频起伏
low frequency radio astronomy 低频射电天文
low-frequency radio telescope 低频射电望远镜
low intensity reciprocity （＝LIR）低强度倒易性
low intensity reciprocity failure （＝LIRF）低强度倒易性失效
low ion 低电离子
low ionization nuclear emission-line region （＝Liner）低电离星系核
lowland 低地
low level 低能级
low light level CCD （＝L3CCD）微光 CCD
low-luminosity galaxy 低光度星系
low-luminosity star 低光度星 又见: *underluminous star*

Lowly concubine of emperor 须女
  又见: *Serving-maid; Waiting-maid*
low-lying state 低能态 又见: *low state*
low-mass binary 小质量双星
low-mass star 小质量星｜低質量星〔台〕
low-mass X-ray binary （＝LMXB）小质量 X 射线双星
low-metallicity 低金属度
low-metallicity cluster ①低金属度星系团｜低金屬豐度星系團〔台〕 又见: *low-metallicity cluster of galaxies* ②低金属度星团｜低金屬豐度星團〔台〕 又见: *low-metallicity cluster of stars*
low-metallicity cluster of galaxies 低金属度星系团｜低金屬豐度星系團〔台〕 又见: *low-metallicity cluster*
low-metallicity cluster of stars 低金属度星团｜低金屬豐度星團〔台〕 又见: *low-metallicity cluster*
low noise 低噪声
low-noise amplifier 低噪声放大器
low pass filter 低通滤波器 又见: *low-band filter*
low-power telescope 低倍率望远镜
low pressure plasma 低压等离子体
low-redshift galaxy （＝low-z galaxy）低红移星系
low-redshift quasar （＝low-z quasar）低红移类星体
low-resolution spectrograph 低分辨率摄谱仪
low-resolution spectroscopy 低分辨率分光
low state 低能态 又见: *low-lying state*
low surface-brightness galaxy （＝LSB galaxy）低表面亮度星系，LSB 星系｜低面亮度星系，LSB 星系〔台〕
low-temperature flare 低温耀斑
low-temperature plasma 低温等离子体
low-temperature star 低温星
low-velocity star 低速星
low water 低潮
low-z galaxy （＝low-redshift galaxy）低红移星系
low-z quasar （＝low-redshift quasar）低红移类星体
loxocosm 地球运行仪
loxodrome 恒向线 又见: *rhumb line*
Loytsianski's theorem 洛伊强斯基定理
LPL （＝Lunar and Planetary Laboratory）月球和行星实验室（美国）
LPV （＝long period variable）长周期变星
Lr （＝lawrencium）铹（103 号元素）
LRG （＝luminous red galaxy）亮红星系

LRO (＝Lunar Reconnaissance Orbiter）月球勘测轨道飞行器（美国探测器）

LRV (＝Lunar Roving Vehicle）月球车 | 月面車〔台〕（美国阿波罗计划）

LSB galaxy (＝low surface-brightness galaxy）低表面亮度星系，LSB 星系 | 低面亮度星系，LSB 星系〔台〕

LS coupling LS 耦合

L-shaped array L 形阵

LSP (＝lightest supersymmetric particle）最轻超对称粒子

LSR (＝local standard of rest）本地静止标准，局域静止标准 | 本地靜止標準〔台〕

LSS (＝large scale structure）大尺度结构

LSST (＝Large Synoptic Survey Telescope）大口径全天巡视望远镜

LST (＝local sidereal time）地方恒星时

L star L 型星

LT ①（＝lag time）滞后时间 ②（＝local time）地方时

LTB model (＝Lemaitre-Tolman-Bondi model）勒梅特－托尔曼－邦迪模型

LTE ①（＝local thermal equilibrium）局部热平衡 ②（＝local thermodynamic equilibrium）局部热动平衡 | 局部热力平衡〔台〕

LTP (＝lunar transient phenomenon）月球暂现现象

Lu (＝lutetium）镥 | 鎦〔台〕（71 号元素）

lucida 最亮星（一个星座内）

lucid star 肉眼可见星

Lucifer 启明星 又见：*Phospherus; Phospher*

luck imaging 幸运成像法

Lucretius crater 卢克莱修环形山（月球）

Lucy 露西（小行星 32605 号）

Lucy algorithm 卢西算法

Luguhu 泸沽湖（小行星 58418 号）

Luichewoo 吕志和（小行星 5538 号）

Lujiaxi 卢嘉锡（小行星 3844 号）

Lulin 鹿林（小行星 145523 号）

luminance 发光率

luminescence 发光

luminosity 光度

luminosity class 光度级

luminosity classification 光度分类

luminosity coefficient 光度系数

luminosity curve 光度曲线

luminosity-decline rate relation 光度－减光率关系

luminosity density 光度密度

luminosity density of the Universe 宇宙光度密度

luminosity distance 光度距离

luminosity distance indicator 光度示距物 又见：*luminosity indicator*

luminosity distribution 光度分布

luminosity evolution 光度演化

luminosity function 光度函数

luminosity indicator 光度示距物 又见：*luminosity distance indicator*

luminosity mass 光度质量

luminosity of night sky 夜天光度

luminosity paradox 光度佯谬 又见：*photometric paradox*

luminosity parallax 光度视差

luminosity profile 光度轮廓

luminosity-rate of variation relation 光度－光变率关系

luminosity segregation 光度分层

luminosity standard 光度标准星

luminosity-volume test 光度－空间体积检验

luminous arc ① 光弧 ② 亮弧

luminous band 光带

luminous blue variable (＝LBV）高光度蓝变星 | 亮藍變星〔台〕

luminous diffuse nebula 亮弥漫星云

luminous dust nebula 亮尘埃星云

luminous efficiency 发光效率

luminous emittance 发光度

luminous energy 光能

luminous flux 光流量 | 光通量〔台〕

luminous flux density [光] 照度 | 照明度〔台〕 又见：*illuminance; illumination*

luminous Galactic nebula 亮银河星云

luminous galaxy nucleus 亮星系核

luminous giant 亮巨星 又见：*bright giant*

luminous high-latitude star 亮高银纬星

luminous infrared galaxy (＝LIRG）亮红外星系

luminous intensity 发光强度

luminous intergalactic matter 亮星系际物质

luminous intergalactic medium 亮星系际介质

luminous interstellar matter 亮星际物质

luminous interstellar medium 亮星际介质

luminous mass 发光质量 | 亮物質〔台〕

luminous nebula 亮星云 | 光星雲〔台〕 又见：*bright nebula*

luminous red galaxy (＝LRG）亮红星系

luminous star 高光度星 | 高光度 [恒] 星〔台〕
又见：*high-luminosity star*
Luna 月球，太阴 又见：*Moon*
lunabase 月海 | [月] 海〔台〕 又见：*marebase*
Luna probe 月球探测器
Luna programme 露娜月球探测项目（苏联）
又见：*Lunik programme*
Lunar and Planetary Laboratory （＝LPL）月球和行星实验室（美国）
lunar appulse ① 月犯星，月星趋近 ② 半影月食
又见：*penumbral lunar eclipse; appulse*
lunar aspect 月相 又见：*phase of the moon; aspect of the moon; lunar phase; moon's phase*
lunar atmosphere 月球大气
Lunar Atmosphere and Dust Environment Explorer （＝LADEE）月球大气及尘埃环境探测器（美国）
lunar aureole 月晕
lunar-based astronomy 月基天文学
lunar-based interferometer 月基干涉仪
lunar-based observatory 月基天文台
lunar-based telescope 月基望远镜
Lunar-based Ultraviolet Telescope （＝LUT）月基紫外天文望远镜
lunar bounce 月球回波
lunar calendar 阴历
lunar chiaroscuro 月相图
lunar circus 月球圆谷 | 月面圆谷〔台〕
lunar crater ① 月面环形山 ② 月面陨击坑 | 月面陨坑〔台〕
Lunar Crater Observation and Sensing Satellite （＝LCROSS）月球环形山观测与遥感卫星（美国）
lunar crust 月壳
lunar cycle 太阴周（19 年） 即：*Metonic cycle*
lunar day 太阴日
lunar dial 月晷 又见：*moondial*
lunar diffraction 月衍射
lunar disk 月面 又见：*lunar surface*
lunar distance 月球距离
lunar dome 月丘
lunar dust 月尘
lunar eclipse 月食 又见：*eclipse of the moon*
lunar ecliptic limit 月食限
lunar ephemeris 月球历表
lunar equation 月离，月行差
lunar excursion 登月
lunar excursion module （＝LEM）登月舱（美国阿波罗计划）

lunar fines 月球岩粉
lunar geology 月质学
lunar grid 月面视栅
lunar highland 月面高地
lunar inequality 月行差 又见：*variational inequality*
lunar interval 月蹉
lunar ionosphere 月球电离层
lunarite 月陆
lunarium 月球运行仪
lunar lander 月球着陆器 | 登月飞行器〔台〕
lunar laser ranging （＝LLR）激光测月 | 月球雷射测距〔台〕
lunar libration 月球天平动
lunar lodge degrees 入宿度 又见：*determinative star distance; equatorial lodge degrees*
lunar mansion 宿 又见：*mansion*
lunar map 月面图 又见：*selenograph; selenographic chart; selenographic map*
lunar meteoroid 月球流星体
lunar module （＝LM）登月舱
lunar month 朔望月，太阴月 又见：*synodic month; moon month; month of the phases; lunation*
lunar mountain 月球山系
lunar node 月轨交点
lunar nodule 月岩体
lunar nomenclature 月面命名 | 月面命名法〔台〕
lunar noon 月球正午
lunar nutation 月球章动
lunar occultation 月掩星，月掩源 | 月掩星〔台〕
lunar occultation technique 月掩星技术
lunar orbit 月球轨道 又见：*moon's orbit*
Lunar Orbiter 月球轨道飞行器（美国探测器）
lunar orbiter 环月飞行器 | 環月衛星，環月太空船〔台〕
lunar orbit rendezvous 绕月会合
lunar parallax 月球视差
lunar phase 月相 又见：*phase of the moon; aspect of the moon; lunar aspect; moon's phase*
lunar pole 月极
lunar primeval element 月球原始元素
lunar probe 月球探测器
Lunar Prospector 月球勘探者 | 月球勘探者號〔台〕（美国探测器）
lunar radar time-synchronization 月球雷达时间同步
lunar radiation 月球辐射
lunar rays 月面辐射纹 | [月面] 輻射紋〔台〕
又见：*rays of the moon*

Lunar Reconnaissance Orbiter （＝LRO）月球勘测轨道飞行器（美国探测器）
lunar rock 月岩　又见：*moonrock*
Lunar Rover 月球巡视器，月球车　又见：*lunar rover vehicle*
lunar rover vehicle 月球巡视器，月球车　又见：*Lunar Rover*
Lunar Roving Vehicle （＝LRV）月球车 | 月面车〔台〕（美国阿波罗计划）
lunar satellite 月球卫星　又见：*selenoid*
lunar seismology 月震学
lunar seismometer 月震计
lunar space 近月空间　又见：*near-lunar space*
lunar surface 月面　又见：*lunar disk*
lunar survey probe 月球探测器
lunar table 月球表
lunar theory 月离理论 | 月球運動說〔台〕
lunar tidal wave 月潮波
lunar tide 月潮
lunar time 太阴时
lunar topography 月志学　即：*selenography*
lunar trajectory 赴月轨道
lunar transient phenomenon （＝LTP）月球暂现现象
lunar year 太阴年
lunar zodiac 二十八宿　又见：*28 lunar mansions; twenty-eight lunar mansions*
lunation 朔望月，太阴月　又见：*synodic month; lunar month; moon month; month of the phases*
lunation number 朔望月数
lunation numerator 朔实
Lundmark crater 伦德马克环形山（月球）
Lundquist number 伦德奎斯特数
Lund Royal University Observatory 伦德皇家大学天文台（瑞典）
Lunik programme 露娜月球探测项目（苏联）　又见：*Luna programme*
lunisolar calendar 阴阳历 | 陰陽 [合] 曆〔台〕
lunisolar diurnal wave 日月周日波
lunisolar nutation 日月章动
lunisolar period 日月周期
lunisolar perturbation 日月摄动
lunisolar precession 日月岁差
lunisolar year 阴阳年
luni-tidal interval 月潮间隔
Lunokhod 月球车 | 月面車〔台〕（苏联）
Luoxiahong 落下闳（小行星 16757 号）
Lup （＝Lupus）豺狼座

Lupus （＝Lup）豺狼座
Lupus Loop 豺狼圈
Lupus Z Dark Cloud 豺狼 Z 暗云
LUT ① （＝look-up table）对照表
② （＝Lunar-based Ultraviolet Telescope）月基紫外天文望远镜
Lutetia 司琴星（小行星 21 号）
lutetium （＝Lu）镥 | 鑥〔台〕（71 号元素）
Lutz-Kelker Bias 卢茨－凯克尔偏差
Luyajia 吕亚佳（小行星 20830 号）
Luyi 鹿邑（小行星 4776 号）
ly （＝light year）光年
Lyapunov characteristic number （＝LCN）李雅普诺夫特征数
Lyapunov exponent 李雅普诺夫指数
Lyapunov function 李雅普诺夫函数
Lyapunov stability 李雅普诺夫稳定性
Lydia 吕女星（小行星 110 号）
Lydia family 吕蒂亚族
Lyman alpha absorption 莱曼 α 吸收　又见：*Lyα absorption*
Lyman alpha emitter （＝LAE）莱曼 α 发射体　又见：*Lyα emitter*
Lyman break galaxy （＝LBG）莱曼断裂星系 | 來曼斷裂星系〔台〕
Lyman break method 莱曼断裂方法
Lyman continuum 莱曼连续区 | 來曼連續區〔台〕
Lyman crater 莱曼环形山（月球）
Lyman discontinuity 莱曼跃变
Lyman ghost 莱曼鬼线
Lyman jump 莱曼跳变
Lyman limit 莱曼系限 | 來曼系限〔台〕
Lyman-limit absorption 莱曼系限吸收
Lyman-limit galaxy 莱曼系限星系
Lyman-limit system 莱曼系限系统
Lyman line 莱曼线
Lyman series 莱曼系 | 來曼系〔台〕
Lyman-Werner photon 莱曼－沃纳光子
Lyman-α blob （＝LAB）莱曼 α 团块
Lyman-α cloud （＝Ly-α cloud）莱曼 α 云
Lyman-α forest （＝Ly-α forest）莱曼 α 森林 | 來曼 α 叢〔台〕
Lyman-α galaxy （＝Ly-α galaxy）莱曼 α 星系 | 來曼 α 星系〔台〕
Lyman-α line （＝Ly-α line）莱曼 α 线
Lyman-β line （＝Ly-β line）莱曼 β 线
Lyn （＝Lynx）天猫座
Lynds catalogue 林茨暗星云表

Lynx (＝Lyn）天猫座
Lyons University Observatory 里昂大学天文台
Lyot coronagraph 李奥星冕仪，李奥日冕仪
Lyot crater 李奥环形山（月球）
Lyot depolarizer 李奥消偏振镜
Lyot division 李奥环缝（土星）
Lyot filter 李奥滤光器 | 利奥濾鏡〔台〕
Lyot-Oehman filter 李奥－欧曼滤光器
Lyot stop 李奥光阑
Lyr (＝Lyra）天琴座
Lyra (＝Lyr）天琴座
Lyrid meteor shower 天琴座流星雨
Lyrids 天琴流星群 | 天琴[座]流星雨〔台〕
Lysithea 木卫十

Lyα absorption 莱曼 α 吸收　又见：*Lyman alpha absorption*
Ly-α cloud (＝Lyman-α cloud）莱曼 α 云
Lyα emitter 莱曼 α 发射体　又见：*Lyman alpha emitter*
Ly-α forest (＝Lyman-α forest）莱曼 α 森林 | 來曼 α 叢〔台〕
Ly-α galaxy (＝Lyman-α galaxy）莱曼 α 星系 | 來曼 α 星系〔台〕
Ly-α line (＝Lyman-α line）莱曼 α 线
Lyα scattering 莱曼 α 散射
Ly-β line (＝Lyman-β line）莱曼 β 线
LZT (＝Large Zenith Telescope）大型天顶望远镜（加拿大）

M

M （=Messier Catalogue）梅西叶星云星团表 | 梅西耳星表〔台〕
Maat Mons 玛阿特山（金星）
Mab 天卫二十六
Mabsuthat 上台增四（天猫座 31）
　　又见：Alsciaukat
Macao 澳门（小行星 8423 号）
Mach crater 马赫环形山（月球）
Machina Electrica 电机座（现已不用）
machine language 机器语言
machine-readable form 机器可读形式
Mach number 马赫数
MACHO （=massive compact halo object）晕族大质量致密天体 | 大質量緻密暈天體〔台〕
Mach principle 马赫原理 又见：Mach's principle
Mach's principle 马赫原理 又见：Mach principle
MacLaurin crater 麦克劳林环形山（月球）
MacLaurin disk 麦克劳林盘
MacLaurin spheroid 麦克劳林球体 | 馬氏球體〔台〕
MacLaurin stage of light curve 麦克劳林光变曲线分段
Macrobius crater 马克罗比乌斯环形山（月球）
macrocosmos 大宇宙
macrolensing 巨引力透镜
macrolensing effect 巨引力透镜效应 | 巨重力透鏡效應〔台〕
macroquake 宏震
macroscopic motion 宏观运动
macroscopic quantity 宏观量
macroscopic scattering cross-section 宏观散射截面
macrospicule 巨针状体
macroturbulence 宏观湍流
macroturbulent motion 宏观湍动
macula 暗斑
Madau plot 马道图

Madrid Astronomical Observatory 马德里天文台
Maffei galaxy 马费伊星系
Maffei group 马费伊星系群
MAG （=Mitteilungen der Astronomischen Gesellschaft）【德】《天文学会会刊》
mag （=magnitude）星等
Magdalena Ridge Optical Interferometer （=MROI）马格达莱纳岭光学干涉仪
Magelhaens crater 麦哲伦环形山（月球）
Magellanic Clouds 麦哲伦云 又见：Nubeculae
Magellanic Cloud-type galaxy 麦哲伦云型星系
Magellanic irregular galaxies 麦哲伦型不规则星系
Magellanic Stream 麦哲伦流
Magellanic System 麦哲伦系
Magellan spacecraft 麦哲伦号金星探测器 | 麥哲倫號[金星探測器]〔台〕（美国）
Magellan Telescopes 麦哲伦望远镜
MAGIC （=Major Atmospheric Gamma-ray Imaging Cherenkov Telescopes）大型大气伽马射线成像切伦科夫望远镜
magic nucleus 幻核
magic number 幻数
magma 岩浆
magma theory 岩浆学说
Magnanimity 宽容（小行星 8992 号）
magnesium （=Mg）镁（12 号元素）
magnetar 强磁星
magnetic active feature 磁活动体
magnetic activity 磁活动性 | 磁活動〔台〕
magnetically driven shock-wave 磁致激波
magnetic annihilation 磁湮灭
magnetic anomaly 磁异常
magnetic arc 磁弧
magnetic arcade ① 磁环 又见：magnetic loop ② 磁拱
magnetic arm 磁臂

magnetic axis 磁轴
magnetic azimuth 磁方位角
magnetic binary 磁双星
magnetic bottle 磁瓶
magnetic braking 磁阻尼
magnetic bremsstrahlung 磁韧致辐射 | 磁製動輻射〔台〕 又见：*magnetobrems; magnetobremsstrahlung*
magnetic buoyancy 磁浮力
magnetic cancellation 磁对消
magnetic canopy 磁蓬
magnetic cataclysmic binary 磁激变双星
magnetic cataclysmic variable 磁激变变星
magnetic cavity 磁穴
magnetic cell 磁胞
magnetic chart 磁图 | 磁[場]强[度]圖〔台〕
　又见：*magnetogram*
magnetic classification 磁场分类，磁分类
magnetic cloud 磁云
magnetic coalescence 磁吞并
magnetic collapse 磁坍缩
magnetic compression 磁压缩
magnetic conductivity 磁导率，导磁性
magnetic confinement 磁约束
magnetic connection 磁联结
magnetic cooling 磁致冷 | 磁[致]冷卻〔台〕
magnetic coupling 磁耦合
magnetic crochet 磁钩 又见：*crochet*
magnetic cumulation 磁累积
magnetic curve 磁变曲线
magnetic cycle 磁周
magnetic declination 磁偏角
magnetic density 磁密度
magnetic diffusion 磁扩散
magnetic diffusivity 磁扩散率
magnetic dip 磁倾角，磁凹陷 又见：*magnetic inclination*
magnetic dipole 磁偶极子 | 磁偶極[子]〔台〕
magnetic dipole moment 磁偶极矩
magnetic dipole radiation 磁偶极辐射
magnetic dipole transition 磁偶极跃迁
magnetic disorder 磁无序
magnetic disturbance 磁扰
magnetic dot 磁点
magnetic dynamo 磁发电机
magnetic element 磁元
magnetic energy 磁能
magnetic equator 磁赤道

magnetic field 磁场
Magnetic Field Satellite （＝Magsat）磁场科学卫星（美国）
magnetic field wind 磁场风
magnetic filament 磁暗条
magnetic flare 磁耀斑
magnetic flux 磁流 | 磁流[量]〔台〕
　又见：*magnetic stream*
magnetic flux conservation 磁流守恒
magnetic flux density 磁流密度
magnetic flux loop 磁流环，磁环
magnetic flux rope 磁流绳，磁绳
magnetic flux tube 磁流管
magnetic focusing 磁致聚焦
magnetic force 磁力
magnetic gradient 磁场梯度 | 磁梯度〔台〕
magnetic helicity 磁螺度
magnetic hill 磁丘 | 磁峰〔台〕
magnetic inclination 磁倾角，磁凹陷
　又见：*magnetic dip*
magnetic index 磁指数
magnetic induction 磁感应 | 磁感應，磁感應强度〔台〕
magnetic instability 磁不稳定性
magnetic intensity 磁场强度
magnetic interaction 磁相互作用
magnetic inversion line 磁反向线
magnetic knot 磁结
magnetic leakage 磁漏
magnetic line 磁力线
magnetic loop 磁环 又见：*magnetic arcade*
magnetic Mach number 磁马赫数
magnetic macroturbulence 磁宏观湍流
magnetic merging 磁并合
magnetic meridian 磁子午线
magnetic microturbulence 磁微观湍流
magnetic mirror 磁镜
magnetic moment 磁矩
magnetic monopole 磁单极
magnetic monopole radiation 磁单极辐射
magnetic motive force 磁动力
magnetic multipole 磁多极
magnetic multipole radiation 磁多极辐射
magnetic neutral line 磁中性线
magnetic north 磁北
magnetic obliquity 磁倾斜
magnetic oscillation 磁振荡

magnetic patch 磁块
magnetic period 磁变周期
magnetic permeability 磁导率　又见：*permeability*
magnetic phase 磁相位
magnetic plasma 磁等离子体
magnetic plasma configuration 磁等离子体位形
magnetic polarity 磁极性
magnetic polarization 磁极化
magnetic pole 磁极
magnetic potential 磁势 | 磁势，磁位〔台〕
magnetic pressure 磁压
magnetic prestorm 前磁暴
magnetic prime vertical 磁卯酉圈
magnetic property 磁特性
magnetic puka 磁洞
magnetic quadrupole 磁四极
magnetic quadrupole radiation 磁四极辐射
magnetic reconnection 磁重联
magnetic region 磁区
magnetic Reynolds number 磁雷诺数
magnetic rigidity 磁刚性
magnetic rope 磁绳
magnetic separator 磁拓扑界线
magnetic separatrix 磁拓扑界面
magnetic shear 磁剪切 | 磁[场]剪切〔台〕
magnetic sheath 磁鞘　又见：*magneto-sheath*
magnetic shell 磁壳
magnetic shield 磁屏蔽
magnetic spectrometer 磁谱仪
magnetic spherule 磁小球体
magnetic spiral 磁螺线
magnetic star 磁星
magnetic storm 磁暴
magnetic stream 磁流 | 磁流[量]〔台〕
　　又见：*magnetic flux*
magnetic strength 磁强
magnetic stress 磁致应力
magnetic tape 磁带
magnetic tape unit 磁带机
magnetic tongue 磁舌
magnetic trap 磁阱
magnetic turbulence 磁致湍流
　　又见：*magneto-turbulence*
magnetic twist 磁扭绞 | 磁[场]扭绞〔台〕
magnetic variable 磁变星
magnetic variometer 磁变计
magnetic viscosity 磁黏性

magnetic wave 磁波
magnetized plasma 磁化等离子体
magnetoacoustic gravity 磁声重力
magnetoacoustic gravity wave 磁声重力波
magnetoacoustic wave 磁声波　又见：*magnetosonic wave; magneto-sound wave*
magnetoactive plasma 磁活动等离子体
magnetobrems 磁韧致辐射 | 磁製動輻射〔台〕
　　又见：*magnetic bremsstrahlung; magnetobremsstrahlung*
magnetobremsstrahlung 磁韧致辐射 | 磁製動輻射〔台〕　又见：*magnetic bremsstrahlung; magnetobrems*
magneto-bremstrahlung radiation 磁韧致辐射，磁阻尼辐射
magneto-convection 磁对流
magnetodynamics 磁动力学
magnetofluid 磁流体
magnetogram 磁图 | 磁[场]强[度]圖〔台〕
　　又见：*magnetic chart*
magnetograph 磁像仪 | 磁[场]强[度]計〔台〕
magnetogravity 磁重力
magnetogravity wave 磁重力波
magnetoheliograph 太阳磁像仪 | 太陽磁[場]强[度]計〔台〕　又见：*solar magnetograph*
magnetohydrodynamic approximation 磁流[体动]力学近似
magnetohydrodynamic compression 磁流体压缩
magnetohydrodynamic instability 磁流[体动]力学不稳定性 | 磁流體不穩定度〔台〕
magnetohydrodynamics （＝MHD）磁流[体动]力学 | 磁流體[動]力學〔台〕
magnetohydrodynamic shock-wave 磁流[体动力]激波
magnetohydrodynamic turbulence 磁流[体动]力学湍流 | 磁流湍流〔台〕
magnetohydrodynamic wave 磁流[体动力]波 | 磁流[體力學]波〔台〕
magnetoid 磁旋体　又见：*spinar*
magneto-ionic theory 磁离子理论 | 磁離說〔台〕
magnetometer 磁强计 | 磁[場]强[度]計〔台〕
magneton 磁子
magneto-optical effect 磁光效应
magneto-optical filter （＝MOF）磁光滤光器
magneto-optics 磁光学
magnetopause 磁层顶
magnetoplasma dynamics 磁等离子体动力学
magnetorotational instability （＝MRI）磁转动不稳定性
magneto-sheath 磁鞘　又见：*magnetic sheath*

magnetosonic mode　磁声模式
magnetosonic wave　磁声波　又见：*magnetoacoustic wave*; *magneto-sound wave*
magneto-sound turbulence　磁声湍流
magneto-sound wave　磁声波　又见：*magnetoacoustic wave*; *magnetosonic wave*
magnetosphere　磁层
magnetospheric accretion　磁层吸积
magnetospheric discontinuity　磁层间断
magnetospheric eternally collapsing object　（＝MECO）磁层永坍缩体
magnetospheric model　磁层模型
magnetospheric storm　磁层暴
magnetospheric substorm　磁层亚暴
magnetospheric tail　磁尾　又见：*magnetotail*
magnetotail　磁尾　又见：*magnetospheric tail*
magneto-turbulence　磁致湍流　又见：*magnetic turbulence*
magnification factor　放大因子　又见：*amplification factor*
magnification tensor　放大率张量
magnifying glass　放大镜
magnifying lens　放大透镜
magnifying power　放大率
magnitude　（＝mag）星等　又见：*stellar magnitude*
magnitude at opposition　冲时星等
magnitude-color-index diagram　星等－色指数图
magnitude-color-index relation　星等－色指数关系
magnitude difference　星等差　又见：*magnitude equation*
magnitude distortion　星等畸变
magnitude effect　星等效应
magnitude equation　星等差　又见：*magnitude difference*
magnitude of an eclipse　食分
magnitude of eclipse　食分　又见：*degree of obscuration*; *eclipse factor*
magnitude of greatest eclipse　最大食分　又见：*maximum eclipse*
magnitude range　星等范围
magnitude-redshift diagram　星等－红移图
magnitude-redshift relation　星等－红移关系
magnitude scale　星等标 | 星等標 [度] 〔台〕
magnitude screen　星等订正屏 | [訂正] 星等網屏〔台〕
magnitude-spectral type diagram　星等－光谱型图
magnitude-spectral type relation　星等－光谱型关系
magnitude system　星等系统
magnon　磁振子
Magsat　（＝Magnetic Field Satellite）磁场科学卫星（美国）
Maia　昴宿四（金牛座20）
Maia sequence　昴宿四变星序
Maid　女宿（二十八宿）　又见：*Girl*; *Woman*; *Damsel*
main asteroid belt　小行星主带 | 主小行星带〔台〕
main beam　主波束，主射束
main-beam efficiency　主束效率
main belt　主带
main-belt asteroid　主带小行星
main-belt comet　主带彗星
main image　主像　又见：*primary image*
main line　主谱线
main lobe　主瓣
main-lobe solid angle　主瓣立体角
main maximum　主极大　又见：*primary maximum*; *principal maximum*
main minimum　主极小　又见：*primary minimum*; *principal minimum*
main pulse　主脉冲
main reflector　主反射镜
main resonance　主共振
Main Ring　主环（木星）
main sequence　主序 | 主星序〔台〕
main sequence band　主序带
main sequence fitting　主序拟合
main sequence lifetime　主序寿命
main sequence star　（＝MS star）主序星
main source　主源
main spot　主黑子
main sunspot　主太阳黑子
main tail　主彗尾
Maja　光神星（小行星66号）
Majorana mass term　马约拉纳质量项
Majorana neutrino　马约拉纳中微子
Majorana particle　马约拉纳粒子
Majorana representation　马约拉纳表象
majorant series　强级数
Major Atmospheric Gamma-ray Imaging Cherenkov Telescopes　（＝MAGIC）大型大气伽马射线成像切伦科夫望远镜
major axis　长径，长轴 | 長軸〔台〕
major flare　大耀斑 | 大閃焰〔台〕　又见：*big flare*

major limit 大限
major merger 主并合
major planet 大行星 又见：*principal planet*
Makemake 鸟神星（小行星136472号，矮行星）
Maksutov camera 马克苏托夫照相机
Maksutov corrector 马克苏托夫改正镜
Maksutov crater 马克苏托夫环形山（月球）
Maksutov telescope 马克苏托夫望远镜
Mallet crater 马利特环形山（月球）
Malmquist bias 马姆奎斯特偏差
Malmquist correction 马姆奎斯特改正
Malus 船桅座（现已不用）
MAMA （=multi-anode microchannel array）多阳极微通道阵
Mandel'shtam crater 曼德尔施塔姆环形山（月球）
maneuvering satellite 机动卫星
Mangalyaan 火星轨道飞行器任务 又见：*Mars Orbiter Mission*
manganese （=Mn）锰（25号元素）
manganese-mercury star 锰汞星
manganese star 锰星
Manicouagan crater 马尼夸根陨星坑（地球）
manifest covariance 显协变性
manifold 流形
Manley-Rowe equation 曼利－罗氏方程
man-made interference 人为干涉
manned balloon 载人气球
manned flight 载人飞行
Manned Orbiting Laboratory （=MOL）载人环地轨道实验室（美国）
manned rocket 载人火箭
manned spacecraft 载人飞船
manned vehicle 载人飞行器
mansion 宿 又见：*lunar mansion*
Manson crater 曼森陨星坑（地球）
mantle 幔，地幔｜函〔台〕
mantle convection 地幔对流
mantle plume 地幔柱
manual control 手控
many-body problem 多体问题 又见：*multiple-body problem; problem of many bodies*
many-particle system 多粒子系统
Maoyisheng 茅以升（小行星18550号）
MAP （=multichannel astrometric photometer）多通道天测光度计
map cleaning 图像洁化

map distortion 图像畸变｜影像扭曲〔台〕 又见：*image distortion*
Mapihsia 马碧霞（小行星216261号）
mapping 测绘，制图
mapping source 测绘源
Marchab 室宿一（飞马座α） 又见：*Markab*
marching pulse 漂移脉冲 又见：*drift pulse; drifting pulse*
marching subpulse 漂移次脉冲 又见：*drift subpulse; drifting subpulse*
Marconi crater 马可尼环形山（月球）
Marco Polo crater 马可·波罗环形山（月球）
mare 【拉】[月]海（月面地貌）
M-area M区 又见：*M region*
Mare Anguis 【拉】蛇海（月球）
Mare Australe 【拉】南海（月球）
marebase 月海｜[月]海〔台〕 又见：*lunabase*
Mare Cognitum 【拉】知海（月球）
Mare Crisium 【拉】危海（月球）
Mare Desiderii 【拉】梦海（月球，已废弃）
Mare Fecunditatis 【拉】丰富海（月球）
Mare Frigoris 【拉】冷海（月球）
Mare Humboldtianum 【拉】洪堡海（月球）
Mare Humorum 【拉】湿海（月球）
Mare Imbrium 【拉】雨海（月球）
Mare Ingenii 【拉】智海（月球）
Mare Insularum 【拉】岛海（月球）
mare-like plain 类海平原
Mare Marginis 【拉】界海（月球）
Mare Moscoviense 【拉】莫斯科海（月球）
Mare Nectaris 【拉】酒海（月球）
Mare Nubium 【拉】云海（月球）
Mare Orientale 【拉】东海（月球）
Mare Serenitatis 【拉】澄海（月球）
Mare Smythii 【拉】史密斯海（月球）
Mare Spumans 【拉】泡沫海（月球）
Mare Tranquillitatis 【拉】静海｜[宁]靜海〔台〕（月球）
Mare Undarum 【拉】浪海（月球）
Mare Vaporum 【拉】汽海（月球）
Marfak ①晋，天市右垣三（武仙座κ） 又见：*Marsik* ②天船三（英仙座α） 又见：*Mirfak; Mirphak*
Marfic 列肆二（蛇夫座λ） 又见：*Marfik*
Marfik 列肆二（蛇夫座λ） 又见：*Marfic*
Margaret 天卫二十三
marginalization 边缘化

marginally bound orbit　边缘束缚轨道
marginally stable orbit　边缘稳定轨道
marginal stability　边界稳定性
Maria family　玛利亚星族
marial rock　海岩（月球）
Marilyn　玛莉琳（小行星 1486 号）
Mariner　水手号（美国行星探测器）
Marinus crater　马里纳斯环形山（月球）
Mariotte crater　马略特环形山（月球）
Marisat　海事通信卫星（美国）
Markab　① 室宿一（飞马座 α）　又见：*Marchab*
　② 天社五（船帆座 κ）　又见：*Markeb*
Markarian galaxy　马卡良星系｜馬氏型星系〔台〕
Markarian object　马卡良天体
Markeb　天社五（船帆座 κ）　又见：*Markab*
marking　斑纹，斑点
Markov chain　马尔可夫链
Markov chain Monte Carlo method　（＝MCMC method）马尔可夫链蒙特卡洛方法
Markov process　马尔可夫过程
Markowitz moon camera　马科维茨月球照相机｜馬科維茨月球相機〔台〕
Mars　火星，荧惑
Mars Atmosphere and Volatile Evolution　（＝MAVEN）专家号（美国火星探测器）
Mars Climate Orbiter　（＝MCO）环火星气候探测器（美国）
Mars-crossing asteroid　越火小行星
Marseilles National Observatory　马赛国立天文台（法国）
Mars Exploration Rover　火星探险漫游者
Mars Express　火星快车（欧洲火星探测器）
Mars Global Surveyor　（＝MGS）火星环球勘测者
Mars globe　火星仪
Marshall Space Flight Center　（＝MSFC）马歇尔空间中心（美国）
Marsik　晋，天市右垣三（武仙座 κ）
　又见：*Marfak*
Mars Observer　火星观测者（美国火星探测器）
Mars Orbiter Mission　火星轨道飞行器任务
　又见：*Mangalyaan*
Mars Pathfinder　火星探路者｜火星拓荒者號〔台〕（美国火星探测器）
Mars Polar Lander　（＝MPL）火星极地着陆器
Mars probe　火星探测器（苏联）
marsquake　火星地震

Mars Reconnaissance Orbiter　（＝MRO）火星勘测轨道飞行器
Mars Science Laboratory　（＝MSL）火星科学实验室
Mars-Trojan asteroid　火星－特洛伊族小行星
Martian atmosphere　火星大气
Martian dust storm　火星尘暴
Martian moon　火卫　又见：*Martian satellite; satellite of Mars*
Martian opposition　火星大冲
Martian probe　火星探测器
Martian satellite　火卫　又见：*Martian moon; satellite of Mars*
Martian year　火星年
Maryland Point Observatory　马里兰射电观测站
mas　（＝milliarcsecond）毫角秒
mascon　质量瘤｜重力异常区〔台〕　又见：*mass concentration*
maser　微波激射，脉泽｜邁射〔台〕
　又见：*microwave amplification by stimulated emission of radiation*
maser amplifier　微波激射器
maser source　微波激射源｜邁射源〔台〕
maser star　微波激射星，脉泽星｜邁射星〔台〕
masking aperture　拦光孔径
mass　质量
mass absorption coefficient　质量吸收系数
Massachusetts Institute of Technology　（＝MIT）麻省理工学院
Massalia　王后星（小行星 20 号）
mass balance　质量平衡
mass concentration　质量瘤｜重力异常区〔台〕
　又见：*mascon*
mass conservation　质量守恒　又见：*conservation of mass*
mass defect　质量亏损
mass density　质量密度
mass discrepancy　质量差异｜質量差〔台〕
mass distribution　质量分布
mass ejection　质量抛射
mass-energy equivalence　质能相当性
mass-energy relation　质能关系
mass equation　质量方程
mass exchange　质量交换　又见：*exchange of mass*
mass extinction　质量消散
mass factor　质量因子
mass flow　质量流
mass fraction　质量百分比

mass function 质量函数
mass-gaining object 增质量天体
mass-gaining star 增质量星
massive black hole 大质量黑洞
massive compact halo object （=MACHO）晕族大质量致密天体 | 大質量緻密暈天體〔台〕
massive galaxy 大质量星系
massive halo 大质量晕
massive neutrino 有质量中微子
massive object 大质量天体
massive star 大质量星 又见：*high-mass star*
massive X-ray binary （=MXRB）大质量 X 射线双星 | 大質量 X 光雙星〔台〕
massless particle 无质量粒子
mass-losing object 损质量天体
mass-losing star 损质量星
mass loss 质量损失
mass-loss rate 质量损失率
mass-luminosity curve 质光曲线 | 質－光曲線〔台〕
mass-luminosity law 质光定律 | 質－光定率〔台〕
mass-luminosity-radius relation 质光半径关系 | 質量－光度－半徑關係〔台〕
mass-luminosity ratio 质光比 又见：*mass-to-light ratio; mass-to-luminosity ratio*
mass-luminosity relation （=M-L relation）质光关系
mass matrix 质量矩阵
mass-metallicity relation 质量－金属丰度关系
mass moment 质量矩
mass motion 整体运动
mass-radius relation 质量半径关系 | 質量－半徑關係〔台〕
mass ratio 质量比
mass scattering 质量散射
mass scattering coefficient 质量散射系数
mass segregation 质量层化
mass shell 质壳
mass-shell condition 质壳条件
mass spectrograph 质谱仪
mass spectrometer 质谱计 | 質譜儀〔台〕
mass spectrometry 质谱分析 | 質譜學，質譜分析〔台〕
mass spectrum 质谱
mass-temperature relation 质温关系 | 質量－溫度關係〔台〕
mass-to-charge ratio 质荷比

mass-to-light ratio 质光比 又见：*mass-luminosity ratio; mass-to-luminosity ratio*
mass-to-luminosity ratio 质光比 又见：*mass-to-light ratio; mass-luminosity ratio*
mass transfer 质量转移
master clock 母钟 又见：*primary clock*
master equation 主方程
master station 主站
Masym 赵，天市左垣二（武仙座 λ） 又见：*Mazym; Mazim*
Matar 离宫四（飞马座 η）
matched filter 匹配滤波器
matching principle 匹配原理
material arm 物质臂
mathematical expectation 数学期望
mathematical horizon 数学地平，理想地平
mathematical mode 数学模式
mathematical model 数学模型
mathematical pendulum 数学摆，单摆
mathematical statistics 数理统计
Mathieu equation 马蒂厄方程
Mathilde 玛蒂尔德（253 号小行星）
matrix inversion 矩阵求逆
matrix representation 矩阵表示
matter 物质
Matterania 马特尔（小行星 883 号）
matter-antimatter cosmology 物质－反物质宇宙学 | 物質－反物質宇宙論〔台〕
matter-antimatter symmetry 物质－反物质对称性
matter dominated era 物质主导期，物质占优期 | 物質主導期〔台〕
matter dominated universe 物质主导宇宙，物质占优宇宙 | 物質主導宇宙〔台〕
matter era 物质期 | 物質時代〔台〕
matter-radiation equality 物质辐射等量
Mattig formula 马蒂希公式
mature prominence 成熟日珥
Mauna Kea Observatory （=MKO）莫纳克亚天文台
Maunder crater 蒙德环形山（月球）
Maunder minimum 蒙德极小期 | 芒得極小期〔台〕
MAVEN （=Mars Atmosphere and Volatile Evolution）专家号（美国火星探测器）
MAXI （=Monitor of All-sky X-ray Image）全天 X 射线图像监视器

MAXIMA （=Millimeter Anisotropy eXperiment IMaging Array）毫米波各向异性实验成像阵（美国）
maximal disk model 最大盘模型
maximum 极大
maximum activity 极大活动性
maximum corona 极大期日冕，极大日冕
maximum eclipse 最大食分　又见：magnitude of greatest eclipse
maximum entropy 最大熵
maximum entropy method （=MEM）最大熵法
maximum likelihood estimation （=MLE）极大似然估计
maximum likelihood method 极大似然法
maximum of eclipse 食甚　又见：middle of eclipse
maximum tide 高潮
Max-Planck-Institute for Astrophysics （=MPA）马普天体物理研究所
Max-Planck-Institut fur Astronomie （=MPIfA）【德】马普天文研究所
Max-Planck-Institut fur Physik und Astrophysik （=MPIfPA）【德】马普物理和天体物理研究所
Max-Planck-Institut fur Radioastronomie （=MPIfR）【德】马普射电天文研究所
Maxwell-Boltzmann distribution 麦克斯韦－玻尔兹曼分布
Maxwell crater 麦克斯韦环形山（月球）
Maxwell distribution 麦克斯韦分布 | 馬克斯威分佈〔台〕
Maxwell equation 麦克斯韦方程 | 馬克斯威方程〔台〕
Maxwell field equation 麦克斯韦场方程
Maxwell force 麦克斯韦力
Maxwell Gap 麦克斯韦环缝 | 馬克斯威爾環縫〔台〕（土星）
Maxwellian plasma 麦克斯韦等离子体
Maxwellian plasma gas 麦克斯韦等离子气体
Maxwellian velocity distribution 麦克斯韦速度分布
Maxwell Montes 麦克斯韦山脉（金星）
Maxwell Observatory 麦克斯韦天文台
Maxwell Telescope 麦克斯韦望远镜
Mayall Reflector 梅奥尔反射望远镜
Mayall Telescope 梅奥尔望远镜（美国）
Mayan astronomy 玛雅天文学 | 馬雅天文學〔台〕
Mayer formula 梅耶公式
Mazim 赵，天市左垣二（武仙座λ）
　又见：Masym; Mazym
Mazym 赵，天市左垣二（武仙座λ）
　又见：Masym; Mazim
MBR （=microwave background radiation）微波背景辐射
MBTS （=moon bounce time synchronization）月球回波时间同步
M burst M 型暴
MC （=meridian circle）子午环
MCAO （=multi-congugate adaptive optics）多重共轭自适应光学
McDonald crater 麦克唐纳环形山（月球）
McDonald Observatory 麦克唐纳天文台
MCF （=mutual coherence function）互相干函数
MCG （=Morphological Catalogue of Galaxies）星系形态表
McGee spectracon 麦基光谱摄像管
McGraw-Hill Observatory 麦格劳－希尔天文台
McIntosh classification 麦金托什分类
McIntosh scheme 麦金托什图式
McKellar crater 麦凯勒环形山（月球）
M-class asteroid M 型小行星
McLaughlin crater 麦克劳林环形山（月球）
McMath crater 麦克马思环形山（月球）
McMath Hulbert Observatory 麦克马思·赫伯特天文台
McMath-Pierce Solar Telescope 麦克马思－皮尔斯太阳望远镜
McMath-Pierce Telescope 麦克马思－皮尔斯望远镜 | 麥克馬斯－皮爾斯望遠鏡〔台〕
McMath Telescope 麦克马思望远镜
MCMC method （=Markov chain Monte Carlo method）马尔可夫链蒙特卡洛方法
McMullan camera 麦克马伦电子照相机
McNaught Comet 麦克诺特彗星
MCO （=Mars Climate Orbiter）环火星气候探测器（美国）
MCP （=micro-channel plate）微通道板
MCP detector （=micro-channel plate detector）微通道板检测器
MCP optics （=microchannel plate optics）微通道板光学系统
MD （=Melchior-Dejaiffe Catalogue）MD 测纬星表
Md （=mendelevium）钔（101 号元素）
M-discontinuity （=Mohorovičić discontinuity）莫霍间断面
MDM （=mixed dark matter）混合暗物质
MDM model （=mixed dark matter model）混合暗物质模型

M dwarf　M 型矮星
Mead crater　米德环形山（金星）
mean absolute magnitude　平均绝对星等
mean absorption coefficient　平均吸收系数
mean anomaly　平近点角
mean contour　平均轮廓　又见：*mean profile*
mean daily motion　平均日运动 | 平均週日運動〔台〕　又见：*mean diurnal motion*
mean day　平日
mean declination　平赤纬
mean density of matter　平均物质密度
mean diurnal motion　平均日运动 | 平均週日運動〔台〕　又见：*mean daily motion*
mean ecliptic sun　黄道平太阳
mean electron concentration　平均电子浓度
mean element　平根数
mean epoch　平均历元
mean equator　平赤道
mean equatorial sun　赤道平太阳
mean equinox　平春分点
mean error　（＝rmse）中误差　又见：*root mean square error*
mean extinction coefficient　平均消光系数
mean fictitious clock　平均假钟
mean free collision time　平均自由碰撞时间
mean free path　平均自由程 | 平均無礙 [路] 程〔台〕
mean free time　平均自由时
mean latitude　平纬
mean life　平均寿命　又见：*average life; mean lifetime*
mean lifetime　平均寿命　又见：*average life; mean life*
mean light curve　平均光变曲线
mean longitude　平经，平黄经
mean luminosity density　平均光度密度
mean magnitude　平均星等　又见：*average magnitude*
mean midnight　平子夜
mean molecular weight　平均分子量
mean moon　平月亮
mean motion　平运动 | 平均運動〔台〕
mean motion resonance　平运动共振
mean noon　平正午
mean obliquity　平黄赤交角 | 平均黄赤交角〔台〕
mean observatory　平均天文台
mean opposition distance　平均冲距 | 平 [均] 衝距〔台〕
mean orbit　平均轨道　又见：*median orbit*

mean orbital element　平轨道根数 | 平均軌道要素，平均軌道根數〔台〕
mean parallax　平均视差，统计视差
　又见：*statistical parallax*
mean photoelectric magnitude　平均光电星等
mean photographic magnitude　平均照相星等
mean photovisual magnitude　平均仿视星等
mean place　平位置　又见：*mean position*
mean pole　平极
mean position　平位置　又见：*mean place*
mean profile　平均轮廓　又见：*mean contour*
mean pulse　平均脉冲
mean radial velocity　平均视向速度
mean radiation intensity　平均辐射强度
mean refraction　平均大气折射 | 平均 [大氣] 折射〔台〕
mean right ascension　平赤经
mean sea-level　平均海平面
mean sidereal day　平恒星日
mean sidereal second　平恒星秒
mean sidereal time　平恒星时
mean sidereal year　平恒星年
mean solar day　平太阳日
mean solar second　平太阳秒
mean solar time　平太阳时 | 平 [太陽] 時〔台〕
mean spheroid　平均椭球体
mean square　均方
mean square density fluctuation　均方密度涨落
mean square deviation　均方偏差
mean square error　均方误差，中误差 | 均方 [誤] 差〔台〕
mean square value　均方值
mean sun　平太阳
mean synodic month　平朔望月
mean synodic motion　平朔望运动
mean synodic period　平朔望周期
mean synodic revolution　平朔望周　又见：*mean synodic rotation*
mean synodic rotation　平朔望周　又见：*mean synodic revolution*
mean synodic year　平朔望年
mean tide　平均潮汐
mean time　平时 | 平 [太陽] 時〔台〕　即：*mean solar time*
mean-time clock　平时钟
mean-time type time-signal　平时式时号
mean wavelength　平均波长

mear solar year　平太阳年
measurability　可测性
measurable quantity　可测量
measured coordinate　量度坐标 | 量度坐標，测定坐標〔台〕
measure-preserving mapping　保测绘制
measures and weights　权度
measuring accuracy　测量精度
measuring error　测量误差
measuring instrument　测量仪器
measuring machine　量度仪
measuring microscope　测微显微镜 | 测微鏡〔台〕
　　又见：*micrometer-microscope*
measuring screw　测微螺旋 | 测微螺絲〔台〕
　　又见：*micrometer screw*
Mebsuta　井宿五（双子座 ε）　又见：*Melboula*; *Melucta*
mechanical ellipticity　动力学椭率
mechanical energy　机械能
mechanical equivalent　功当量
mechanical equivalent of heat　热功当量
mechanical equivalent of light　光功当量
mechanical feedback　力学反馈
mechanical property　力学特性
mechanical quantity　力学量
mechanics　力学
mechanism　机制
MECO　（＝magnetospheric eternally collapsing object）磁层永坍缩体
Media　箕宿二（人马座 δ）　又见：*Kaus Meridionalis*
median　中位数 | 中 [位] 數〔台〕
median error　中间误差
median magnitude　中位星等
median orbit　平均轨道　又见：*mean orbit*
median period　中位周期
median photoelectric magnitude　中位光电星等
median photographic magnitude　中位照相星等
median photovisual magnitude　中位仿视星等
median radius　中位半径
median relaxation time　中位弛豫时标
medieval maximum　中世纪极大期
medieval minimum　中世纪极小期
medium　介质
medium-altitude satellite　中高度卫星
medium-band photometry　中带测光
　　又见：*intermediate band photometry*
medium-band width　中带宽度

Medium Scale Anisotropy Measurement Experiment　（＝MSAM experiment）中等尺度各向异性测量实验
medium speed emulsion　中度乳胶
medium speed plate　中度底片
MEDOC　（＝Motion of the Earth by Doppler Observation Campaign）多普勒观测地球运动计划
Medusa Nebula　美杜莎星云（Abell 21）
Mee crater　米环形山（月球）
Mees crater　米斯环形山（月球）
Megaclite　木卫十九
mega-electron-volt　（＝MeV）兆电子伏
Megalithic astronomy　巨石天文学
megamaser　巨微波激射，巨脉泽 | 巨邁射〔台〕
megaparsec　（＝Mpc）兆秒差距 | 百萬秒差距〔台〕
Megaregolith　巨石表土
Meggers crater　梅格斯环形山（月球）
Megrez　天权，北斗四（大熊座 δ）
Meinel band　迈内尔谱带
Meinel optics　迈内尔光学系统
Meissa　觜宿一（猎户座 λ）
Meitner crater　迈特纳环形山（月球）
Meizhou　梅州（小行星 3239 号）
Mekbuda　井宿七（双子座 ζ）
Melas Chasma　米拉斯深谷（火星）
Melboula　井宿五（双子座 ε）　又见：*Mebsuta*; *Melucta*
Melchior-Dejaiffe Catalogue　（＝MD）MD 测纬星表
Melete　中神星（小行星 56 号）
melon-seed effect　瓜籽效应
Melpomene　司曲星（小行星 18 号）
Melucta　井宿五（双子座 ε）　又见：*Mebsuta*; *Melboula*
MEM　（＝maximum entropy method）最大熵法
member galaxy　成员星系
member object　成员天体
member star　成员星
membrane mirror　薄膜镜
membrane paradigm　膜范式
Memoirs of the Royal Astronomical Society　（＝MmRAS）《皇家天文学会论文集》（英国期刊）
Men　① 骑官十（豺狼座 α）②（＝Mensa）山案座
Menchib　卷舌三（英仙座 ξ）　又见：*Menkib*; *Menkhib*

Mendel crater　孟德尔环形山（月球）
Mendeleev crater　门捷列夫环形山（月球）
mendelevium　（＝Md）钔（101号元素）
Menghuan　孟奂（小行星12106号）
meniscus　弯月形透镜 | 彎月形[透鏡]〔台〕
　　又见：*meniscus lens*
meniscus correcting plate　弯月形改正板
meniscus corrector　弯月形改正透镜
meniscus lens　弯月形透镜 | 彎月形[透鏡]〔台〕
　　又见：*meniscus*
meniscus mirror　弯月形反射镜
meniscus Schmidt telescope　弯月形改正镜施密特望远镜
meniscus telescope　弯月形镜望远镜 | 彎月形[透鏡]望遠鏡〔台〕
meniscus transit　弯月形镜中星仪
Menkab　天囷一（鲸鱼座α）　又见：*Menkar; Monkar*
Menkalina　五车三（御夫座β）
Menkar　①天囷一（鲸鱼座α）　又见：*Menkab; Monkar* ②天囷三（鲸鱼座λ）
Menkent　库楼三（半人马座θ）
Menkhib　卷舌三（英仙座ξ）　又见：*Menkib; Menchib*
Menkib　①卷舌三（英仙座ξ）　又见：*Menkhib; Menchib* ②室宿二（飞马座β）　又见：*Scheat*
Mensa　（＝Men）山案座
menstrual epact　月闰余
Merak　天璇，北斗二（大熊座β）
Mercator crater　墨卡托环形山（月球）
Mercurial magnetosphere　水星磁层
Mercurial precession　水星进动
Mercurius crater　墨丘利环形山（月球）
Mercury　水星，辰星
mercury　（＝Hg）汞（80号元素）
mercury horizon　水银地平
mercury-manganese star　汞锰星
Mercury Spacecraft　水星号飞船（美国）
Mercury Surface Space Environment Geochemistry and Ranging　（＝MESSENGER）信使号（美国水星探测器）
Merga　玄戈增二（牧夫座38）　又见：*Falx Italica*
merge of galaxy　星系并合　又见：*galaxy merging; galactic merging*
merger rate　并合率
merger remnent　并合后星系
merger scenario　并合图景
merger tree　并合树
merging galaxy　并合星系　又见：*galaxy merger*

merging star　并合恒星
meridian　①子午圈 ②子午线　又见：*meridian line*
meridian altitude　子午圈高度
meridian angle　子午角距
meridian astrometry　子午天体测量 | 子午天體測量[學]〔台〕
meridian astronomy　子午天文学
meridian catalogue　子午星表
meridian circle　（＝MC; TC）子午环　又见：*transit circle*
meridian circulation　子午环流 | 子午圈環流〔台〕
meridian instrument　①子午仪，中星仪 ②中星仪　又见：*transit instrument*
meridian interferometer　中星仪式干涉仪
Meridiani Planum　子午高原（火星）
meridian line　子午线　又见：*meridian*
meridian mark　子午标 | 子午[線]標〔台〕
meridian observation　①中天观测 ②子午观测
meridian of Greenwich　格林尼治子午线
meridian passage　中天　又见：*culmination; meridian transit*
meridian photometer　中天光度计
meridian photometry　中天测光
meridian plane　子午面　又见：*meridional plane*
meridian transit　中天　又见：*culmination; meridian passage*
meridian wire　子午丝 | 子午線〔台〕
meridian zenith distance　子午天顶距 | [過]子午圈天頂距〔台〕
meridiem　【拉】正午　即：*midday*
meridional circulation　子午环流
meridional component　子午分量
meridional flow　子午流 | 子午圈環流〔台〕
meridional plane　子午面　又见：*meridian plane*
MERIT　（＝Monitoring of Earth-Rotation and Intercomparison of the Techniques of Observation and Analysis）MERIT 地球自转监测
MERLIN　（＝Multi-Element Radio Linked Interferometer Network）默林[多元射电联合干涉网] | 梅林[多元電波聯合干涉網]〔台〕
Merope　昴宿五（金牛座23）
Merope Nebula　昴宿星云（NGC 1435）
Merrill crater　梅里尔环形山（月球）
merrillite　磷钙钠石
Mersenius crater　梅森环形山（月球）
mesa　方山
Mesarthim　娄宿二（白羊座γ）　又见：*Mesartim*
Mesartim　娄宿二（白羊座γ）　又见：*Mesarthim*

Meshcherskiy crater 梅谢尔斯基环形山（月球）
mesogranulation 中米粒组织
mesogranule 中米粒
meson 介子 又见：baryton; mesotron
mesopause 中间层顶 | 中氣層頂〔台〕
Mesopotamian astronomy 美索不达米亚天文学
mesosiderite 中陨铁
mesosphere 中间层 | 中氣層〔台〕
mesotron 介子 又见：baryton; meson
MESSENGER （=Mercury Surface Space Environment Geochemistry and Ranging）信使号（美国水星探测器）
Messier Catalogue （=M）梅西叶星云星团表 | 梅西耳星表〔台〕
Messier crater 梅西叶环形山（月球）
Messier number 梅西叶编号 | 梅西耳編號〔台〕
Messier object 梅西叶天体 | 梅西耳天體〔台〕
Me star M 型发射线星, Me 型星 | Me 型星〔台〕
Mestel's disk 梅斯特尔盘
metabolism 新陈代谢
metagalactic astronomy 总星系天文学
metagalactic space 总星系空间
Metagalaxy 总星系 | 總銀河系〔台〕
metal absorption line system 金属吸收线系统
metal abundance 金属丰度
metal-deficient star 贫金属星 又见：metal-poor star
metal enrichment 金属增丰
metallic asteroid 金属小行星
metallic helium 金属氦
metallic hydrogen 金属氢
metallicity 金属度
metallicity gradient 金属度梯度
metallicity-luminosity relation 金属丰度－光度关系
metallicity parameter 金属度参数
metallic-line star 金属线星 | 金屬[譜]線星〔台〕 又见：metal-line star
metallic prominence 金属日珥 | 金屬[譜]線日珥〔台〕
metallic star 金属星
metal line 金属谱线
metal-line star 金属线星 | 金屬[譜]線星〔台〕 又见：metallic-line star
metal-poor binary 贫金属双星
metal-poor cluster 贫金属星团
metal-poor dwarf 贫金属矮星
metal-poor giant 贫金属巨星
metal-poor globular cluster 贫金属球状星团
metal-poor object 贫金属天体
metal-poor star 贫金属星 又见：metal-deficient star
metal ratio 金属比
metal-rich binary 富金属双星
metal-rich cluster 富金属星团
metal-rich dwarf 富金属矮星
metal-rich giant 富金属巨星
metal-rich globular cluster 富金属球状星团
metal-rich object 富金属天体
metal-rich star 富金属星
metamorphic rock 变质岩
metamorphism 变质
metastability 亚稳度
metastable atom 亚稳原子
metastable level 亚稳能级 | 暫穩能階〔台〕
metastable state 亚稳态 | 暫穩態〔台〕
metastable-state transition 亚稳态跃迁
meteor 流星 又见：shooting star; falling star
meteor astronomy 流星天文学
meteor brightness 流星亮度
meteor camera 流星照相机
meteor echo 流星回波
meteor flare 流星耀发 | 流星爆發〔台〕
meteoric apex 流星向点
meteoric body 流星体 又见：meteoroid
meteoric collision 流星体碰撞
meteoric dust 流星尘
meteoric hypothesis 流星假说
meteoric iron 铁陨星，铁陨石，陨铁 | 隕鐵，鐵[質]隕石〔台〕 又见：aerosiderite; iron meteorite; siderite
meteoric material 流星物质 又见：meteoric matter
meteoric matter 流星物质 又见：meteoric material
meteoric seism 陨致地震 | 隕[石]致地震〔台〕
meteoric shower 流星雨 又见：meteor shower
meteoric stream 流星群 又见：meteoroid stream; meteor stream; meteoric swarm; meteor swarm
meteoric swarm 流星群 又见：meteoric stream; meteoroid stream; meteor stream; meteor swarm
meteoric year 流星年
meteorist 流星学家
meteorite 陨星 | 隕石〔台〕 又见：meteorolite
meteorite astronomy 陨星天文学 | 隕石天文學〔台〕
meteorite crater 陨星坑 | 隕石坑〔台〕

meteorite dust　陨星尘 | 隕石塵〔台〕
　　又见：*meteoroid dust*
meteorite hypothesis　陨星假说 | 隕石假說〔台〕
　　又见：*meteoritic hypothesis*
meteorite impact　陨击　又见：*cratering*
meteorite matter　陨星物质 | 隕石物質〔台〕
meteorite shower　陨星雨 | 隕石雨〔台〕
meteoritic hypothesis　陨星假说 | 隕石假說〔台〕
　　又见：*meteorite hypothesis*
meteoriticist　陨星学家
meteoritics　陨星学，陨石学 | 隕石學〔台〕
　　又见：*astrolithology*
meteoritic theory　陨星学说
meteoroid　流星体　又见：*meteoric body*
meteoroid dust　① 陨星尘 | 隕星塵〔台〕
　　又见：*meteorite dust* ② 微陨星尘 | 微隕石塵〔台〕
　　又见：*micrometeoritic dust*
meteoroid dust cloud　微陨星尘云
meteoroid stream　流星群　又见：*meteoric stream; meteor stream; meteoric swarm; meteor swarm*
meteorolite　陨星 | 隕石〔台〕　又见：*meteorite*
Meteorological Satellite　（＝Meteosat）气象科学卫星
meteorological satellite　气象卫星　又见：*weather satellite*
meteorology　气象学
meteor patrol　流星巡天　又见：*meteor survey*
meteor patrol camera　流星巡天照相机
meteor shower　流星雨　又见：*meteoric shower*
meteor spectrograph　流星摄谱仪
meteor station　流星观测站
meteor storm　流星暴
meteor stream　流星群　又见：*meteoric stream; meteoroid stream; meteoric swarm; meteor swarm*
meteor survey　流星巡天　又见：*meteor patrol*
meteor swarm　流星群　又见：*meteoric stream; meteoroid stream; meteor stream; meteoric swarm*
meteor trail　流星余迹　又见：*meteor train*
meteor trail communication　流星余迹通信
meteor trail radar　流星尾迹雷达
meteor train　流星余迹　又见：*meteor trail*
meteor trajectory　流星轨迹
Meteosat　（＝Meteorological Satellite）气象科学卫星
meter component　米波分量
meter-wave astronomy　米波天文学
meter-wave burst　米波暴 | 米波爆發〔台〕

meter-wavelength radiation　米波辐射
　　又见：*meter-wave radiation*
meter-wave radiation　米波辐射
　　又见：*meter-wavelength radiation*
meter-wave radio astronomy　米波射电天文学
meter-wave radio telescope　米波射电望远镜
meter-wave telescope　米波望远镜
methanal　甲醛（H_2CO）　又见：*formaldehyde*
methane　甲烷（CH_4）
methanol　甲醇（CH_3OH）　又见：*methyl alcohol*
methanol maser　甲醇微波激射
method of coincidence　切拍法
method of contiguity　偕日法
method of dependences　依数法　又见：*dependence method*
method of least square　最小二乘法　又见：*least square method; method of minimum square*
method of minimum square　最小二乘法
　　又见：*least square method; method of least square*
method of observation　观测方法
method of opposability　冲日法
method of scale　尺度法
method of successive approximation　逐步近似法
method of successive generations　逐代法
method of surface of section　截面法
Methone　土卫三十二
methyl-acetylene　丙炔，甲基乙炔（C_3H_4）
　　又见：*propyne*
methyl alcohol　甲醇（CH_3OH）　又见：*methanol*
methylamine　甲胺（CH_3NH_2）
methyl cyanide　乙腈（CH_3CN）　又见：*acetonitrile*
methylenimine　甲亚胺（CH_2NH）
methyl formate　甲酸甲酯（$HCOCH_3$）
methyl mercaptan　甲硫醇（CH_3SH）
Metis　① 木卫十六 ② 颖神星（小行星 9 号）
Metius crater　梅修斯环形山（月球）
Metonic cycle　默冬章（19 年）
metric　度规
metric field　度规场
metric perturbation　度规扰动
metric system　米制
metric tensor　度规张量
metric wave　米波
metrology　计量学
Meudon Observatory　墨东天文台
MeV　（＝mega-electron-volt）兆电子伏
Me variable　Me 型变星

Mexican hat potential 墨西哥帽势
Mezentsev crater 梅津采夫环形山（月球）
Mg （=magnesium）镁（12号元素）
MGC （=Millennium Galaxy Catalogue）千亿星系表
M giant M型巨星
MGIO （=Mount Graham International Observatory）格雷厄姆山国际天文台
MGS （=Mars Global Surveyor）火星环球勘测者
MHD （=magnetohydrodynamics）磁流[体动]力学 | 磁流體[動]力學〔台〕
Miaplacidus 南船五（船底座β）
Mic （=Microscopium）显微镜座
Mice Galaxies 双鼠星系（NGC 4676）
Michelson crater 迈克耳孙环形山（月球）
Michelson interferometer 迈克耳孙干涉仪
Michelson interferometry 迈克耳孙干涉测量
Michelson-Morley experiment 迈克耳孙－莫雷实验
Michelson stellar interferometer 迈克耳孙恒星干涉仪 | 邁克爾孫恆星干涉儀〔台〕
Michie model 米基模型
Michigan State University Observatory 密歇根州立大学天文台
Michigan University Radio Astronomy Observatory 密歇根大学射电天文台
microarcsec （=micro-arcsecond）微角秒
microarcsec astrometry 微角秒天体测量
micro-arcsecond （=microarcsec）微角秒
microbeam 微射束
microcalorimeter 微量能器
micro-channel electron multiplier 微通道电子倍增管
micro-channel plate （=MCP）微通道板
micro-channel plate detector （=MCP detector）微通道板检测器
microchannel plate optics （=MCP optics）微通道板光学系统
micro-chronometer 测微时计
microcrater 微陨坑
microdensitometer 测微密度计
microdensitometry 显微密度测量
micro-electro-mechanical deformable mirrors 微机电变形镜
microflare 微耀斑 | 微閃焰〔台〕
microgravitational lens 微引力透镜 | 微重力透镜〔台〕 又见：gravitational micro-lens; microlens

microgravitational lensing 微引力透镜效应
又见：microlensing; mini gravitational lensing; mini lensing
microgravity 微重力
microlens 微引力透镜 | 微重力透镜〔台〕
又见：gravitational micro-lens; microgravitational lens
microlensed quasar 微引力透镜化类星体
microlensing 微引力透镜效应
又见：microgravitational lensing; mini gravitational lensing; mini lensing
microlensing galaxy 微引力透镜效应星系 | 微重力透镜星系〔台〕
microlensing light curve 微引力透镜光变曲线
Microlensing Planet Finder 微引力透镜行星探测器
microlensing probability 微引力透镜概率
microlensing survey 微引力透镜巡天
micromechanism 微观机制
micrometeor 微流星 又见：ultra-telescopic meteor
micrometeorite 微陨星 | 微隕石〔台〕
micrometeoritic dust 微陨星尘 | 微隕石塵〔台〕
又见：meteoroid dust
micrometeoroid 微流星体
micrometer 测微计 | 測微計，測微器〔台〕
micrometer constant 测微计常数 | 測微常數〔台〕
micrometer drum 测微鼓
micrometer-microscope 测微显微镜 | 測微鏡〔台〕
又见：measuring microscope
micrometer screw 测微螺旋 | 測微螺絲〔台〕
又见：measuring screw
micrometer value 测微计周值
microphone sensor 微声传感器
microphotogram 显微光度图 | 微光光度圖〔台〕
microphotometer 显微光度计 | 測微光度計〔台〕
microphotometric tracing 显微光度描迹 | 微光光度描跡圖〔台〕
microphotometry 显微光度测量
microphysics 微观物理学
micropore optics （=MPO）微孔光学系统
microprocessor 微处理机
micro-programmed control 微程序控制
micropulse 微脉冲
microquake 微震 又见：microseism
microquasar 微类星体
micro-satellite 微卫星
microscope-micrometer 显微测微计
Microscopium （=Mic）显微镜座
microsecond 微秒
microsecond pulsar 微秒脉冲星

microsecond pulse 微秒脉冲
microseism 微震　又见：*microquake*
microspot 微黑子 | 小黑點〔台〕　又见：*pore*
microstate 微观状态
micro-strip 微带
microstructure 微观结构
microturbulence 微观湍流 | 微[觀]湍流〔台〕
microturbulent motion 微观湍动
microturbulent velocity 微观湍速
Microvariability and Oscillations of Stars telescope （＝MOST）恒星微变和振荡望远镜
microwave 微波
microwave absorption 微波吸收
microwave amplification by stimulated emission of radiation 微波激射，脉泽 | 邁射〔台〕
　又见：*maser*
microwave antenna 微波天线
microwave background 微波背景
microwave background radiation （＝MBR）微波背景辐射
microwave background rest frame 微波背景静止框架
microwave burst 微波暴 | 微波爆發〔台〕
microwave component 微波分量
microwave emission 微波发射
microwave heliograph 微波日像仪
microwave imaging 微波成像
microwave kinetic induction detector 微波动态电感探测器
microwave line 微波谱线
microwave radiation 微波辐射
microwave spectroscopy 微波频谱测量
microwave spectrum 微波频谱
microwave storm 微波暴
mid-Atlantic ridge 大西洋中脊
Midcourse Space Experiment （＝MSX）太空中途红外实验
midday 正午
middle atmosphere 中层大气
middle corona 中冕 | 中[日]冕〔台〕
middle of eclipse 食甚　又见：*maximum of eclipse*
middle ring 中环
middle weight star 中等质量星
mid-infrared 中红外
midnight 子夜
mid-oceanic ridge 洋中脊 | 中洋脊〔台〕
midsize black hole 中型黑洞

mid-term 中气
Mie scattering 米氏散射
Mie theory 米氏理论
Mikheyev-Smirnov-Wolfenstein effect （＝MSW effect）MSW 效应
Milagro telescope 奇迹伽马射线望远镜
Milanković crater 米兰科维奇环形山（月球）
Milankovich theory 米兰科维奇理论
military satellite 军用卫星
Milk Dipper 南斗（星组）　又见：*Southern Dipper*
Milkomeda 银河仙女星系
Milky Way （＝MW）银河，天河 | 銀河〔台〕
Milky Way galaxy 银河系　又见：*Galaxy; Galactic System*
millennium 千禧年，千年纪 | 千年紀〔台〕
Millennium Galaxy Catalogue （＝MGC）千僖星系表
Millennium simulation 千禧模拟
Miller crater 米勒环形山（月球）
milliarcsec 毫角秒　又见：*milliarcsecond*
milliarcsecond （＝mas）毫角秒　又见：*milliarcsec*
milli-jansky （＝mJy）毫央
Millikan crater 密立根环形山（月球）
Millimeter Anisotropy eXperiment IMaging Array （＝MAXIMA）毫米波各向异性实验成像阵（美国）
Milli-Meter Array （＝MMA）毫米波射电望远镜阵 | 毫米波[電波望遠鏡]陣〔台〕
millimeter astronomy 毫米波天文学
　又见：*millimeter-wave astronomy*
millimeter radiation 毫米波辐射
millimeter-wave 毫米波
millimeter-wave astronomy 毫米波天文学
　又见：*millimeter astronomy*
millimeter-wave observatory 毫米波天文台
millimeter-wave radio astronomy 毫米波射电天文学
millimeter-wave radio telescope 毫米波射电望远镜
millimeter-wave spectrum 毫米波谱
Millimeter-Wave Telescope （＝MWT）MWT 毫米波望远镜
millimeter-wave telescope 毫米波望远镜
millimicrosecond 毫微秒
million years （＝Myr）百万年
millisecond 毫秒
millisecond pulsar （＝MSP）毫秒脉冲星
millisecond pulse 毫秒脉冲

Mills Cross　米尔斯十字 | 密耳式十字天线阵〔台〕（射电）

Mills-Slee-Hill Catalogue of Radio Sources　（＝MSH）MSH 射电源表

Mills-Strurrock model　米尔斯－斯特鲁洛克模型

Milne crater　米尔恩环形山（月球）

Milne-Eddington approximation　米尔恩－爱丁顿近似

Milne-Eddington model　米尔恩－爱丁顿模型

Milne model　米尔恩模型

Milne relation　米尔恩关系

Milne universe　米尔恩宇宙

Mimas　土卫一

Mimosa　十字架三（南十字座 β）　又见：*Becrux*

mineral deposit　矿床

Minerva　慧神星（小行星 93 号）

Mineur crater　米纳尔环形山（月球）

Mingantu　明安图（小行星 28242 号）

mini black-hole　超小黑洞 | 小黑洞〔台〕

mini-computer　微机

mini gravitational lensing　微引力透镜效应
　　又见：*microgravitational lensing; microlensing; mini lensing*

minihalo　超小晕

mini lensing　微引力透镜效应
　　又见：*microgravitational lensing; microlensing; mini gravitational lensing*

Mini Magellanic Cloud　（＝MMC）微麦哲伦云

minimal supersymmetric standard model　（＝MSSM）最小超对称模型

mini-moon　超小卫星 | 小衛星；小月亮〔台〕
　　又见：*moonlet*

minimum　极小

minimum corona　极小期日冕，极小日冕

minimum detectable brightness　最小可检测亮度

minimum detectable error　最小可检测误差

minimum detectable flux　最小可检测流量

minimum detectable flux density　最小可检测流量密度

minimum detectable signal　最小可检测信号

minimum deviation　最小偏向

minimum deviation angle　最小偏向角

minimum deviation grating　最小偏向光栅

minimum e-folds　最小 e 叠数

minimum-energy orbit　最小耗能轨道

minimum mass　质量下限

minimum-redundancy array　最小冗余 [天线] 阵 | 最小重複天線陣〔台〕

minimum resolution　最小分辨率

minimum resolvable angle　最小可分辨角 | 最小可鑒角〔台〕

minimum variance　最小方差

minimum visibility　最小能见度

mini quasar　超小类星体

Minisat 01　超小卫星 1 号（西班牙）

mini-satellite　超小卫星

Minkar　轸宿二（乌鸦座 ε）

Minkowski crater　闵科夫斯基环形山（月球）

Minkowski functional　闵可夫斯基泛函

Minkowski metric　闵可夫斯基度规

Minkowski space　闵可夫斯基空间 | 明氏空間〔台〕

Minkowski spacetime　闵可夫斯基时空

Minkowski world　闵可夫斯基宇宙 | 明氏宇宙〔台〕

Minnaert crater　米奈尔环形山（月球）

minor axis　短轴

minor limit　小限

minor lobe　副瓣

minor-lobe radiation　副瓣辐射

minor-lobe solid angle　副瓣立体角

minor merger　次并合

minor planet　小行星　又见：*asteroid*

Minor Planet Center　（＝MPC）小行星中心

minor-planet family　小行星族 | 小行星 [家] 族〔台〕　又见：*family of asteroids; asteroid family*

Mintaka　参宿三（猎户座 δ）　又见：*Mintika*

Mintika　参宿三（猎户座 δ）　又见：*Mintaka*

minute　分 | 分 [鐘]〔台〕

minute of arc　（＝arcmin）角分　又见：*arc minute*

minute of time　时分

Mir　和平号空间站（苏联）

Mira　刍藁增二（鲸鱼座 o）　又见：*Stella Mira*

Mirac　奎宿九（仙女座 β）　又见：*Mirach*

Mira Ceti variable　刍藁 [型] 变星 | 蒭藁 [增二] 型星，米拉 [型] 變星〔台〕　又见：*Mira variable; Mira-type variable*

Mirach　① 梗河一（牧夫座 ε）　又见：*Izar; Mirak; Mirrak; Pulcherrima* ② 奎宿九（仙女座 β）　又见：*Mirac*

mirage　海市蜃楼 | 蜃景〔台〕

Mirak　梗河一（牧夫座 ε）　又见：*Izar; Mirach; Mirrak; Pulcherrima*

Miralaidjii Corona　米拉莱狄地冕（金星）

Miram　天船一（英仙座 η）

Miranda　天卫五 | 天 [王] 衛五〔台〕

Mira star 刍藁型星 | 蒭藁 [增二] 型星〔台〕
 即：*Mira Ceti variable; Mira variable; Mira-type variable*；
 又见：*Mira-type star*

Mira-type star 刍藁型星 | 蒭藁 [增二] 型星〔台〕
 即：*Mira Ceti variable; Mira variable; Mira-type variable*；
 又见：*Mira star*

Mira-type variable 刍藁 [型] 变星 | 蒭藁 [增二] 型星，米拉 [型] 變星〔台〕 又见：*Mira Ceti variable; Mira variable*

Mira variable 刍藁 [型] 变星 | 蒭藁 [增二] 型星，米拉 [型] 變星〔台〕 又见：*Mira Ceti variable; Mira-type variable*

mire 照准标 | 標點〔台〕

Mirfak 天船三（英仙座 α） 又见：*Marfak; Mirphak*

Miriam 圣女星（小行星 102 号）

Mirphak 天船三（英仙座 α） 又见：*Marfak; Mirfak*

Mirrak 梗河一（牧夫座 ε） 又见：*Izar; Mirach; Mirak; Pulcherrima*

mirror 反射镜 又见：*reflector*

mirror actuator 镜面致动器

mirror blank 镜坯

mirror cell 镜室

mirror cover 镜盖

mirror figuring 镜面修琢

mirror grating 镜栅

mirror image 镜像

mirror instability 镜面不稳定性 | 反射鏡不穩定 [性]〔台〕

mirror-lenses system 折反射光学系统

mirror plane 镜平面

mirror reflection 镜式反射

mirror seeing 镜面视宁度

mirror sextant 反射式六分仪 | 反射鏡六分儀〔台〕

mirror support 镜面支承

mirror surface 镜面

mirror telescope 反射望远镜 又见：*reflecting telescope; catoptric telescope; reflector*

mirror transit circle 反射式子午环

mirror transit instrument 反射式中星仪 | 地平式中星儀〔台〕

mirror trap 镜俘获

Mirzam 军市一（大犬座 β） 又见：*Murzim*

miscellaneous correlation 混杂相关

misidentification 误认

missing asteroid 丢失小行星

missing baryon problem 重子缺失问题

missing comet 丢失彗星

missing mass 短缺质量 | 無蹤質量〔台〕

missing mass problem 质量缺失问题

missing satellite problem 卫星星系缺失问题

Mistastin crater 米斯塔斯汀湖陨星坑（地球）

MIT （＝Massachusetts Institute of Technology）麻省理工学院

Mitra crater 米特拉环形山（月球）

Mitteilungen der Astronomischen Gesellschaft （＝MAG）【德】《天文学会会刊》

Mitteilungen über Veränderliche Sterne （＝MVS）【德】《变星通报》

Mittenzwey eyepiece 米腾兹威目镜

mixed dark matter （＝MDM）混合暗物质

mixed dark matter model （＝MDM model）混合暗物质模型

mixed perturbation 混合摄动

mixed-polarity magnetic field 极性混合磁场

mixed state 混合态

mixed term 混合项

mixer 混频器

mixing layer 混合层

mixing length 混合长 | 混合長度〔台〕

mixing length theory 混合长理论 | 混合長度理論〔台〕

mixmaster universe 混合主控宇宙

mixture 混合物

Mizar 北斗六，开阳（大熊座 ζ）

MJD （＝modified Julian date）简化儒略日期 | 約簡儒略日〔台〕

Mjølnir crater 米约涅尔陨星坑（地球）

MJUO （＝Mount John University Observatory）约翰山大学天文台（新西兰）

mJy （＝milli-jansky）毫央

MK classification MK 光谱分类
 又见：*Morgan-Keenan classification*

MKK classification （＝Morgan-Keenan-Kellman classification）MKK 光谱分类

MKK system （＝Morgan-Keenan-Kellman system）MKK 分类系统

MK luminosity class （＝Morgan-Keenan luminosity class）MK 光度级

MKO （＝Mauna Kea Observatory）莫纳克亚天文台

MK system （＝Morgan-Keenan system）MK 分类系统

MLE （＝maximum likelihood estimation）极大似然估计

MLO （＝Mount Lemmon Observatory）莱蒙山天文台（美国）

M-L relation (=mass-luminosity relation) 质光关系

MMA (=Milli-Meter Array) 毫米波射电望远镜阵 | 毫米波 [電波望遠鏡] 陣〔台〕

M-magnitude M 星等

MMC (=Mini Magellanic Cloud) 微麦哲伦云

MMIC (=monolithic microwave integrated circuit) 微波单片集成电路

MmRAS (=Memoirs of the Royal Astronomical Society)《皇家天文学会论文集》（英国期刊）

MMS (=multimission modular spacecraft) 多任务模块化空间飞行器

MMT (=Multiple-Mirror Telescope) MMT 望远镜（美国，1998 年改为单面 6.5 米主镜，全称现已不用）

MN (=Monthly Notices of the Royal Astronomical Society)《皇家天文学会月刊》

Mn (=manganese) 锰（25 号元素）

Mneme 木卫四十

Mnemosyne 龙女星（小行星 57 号）

MNRAS (=Monthly Notices of the Royal Astronomical Society)《皇家天文学会月刊》

Mo (=molybdenum) 钼（42 号元素）

MOAO (=multi-object adaptive optics) 多目标自适应光学

mobility 活动性 | 遷移率〔台〕

Möbius crater 麦比乌斯环形山（月球）

mock gravitational force 伪引力

mock lunar-ring 幻月环　又见：*paraselenic circle*

mock moon 幻月　又见：*moon dog*

mock solar-ring 幻日环　又见：*paranthelionic circle; parhelic circle*

mock sun 幻日　又见：*parhelion; sundog*

mode change 模式变化

mode-hopping 模式耀变

model atmosphere 模型大气 | 大氣模型〔台〕

model atmosphere analysis 模型大气分析

model chromosphere 模型色球

model corona 模型日冕

model photosphere 模型光球

model solar atmosphere 模型太阳大气

model stellar atmosphere 模型恒星大气

moderate-resolution spectrograph 中分辨摄谱仪

moderate-resolution spectroscopy 中分辨分光

modern astronomy 现代天文学

modest telescope 中型望远镜

mode subtraction 模减除

modified Bessel function 修正贝塞尔函数

modified English mounting 改进型英国式装置

modified fork mounting 改进型叉式装置

modified German mounting 改进型德国式装置

modified gravity 修改引力

modified horseshoe mounting 改进型马蹄式装置

modified Hubble profile 修正哈勃分布

modified isochrone method 改进等龄线法

modified isothermal sphere 修正等温球

modified Julian date (=MJD) 简化儒略日期 | 約簡儒略日〔台〕

Modified Newtonian Dynamics (=MOND) 修改的牛顿动力学，MOND 理论

modified normal equation 约化法方程

modified value 修正值

modular field 模场

modulated receiver 调制接收机

modulated time signal 调制式时号 | 調變時號〔台〕

modulation 调制 | 調變，調幅〔台〕

modulation collimator 调制式准直器

modulation effect 调制效应

modulation index 调制指数

modulation interferometer 调制型干涉仪 | 调制[型]干涉儀〔台〕

modulation interferometry 调制型干涉测量

modulation transfer function (=MTF) 调制传递函数

modulator 调制器 | 調變器，調幅器〔台〕

moduli space 模空间

modulus of deformation 形变模量

modulus of distance 距离模数　又见：*distance modulus*

modulus of elasticity 弹性模量

Moessbauer effect 穆斯堡尔效应 | 梅斯堡效應〔台〕

MOF (=magneto-optical filter) 磁光滤光器

Mögel-Dellinger effect 默格尔－德林格尔效应

Mohammedan calendar 伊斯兰历 | 回曆〔台〕　又见：*islamic calendar*

Mohammedan era 穆罕默德纪元

Mohorovičić crater 莫霍洛维奇环形山（月球）

Mohorovičić discontinuity (=M-discontinuity) 莫霍间断面

Mohorovičić surface 莫霍面　又见：*Moho surface*

Moho surface 莫霍面　又见：*Mohorovičić surface*

Moiseev crater 莫伊谢耶夫环形山（月球）

MOL (=Manned Orbiting Laboratory) 载人环地轨道实验室（美国）

moldavite 暗绿玻璃
molecular astronomy 分子天文学
molecular astrophysics 分子天体物理
molecular band 分子谱带 | 分子 [谱] 帶〔台〕
molecular clock 分子钟
molecular cloud 分子云
molecular clump 分子团块
molecular core 分子核
molecular hydrogen 分子氢
molecular line 分子谱线 | 分子 [谱] 線〔台〕
molecular-line radio astronomy 分子谱线射电天文学
molecular maser 分子微波激射，分子脉泽
molecular outflow 外向分子流
molecular scattering 分子散射
molecular shock 分子激波
molecular spectroscopy 分子谱测量
molecular spectrum 分子谱
molecule 分子
Møller scattering 摩勒散射
Molonglo Observatory Synthesis Telescope （=MOST）莫隆格勒综合孔径望远镜
Molonglo pulsar （=MP）莫隆格勒脉冲星
Molonglo Radio Observatory 莫隆格勒射电天文台
Moloniya 闪电（苏联通信卫星）
molybdenum （=Mo）钼（42 号元素）
moment actuator 力矩致动器
momental ellipse 惯量椭圆 又见: ellipse of inertia
momental ellipsoid 惯量椭球 又见: ellipsoid of inertia
moment arm 矩臂
moment equation 矩方程
moment of couple 力耦矩
moment of dipole 偶极矩 又见: dipole moment
moment of force 力矩
moment of inertia 惯性矩 | 轉動慣量〔台〕
moment of inertia tensor 惯性矩张量
moment of momentum 动量矩 | 動量矩，角動量〔台〕
momentum 动量
momentum current 动量流
momentum-driven wind 动量驱动风
momentum equation 动量方程
Mon （=Monoceros）麒麟座
MOND （=Modified Newtonian Dynamics）修改的牛顿动力学，MOND 理论

Mongmanwai 蒙民伟（小行星 3678 号）
Monitoring of Earth-Rotation and Intercomparison of the Techniques of Observation and Analysis （=MERIT）MERIT 地球自转监测
Monitoring of the Sun-Earth Environment （=MONSEE）MONSEE 日地环境监测
Monitor of All-sky X-ray Image （=MAXI）全天 X 射线图像监视器
Monkar 天囷一（鲸鱼座 α） 又见: Menkab; Menkar
monoatomic gas 单原子气体
monocentric eyepiece 单心目镜
Monocerids 麒麟流星群 | 麒麟 [座] 流星雨〔台〕
Monoceros （=Mon）麒麟座
Monoceros Hot Spot 麒麟热斑
Monoceros Loop 麒麟圈 又见: Monogem Ring
monochromatic aberration 单色像差
monochromatic absorption 单色吸收
monochromatic amplitude 单色光变幅 | 單色 [光] 變幅〔台〕
monochromatic corona 单色日冕，单色星冕
monochromatic filter ① 单色滤光器 ② 单色滤光片 | 單色濾 [光] 鏡〔台〕
monochromatic flux 单色流量
monochromatic image 单色像
monochromatic light 单色光 又见: homogeneous light
monochromatic radiation 单色辐射
monochromatic wave 单色波
monochromator 单色器
monochrometer 单色计
monocular 单筒望远镜
monodromy 单值性
monoenergetic electron 单能电子
monoenergetic spectrum 单能谱
Monogem Ring 麒麟圈 又见: Monoceros Loop
monolithic CMOS 单片型 CMOS
monolithic collapse 整体坍缩
monolithic microwave integrated circuit （=MMIC）微波单片集成电路
monolithic-mirror telescope 整单镜面望远镜
monolithic model 单块模型
monomolecular radiation 单分子辐射
monomolecule 单分子
monopolar radiation 单极辐射 又见: monopole radiation
monopole 单极
monopole emission 单极发射
monopole problem 单极问题

monopole radiation　单极辐射　又见：*monopolar radiation*
monopole term　单极项
monostatic configuration　单元组态
monotonic function　单调函数
monotonic model　单调宇宙模型 | 單調 [宇宙] 模型〔台〕
monotonic model of the first kind　第一类单调宇宙模型
monotonic model of the second kind　第二类单调宇宙模型
Mons Ampère　【拉】安培山（月球）
Mons Argaeus　【拉】阿尔加山（月球）
MONSEE　（＝Monitoring of the Sun-Earth Environment）MONSEE 日地环境监测
Mons Huygens　【拉】惠更斯山（月球）
Mons La Hire　【拉】拉希尔山（月球）
Mons Pico　【拉】皮科山（月球）
Mons Piton　【拉】皮通山（月球）
Mons Rümker　【拉】吕姆克山 | 呂姆克爾山〔台〕（月球）
mons（复数：montes）【拉】山 | 山系〔台〕
Montagnais crater　蒙塔格尼陨星坑（地球）
Montanari crater　蒙塔纳里环形山（月球）
Mont Blanc　【拉】勃朗峰（月球）
Monte-Carlo method　蒙特卡罗方法 | 蒙地卡羅法〔台〕　又见：*Monte-Carlo technique*
Monte-Carlo technique　蒙特卡罗方法 | 蒙地卡羅法〔台〕　又见：*Monte-Carlo method*
Montes Agricola　【拉】阿格里科拉山脉（月球）
Montes Alpes　【拉】阿尔卑斯山脉（月球）
Montes Apenninus　【拉】亚平宁山脉（月球）
Montes Archimedes　【拉】阿基米德山脉（月球）
Montes Carpatus　【拉】喀尔巴阡山脉 | 喀爾巴仟山脈〔台〕（月球）
Montes Caucasus　【拉】高加索山脉（月球）
Montes Cordillera　【拉】科迪勒拉山脉（月球）
Montes Haemus　【拉】海玛斯山脉（月球）
Montes Harbinger　【拉】前锋山脉（月球）
Montes Jura　【拉】侏罗山脉（月球）
Montes Pyrenaeus　【拉】比利牛斯山脉 | 庇里牛斯山脈〔台〕（月球）
Montes Recti　【拉】直列山脉（月球）
Montes Riphaeus　【拉】里菲山脉（月球）
Montes Rook　【拉】鲁克山脉 | 盧克山脈〔台〕（月球）
Montes Secchi　【拉】塞奇山脉（月球）

Montes Spitzbergen　【拉】施皮茨贝尔根山脉（月球）
Montes Taurus　【拉】金牛山脉 | 托鲁斯山脈〔台〕（月球）
Montes Teneriffe　【拉】特内里费山脉（月球）
Montgolfier crater　蒙戈尔费埃环形山（月球）
month　月
monthly mean　月平均
Monthly Notices of the Royal Astronomical Society　（＝MNRAS; MN）《皇家天文学会月刊》
monthly nutation　周月章动
month of the phases　朔望月，太阴月　又见：*synodic month; lunar month; moon month; lunation*
moom　行星环卫星　又见：*ring-moon*
Moon　月球，太阴　又见：*Luna*
moon bounce time synchronization　（＝MBTS）月球回波时间同步
moon-crossing asteroid　越月小行星
moon-crossing object　越月天体
moondial　月晷　又见：*lunar dial*
moon dog　幻月　又见：*mock moon*
moon enters penumbra　半影月食始
moon enters umbra　本影月食始 | 初虧〔台〕
moon leaves penumbra　半影月食终
moon leaves umbra　本影月食终 | 復圓〔台〕
moonlet　超小卫星 | 小衛星；小月亮〔台〕　又见：*mini-moon*
Moon Mineralogy Mapper　月球矿藏勘测器（印度月船号上的科学设备）
moon month　朔望月，太阴月　又见：*synodic month; lunar month; month of the phases; lunation*
moonquake　月震
moonrise　月出
moonrock　月岩　又见：*lunar rock*
moon rocket　月球火箭
moon's age　月龄　又见：*age of the moon*
moon scraper　月岩取样器
moon seismograph　月震仪
moonset　月没
moon's evection　出差　又见：*evection*
moon's orbit　月球轨道　又见：*lunar orbit*
moon's path　白道
moon's phase　月相　又见：*phase of the moon; aspect of the moon; lunar aspect; lunar phase*
moon's phase law　月相律
moon's tide　月球潮汐
moon's variation　二均差　又见：*variation of the moon*
moonwatch　人卫监测　又见：*moon watching*

moon watching　人卫监测　又见：*moonwatch*
moonwatch telescope　人卫广角镜
Moore crater　穆尔环形山（月球）
Moore density profile　穆尔密度轮廓
Mopra Observatory　莫普拉天文台
Moreton wave　莫尔顿波
Moretus crater　莫雷环形山（月球）
Morgan-Keenan classification　MK 光谱分类
　又见：*MK classification*
Morgan-Keenan-Kellman classification
　（＝MKK classification）MKK 光谱分类
Morgan-Keenan-Kellman system　（＝MKK system）MKK 分类系统
Morgan-Keenan luminosity class　（＝MK luminosity class）MK 光度级
Morgan-Keenan system　（＝MK system）MK 分类系统
Morgan's classification　摩根星系分类
morning glow　晨辉
morning group　晨星组 | 早晨 [星] 组〔台〕
morning star　晨星
morning twilight　曙光
Morokweng crater　莫罗昆陨星坑（地球）
morphological astronomy　形态天文学
Morphological Catalogue of Galaxies　（＝MCG）星系形态表
morphological classification　形态分类
morphological peculiarity　形态特殊性
morphological property　形态特征
morphological type　形态型
morphology　形态学 | 形態學，形態〔台〕
morphology-density relation　形态－密度关系
morphology segregation　形态层化
Morse crater　莫尔斯环形山（月球）
mosaicing　图像拼接法
mosaic mirror　镶嵌镜面
mosaic mirror telescope　镶嵌镜面望远镜
Moseley crater　莫塞莱环形山（月球）
MOST　① （＝Microvariability and Oscillations of Stars telescope）恒星微变和振荡望远镜　② （＝Molonglo Observatory Synthesis Telescope）莫隆格勒综合孔径望远镜
most probable error　最概然误差 | 最可能誤差〔台〕
most probable value　最概然值 | 最可能值〔台〕
most probable velocity　最概然速度

motion of earth pole　极移　又见：*polar motion; polar displacement; polar shift; polar wandering; wandering of the pole*
Motion of the Earth by Doppler Observation Campaign　（＝MEDOC）多普勒观测地球运动计划
motive magnetic structure　运动磁结构
　又见：*moving magnetic feature*
mottle　日芒 | 日芒，日斑〔台〕　又见：*solar mottle*
Mott scattering　莫特散射
Mouchez crater　穆谢环形山（月球）
mountain-building activity　造山运动
mountain range　山脉
mounted standard instrument　座正仪
Mount Graham International Observatory　（＝MGIO）格雷厄姆山国际天文台
Mount Hopkins Observatory　霍普金斯山天文台
　即：*Fred Lawrence Whipple Observatory; Whipple Observatory*；又见：*Mt. Hopkins Observatory*
mounting error　安装误差　又见：*set-up error*
mounting of telescope　望远镜机架 | 望遠鏡架台〔台〕
mounting system　支架系统
Mount John University Observatory　（＝MJUO）约翰山大学天文台（新西兰）　又见：*Mt. John University Observatory*
Mount Lemmon Observatory　（＝MLO）莱蒙山天文台（美国）　又见：*Mt. Lemmon Observatory*
Mount Palomar Observatory　帕洛玛山天文台（美国）　又见：*Mt. Palomar Observatory*
Mount Stromlo and Siding Spring Observatory　（＝MSSSO）斯特朗洛山和赛丁泉天文台（澳大利亚）
Mount Stromlo Observatory　（＝MSO）斯特朗洛山天文台（澳大利亚）　又见：*Mt. Stromlo Observatory*
Mount Wilson Catalogue of Early Type Emission Stars　（＝MWC）MWC 早型发射线星表 | 威爾遜山早型發射線星表〔台〕
Mount Wilson Observatory　威尔逊山天文台
　又见：*Mt. Wilson Observatory*
moustache　胡须状结构 | 鬍鬚〔台〕　即：*Ellerman bomb*
Mouth　觜宿（二十八宿）　又见：*Turtle beak*
movable antenna　可移动天线
movable ecliptic　移动黄道
movable sighting set　四游仪　又见：*sighting-tube ring; component of the four displacements*
movable telescope　可移动望远镜
movable wire　动丝　又见：*traveling wire*

moving atmosphere 运动大气
moving-blade shutter 动片快门
moving cluster 移动星团
moving cluster parallax 移动星团视差
moving element 移动部件
moving group 移动星群
moving magnetic feature 运动磁结构
  又见：*motive magnetic structure*
MP (＝Molonglo pulsar) 莫隆格勒脉冲星
MPA (＝Max-Planck-Institute for Astrophysics) 马普天体物理研究所
MPC (＝Minor Planet Center) 小行星中心
Mpc (＝megaparsec) 兆秒差距 | 百万秒差距〔台〕
MPIfA (＝Max-Planck-Institut fur Astronomie)【德】马普天文研究所
MPIfPA (＝Max-Planck-Institut fur Physik und Astrophysik)【德】马普物理和天体物理研究所
MPIfR (＝Max-Planck-Institut fur Radioastronomie)【德】马普射电天文研究所
MPL (＝Mars Polar Lander) 火星极地着陆器
MPO (＝micropore optics) 微孔光学系统
MPP CCD (＝multi-pinned phase CCD) 多相钉扎型 CCD
MRAO (＝Mullard Radio Astronomy Observatory) 玛拉德射电天文台
M region M 区 又见：*M-area*
M-region storm M 区暴
MRI (＝magnetorotational instability) 磁转动不稳定性
MRO (＝Mars Reconnaissance Orbiter) 火星勘测轨道飞行器
MROI (＝Magdalena Ridge Optical Interferometer) 马格达莱纳岭光学干涉仪
MSAM experiment (＝Medium Scale Anisotropy Measurement Experiment) 中等尺度各向异性测量实验
MSFC (＝Marshall Space Flight Center) 马歇尔空间中心（美国）
MSH (＝Mills-Slee-Hill Catalogue of Radio Sources) MSH 射电源表
MSL (＝Mars Science Laboratory) 火星科学实验室
MSO (＝Mount Stromlo Observatory) 斯特朗洛山天文台（澳大利亚）
MSP (＝millisecond pulsar) 毫秒脉冲星
MSS (＝multi-spectral scanner) 多谱扫描仪
MSSM (＝minimal supersymmetric standard model) 最小超对称模型
MSSSO (＝Mount Stromlo and Siding Spring Observatory) 斯特朗洛山和赛丁泉天文台（澳大利亚）
MS star (＝main sequence star) 主序星
M star M 型星 又见：*M-type star*
M subdwarf M 型亚矮星 | M 型次矮星〔台〕
M subgiant M 型亚巨星 | M 型次巨星〔台〕
M supergiant M 型超巨星
MSW effect (＝Mikheyev-Smirnov-Wolfenstein effect) MSW 效应
MSX (＝Midcourse Space Experiment) 太空中途红外实验
MTF (＝modulation transfer function) 调制传递函数
M-theory M 理论
Mt. Hopkins Observatory 霍普金斯山天文台
  即：*Fred Lawrence Whipple Observatory*; *Whipple Observatory*; 又见：*Mount Hopkins Observatory*
Mt. John University Observatory 约翰山大学天文台（新西兰） 又见：*Mount John University Observatory*
Mt. Lemmon Observatory 莱蒙山天文台（美国）
  又见：*Mount Lemmon Observatory*
Mt. Palomar Observatory 帕洛玛山天文台（美国） 又见：*Mount Palomar Observatory*
MTRLI (＝Multi-Telescope Radio-Linked Interferometer) 多望远镜射电联合干涉仪
  即：*MERLIN*
Mt. Stromlo Observatory 斯特朗洛山天文台（澳大利亚） 又见：*Mount Stromlo Observatory*
Mt. Wilson Observatory 威尔逊山天文台
  又见：*Mount Wilson Observatory*
M-type asteroid M 型小行星
M-type star M 型星 又见：*M star*
Mufrid 右摄提一（牧夫座 η） 又见：*Mufride*; *Muphrid*
Mufride 右摄提一（牧夫座 η） 又见：*Mufrid*; *Muphrid*
Muhlifain 库楼七（半人马座 γ）
Mukhanov-Sasaki equation 穆哈诺夫－佐佐木方程
Muliphain 天狼增四（大犬座 γ） 又见：*Muliphein*; *Muliphen*
Muliphein 天狼增四（大犬座 γ） 又见：*Muliphain*; *Muliphen*
Muliphen ① 天狼增四（大犬座 γ）
  又见：*Muliphain*; *Muliphein* ② 宗正二（蛇夫座 γ）
Mullard Radio Astronomy Observatory (＝MRAO) 玛拉德射电天文台
Mulliken band 马勒肯谱带

multi-anode microchannel array （=MAMA）多阳极微通道阵

multi-antenna array 多天线阵　又见：*multi-element array*

multi-antenna radio telescope 多天线射电望远镜　又见：*multi-element radio telescope*

multi-band 多波段

multiband color-index 多波段色指数

multiband photometer 多波段光度计

multiband photometry 多波段测光

multibay truss 多层桁架

multibeam receiver 多波束接收机

multi-channel 多通道

multichannel astrometric photometer （=MAP）多通道天测光度计

multichannel coronagraph 多通道日冕仪　又见：*multichannel coronagraph*

multichannel coronal spectrometer 多通道日冕分光仪

multichannel coronograph 多通道日冕仪　又见：*multichannel coronagraph*

multichannel detector 多通道检测器

multichannel filter spectrometer 多通道滤波型频谱仪

multichannel magnetograph 多通道磁像仪 | 多道磁场儀〔台〕

multichannel photometer 多通道光度计

multichannel photometry 多通道测光

multichannel receiver 多通道接收器 | 多道接收機〔台〕

multichannel solar telescope 多通道太阳望远镜

multichannel spectrometer 多通道分光仪

multicolor photometer 多色光度计

multicolor photometry 多色测光

multicomponent spectrum 多组元波谱

multi-congugate adaptive optics （=MCAO）多重共轭自适应光学

multidimensional space 多维空间　又见：*hyperspace*

multidirectional beam 多向射束

multi-element array 多天线阵　又见：*multi-antenna array*

multi-element interferometer 多天线干涉仪 | 多元干涉儀，多天線干涉儀〔台〕

multi-element interferometry 多天线干涉测量

Multi-Element Radio Linked Interferometer Network （=MERLIN）默林 [多元射电联合干涉网] | 梅林 [多元電波聯合干涉網]〔台〕

multi-element radio telescope 多天线射电望远镜　又见：*multi-antenna radio telescope*

multifibre spectrometer 多光纤分光仪

multifibre spectroscopy 多光纤分光

multifield inflation 多场暴胀

multilayer coating 多层镀膜

multilayer mirror 多层膜镜

multilayer X-ray telescope 多层膜X射线望远镜

multi-mirror telescope 多镜面望远镜 | 多面鏡望遠鏡〔台〕　又见：*multiple mirror telescope*

multimission modular spacecraft （=MMS）多任务模块化空间飞行器

multimode horn 多模喇叭

multi-object adaptive optics （=MOAO）多目标自适应光学

multi-object spectrograph 多天体摄谱仪

multi-object spectroscopy 多天体分光

multi-period 多重周期

multi-periodic variable 多重周期变星

multiphase IGM 多相星系际介质

multi-phase interferometer 多相干涉仪

multi-phase interferometry 多相干涉测量

multi-pinned phase CCD （=MPP CCD）多相钉扎型CCD

multiple-arc method 复弧法

multiple beam 多波束

multiple-body problem 多体问题　又见：*many-body problem; problem of many bodies*

multiple fly-bys 多重飞掠

multiple galaxy 多重星系　又见：*multiple system*

multiple grating 复光栅

multiple interferometer 多元干涉仪

multiple interferometry 多元干涉测量

multiple meteor 多重流星

Multiple-Mirror Telescope （=MMT）MMT望远镜（美国，1998年改为单面6.5米主镜，全称现已不用）

multiple mirror telescope 多镜面望远镜 | 多面鏡望遠鏡〔台〕　又见：*multi-mirror telescope*

multiple quasars 多重类星体

multiple radio source 多重射电源 | 多重電波源〔台〕　又见：*radio multi-source*

multiple redshift 多重红移

multiple scattering 多次散射

multiple star 聚星 | 聚星，多重星〔台〕

multiple system 多重星系　又见：*multiple galaxy*

multiplet 多重线 | 多重[譜]線〔台〕

multiple tail 多重彗尾

multiplet component 多重子线
multiplet structure 多重结构
multiplication 倍增
multiplication constant 倍增常数
multiplication factor 倍增因子
multiplicative noise 可乘噪声
multiplicity 多重性
multiplicity function 多重度函数
multiplier 倍增管，乘法器
multiplying interferometer 相乘干涉仪
multiplying interferometry 相乘干涉测量
multipolar radiation 多极辐射 又见：*multipole radiation*
multipole 多极
multipole coefficients 多极系数
multipole component 多极分量
multipole emission 多极发射
multipole expansion 多极展开
multipole moments 多极矩
multipole radiation 多极辐射 又见：*multipolar radiation*
multipole term 多极项
multiradix index 多基数指数
multi-reflecting antenna 多反射面天线
multi-reflection 多重反射
multi-reflector 多重反射镜，多重反射器
multi-slit mask 多缝掩模
multi-slit spectrograph 多缝摄谱仪
multi-spectral scanner （=MSS）多谱扫描仪
multi-stage rocket 多级火箭 又见：*rocket train; step rocket*
Multi-Telescope Radio-Linked Interferometer （=MTRLI）多望远镜射电联合干涉仪 即：*MERLIN*
multivariate distribution 多变量分布
multiverse 多重宇宙
multiwave interaction 多波相互作用
multiwavelength astronomy 多波段天文学
multiwavelength astrophysics 多波段天体物理
multiwavelength digital sky survey 多波段数字化巡天
multiwavelength sky survey 多波段巡天 又见：*multiwavelength survey*
multiwavelength survey 多波段巡天 又见：*multiwavelength sky survey*
multiwave-particle interaction 多波—粒子相互作用
multi-wire counter 多线计数器
multiwire proportional counter 多丝正比计数器
Mundilfari 土卫二十五
Mundrabilla meteorite 曼德拉比拉陨星
Munich University Observatory 慕尼黑大学天文台
muon μ介子 又见：*μ-meson*
Muphrid 右摄提一（牧夫座 η） 又见：*Mufrid; Mufride*
mural circle 墙仪
mural quadrant 墙象限仪
Murchison crater 默奇森环形山（月球）
Murchison meteorite 默奇森陨石
Murchison Widefield Array （=MWA）默奇森大视场[射电]阵
Murgab crater 默加布陨星坑（地球）
Murray crater 默里陨星坑（地球）
Murzim 军市一（大犬座 β） 又见：*Mirzam*
Mus （=Musca）苍蝇座
Musca （=Mus）苍蝇座
Musca Borealis 北蝇座（现已不用）
Muscida ①内阶一（大熊座 o）②内阶增七（大熊座 $π^2$）
Muses-A 缪斯A月球探测器
mutual coherence 互相干
mutual coherence function （=MCF）互相干函数
mutual induction 互感性 | 互感應〔台〕
Mutus crater 穆图什环形山（月球）
M variable M型变星
MVS （=Mitteilungen über Veränderliche Sterne）【德】《变星通报》
MW （=Milky Way）银河，天河 | 銀河〔台〕
MWA （=Murchison Widefield Array）默奇森大视场[射电]阵
MWC （=Mount Wilson Catalogue of Early Type Emission Stars）MWC 早型发射线星表 | 威爾遜山早型發射線星表〔台〕
MWT （=Millimeter-Wave Telescope）MWT 毫米波望远镜
MXRB （=massive X-ray binary）大质量X射线双星 | 大質量X光雙星〔台〕
Myr （=million years）百万年

N

N （=nitrogen）氮（7 号元素）
Na （=sodium）钠（11 号元素）
nacreous cloud 珠母云
nadir 天底
nadir distance 天底距
nadir reading 天底读数
Nagler eyepiece 纳格勒目镜
Naiad 海卫三
NAIC （=National Astronomy and Ionosphere Center）美国国家天文和电离层中心
NaI detector （=sodium-iodide detector）碘化钠探测器 | 碘化钠侦测器〔台〕
nail-bed function 钉板函数
Nair al Zaurak 火鸟六（凤凰座 α） 又见：Ankaa
Nairc 耐尔思（小行星 4245 号）
naked eye 肉眼 又见：human eye
naked-eye brightness 肉眼亮度
naked-eye nova 肉眼新星
naked-eye object 肉眼天体
naked-eye observation 肉眼观测
naked-eye supernova 肉眼超新星
naked-eye variable 肉眼变星
naked singularity 裸奇点
naked T-Tauri star 显露金牛 T 型星 | 顯露金牛[座]T 型星〔台〕
naked T-Tauri variable 显露金牛 T 型变星
Nakhla and Chassigny meteorite SNC 陨星 又见：Shergotty
Nakhla meteorite 纳赫利赫陨星
nakhlite 透辉橄无球粒陨石
Nakkar 七公增五（牧夫座 β） 又见：Nekkar
Nakshatra 月站（印度二十七宿）
nakshatra month 星宿月
Namaka 妊卫二（矮行星 Haumea 卫星）
Nambu-Goldstone boson （=NGB）南部－戈德斯通玻色子
naming 命名

Nançay Radio Astronomy Facility 南赛射电天文观测站（法国）
Nançay radio telescope （=NRT）南赛射电望远镜（法国）
Nanedi Valles 纳内迪谷（火星）
Nanjingdaxue 南京大学（小行星 3901 号）
Nanking 南京（小行星 2078 号）
nanoflare 纳耀斑 | 奈闪焰〔台〕
Nano-Japan Astrometry Satellite Mission for INfrared Exploration （=Nano-JASMINE）纳茉莉天体测量卫星
Nano-JASMINE （=Nano-Japan Astrometry Satellite Mission for INfrared Exploration）纳茉莉天体测量卫星
nanosecond 纳秒 | 奈秒，毫微秒〔台〕
Nansen crater ①南森环形山（月球）②南森环形山（火星）
Nantong 南通（小行星 3051 号）
Nantou 南投（小行星 160493 号）
Nanyang 南阳（小行星 9092 号）
NAO （=national astronomical observatory）国家天文台
NAOC （=National Astronomical Observatories, Chinese Academy of Sciences）中国科学院国家天文台
Naos 弧矢增二十二（船尾座 ζ）
Narrabri intensity interferometer 纳拉布赖强度干涉仪
narrow-band filter 窄波段滤光片，窄带滤波器
narrow-band magnitude 窄波段星等
narrow-band photometer 窄带光度计
narrow-band photometry 窄带测光
narrow-band width 窄带宽度
narrow-beam radiation 窄束辐射
narrow-line radio galaxy （=NLRG）窄线射电星系 | 窄線電波星系〔台〕
narrow-line region （=NLR）窄线区
narrow-line spectrum 窄线光谱

narrow linewidth 窄线宽
narrow resonance 窄共振
Narvi 土卫三十一
NASA （＝National Aeronautics and Space Administration）美国[国家]航天局｜美國[國家]航太總署〔台〕
NASA Infrared Telescope Facility （＝IRTF）[美国]航天局红外望远镜｜NASA 紅外望遠鏡〔台〕
NASA/IPAC Extragalactic Database （＝NED）美国航天局河外数据库
Nashira 壁垒阵三（摩羯座 γ）
Nasmyth crater 内史密斯环形山（月球）
Nasmyth focus 内氏焦点
Nasmyth spectrograph 内氏焦点摄谱仪｜內史密斯攝譜儀〔台〕
Nassau crater 纳索环形山（月球）
Nata 娜塔（小行星 1086 号）
Nath 玉井三（波江座 β）又见：*Cursa*
National Aeronautics and Space Administration （＝NASA）美国[国家]航天局｜美國[國家]航太總署〔台〕
National Astronomical Observatories, Chinese Academy of Sciences （＝NAOC）中国科学院国家天文台
national astronomical observatory （＝NAO）国家天文台
National Astronomy and Ionosphere Center （＝NAIC）美国国家天文和电离层中心
National Bureau of Standards （＝NBS）美国国家标准局
National Geodetic Survey （＝NGS）美国测地局
National New Technology Telescope （＝NNTT）美国新技术望远镜
National Oceanic and Atmospheric Administration （＝NOAA）美国国家海洋和大气管理局
National Oceanic Survey （＝NOS）美国海洋测量局
National Optical Astronomy Observatory （＝NOAO）美国国家光学天文台｜[美國]國家光學天文台〔台〕
National Radio Astronomy Observatory （＝NRAO）美国国家射电天文台｜[美國]國家電波天文台〔台〕
National Science Foundation （＝NSF）国家科学基金会（美国）
National Solar Observatory （＝NSO）美国国家太阳观测台｜[美國]國家太陽觀測台〔台〕
National Space Science Data Center （＝NSSDC）国家空间科学数据中心（美国）

National Virtual Observatory, United States （＝NVO）美国国家虚拟天文台
natural bias 自然偏差
natural broadening 自然展宽｜自然致寬〔台〕
natural damping 自然阻尼
natural direction 自然方向
natural frequency 固有频率
natural guide star 自然引导星
natural half-width 自然半宽
natural inflation 自然暴胀
natural line-broadening 谱线自然展宽
natural line-width 自然线宽
natural optical background 天然光学背景
natural oscillation 固有振荡｜固有振動,固有振盪〔台〕
natural period 固有周期
natural quartz 天然石英
natural radiation belt 天然辐射带
natural radio background 天然射电背景
natural reference frame 自然参考架
natural reference system 自然参考系
natural satellite 天然卫星
natural seeing 自然视宁度
natural selection 自然选择
natural termination 自然终止
natural tetrad 自然基,自然四维标架｜自然基〔台〕
natural unit 自然单位
natural unit system 自然单位系统
natural vibration 固有振动
natural weighting 自然权重
natural width 自然宽度
natural X-ray background 天然 X 射线背景
natural year 回归年 又见：*tropical year*
natural γ-ray background 天然 γ 射线背景
nature 本原
nautical almanac 航海历书｜航海[曆]書〔台〕
nautical astronomy 航海天文学
nautical mile 海里
nautical star 导航星 又见：*navigation star; navigational star*
nautical table 航海用表
nautical triangle 航海三角形
nautical twilight 航海晨昏蒙影｜航海曙暮光〔台〕
Navagraha 九执历（印度历数）
Naval Research Laboratory （＝NRL）美国海军科研实验室

Navarro-Frenk-White density profile （=NFW profile）NFW 密度轮廓
Navi 阁道二（仙后座 ε） 又见：Segin
Navier-Stokes equation 纳维耶－斯托克斯方程
navigational object 导航天体
navigational planet 导航行星
navigational satellite 导航卫星 又见：navigation satellite
navigational star 导航星 又见：navigation star; nautical star
navigational star number 导航星号
navigational time scale 导航时标
navigational triangle 导航三角形 | 航海三角形〔台〕
navigation by earth reference 地球基准导航
navigation by space reference 空间基准导航
navigation clock 导航时钟 | 航海鐘〔台〕
navigation satellite 导航卫星 又见：navigational satellite
navigation star 导航星 又见：nautical star; navigational star
Navigation Technology Satellite （=NTS）NTS 系列卫星，导航技术卫星（美国）
Navy Precision Optical Interferometer （=NPOI）海军精确光学干涉仪
Nb （=niobium）铌（41 号元素）
N-body problem n 体问题 又见：problem of n bodies
N-body simulation n 体模拟
NBS （=National Bureau of Standards）美国国家标准局
Nd （=neodymium）钕（60 号元素）
NDRO （=non-destroy readout）无损示值读数
Ne （=neon）氖（10 号元素）
NEA （=near-earth asteroid）近地小行星
Neander crater 内安德环形山（月球）
neap tide 小潮
NEAR （=Near Earth Asteroid Rendezvous）近地小行星探测器
nearby dwarf 近距矮星
nearby galaxy 近邻星系 | 鄰近星系〔台〕
nearby object 近距天体
nearby space 近距空间
nearby star 近距恒星
Nearch crater 奈阿尔科环形山（月球）
near-circular orbit 近圆轨道
near-collision nebula 近碰撞星云
near-contact binary 近相接双星

near-earth asteroid （=NEA）近地小行星 又见：earth-approaching asteroid
near-earth asteroid belt 近地小行星带
Near Earth Asteroid Rendezvous （=NEAR）近地小行星探测器
Near-Earth Asteroid Tracking （=NEAT）近地小行星跟踪
near-earth comet 近地彗星 又见：earth-approaching comet
near-Earth flyby 近地飞掠
near-earth object （=NEO）近地天体 又见：earth-approaching object
Near-Earth Object Surveillance Satellite （=NEOSSat）近地天体监测卫星
near-earth satellite 近地卫星
near-earth space （=NES）近地空间
nearest star 最近距恒星
near-field cosmology 近场宇宙学
near-field pattern 近场方向图
near-infrared （=NIR）近红外 又见：near-IR
near-infrared glow 近红外天光
near-in sidelobe 内旁瓣
near-IR 近红外 又见：near-infrared
near-lunar space 近月空间 又见：lunar space
nearly parabolic comet 近抛物线彗星
nearly parabolic orbit 近抛物线轨道
near real-time 近实时
NEAR Shoemaker probe NEAR－舒梅克号探测器
nearside of the Moon 月球正面
near-solar space 近日空间
near-ultraviolet 近紫外 又见：near-UV
near-UV 近紫外 又见：near-ultraviolet
NEAT （=Near-Earth Asteroid Tracking）近地小行星跟踪
NEB ①（=noise equivalent bandwidth）噪声等效带宽 ②（=north equatorial belt）[木星] 北赤道带
Nebra Disc 内布拉星象盘
nebula filter 星云滤镜
nebular envelope 星云包层 | 星雲狀外殼〔台〕
nebular hypothesis 星云假说
nebular line 星云谱线 | 星雲[譜]線〔台〕
nebular luminosity 星云光度
nebular physics 星云物理 | 星雲物理[學]〔台〕
nebular spectrograph 星云摄谱仪
nebular spectrum 星云光谱
nebular stage 星云阶段
nebular transition 星云跃迁

nebular variable 星云变星

nebula（复数：**nebulae** 或 **nebulas**） 星云

nebulium 氜（yún，假想元素）

nebulosity 星云状物质

nebulous bridge 星云态桥状结构

nebulous cluster 伴云星团

nebulous envelope 星云状包层

nebulous object 星云状天体

nebulous patch 星云态块状结构

nebulous ring 星云态环状结构 | 星雲環〔台〕

NEC （＝noise equivalent charge）噪声等效电荷

Neck 亢宿（二十八宿）

Necklace Nebula 项链星云（PN G054.2−03.4）

Nectarian System 酒海地层系统

Nectaris Basin 酒海盆地（月球）

NED （＝NASA/IPAC Extragalactic Database）美国航天局河外数据库

Needham 李约瑟（小行星 2790 号）

Needle Galaxy 针状星系（NGC 4564）

needlet 针状小波

negative 负片

negative absorption 负吸收

negative charge 负电荷

negative conductance amplifier 负导放大器

negative curvature 负曲率

negative electricity 负电性

negative electron 负电子　又见：*negatron*

negative energy 负能 [量]

negative energy state 负能态

negative energy wave 负能波

negative eyepiece 负目镜

negative hydrogen ion 负氢离子　又见：*H-ion*

negative image 负像

negative ion 负离子，阴离子　又见：*anion*

negative leap second 负闰秒　又见：*rubber second*

negative lens 负透镜

negative molecule 负分子

negative parallax 负视差

negative pressure 负压

negative resistance amplifier 负阻放大器

negative response 负响应

negatron 负电子　又见：*negative electron*

neighborhood 邻域

Nei Monggol 内蒙古（小行星 2355 号）

Nekkar 七公增五（牧夫座 β）　又见：*Nakkar*

Nemausa 禽神星（小行星 51 号）

Nemesis 涅墨西斯（假想的太阳伴星）

NEMO ①（＝Neutrino Mediterranean Observatory）地中海中微子天文台
②（＝Neutrino Ettore Majorana Observatory）马约拉纳中微子天文台

Nengshun 能顺（小行星 142106 号）

NEO （＝near-earth object）近地天体

neocosmos 新宇宙

neodymium （＝Nd）钕（60 号元素）

neon （＝Ne）氖（10 号元素）

neon nova 氖新星

neon-rich star 富氖星

NEOSSat （＝Near-Earth Object Surveillance Satellite）近地天体监测卫星

NEP （＝noise equivalent power）噪声等效功率

nephelimeter 能见度测定计 | 天霧測定計〔台〕

Neptune 海王星

Neptune's ring 海王星环　又见：*Neptunian ring; ring of Neptune*

Neptune's satellite 海卫 | 海 [王] 衛〔台〕
又见：*Neptunian satellite; satellite of Neptune*

Neptune-Trojan asteroid 海王星−特洛伊族小行星

Neptunian ring 海王星环　又见：*Neptune's ring; ring of Neptune*

Neptunian satellite 海卫 | 海 [王] 衛〔台〕
又见：*Neptune's satellite; satellite of Neptune*

neptunium （＝Np）镎 | 錼〔台〕（93 号元素）

Nereid 海卫二 | 海 [王] 衛二〔台〕

Nereidum Montes 奈瑞杜山脉（火星）

Nernst crater 能斯特环形山（月球）

Nernst theorem 能斯特定理

NES （＝near-earth space）近地空间

Neso 海卫十三

nested X-ray telescope 嵌套式 X 射线望远镜

NESTOR （＝Neutrino Extended Submarine Telescope with Oceanographic Research project）涅斯托耳中微子望远镜

Nestor 内斯托（小行星 659 号）

Net 毕宿（二十八宿）

net current 净电流

net flux 净流量

net gain 净增益

network diaphragm 网状光阑 | 網狀光欄〔台〕

network field 网络场 | 網狀場〔台〕

network magnetic field 网络磁场 | 網狀磁場〔台〕

Network Nebula 天鹅圈（NGC 6960）
又见：*Cygnus Loop*

network nebula 网状星云（NGC 6992）
network structure 网状结构
Neugebauer-Martz-Leighton object （＝NML object）NML 天体
Neugebauer-Martz-Leighton source （＝NML source）NML 红外源
Neumann band 诺伊曼条纹 | 諾以曼條紋〔台〕
Neumann function 诺伊曼函数
Neumann line 诺伊曼线
Neumayer crater 诺伊迈尔环形山（月球）
neutral atmosphere 中性大气
neutral atom 中性原子
neutral current 中性流
neutral current sheet 中性流片
neutral equilibrium 随遇平衡 又见：*indifferent equilibrium*
neutral field 中性场
neutral field line 中性场线
neutral filter 中性滤光片 | 中性濾 [光] 鏡〔台〕
neutral gas 中性气体
neutral helium 中性氦
neutral hydrogen 中性氢
neutral hydrogen distribution 中性氢分布
neutral hydrogen region 中性氢区 | HI 區，氫原子區〔台〕 又见：*HI region; neutral hydrogen zone*
neutral hydrogen zone 中性氢区 | HI 區，氫原子區〔台〕 又见：*HI region; neutral hydrogen region*
neutralino 中性微子
neutral line 中性线
neutral particle 中性粒子
neutral point 中性点 | 中和點〔台〕
neutral region 中性区
neutral sheet 电流片，中性片 又见：*current sheet*
neutral solution 临界解
neutral step filter 中性阶梯滤光片
neutral step weakener 中性阶梯减光片 | 中和階梯減光板〔台〕
neutrino 中微子
neutrino astronomy 中微子天文学 | 微中子天文學〔台〕
neutrino astrophysics 中微子天体物理学 | 微中子天文物理學〔台〕
neutrino background 中微子背景
neutrino bremsstrahlung 中微子韧致辐射
neutrino decoupling 中微子退耦
neutrino deficiency 中微子亏缺
neutrino degeneracy 中微子简并
neutrino detection 中微子检测

neutrino discrepancy 中微子量差
Neutrino Ettore Majorana Observatory （＝NEMO）马约拉纳中微子天文台
Neutrino Extended Submarine Telescope with Oceanographic Research project （＝NESTOR）涅斯托耳中微子望远镜
neutrino freeze-out 中微子冻结
neutrino loss 中微子损耗
neutrino loss rate 中微子损耗率
Neutrino Mediterranean Observatory （＝NEMO）地中海中微子天文台
neutrino mixing 中微子混合
neutrino oscillation 中微子振荡
neutrino process 中微子过程
neutrino scattering 中微子散射
neutrino telescope 中微子望远镜 | 微中子望遠鏡〔台〕
neutron capture 中子俘获
neutron drop 中子滴
neutron excess 中子过剩 | 中子盈餘〔台〕
neutronization 中子化
neutron matter 中子物质
neutron-proton ratio 中子－质子比 又见：*neutron-to-proton ratio*
neutron pulse 中子脉冲
neutron radiation 中子辐射
neutron spectrum 中子能谱
neutron star 中子星
neutron-to-proton ratio 中子－质子比 又见：*neutron-proton ratio*
New Astronomy 《新天文学》（荷兰期刊）
New Astronomy Reviews 《新天文学评论》（英国期刊）
Newcomb's fundamental constants 纽康基本常数
Newcomb theory 纽康理论
New General Catalogue of Nebulae and Clusters of Stars （＝NGC）星云星团新总表 | 新總表〔台〕
New Horizons 新视野号（美国冥王星探测器）
New Horizons mission 新视野任务
new inflation 新暴胀
newly formed star 新生星 | 新見星〔台〕 又见：*new star*
new moon 朔，新月 又见：*beginning of lunation*
New Solar Telescope （＝NST）新太阳望远镜
new star 新生星 | 新見星〔台〕 又见：*newly formed star*
New Style （＝N.S.）新历

New Technology Telescope （＝NTT）新技术望远镜 | [欧南台] 新技術望遠鏡〔台〕
Newton crater 牛顿环形山（月球）
Newtonian-Cassegrain telescope 牛顿－卡塞格林望远镜
Newtonian cosmology 牛顿宇宙学 | 牛頓宇宙論〔台〕
Newtonian focus 牛顿焦点
Newtonian gauge 牛顿规范
Newtonian gravitation 牛顿引力
Newtonian limit 牛顿极限
Newtonian reflector 牛顿反射望远镜 | 牛頓[式]反射望遠鏡〔台〕
Newtonian telescope 牛顿望远镜 | 牛頓型望遠鏡〔台〕
Newtonian time 牛顿时
Newton-Raphson method 牛顿－拉弗森方法
next generation telescope （＝NGT）下一代望远镜 | 次世代望遠鏡〔台〕
Ney-Allen nebula 奈伊－艾伦星云
NFW profile （＝Navarro-Frenk-White density profile）NFW 密度轮廓
N galaxy N 星系
NGB （＝Nambu-Goldstone boson）南部－戈德斯通玻色子
NGC （＝New General Catalogue of Nebulae and Clusters of Stars）星云星团新总表 | 新總表〔台〕
NGS （＝National Geodetic Survey）美国测地局
NGT （＝next generation telescope）下一代望远镜 | 次世代望遠鏡〔台〕
Ngwaikin 吴伟坚（小行星 20887 号）
NH （＝northern hemisphere）北半球
Ni （＝nickel）镍（28 号元素）
Nice Observatory 尼斯天文台
Nicholson Regio 尼科尔森地区（木卫三）
nickel （＝Ni）镍（28 号元素）
nickel-iron meteorite 镍铁陨星
Nicol prism 尼科尔棱镜 | 尼克耳稜鏡〔台〕
Niehaisheng 聂海胜（小行星 9517 号）
Niépce crater 尼埃普斯环形山（月球）
night arc 夜弧
night assistant 夜间观测助理
night brightness 夜天亮度　又见：night sky brightness
night dial 夜晷
night glow 夜天光 | 夜[間天]光〔台〕　又见：night sky light; light of the night sky
night glow emission 夜天光发射
night sky 夜天

night sky background 夜天背景
night sky brightness 夜天亮度　又见：night brightness
night sky light 夜天光 | 夜[間天]光〔台〕
　　又见：night glow; light of the night sky
night sky radiation 夜天辐射 | 夜間天空輻射〔台〕
night sky spectrum 夜天光谱
night-time seeing 夜间视宁度
night transparency 夜天透明度 | 夜間天空透明度〔台〕
night value 夜间值
night vision 夜视
Nihal 厕二（天兔座 β）
Nimbus 雨云系列气象卫星（美国）
nimbus phenomena 雨云现象
Ningbo 宁波（小行星 3543 号）
Ningxia 宁夏（小行星 2539 号）
Niobe 司石星（小行星 71 号）
niobium （＝Nb）铌（41 号元素）
NIR （＝near-infrared）近红外
Nishina crater 仁科芳雄环形山（月球）
nitric oxide 一氧化氮（NO）
nitric sulfide 一硫化氮（NS）
nitrogen （＝N）氮（7 号元素）
nitrogen branch 氮支
nitrogen sequence 氮序 | 氮分支〔台〕
nitrogen star 氮星
Nix 冥卫二
NLC （＝noctilucent cloud）夜光云
N line N 谱线
NLR （＝narrow-line region）窄线区
NLRG （＝narrow-line radio galaxy）窄线射电星系 | 窄線電波星系〔台〕
NLTE （＝non-local thermodynamic equilibrium）非局部热动平衡
N-magnitude N 星等 | N－星等〔台〕
NML Cygni 天鹅 NML 红外源
NML object （＝Neugebauer-Martz-Leighton object）NML 天体
NML source （＝Neugebauer-Martz-Leighton source）NML 红外源
NML Tauri 金牛 NML 红外源
NMR （＝nuclear magnetic resonance）核磁共振
NNTB （＝north north temperate belt）[木星] 北北温带
NNTT （＝National New Technology Telescope）美国新技术望远镜
No （＝nobelium）锘（102 号元素）

NOAA （＝National Oceanic and Atmospheric Administration）美国国家海洋和大气管理局
Noachis Terra Crater 诺亚高地陨击坑（火星）
NOAO （＝National Optical Astronomy Observatory）美国国家光学天文台 | [美國] 國家光學天文台〔台〕
Nobel crater 诺贝尔环形山（月球）
nobelium （＝No）锘（102号元素）
Nobeyama Radio Observatory 野边山射电天文台 | 野邊山電波天文台〔台〕（日本）
Nobile crater 诺毕尔环形山（月球）
noble gas 稀有气体 又见：*rare gas*
no-boundary conjecture 无边界猜想
noctilucent cloud （＝NLC）夜光云
noctilucent train 夜光余迹 | 夜光[流星]餘跡〔台〕
Noctis Labyrinthus 诺克提斯沟网（火星）
Noctua 猫头鹰座（现已不用）
nocturnal 夜间定时仪
nocturnal arc 地平下弧 | 夜間弧〔台〕
nodal line 交点线，节线 | 交點線〔台〕
nodal plane 节平面
nodal point 交点，节点
nodal precession 交点进动
nodal regression 交点退行 又见：*regression of nodes; regression of the node*
nodding technique 点头技术
node 交点
node surface 节面
nodical month 交点月 又见：*draconic month*
nodical year 交点年 又见：*draconic year*
Nodus I 上弼，紫微右垣四（天龙座 ζ）
 又见：*Nodus Primus*
Nodus II 天厨一（天龙座 δ） 又见：*Altain; Nodus Secundus*
Nodus Primus 上弼，紫微右垣四（天龙座 ζ）
 又见：*Nodus I*
Nodus Secundus 天厨一（天龙座 δ）
 又见：*Altain; Nodus II*
Noether's theorem 纳脱定理
no-evolution galaxy model 无演化星系模型
no-hair theorem for black hole 黑洞无毛定理
noise 噪声 | 噪音，雜訊〔台〕
noise-adding receiver 噪声注入接收器
noise bandwidth 噪声带宽
noise equivalent bandwidth （＝NEB）噪声等效带宽
noise equivalent charge （＝NEC）噪声等效电荷
noise equivalent power （＝NEP）噪声等效功率

noise factor 噪声因子 | 噪音因數〔台〕
noise figure 噪声系数
noise fluctuation 噪声起伏
noise-free receiver 无噪声接收器
noise generator 噪声发生器
noise level 噪声电平 | 雜訊級〔台〕
noise power 噪声功率
noise power spectrum 噪声功率谱 | 雜訊功率譜〔台〕
noise source 噪声源
noise spectrum 噪声谱
noise storm 噪暴 | 雜訊爆〔台〕 又见：*storm burst*
noise temperature 噪声温度 | 雜訊溫度〔台〕
noise-temperature ratio 噪声-温度比
nomenclature 命名，命名法
nominal error 标称误差
nominal frequency 标称频率
nominal time 标称时刻
nomogram 列线图 又见：*nomograph; alignment chart*
nomograph 列线图 又见：*nomogram; alignment chart*
nomography 列线图法
non-adiabatic pulsation 非绝热脉动
non-affine deformation 非仿射形变
non-agesimal 黄道地平最高点
non-autonomous system 非自治系统
non-axisymmetrical configuration 非轴对称组态
non-axisymmetrical distribution 非轴对称分布
non-axisymmetric disk 非轴对称盘
non-baryonic dark matter 非重子暗物质
non-baryonic matter 非重子物质
non-central collision 非中心碰撞
non-circularity 非正圆性
non-cluster galaxy 非属团星系 | 非[星系]團星系〔台〕
non-cluster star 非属团恒星 | 非星團星〔台〕
non-coherent emission 非相干发射
 又见：*incoherent emission*
non-coherent radiation 非相干辐射
non-coherent re-emission 非相干再发射 | 非同調再發射〔台〕
non-coherent re-radiation 非相干再辐射
non-coherent scattering 非相干散射 | 非同調散射〔台〕 又见：*incoherent scattering*
non-coherent synchrotron radiation 非相干同步加速辐射
non-conservative scattering 非守恒散射
non-coplanar axes 非共面轴

non-current field 无电流场
non-cyclic variation 非周期变化
 又见：*non-periodic variation*
non-destroy readout （=NDRO）无损示值读数
non-diamagnetic plasma 非抗磁等离子体
non-dimensional number 无量纲数
 又见：*dimensionless number*
non-directional antenna 非定向天线
non-dissipative interaction 非耗散相互作用
non-elastic collision 非弹性碰撞 又见：*inelastic collision; inelastic impact*
non-empty model 非空虚模型
non-equilibrium flow 非平衡态流
non-equilibrium plasma 非平衡态等离子体
non-equilibrium plasmon 非平衡态等离激元
non-equilibrium state 非平衡态
non-Fraunhofer component 非夫琅和费分量
non-Gaussian density field 非高斯密度场
non-Gaussian effect 非高斯效应
non-Gaussian field 非高斯场
non-Gaussian fluctuation 非高斯涨落
non-Gaussianity 非高斯性
non-gravitational effect 非引力效应
non-gravitational force 非引力
non-gravitational motion 非引力运动
non-gray atmosphere 非灰大气 | 非灰色大氣〔台〕
 又见：*non-grey atmosphere*
non-grey atmosphere 非灰大气 | 非灰色大氣〔台〕
 又见：*non-gray atmosphere*
non-holonomic constraint 不完整约束
non-homogeneous model 非均匀模型
 又见：*non-uniform model*
non-homogeneous universe 非均匀宇宙
 又见：*non-uniform universe*
non-interacting binary galaxy 非互扰双重星系
non-interacting binary star 非互扰双星
non-interacting close binary 非互扰密近双星
non-isolating integral 非孤立积分
Nonius crater 努内斯环形山（月球）
non-leap year 平年（非闰年） 又见：*common year*
non-linear astronomy 非线性天文学
nonlinear bias 非线性偏袒
non-linear dispersion 非线性色散
non-linear infall 非线性下落
non-linear interaction 非线性相互作用
non-linear molecule 非线型分子
non-linear optics 非线性光学

non-linear scattering 非线性散射
non-linear stability 非线性稳定性
non-linear stabilization 非线性稳定化
non-linear term 非线性项
non-linear transfer 非线性转移
nonliving planet 无生命行星
non-local thermodynamic equilibrium （=NLTE）非局部热动平衡
non-magnetic particle 非磁化粒子
non-main-sequence star 非主序星
non-nucleated dwarf elliptical galaxy 无核矮椭圆星系
non-nucleated dwarf galaxy 无核矮星系
non-periodic comet 非周期彗星 | 非週期[性]彗星〔台〕 又见：*aperiodic comet*
non-periodic variable 非周期变星 | 非周期[性]變星〔台〕
non-periodic variation 非周期变化
 又见：*non-cyclic variation*
non-polar variation of latitude 非极纬度变化
non-potentiality 非势[场]性
non-potential motion 无势运动
non-proton flare 非质子耀斑 | 非質子閃焰〔台〕
non-radial distribution 非径向分布
non-radial motion 非径向运动
non-radial oscillation 非径向振荡
non-radial pulsation 非径向脉动
non-radial pulsator 非径向脉动体
non-random deviation 非随机偏离
non-random distribution 非随机分布
non-reciprocity 非倒易性
non-recurrent burst 非复现暴 | 非復現爆〔台〕
non-recurrent disturbance 非复现扰动
nonredundant aperture masking 非冗余孔径掩模法
non-redundant array 非冗余阵
non-redundant masking 非过剩遮幅成像
nonrelativistic compton scattering 非相对论康普顿散射
non-relativistic cosmology 非相对论性宇宙学 | 非相對性宇宙論〔台〕
non-relativistic degeneracy 非相对论性简并性 | 非相對論簡併性〔台〕
non-relativistic model 非相对论性模型
non-relativistic particle 非相对论性粒子
non-relativistic universe 非相对论性宇宙 | 非相對論[性]宇宙〔台〕
non-relativistic zone 非相对论性区

non-resonant diffusion 非共振扩散
non-resonant particle 非共振粒子
non-reversibility 不可逆性 又见：irreversibility
non-rotating origin 无旋转原点
non-rotating potential 无转动势
non-rotating white dwarf 无自转白矮星
non-selective absorption 非选择性吸收
non-selective emission 非选择性发射
non-selective radiation 非选择性辐射
non-singular lens 非奇异透镜
non-stable star 不稳定星 又见：unstationary star
non-standard cosmological model 非标准宇宙模型
non-standard cosmology 非标准宇宙学，备择宇宙学 又见：alternative cosmology
non-standard model 非标准模型
non-standard solar model 非标准太阳模型
non-static model 非静态模型 | 非静止模型〔台〕
non-static universe 非静态宇宙
non-stationary model 非稳态模型 | 非穩態[宇宙]模型〔台〕
non-stationary universe 非稳态宇宙
non-steady state 非稳态
non-stellar object 非星天体
non-synchrotron radiation 非同步加速辐射
non-thermal background 非热背景
non-thermal bremsstrahlung 非热韧致辐射
non-thermal electron 非热电子
non-thermal emission 非热发射
non-thermal particle 非热粒子
non-thermal process 非热过程
non-thermal proton 非热质子
non-thermal radiation 非热辐射
non-thermal radio halo 非热射电晕
non-thermal radio radiation 非热射电
non-thermal radio source 非热射电源
non-thermal source 非热辐射源
non-thermal spectrum 非热频谱
non-thermal velocity 非热速度
non-thermodynamic equilibrium （=NTE）非热动平衡
non-turbulent diffusion 非湍动扩散
non-uniform compact source 非均匀致密源
non-uniformity 不均匀性 又见：inhomogeneity; unevenness
non-uniform magnetic field 非均匀磁场

non-uniform model 非均匀模型 又见：non-homogeneous model
non-uniform universe 非均匀宇宙 又见：non-homogeneous universe
non-variable source 不变源
non-velocity redshift 非速度红移
non-volatile substance 非挥发性物质
noon 午 | [正]午〔台〕
noon interval 正午间隔
noon-mark 正午标 | [正]午標，[正]午線〔台〕
Nor （=Norma）矩尺座
Nordic Optical Telescope （=NOT）北欧光学望远镜
Nördlinger Ries crater 诺德林格里斯陨星坑（地球）
Nordtvedt effect 诺特维特效应
norm 范数
Norma （=Nor）矩尺座
Norma arm 矩尺臂
normal acceleration 法向加速度 | 法線加速度〔台〕
normal astrograph 标准天体照相仪
normal distribution 正态分布 | 常態分佈〔台〕
normal equation 法方程 | 法方程，常態方程〔台〕
normal force 法向力
normal form 范式
normal galaxy 正常星系
normal gravity 正常重力
normal incidence telescope 正入射式望远镜
normalization 归一化
normalized coordinate 标准化坐标
normalized power pattern 归一化功率方向图
normalized unit 归一化单位
normalizing 归一
normal latitude 标准点纬度 | 標準[點]緯度〔台〕
normal magnitude 正常星等
normal mode 简正模
normal nebula 正常星云
normal place 标准位置
normal point 基准点，归组点 | 標準點〔台〕
normal profile 正常轮廓
normal shock wave 正激波
normal spectrum 匀排光谱
normal spiral galaxy 正常旋涡星系 | 正常螺旋星系〔台〕
normal star 正常恒星
normal stellar wind 正常星风

normal tail 正常彗尾
normal time 标准时间
normal velocity 法向速度
normal vibration 简正振动
normal Zeeman effect 正常塞曼效应 | 正常則曼效應〔台〕
north 北
North America Nebula 北美星云 | 北美洲星雲〔台〕（NGC 7000）
North America plate 北美板块
northbound node 升交点 又见：ascending node
north celestial pole 北天极
north ecliptic pole 北黄极
north equatorial belt （＝NEB）[木星] 北赤道带
northern celestial hemisphere 北天 [区]
 又见：northern sky
Northern Coalsack 北煤袋
Northern Cross 北十字（星组）
Northern Dipper 北斗 [七星]（星组）
 又见：Triones; Big Dipper; Charles' Wain; Wain; Plough; Plow
northern elongation 北大距
northern galaxy 北天星系
northern hemisphere （＝NH）北半球
northern latitude 北纬
northern light 北极光 又见：aurora borealis
northern limit 北界限
northern limit of eclipse 北食限
Northern Palace Black Tor 北宫玄武
 又见：Northern Palace Black Warrior
Northern Palace Black Warrior 北宫玄武
 又见：Northern Palace Black Tor
northern polarity 北极性 | [磁] 北極性〔台〕
Northern Proper Motion Catalog 北天自行星表
Northern PZT program （＝NPZT）北天 PZT 星纲要
northern reference star 北天参考星
northern sky 北天 [区] 又见：northern celestial hemisphere
northern star 北天恒星
Northern tropic 北回归线 又见：Tropic of Cancer
north following star 东北星
north frigid zone 北寒带
North Galactic Cap 北银冠 | 北银 [极] 冠〔台〕
north galactic-polar spur 北银极支 又见：north galactic spur; north polar spur
North Galactic Pole 北银极

north galactic spur 北银极支 又见：north galactic-polar spur; north polar spur
north heliographic pole 日面北极
north north temperate belt （＝NNTB）[木星] 北北温带
north point 北点
north polar distance （＝NPD）北极距
North Polar Region 北极区（火星）
north polar sequence （＝NPS）北极星序
north polar spur 北银极支 又见：north galactic spur; north galactic-polar spur
north pole 北极
north preceding star 西北星
north-south asymmetry 南北不对称性
 又见：south-north asymmetry
North Star 勾陈一，北极星（小熊座 α）
 又见：Polaris
north temperate belt （＝NTB）木星北温带
north temperate zone 北温带
NOS （＝National Oceanic Survey）美国海洋测量局
nose frequency 鼻频
nose whistler 鼻啸
NOT （＝Nordic Optical Telescope）北欧光学望远镜
notch filter 点阻滤波器
nova 新星
Nova Cygni 天鹅座新星
nova explosion 新星爆发 又见：nova outburst
nova frequency 新星出现率 又见：nova rate
nova-like star 类新星
nova-like variable 类新星变星
nova-like X-ray source 类新星 X 射线源
nova outburst 新星爆发 又见：nova explosion
nova rate 新星出现率 又见：nova frequency
nova spectrum 新星光谱
November meteors 十一月流星雨 又见：November swarm
November swarm 十一月流星雨 又见：November meteors
Nozomi 希望号（日本火星探测器）
NP （＝NRAO pulsar）NRAO 脉冲星
Np （＝neptunium）镎 | 錼〔台〕（93 号元素）
N-particle distribution function N 粒子分布函数
NPD （＝north polar distance）北极距
NPOI （＝Navy Precision Optical Interferometer）海军精确光学干涉仪
n-point correlation function n 点相关函数

NPS (＝north polar sequence) 北极星序
NPZT (＝Northern PZT program) 北天 PZT 星纲要
NQR (＝nuclear quadrupole resonance) 核四极共振
NRAO (＝National Radio Astronomy Observatory) 美国国家射电天文台 | [美國] 國家電波天文台〔台〕
NRAO pulsar (＝NP) NRAO 脉冲星
NRAO VLA Sky Survey (＝NVSS) NVSS 巡天
NRL (＝Naval Research Laboratory) 美国海军科研实验室
NRT (＝Nançay radio telescope) 南赛射电望远镜（法国）
N.S. (＝New Style) 新历
NSF (＝National Science Foundation) 国家科学基金会（美国）
NSO (＝National Solar Observatory) 美国国家太阳观测台 | [美國] 國家太陽觀測台〔台〕
NSSDC (＝National Space Science Data Center) 国家空间科学数据中心（美国）
NST (＝New Solar Telescope) 新太阳望远镜
N star N 型星 又见: *N-type star*
NTB (＝north temperate belt) 木星北温带
NTE (＝non-thermodynamic equilibrium) 非热动平衡
NTS (＝Navigation Technology Satellite) NTS 系列卫星，导航技术卫星（美国）
NTT (＝New Technology Telescope) 新技术望远镜 | [歐南台] 新技術望遠鏡〔台〕
N-type star N 型星 又见: *N star*
Nubeculae 麦哲伦云 又见: *Magellanic Clouds*
Nubecula Minor 小麦哲伦云 | 小麥[哲倫] 雲〔台〕
nuclear activity 核活动
nuclear astrophysics 核天体物理 | 核天文物理學〔台〕
nuclear bulge 核球 又见: *bulge*
nuclear burning 核燃烧
nuclear density 核密度
nuclear disk 核盘
nuclear emulsion 核乳胶
nuclear evolution time 核演化时标
nuclear fission 核裂变 | 核分裂，核裂變〔台〕
nuclear fusion 核聚变 | 核融合，核聚合〔台〕
nuclear magnetic resonance (＝NMR) 核磁共振
nuclear magneton 核磁子
nuclear matter 核物质
nuclear physics 核物理
nuclear process 核过程
nuclear quadrupole resonance (＝NQR) 核四极共振
nuclear radiation 核辐射
nuclear reaction 核反应
Nuclear Spectroscopic Telescope Array (＝NuSTAR) 核分光望远镜阵
nuclear spin 核自旋
nuclear star 核心星
nuclear star cluster 核星团
nuclear statistical equilibrium 核统计平衡
nuclear time-scale 核反应时标
nuclear-track emulsion 核径迹乳胶
nuclear wind 星系核风
nucleated dwarf elliptical galaxy 有核矮椭圆星系
nucleated dwarf galaxy 有核矮星系
nucleid 核类
nucleochronology 核年代测定
nucleocosmochronology 核纪年法
nucleogenesis 核起源，核形成 | [原子] 核起源〔台〕
nucleon 核子
nucleosynthesis 核合成 | [原子] 核合成〔台〕
nucleosynthesis era 核合成时期
nucleus band 核带
nucleus of comet 彗核 又见: *cometary nucleus*
nucleus of condensation 凝聚核 | 凝結核〔台〕
nucleus of galaxy 星系核 又见: *galactic nucleus; galaxy nucleus*
nucleus population 核星系
nuclide 核素 | 核種〔台〕
Nuffield Radio Astronomy Laboratories 纳菲尔德射电天文实验室
null-balanced receiver 零位接收器
null cone 零锥
null field 零场
null geodesic 零测地线 | 零短程線〔台〕
null infinity 零无限远，类光无限远
nulling coronagraph 消零星冕仪
nulling interferometer 消零干涉仪
nulling interferometry 消零干涉测量
nulling phenomenon 脉冲消失现象 又见: *pulse nulling phenomenon*
null point 零点 又见: *zero point*
null surface 零曲面
number counting 计数 又见: *count; counting*
number counts of galaxy 星系计数

number density　数密度
number density profile　数密度轮廓
numbered asteroid　编号小行星
numbered supernova　编号超新星
number-flux relation　计数—流量关系
number-magnitude distribution　计数—星等分布
numerator　计数器　又见：*counter*
numerical analysis　数值分析
numerical approximation　数值近似
numerical cosmology　数值宇宙学
numerical differentiation　数值微分
numerical estimate　数值估计
numerical filtering　数值滤波
numerical integration　数值积分
numerical invariant　不变数
numerical method　数值方法
numerical model　数值模型
numerical simulation　数值模拟
numerical solution　数值解
numerical technique　数值技术
Numerov crater　努梅罗夫环形山（月球）
Nunki　斗宿四（人马座 σ）
Nuovo Cimento C　【意】《新试验 C 辑》（意大利期刊）
Nusakan　贯索三（北冕座 β）
Nushaba　箕宿一（人马座 γ）　又见：*Al Nasl*
Nüsl crater　努斯尔环形山（月球）
NuSTAR　（＝Nuclear Spectroscopic Telescope Array）核分光望远镜阵

nutation　章动
nutational constant　章动常数　又见：*nutation constant; constant of nutation*
nutational ellipse　章动椭圆　又见：*nutation ellipse*
nutation constant　章动常数　又见：*nutational constant; constant of nutation*
nutation ellipse　章动椭圆　又见：*nutational ellipse*
nutation in longitude　黄经章动　又见：*nutation of longitude*
nutation in obliquity　黄赤交角章动｜傾角章動〔台〕　又见：*nutation of inclination*
nutation in right ascension　赤经章动　即：*equation of the equinoxes*
nutation of inclination　黄赤交角章动｜傾角章動〔台〕　又见：*nutation in obliquity*
nutation of longitude　黄经章动　又见：*nutation in longitude*
nutation period　章动周期
Nuwa　女娲星（小行星 150 号）
NVO　（＝National Virtual Observatory, United States）美国国家虚拟天文台
NVSS　（＝NRAO VLA Sky Survey）NVSS 巡天
Nyquist criterion　尼奎斯特判据
Nyquist formula　尼奎斯特公式
Nyquist frequency　尼奎斯特频率
Nyquist limit　尼奎斯特极限
Nyquist sampling　尼奎斯特采样
Nysa　侍神星（小行星 44 号）
Nysa family　侍神星族

O

O （=oxygen）氧（8号元素）
OAO （=Orbiting Astronomical Observatory）轨道天文台（美国）
OAO-3 哥白尼天文卫星（美国） 又见: *Copernicus*
O association O 星协
OB association OB 星协
OB cluster OB 星团
Oberon 天卫四 | 天 [王] 衛四, 奥伯朗〔台〕
Oberth crater 奥伯特环形山（月球）
object glass 物镜 又见: *objective*
objective （=OG）物镜 又见: *object glass*
objective catalog 客观星表
objective classification 客观分类
objective grating 物端光栅
objective lens 物端透镜
objective prism 物端棱镜
objective-prism spectrum 物端棱镜光谱
object plane 物平面
object prism spectrograph 物端棱镜摄谱仪
object space 物方, 物空间 | 物方域〔台〕
oblate ellipsoid 扁椭球体
oblate isotropic rotator 各向同性扁转子
oblateness 扁率 又见: *flattening factor*
oblate spheroid 扁球体
oblate spheroidal coordinate ① 扁球面坐标 ② 等轴扁椭球坐标
oblique ascension 斜升
oblique coordinate 斜交坐标
oblique descension 斜降
oblique hydromagnetic shock wave 斜磁流激波
oblique impact 斜碰
oblique pulsator 斜脉动星
oblique rotator 斜转子
oblique rotator model 斜转子模型
oblique rotator theory 斜转子理论
oblique sphere 斜交天球 | 傾斜球〔台〕

obliquity ① 斜交 ② 黄赤交角 又见: *obliquity of the ecliptic; ecliptic obliquity* ③ 倾角 又见: *tilt angle; angle of tilt*
obliquity of the ecliptic 黄赤交角 又见: *obliquity; ecliptic obliquity*
Obolon' crater 奥博隆陨星坑（地球）
Obruchev crater 奥布鲁切夫环形山（月球）
obscuration 掩 | 掩 [星]〔台〕 又见: *occultation*
obscured variable 屏蔽变星 又见: *shielding variable*
obscuring nebula 遮掩星云
obscuring torus 遮蔽环
observable object 可观测天体
observable quantity 可观测量
observable universe 可观测宇宙
observational astronomy 观测天文学, 实测天文学 | 觀測天文學〔台〕
observational astrophysics 实测天体物理学 | 觀測天文物理學〔台〕
observational catalogue 观测星表
observational cosmology 观测宇宙学
observational data 观测数据 又见: *observed data*
observational dispersion 观测弥散度
observational error 观测误差
observational material 观测资料
observational network 观测网
observational program 观测程序
observational project 观测计划
observational quantity 观测量 又见: *observed quantity*
observational sampling 观测采样
observational selection 观测选择
observation equation 观测方程
observation platform 观测平台
Observatoire de Paris （=OP）【法】巴黎天文台
Observatoire Royal de Belgique （=ORB）【法】比利时皇家天文台

Observatoire Virtuel France （＝OV-France）【法】法国虚拟天文台
Observatorio del Roque de los Muchachos （＝ORM）【西】穆查丘斯罗克天文台（西班牙）
即：*Roque de los Muchachos Observatory*
Observatorio del Teide 【西】泰德峰天文台（西班牙） 即：*Teide Observatory*
Observatory 《天文台》（英国期刊）
observatory 天文台 又见：*astronomical observatory*
observatory dome 天文台圆顶
observed data 观测数据 又见：*observational data*
observed distribution 观测分布
observed place 观测位置
observed quantity 观测量 又见：*observational quantity*
observed value 观测值
observer 观测者 | 觀測者，觀測員〔台〕
observing cage 观测小室
observing frequency 观测频率
observing season 观测季
observing station 观测站
observing table 窥几
observing technique 观测技术
obsidianite 似曜石
OB star OB 型星
OB supergiant OB 型超巨星
obtuse angle 钝角
occultation 掩 | 掩[星]〔台〕 又见：*obscuration*
occultation angle 掩食角
occultation band 掩带
occultation curve 掩食曲线
occultation of planet 掩食行星
occultation of satellite 掩食卫星
occultation of source 掩源
occultation of star 掩星
occultation time 掩食时刻
occultation variable 掩食变星 | [掩]食變星〔台〕
又见：*veil variable star*
occulting disk 遮掩圆面
occulting mask 星掩模
occulting object 遮掩天体
occulting source 遮掩源
occultor 掩星器
occupation number 占据数
occupation ratio 占有比
occurrence rate 出现率

OCDM model （＝Open Cold Dark Matter model） 冷暗物质开模型
ocean-floor spreading 洋底扩张
oceanographic satellite 海洋科学卫星
oceanography 海洋学
ocean ridge 洋脊
ocean tide 洋潮
ocean trench 洋沟
ocean trough 洋槽
oceanus 【拉】洋（行星地貌）
Oceanus Procellarum 【拉】风暴洋（月球）
O cluster O 星团
Oct （＝Octans）南极座
Octahedral Research Satellite （＝ORS）八面体系列科学卫星（美国）
octahedrite 八面体陨星 | 八面體[式]隕鐵〔台〕
Octans （＝Oct）南极座
octant 八分仪，45° 视角
octave 倍频程 | 八[音]度，倍頻程〔台〕
octet ① 八重线 ② 八重态
ocular 目镜 又见：*eyepiece*
ocular circle 出射光瞳 | 出射[光]瞳〔台〕
又见：*exit pupil; eye-circle; emergent pupil*
ocular micrometer 目镜测微计 又见：*eyepiece micrometer*
Oculus Boreus 毕宿一（金牛座ε） 又见：*Ain*
OD （＝optical density）光学密度 | 光密度〔台〕
O'Day crater 奥戴环形山（月球）
odd coupling 奇耦合
odd cycle 奇数周
odd-even nucleus 奇偶核
odd-number theorem 奇数定理
odd-odd nucleus 奇奇核
odd-parity 奇宇称
odd state 奇态 又见：*singular state*
Odessa Astronomical Observatory 敖德萨天文台（乌克兰）
Odessa crater 敖德萨陨星坑（地球）
Odin 奥丁号（瑞典天文与高层大气卫星）
O dwarf O 型矮星
Odysseus 奥德赛（小行星 1143 号）
Odysseus crater 奥德赛环形山（土卫三）
Oef star Oef 型星
Oe star Oe 型星
off-axis aberration 轴外像差 又见：*abaxial aberration*
off-axis Cassegrain telescope 离轴卡塞格林望远镜

off-axis distribution　离轴分布
off-axis effect　偏轴效应
off-axis English mounting　偏轴英国式装置
off-axis Gregorian telescope　离轴格里高利望远镜
off-axis guider　离轴导星镜
off-axis image　轴外像 | 偏轴像〔台〕
off-axis imaging　轴外成像
off-axis mounting　偏轴式装置
off-axis observation　轴外观测
off-band observation　偏带观测
off-center interferogram　离心干涉图谱
off-center interferometry　离心干涉测量
off-focus detection　离焦探测
off-focus image　离焦像
off-focus imaging　离焦成像
off-focus observation　离焦观测
Offner relay system　奥夫纳中继光学系统
offset　偏置
offset coordinated time　补偿协调时
offset guiding　偏置导星
offset guiding device　偏置导星装置
offset guiding telescope　偏置导星镜
offset in right ascension　赤经偏差
offset local time　补偿地方时
offset observation　偏置观测
Of star　Of 型星
OG　(＝objective) 物镜
O giant　O 型巨星
OGLE　(＝Optical Gravitational Lensing Experiment) 光学引力透镜实验
OGO　(＝Orbiting Geophysical Observatory) 轨道地球物理台(美国)
OH absorption　羟基吸收，OH 吸收
OHANA　(＝Optical Hawaiian Array for Nanoradian Astronomy) 夏威夷纳弧度天文光学阵
OH cloud　羟基云，OH 云
OH emission　羟基发射，OH 发射
Ohio Sky Survey　俄亥俄射电巡天
Ohio State University Radio Observatory　俄亥俄州立大学射电天文台
OH-IR source　羟基红外源
OH/IR source　羟基红外源，OH/IR 源
OH/IR star　羟基红外星，OH/IR 星
OH line　羟基谱线，OH 谱线
Ohlson galactic plane　奥尔森银道面
Ohlson galactic pole　奥尔森银极

OH maser　(＝hydroxyl maser) 羟基微波激射 | 羟基邁射〔台〕
Ohm crater　欧姆环形山(月球)
ohmic dissipation　欧姆耗散
ohmic heating　欧姆致热
ohmic loss　欧姆损耗 | 歐姆損耗，電阻損耗〔台〕
OHP　(＝Haute Provence Observatory) 上普罗旺斯天文台
OH radical　(＝hydroxyl radical) 羟基
OH source　羟基源，OH 源 | OH 源〔台〕
OH star　羟基星，OH 星
Ohsumi　大隅号(日本卫星)　又见：Ōsumi
OIFITS　(＝Optical Interferometry Exchange Format) 光学干涉 FITS 格式
Okda　外屏七(双鱼座 α)　又见：Alrescha; Kaitain
Oken crater　奥肯环形山(月球)
Olbers crater　奥伯斯环形山(月球)
Olbersia　奥伯斯(小行星 1002 号)
Olbers' method　奥伯斯方法
Olbers' paradox　奥伯斯佯谬
Olcott crater　奥尔科特环形山(月球)
old disk star　老年盘族星
older population　老年星族
old inflation　旧暴胀
old nova　老新星　即：faded nova; ex-nova; postnova
old star　老年恒星
Old Style　(＝O.S.) 旧历
oligarchic growth　寡头式生长
olivine　橄榄石
Oljato　奥加托(小行星 2201 号)
Olympus Mons　奥林匹斯山(火星)
O magnitude　O 星等
Omar Khayyam crater　欧玛尔·海亚姆环形山(月球)
OMC　(＝Orion Molecular Cloud) 猎户座分子云 | 獵戶[座]分子雲〔台〕
Omega Nebula　ω 星云(M17)　又见：ω Nebula
omni-directional antenna　全向天线
omni-directional effect　全向效应
omni-directional flux　全向流量
on-band observation　在带观测
on-board clock　在载钟
on-board telescope　在载望远镜
once-through correcting plate　单程改正片
once-through corrector　单程改正镜
Ondrejov Astrophysical Observatory　昂德列约夫天体物理台

one-dimensional aperture synthesis 一维综合孔径
又见：*one-dimensional synthesis*

one-dimensional classification 一元分类

one-dimensional synthesis 一维综合孔径
又见：*one-dimensional aperture synthesis*

O-Ne-Mg white dwarf 氧氖镁白矮星

One-Mile Telescope 一英里射电望远镜 | 一英里電波望遠鏡〔台〕

one-parameter method 单参数法

one-particle distribution function 单粒子分布函数

one-way Doppler 单程多普勒

one-way light time 单程光行时

one-way membrane 单向膜

one-way satellite system 单路卫星系统

one-zone model 单区模型

onion skin model 洋葱皮模型

on-line clock 联机钟

on-line data handling 联机数据处理 又见：*on-line data processing*

on-line data processing 联机数据处理
又见：*on-line data handling*

on-line filtering 联机滤波

on-line system 联机系统

on-off measurement 通断测量，离合测量

Onsala Space Observatory 翁萨拉天文台

on-site clock visit 到站钟访，现场钟访

on-site measurement 就地测量

on-the-fly mapping 飞扫成图法

On the Revolution of the Heavenly Spheres 《天体运行论》（哥白尼著作）

Oort cloud 奥尔特云 | 歐特[彗星]雲〔台〕

Oort cloud comet 奥尔特云彗星

Oort cloud object 奥尔特云天体

Oort comet cloud 奥尔特彗星云

Oort constant 奥尔特常数 | 歐特常數〔台〕
又见：*Oort's constant*

Oort cycle 奥尔特周

Oort formulae 奥尔特公式 | 歐特公式〔台〕
又见：*Oort's formulae*

Oort limit 奥尔特极限

Oort's constant 奥尔特常数 | 歐特常數〔台〕
又见：*Oort constant*

Oort's formulae 奥尔特公式 | 歐特公式〔台〕
又见：*Oort formulae*

Oort-Spitzer hypothesis 奥尔特-斯皮策假说

Oort-Spitzer mechanism 奥尔特-斯皮策机制

Oosterhoff class 奥斯特霍夫类型

Oosterhoff group 奥斯特霍夫星群

Ootacamund Radio Astronomy Centre 乌塔卡蒙德射电天文中心

Ooty Radio Telescope （=ORT）乌塔卡蒙德射电望远镜（印度）

OP ① （=Observatoire de Paris）【法】巴黎天文台 ② （=Paris Observatory）巴黎天文台

opacity 不透明度

opacity coefficient 不透明系数 又见：*coefficient of opacity*

opaque atmosphere 不透明大气

Opaque Era 不透明时期

open-air radio telescope 露天式射电望远镜

open-air telescope 露天望远镜

open arc crater 开环形山弧

open chain crater 开环形山链

open cluster ① 疏散星系团 又见：*open cluster of galaxies* ② 疏散星团 | 疏散星團，銀河星團〔台〕
又见：*open cluster of stars; Galactic cluster; open star cluster*

open cluster of galaxies 疏散星系团 又见：*open cluster*

open cluster of stars 疏散星团 | 疏散星團，銀河星團〔台〕 又见：*open cluster; Galactic cluster; open star cluster*

Open Cold Dark Matter model （=OCDM model）冷暗物质开模型

open configuration 开组态

open inflation 开暴胀

open model 开模型

openness 松紧度

open solar telescope 开放式太阳望远镜

open spiral galaxy 疏散旋涡星系

open star cluster 疏散星团 | 疏散星團，銀河星團〔台〕 又见：*open cluster of stars; open cluster; Galactic cluster*

open string 开弦

open universe 开宇宙 | 開放宇宙〔台〕

operation 运作

operational amplifier 运算放大器

Oph （=Ophiuchus）蛇夫座

Ophelia ① 天卫七 | 天[王]衛七〔台〕 ② 奥菲丽亚（小行星171号）

Ophir Chasma 俄斐深谷（火星）

Ophiuchid meteor shower 蛇夫流星雨

Ophiuchids 蛇夫流星群

Ophiuchus （=Oph）蛇夫座

Ophiuchus Ridge 蛇夫脊

Opik-Oort cloud 欧皮克-奥尔特云

Opik theory　欧皮克理论
Oppenheimer crater　奥本海默环形山（月球）
Oppenheimer-Volkoff limit　奥本海默－沃尔科夫极限 | 歐本海默－沃科夫極限〔台〕
Opportunity Mars Exploration Rover　机遇号火星探测车
opposition　冲
opposition in right ascension　赤经冲
optical aperture-synthesis imaging　光学综合孔径成像 | 光學孔徑合成成像〔台〕
optical aperture-synthesis imaging technique　光学综合孔径成像技术 | 光學孔徑合成成像技術〔台〕
optical appearance　光学外观
optical arm　光学臂
optical astrometry　光学天体测量
optical astronomy　光学天文学 | 光學天文 [學]〔台〕
optical astrophysics　光学天体物理 | 光學天文物理學〔台〕
optical axis　光轴
optical binary　光学双星　又见：*optical double star; optical pair*
optical brightness　光学波段亮度
optical burst　光学波段暴
optical center　光心　又见：*photocenter; center of light*
optical clock　光学钟
optical component　光学子星
optical counterpart　光学对应体
optical density　（＝OD）光学密度 | 光密度〔台〕
optical depth　光深 | 光 [學] 深 [度]〔台〕
optical design　光学设计
optical disk　光学盘
optical double　视双星，光学双星　又见：*apparent binary*
optical double galaxy　光学双星系
optical double star　光学双星　又见：*optical binary; optical pair*
optical emission　光学发射
optical equipment　光学设备
optical fiber interferometer　光纤干涉仪
optical fibre　光纤
optical film　光膜
optical filter　滤光片
optical flare　光学耀斑
optical flash　光学闪
optical galaxy　光学星系
optical glass　光学玻璃
optical glow　光学天光

Optical Gravitational Lensing Experiment　（＝OGLE）光学引力透镜实验
optical guidance　光学导航
Optical Hawaiian Array for Nanoradian Astronomy　（＝OHANA）夏威夷纳弧度天文光学阵
optical horizon　光视界
optical identification　光学证认 | 光學識別〔台〕
optical image　光学像
optical imaging　光学成像
optical instrument　光学仪器
optical interferometer　光学干涉仪 | 光干涉儀〔台〕
optical interferometry　光干涉测量
Optical Interferometry Exchange Format　（＝OIFITS）光学干涉 FITS 格式
optical laser radar　激光雷达　又见：*laser radar*
optical libration　光学天平动
optical light　可见光　又见：*visible light*
optical luminosity function　光学光度函数
optically identified object　光学证认天体
optically identified source　光学证认源
optically thick　光学厚
optically thick medium　光厚介质
optically thin　光学薄
optically thin absorption　光学薄吸收
optically thin emission　光学薄发射
optically thin medium　光薄介质
optically violently variable quasar　光剧变类星体，OVV 类星体 | 光巨變類星體，OVV 類星體〔台〕　又见：*optically violent variable quasar*
optically violent variable quasar　（＝OVV quasar）光剧变类星体，OVV 类星体 | 光巨變類星體，OVV 類星體〔台〕　又见：*optically violently variable quasar*
optically visible object　光学可见天体
optical material　光学材料
optical micrometer　光学测微计
optical object　光学天体
optical observation　光学观测
optical observatory　光学天文台
optical outburst　光学波段爆发
optical pair　光学双星　又见：*optical binary; optical double star*
optical path　光程
optical path length　光程长
optical path length equalizer　光程补偿器
optical phase　光学相位（耀斑）
optical picture　光学图

optical pointing telescope　光学指向望远镜
optical position　光学位置
optical processor　光学处理机
optical property　光学特性
optical pulsar　光学脉冲星
optical pulsation　光学脉动
optical pulse　光学脉冲
optical pumping　光学抽运
optical radar　光波雷达
optical radiation　光学辐射
optical scalar equation　光学标量方程
optical scanner　光学扫描仪
optical scanning　光学扫描
optical sensibilization　光学增敏
optical spectrometer　光波分光计
optical spectroscopy　光波分光
optical spectrum　光谱　又: *spectrum*
optical sun　光学太阳
optical surface　光学表面
optical system　光学系统，光具组
optical telescope　光学望远镜
optical temperature　光学温度
optical testing　光学检验
optical theodolite　光学经纬仪
optical thickness　光学厚度
optical tracking　光学跟踪 | 光學追蹤〔台〕
optical tracking theodolite　光学跟踪经纬仪
optical transfer function　（=OTF）光学转移函数
optical truss　光学桁架
optical variable　光学变星
optical visual binary　光学目视双星
optical waveguide　光学波导
optical wedge　光学尖劈
optical window　光学窗口 | 光窗〔台〕
optics　光学
optimization　最优化
optimum　最优
optimum filter　最佳滤光片
optimum transfer　最佳过渡
opto-acoustical processor　光声处理机
ORB　（=Observatoire Royal de Belgique）【法】比利时皇家天文台
orb　天体　又见: *aster; celestial body; celestial object*
orbifold　轨形
orbit　轨道
orbital acceleration　轨道加速度

orbital angular momentum　轨道角动量
orbital angular velocity　轨道角速度
orbital anisotropy　轨道各向异性
orbital chaos　轨道混沌
orbital circularization　轨道圆化
orbital decay　轨道衰减
orbital determination　定轨，轨道测定
　又见: *determination of orbit*
orbital eccentricity　轨道偏心率
orbital electron capture　轨道电子俘获
orbital element　轨道根数 | 轨道要素，轨道根數〔台〕　又见: *element of orbit*
orbital evolution　轨道演化
orbital frequency　轨道频率
orbital inclination　轨道倾角
orbital instability　轨道不稳定性
orbital latitude　轨道纬度
orbital longitude　轨道经度
orbital motion　轨道运动
orbital node　轨道交点
orbital period　轨道周期
orbital phase　轨道相位
orbital plane　轨道面　又见: *orbit plane*
orbital stability　轨道稳定性
orbital theory　轨道理论
orbital velocity　轨道速度
orbital workshop　轨道工作室
orbit averaged approximation　轨道平均近似
orbit-averaged diffusion coefficient　轨道平均扩散函数
orbit-averaged Fokker-Planck equation　轨道平均福克－普朗克方程
orbit control　轨道控制
orbit correction　轨道改正 | 軌道修正〔台〕
orbit determination　定轨 | 軌道測定〔台〕
orbit dynamics　轨道力学
orbiter　轨道飞行器 | 軌道衛星，軌道太空船〔台〕
orbit improvement　轨道改进　又见: *improvement of orbit*
Orbiting Astronomical Observatory　（=OAO）轨道天文台（美国）
orbiting collision　轨道碰撞
Orbiting Geophysical Observatory　（=OGO）轨道地球物理台（美国）
Orbiting Satellite Carrying Amateur Radio　（=OSCAR）奥斯卡系列卫星（美国）
Orbiting Solar Observatory　（=OSO）轨道太阳观测站 | 軌道太陽觀測台〔台〕（美国）

orbiting telescope 轨道望远镜
Orbiting Vehicle （＝OV）OV 轨道飞行器
Orbiting Wide-angle Light Collectors （＝OWL）广角集光卫星（美国高能卫星）
orbit-orbit resonance 轨轨共振 | 與軌共振〔台〕
orbit plane 轨道面　又见：*orbital plane*
orbit prediction 轨道预报
orbit resonance 轨道共振
orbit-rotation resonance 轨旋共振　又见：*spin-orbit resonance*
orbit-superposition method 轨道交叠法
orbit transfer 轨道转移 | 軌道變換〔台〕　又见：*transfer between orbits*
Orcus 亡神星（小行星 90482 号）
ordered orbit 有序轨道，规则轨道 | 有序軌道〔台〕　又见：*regular orbit*
order of magnitude 数量级 | [數] 量級〔台〕
order of perturbation 摄动阶
order of spectrum 光谱级
order parameter 有序参数
order region 有序区
ordinary chondrite 普通球粒陨星
ordinary periodic solution 常周期解
ordinary ray 寻常光线 | 尋常射線〔台〕
ordinary wave 寻常光波 | 尋常波〔台〕
Oreol 日晕号（苏联-法国科学卫星）　又见：*Aureole*
Oresme crater 奥雷姆环形山（月球）
organic matter 有机物质
organic molecule 有机分子
organic scintillator 有机闪烁体
organon parallacticon 星位尺 | 星位儀〔台〕
Orgueil meteorite 奥盖尔陨星 | 奧蓋爾隕石〔台〕
Ori （＝Orion）猎户座
Orientale Basin 东海盆地（月球）
orientation 定向，定位 | 取向〔台〕
orientation angle 定向角，定位角
orientation correction 定向改正，定位改正
orientation parameter 定向参数，定位参数
orientation star 定向星，定位星 | 定向星〔台〕
oriented meteorite 有向陨星
Origem Loop 猎户双子圈
original atmosphere 原大气
Original Gas hypothesis 原气说　又见：*Primary Atmosphere hypothesis*
original planet 原行星　又见：*proto-planet*
original star 原恒星　又见：*protostar; primordial star*

original sun 原太阳　又见：*protosun*
origin of elements 元素起源
origin of the universe 宇宙起源
Origins Spectral Interpretation Resource Identification Security Regolith Explorer （＝OSIRIS-REx）奥西里斯王号小行星探测器
Orion （＝Ori）猎户座
Orion aggregate 猎户星集
Orion Arm 猎户臂
Orion association 猎户星协
Orionid meteor shower 猎户座流星雨
Orionids 猎户流星群 | 獵戶 [座] 流星雨〔台〕
Orion Loop 猎户圈
Orion Molecular Cloud （＝OMC）猎户座分子云 | 獵戶 [座] 分子雲〔台〕
Orion Nebula 猎户 [大] 星云 | 獵戶 [座] 大星雲〔台〕（M42）　又见：*Great Nebula in Orion; Great Orion Nebula*
Orion population star 猎户族恒星
Orion's belt 猎户腰带（星组）　又见：*Belt of Orion*
Orion spectrum 猎户光谱
Orion spur 猎户射电支 | 獵戶座分支〔台〕
Orion's sword 猎户腰刀（星组）
Orion-type flare star 猎户型耀星 | 獵戶 [座型閃] 焰星〔台〕
Orion variable 猎户型变星
Orlov crater 奥尔洛夫环形山（月球）
ORM （＝Observatorio del Roque de los Muchachos）【西】穆查丘斯罗克天文台（西班牙）　即：*Roque de los Muchachos Observatory*
orogenetic process 造山过程
orphan afterglow 孤立余辉
orrery 太阳系仪，七政仪 | 太陽系儀〔台〕
ORS （＝Octahedral Research Satellite）八面体系列科学卫星（美国）
ORT （＝Ooty Radio Telescope）乌塔卡蒙德射电望远镜（印度）
orthicon 正析摄像管 | 超正析像管〔台〕
orthochromatic plate 正色底片
orthodrome 大圆　又见：*great circle*
orthogonal axes 正交轴
orthogonal coordinate 正交坐标
orthogonal function 正交函数
orthogonality 正交性
orthogonal polarization 正交偏振
orthogonal space-time 正交时空
orthogonal transfer CCD （＝OTCCD）正交转移 CCD

orthogonal transformation　正交变换
orthographic projection　正投影 | 正射投影〔台〕
ortho-hydrogen　正氢
orthometric correction　正交改正
orthomode　正交模式
orthonormal tetrad　正交四维标架
orthopyroxene　斜方辉石
orthoscopic eyepiece　无畸变目镜
Orthosie　木卫三十五
ortho-spectrum　正谱
orthostate　正态
O.S.　（＝Old Style）旧历
Os　（＝osmium）锇（76 号元素）
OSCAR　（＝Orbiting Satellite Carrying Amateur Radio）奥斯卡系列卫星（美国）
Oschin Telescope　奥欣望远镜（美国）
oscillating mode　振动模
oscillating model　振动模型，振荡模型 | 振盪 [宇宙] 模型〔台〕
oscillating satellite　振动卫星
oscillating universe　振荡宇宙　又见：pulsating universe
oscillation　振动，振荡
oscillation phase　振动相位
oscillation pile-up　振荡堆垒
oscillator　振子，振荡器
oscillator strength　振子强度
oscillatory motion　振荡运动
oscillogram　波性图
oscillograph　示波仪
oscilloscope　示波器
osculating element　吻切根数 | 密切轨道要素，密切轨道根数〔台〕
osculating ellipse　吻切椭圆 | 密切橢圓〔台〕
osculating epoch　吻切历元 | 密切曆元〔台〕
osculating orbit　吻切轨道 | 密切軌道〔台〕
osculating orbital elements　吻切轨道根数
osculating plane　吻切平面 | 密切平面〔台〕
osculating point　吻切点 | 密切點〔台〕
OSIRIS-REx　（＝Origins Spectral Interpretation Resource Identification Security Regolith Explorer）奥西里斯王号小行星探测器
Oslo University Observatory　奥斯陆大学天文台
osmium　（＝Os）锇（76 号元素）
OSO　（＝Orbiting Solar Observatory）轨道太阳观测站 | 軌道太陽觀測台〔台〕（美国）
O star　O 型星　又见：O-type star

Ostriker-Peebles criterion　奥斯特里克－皮布尔斯判据
Ostriker-Vishniac effect　奥斯特里克－维希尼克效应
Ostwald crater　奥斯特瓦尔德环形山（月球）
O subdwarf　O 型亚矮星
Ōsumi　大隅号（日本卫星）　又见：Ohsumi
O supergiant　O 型超巨星
OTCCD　（＝orthogonal transfer CCD）正交转移 CCD
OTF　（＝optical transfer function）光学转移函数
Ottawa River Solar Observatory　渥太华河太阳观测台
Otto Struve Telescope　斯特鲁维望远镜
O-type star　O 型星　又见：O star
Oujianquan　区鉴泉（小行星 3089 号）
outbreak　爆发　又见：eruption; outburst
outburst　爆发　又见：eruption; outbreak
Outer Arm　外缘旋臂
outer asteroid belt　外小行星带
outer atmosphere　外层大气
outer-belt asteroid　外带小行星
outer core　外核
outer corona　外冕 | 外 [日] 冕〔台〕
outer gap　外围隙
outer halo　外晕
outer halo cluster　外晕族星团
outer Lagrangian point　外拉格朗日点
outer Lindblad radius　外林德布拉德半径
outer Lindblad resonance　外林德布拉德共振
outer planet　带外行星 | [带] 外行星〔台〕
outer product　外积　又见：wedge product
outer ring　外环
outer-shell electron　外壳电子
outer solar system　外太阳系
outer space　外层空间 | 外太空〔台〕
outgassing　排气 | 釋氣〔台〕
outline curve　轮廓线
out-of-equilibrium decay　失衡衰变
out of phase　异相
output　输出
output device　输出装置
output impedance　输出阻抗
outside-eclipse variation　食外变光
outward winding　旋开
OV　（＝Orbiting Vehicle）OV 轨道飞行器
oval nebula　卵形星云 | 蛋形星雲〔台〕

oval-shaped curve 卵形曲线
oval-shaped curve of zero velocity 卵形零速度曲线
Ovda Regio 奥华特地区（金星）
overall bandwidth 总带宽
overall magnetic field 总体磁场
overall noise 总噪声
overall noise temperature 总噪声温度
overcast sky 阴天
overcontact binary 过接双星
overcooling problem 过冷却问题
overdense matter 超密物质
overdense plasma 超密等离子体
overdensity 过密度
over-exposure 曝光过度 | 曝光過度，露光過度〔台〕
over-flux 超额流量
over-heating 过热
over-heating instability 过热不稳定性
overlap method 重叠法
overlap of spectral lines 谱线重叠
overlapping region 重叠区
overluminous object 超大光度天体 | 光度[特]大天體〔台〕
overluminous star 超大光度恒星 | 光度[特]大恆星〔台〕
overlying region 叠加区
overmassive object 超大质量天体
overmassive star 超大质量恒星
overshooting ①超射，过冲 ②超射 即：convective overshooting
overstability 超稳定性
overstable mode 超稳定模
overtaking collision 追撞
overtone 谐波 又见：harmonic overtone
overtone frequency 谐波频率
overtone mode 谐波模
OV-France （=Observatoire Virtuel France）【法】法国虚拟天文台
OVRO （=Owens Valley Radio Observatory）欧文斯谷射电天文台
OVV quasar （=optically violent variable quasar）光剧变类星体，OVV类星体 | 光巨變類星體，OVV類星體〔台〕
Owens Valley Radio Observatory （=OVRO）欧文斯谷射电天文台
OWL （=Orbiting Wide-angle Light Collectors）广角集光卫星（美国高能卫星）
Owl Cluster 夜枭星团（NGC 457）
Owl Nebula 夜枭星云（NGC 3587）
Ox 牛宿（二十八宿）
oxygen （=O）氧（8号元素）
oxygen-rich star 富氧星
oxygen sequence 氧序
oxygen star 氧星
oxysphere 岩石圈 又见：lithosphere
Ozma project 奥兹玛计划 | 奥斯瑪計劃〔台〕
ozone layer 臭氧层 又见：ozonosphere
ozonosphere 臭氧层 又见：ozone layer
Ozza Mons 奥扎山（金星）

P

P （＝phosphorus）磷（15 号元素）
PA （＝position angle）位置角 | 方位角〔台〕
Pa （＝protactinium）镤（91 号元素）
Paaliaq 土卫二十
PAAS （＝Publications of the American Astronomical Society）《美国天文学会会刊》
Pabu 曝卫（小行星 Borasisi 的卫星）
Pacific plate 太平洋板块
Pacific standard time （＝PST）太平洋标准时
packing fraction 敛集率
Padua University Astrophysical Observatory 帕多瓦大学天体物理台
PAH （＝polycyclic aromatic hydrocarbon）多环芳烃
pair annihilation 对湮灭 | 對滅〔台〕
pair annihilation cross-section 对湮灭截面
pair annihilation neutrino process 对湮灭中微子过程
pair creation 粒子对产生 | 對生〔台〕 又见：pair production
pairing energy 成对能量
pair-instability supernova 生对不稳定性超新星
pair of galaxies 星系对
pair of radio sources 射电源对 | 電波源對〔台〕
pair of stars 星对
pair potential 对势
pair production 粒子对产生 | 對生〔台〕 又见：pair creation
pair-production instability 生对不稳定性
pair production telescope 正负电子对生成望远镜
pair tracker 正负电子对示踪器
pairwise peculiar velocity dispersion 成对本动速度弥散
pairwise velocity dispersion （＝PVD）成对速度弥散
palaeoclimate 古气候 又见：paleoclimate
palaeogeography 古地理 又见：paleogeography
palaeogeology 古地质 又见：paleogeology
paleoclimate 古气候 又见：palaeoclimate
paleogeography 古地理 又见：palaeogeography
paleogeology 古地质 又见：palaeogeology
paleointensity 古磁场强度
paleomagnetism 古地磁 | 古地磁 [學]〔台〕
paleotectonics 古地质构造
paleotopography 古地形
Palermo Circle 巴勒莫环形仪
Pales 牧神星（小行星 49 号）
palladium （＝Pd）钯（46 号元素）
Pallas 智神星（小行星 2 号）
pallasite 橄榄陨铁
Pallene 土卫三十三
Palomar Green survey （＝PG survey）帕洛玛－格林巡天
Palomar-Haro-Luyten Catalogue of Blue Stellar Objects （＝PHL）PHL 蓝星体表
Palomar Observatory 帕洛玛天文台
Palomar Observatory Sky Survey （＝POSS）帕洛玛天图 | 帕洛瑪星圖〔台〕
Palomar Sky Survey （＝PSS）帕洛玛天图 | 帕洛瑪星圖〔台〕 即：POSS
palus 沼
Palus Epidemiarum 【拉】疫沼（月球）
Palus Nebularum 【拉】雾沼（月球，非正式名称）
Palus Putredinis 【拉】腐沼（月球）
Palus Somni 【拉】梦沼（月球）
PAMELA （＝Payload for Antimatter Matter Exploration and Light-nuclei Astrophysics）PAMELA 实验
Pan 土卫十八
pancake model 薄饼模型
pancake theory 薄饼理论
Pancha-Siddhantika 【梵】《五种历数全书》（印度古历）

panchromatic plate 全色底片
pancosmic theory 泛宇宙学说
pancosmism 泛宇宙论
pancosmos 泛宇宙
Pandora ① 土卫十七 ② 祸神星（小行星 55 号）
Paneth crater 帕内特环形山（月球）
Panopaea 蟹神星（小行星 70 号）
panoramic camera 全景照相机
panoramic detection 全景探测
panoramic receiver 扫频接收机
  又见：*sweep-frequency receiver*
Panoramic Survey Telescope and Rapid Response System （=Pan-STARRS）泛星计划 | 泛星計畫〔台〕
Pan-STARRS （=Panoramic Survey Telescope and Rapid Response System）泛星计划 | 泛星計畫〔台〕
PAPA （=precision analog photon address）精密模拟光子定位
Papaleksi crater 帕帕列克西环形山（月球）
paper clock 纸钟（平均假钟）
parabola 抛物线 又见：*para-curve*
parabolic antenna 抛物面天线 又见：*paraboloid antenna; paraboloidal antenna*
parabolic comet 抛物线轨道彗星 | 抛物線[軌道]彗星〔台〕
parabolic coordinate 抛物坐标
parabolic cylinder antenna 抛物柱面天线
parabolic mirror 抛物面镜 又见：*paraboloidal mirror; paraboloid mirror*
parabolic orbit 抛物线轨道
parabolic radio telescope 抛物面射电望远镜
  又见：*paraboloid radio telescope; paraboloidal radio telescope*
parabolic reflector 抛物面反射镜 又见：*paraboloid reflector; paraboloidal reflector*
parabolic space 抛物空间
parabolic velocity 抛物线速度
paraboloid 抛物面 | 抛物體〔台〕
paraboloidal antenna 抛物面天线
  又见：*paraboloid antenna; parabolic antenna*
paraboloidal mirror 抛物面镜 又见：*paraboloid mirror; parabolic mirror*
paraboloidal radio telescope 抛物面射电望远镜
  又见：*paraboloid radio telescope; parabolic radio telescope*
paraboloidal reflector 抛物面反射镜
  又见：*parabolic reflector; paraboloid reflector*
paraboloid antenna 抛物面天线 又见：*paraboloidal antenna; parabolic antenna*

paraboloid mirror 抛物面镜 又见：*paraboloidal mirror; parabolic mirror*
paraboloid radio telescope 抛物面射电望远镜
  又见：*paraboloidal radio telescope; parabolic radio telescope*
paraboloid reflector 抛物面反射镜
  又见：*parabolic reflector; paraboloidal reflector*
para-curve 抛物线 又见：*parabola*
paradox 佯谬
paragravity 类重力
para-hydrogen 仲氢
parallactic angle 星位角
parallactic displacement 视差位移
  又见：*parallactic shift*
parallactic ellipse 视差椭圆
parallactic equation 月角差 又见：*parallactic inequality*
parallactic error 视差误差
parallactic inequality 月角差 又见：*parallactic equation*
parallactic libration 视差天平动
parallactic motion 视差动
parallactic mounting 赤道式装置 | 赤道[式]裝置〔台〕
parallactic orbit 视差轨道
parallactic refraction 折射视差
parallactic rule 视差尺
parallactic shift 视差位移 又见：*parallactic displacement*
parallactic triangle 视差三角形
parallax 视差
parallax factor 视差因子
parallax second （=pc; parsec）秒差距
parallax star 视差星
parallel 纬线，纬圈，平行圈 | 緯圈，平行圈，緯線〔台〕
parallel circle ① 纬[度]圈 又见：*latitude circle; parallel of latitude* ② 纬圈，平行圈
parallel connection 并联
parallel displacement 平行位移
parallel drift 平行漂移
parallel drifting bands 平行漂移带 又见：*parallel drifting zebras*
parallel drifting zebras 平行漂移带 又见：*parallel drifting bands*
parallel flow 层流
parallel of altitude 地平纬圈，平行圈 | 地平緯圈，等高圈〔台〕 又见：*altitude circle; circle of altitude; almucantar*

parallel of declination 赤纬圈　又见：*declination circle; circle of declination; declination parallel*
parallel of latitude ①纬[度]圈　又见：*latitude circle; parallel circle* ②黄纬圈　又见：*latitude circle*
parallel plate 平行片
parallel sphere 平行球
parallel transport 平行移动
paramagnet 顺磁体
paramagnetic resonance 顺磁共振
parameterized post-Newtonian formalism （=PPN formalism）参数化后牛顿形式
parameter variance 参数方差
parametric amplification 参量放大
parametric amplifier 参量放大器 | 参數放大器〔台〕
parametric converter 参量变频器
parametric latitude 归化纬度，归心纬度　又见：*reduced latitude*
parametric oscillation 参量振荡
parametric oscillator 参量振荡器
parametric resonance 参数共振
Paranal Observatory 帕拉纳尔天文台（智利）
paranthelionic circle 幻日环　又见：*parhelic circle; mock solar-ring*
paranthelion（复数：**paranthelia**） 远幻日 | 幻日〔台〕
parantiselene 远幻月 | 幻月〔台〕
paraselene（复数：**paraselenae**） 近幻月 | 幻月〔台〕
paraselenic circle 幻月环　又见：*mock lunar-ring*
parasite 天电干扰　又见：*atmospherics; atmospheric interference*
parasitic image 寄生像
parasitic light 寄生光
parasitics 寄生效应
parasitic signal 寄生信号
parasitic star 寄生星
para-spectrum 仲谱
para-state 仲态
paraxial image 近轴像
paraxial ray 近轴光线
parent body 母体
parent cloud 母云
parent comet 母彗星
parent galaxy 母星系
parent object 母天体
parent planet 母行星

parhelic circle 幻日环　又见：*paranthelionic circle; mock solar-ring*
parhelion 幻日　又见：*mock sun; sundog*
Paris 帕里斯（小行星3317号）
Paris classification 巴黎光谱分类
Paris Observatory （=OP）巴黎天文台
parity 宇称
parity conservation 宇称守恒
parity nonconservation 宇称不守恒
parity violation 宇称破坏
Parker instability 帕克不稳定性
Parker model 帕克模型
Parkes Catalogue of Radio Sources （=PKS）帕克斯射电源表 | 帕克斯電波源表〔台〕
Parkes Observatory 帕克斯天文台
Parkes Radio Telescope 帕克斯射电望远镜 | 帕克斯電波望遠鏡〔台〕
Parkhurst crater 帕克赫斯特环形山（月球）
parking orbit 停泊轨道
Parrot crater 帕罗特环形山（月球）
parry arc 彩晕弧
parsec （=parallax second）秒差距
Parthenope 海妖星（小行星11号）
partial coherence 部分相干性
partial degeneracy 部分简并性 | 部份簡併性〔台〕
partial density 部分密度 | 部份密度〔台〕
partial eclipse 偏食
partial eclipse beginning 偏食始
partial eclipse end 偏食终 | 偏食終，復圓〔台〕
partial eclipse solution 偏食解
partial lunar eclipse 月偏食
partially eclipsing binary 偏食双星
partially filled aperture 部分连续孔径
partially ionization 部分电离
partially ionized plasma 部分电离等离子体
partially polarization 部分偏振　又见：*fractional polarization*
partially polarized radiation 部分偏振辐射 | 部份偏振輻射〔台〕
partial pressure 分压
partial solar eclipse 日偏食
partial wave amplitude 分波强度
partial wave decomposition 分波分解
particle 粒子
particle acceleration 粒子加速
particle accretion 粒子吸积

particle astronomy 粒子天文学 | [高能]粒子天文學〔台〕
particle astrophysics 粒子天体物理学 | 粒子天文物理學〔台〕
particle beam 粒子束　又见：*bunch of particles*
particle confinement 粒子囚禁
particle creation 粒子创生
particle detection 粒子探测
particle distribution function 粒子分布函数
particle encounter 粒子碰撞
particle escape 粒子逃逸
particle-field interaction 粒子－场相互作用
particle flux 粒子流量
particle horizon 粒子视界
particle interaction 粒子相互作用
particle-mesh algorithm （＝PM algorithm）质点－网格算法
Particle Mesh code （＝PM code）质点－网格程序
particle-particle algorithm （＝PP algorithm）质点－质点算法
particle production 粒子产生
particular solution 特解
particulate 微粒　又见：*corpuscle*
partition function 配分函数
partner star 伴星　又见：*companion star*
parton 部分子 | 成子，部分子〔台〕
PASA （＝Proceedings of the Astronomical Society of Australia）《澳大利亚天文学会会刊》
Pascal crater 帕斯卡环形山（月球）
Pascal triangle 帕斯卡三角
Paschal table 复活节历表
Paschen-Back effect 帕邢－巴克效应 | 帕申－贝克效應〔台〕
Paschen continuum 帕邢连续区
Paschen crater 帕邢环形山（月球）
Paschen limit 帕邢系限
Paschen lines 帕邢谱线
Paschen series 帕邢线系 | 帕申譜系〔台〕
Pasiphae 木卫八
Pasithee 木卫三十八
PASJ （＝Publications of the Astronomical Society of Japan）《日本天文学会会刊》
PASP （＝Publications of the Astronomical Society of the Pacific）《太平洋天文学会会刊》
passband 通带
passband width 带宽 | 通帶寬〔台〕
passive device 无源器件

passive evolution 被动演化
passive gravitational mass 被动引力质量
passive network 无源网络
passive satellite 无源卫星
passive shielding 被动屏蔽
Pasteur crater 巴斯德环形山（月球）
past light-cone 昔日光锥
patera 托边火山结构（行星地貌）
path delay 路径延迟
path difference 程差
path length 程长 | 軌跡長度〔台〕
path length correction 程长改正
path loss 路径损耗
path of annular eclipse 环食带　又见：*zone of annularity; path of annular phase; path of annularity*
path of annularity 环食带　又见：*zone of annularity; path of annular phase; path of annular eclipse*
path of annular phase 环食带　又见：*zone of annularity; path of annularity; path of annular eclipse*
path of eclipse 食带　又见：*zone of eclipse*
path of total eclipse 全食带　又见：*zone of totality; belt of totality; path of totality*
path of totality 全食带　又见：*zone of totality; path of total eclipse; belt of totality*
Patientia 忍神星（小行星451号）
Patroclus 帕特洛克鲁斯 | 派特洛克魯斯〔台〕（小行星617号）
Patroclus group 帕特洛克鲁斯群 | 派特洛克魯斯群〔台〕
patrol camera 巡天照相机
patrol survey 巡天　又见：*survey; all-sky patrol*
patterned ground 月面花样表层
pattern flattening 方向图平化
pattern function 方向图函数
pattern multiplication 方向图相乘
pattern smoothing 方向图平滑
pattern solid angle 方向图立体角
pattern speed 图案速度　又见：*pattern velocity*
pattern velocity 图案速度　又见：*pattern speed*
Paul-Baker telescope 保罗－贝克望远镜
Pauli crater 泡利环形山（月球）
Paul trap 保罗俘获 | 保羅阱〔台〕
Paul Wild Observatory 怀尔德天文台
Pav （＝Pavo）孔雀座
Pavlov crater 巴甫洛夫环形山（月球）
Pavo （＝Pav）孔雀座
Pavonis Mons 帕弗尼斯山（火星）

Pawsey crater 波西环形山（月球）
payload 有效载荷，仪表舱 | 酬载〔台〕
Payload for Antimatter Matter Exploration and Light-nuclei Astrophysics （=PAMELA）PAMELA 实验
Pb （=lead）铅（82 号元素）
P branch P 谱带 | P－支〔台〕
p brane p 膜
pc （=parallax second）秒差距
PCA ①（=polar cap absorption）极冠吸收 ②（=principal component analysis）主成分分析
P-class asteroid P 型小行星
P-Cygni line profile 天鹅 P 谱线轮廓 又见：*P-Cygni profile*
P-Cygni phenomenon 天鹅 P 现象
P-Cygni profile 天鹅 P 谱线轮廓 又见：*P-Cygni line profile*
P-Cygni star （=P Cyg star）天鹅 P 型星 | 天鹅[座]P 型星〔台〕又见：*P-Cygni type star*
P-Cygni type star 天鹅 P 型星 | 天鹅[座]P 型星〔台〕又见：*P-Cygni star*
P-Cygni variable 天鹅 P 型变星
P Cyg star （=P-Cygni star）天鹅 P 型星 | 天鹅[座]P 型星〔台〕
PD （=Potsdamer Durchmusterung）波茨坦巡天星表
Pd （=palladium）钯（46 号元素）
PdBI （=Plateau de Bure Interferometer）比尔高原干涉仪（法国）
PDE （=pure density evolution）纯密度演化
PDF （=probability distribution function）概率分布函数
PDR （=photo-dissociation region）光致离解区
PDS ①（=photo-digitizing system）图像数字仪 | 相片數位化系統〔台〕②（=photometric data system）测光数据仪 | 測光資料系統〔台〕
p.e. （=probable error）概差，可几误差 | 概差〔台〕
peak aphelion 最大远日点
peak apogee 最大远地点
peak-background split 峰－背景分离
peak energy 峰值能量
peak load 峰值载荷
peakology 峰值分析
peak value 峰值 又见：*crest value*
pear-shaped curve 梨形曲线
pear-shaped curve of zero velocity 梨形零速度线
pear-shaped Earth 梨形地球

Peary crater 皮里环形山（月球）
Pecci-Quinn symmetry 佩奇－奎因对称性
peculiar A star （=Ap star）Ap 星，A 型特殊星
peculiar B-A star B－A 型特殊星
peculiar B star （=Bp star）B 型特殊星 | B 型特殊星，Bp 星〔台〕
peculiar galaxy 特殊星系
peculiar giant star 特殊巨星
peculiar motion 本动 又见：*peculiar proper motion*
peculiar motion component 本动分量
peculiar nebula 特殊星云
peculiar object 特殊天体
peculiar proper motion 本动 又见：*peculiar motion*
peculiar spectrum 特殊光谱
peculiar speed 本动速度 又见：*peculiar velocity*
peculiar star 特殊恒星 | 特殊[恆]星〔台〕
peculiar variable 特殊变星
peculiar velocity 本动速度 又见：*peculiar speed*
peculiar velocity field 本动速度场
peculiar white dwarf 特殊白矮星
pedestal 支柱（望远镜）
pedestal crater 基座陨击坑
Peekskill meteorite 皮克斯基尔陨星
Peg （=Pegasus）飞马座
Pegasus ①（=Peg）飞马座 ②飞马座 | 飛馬[星]座〔台〕（科学卫星）
Pegasus Irregular Galaxy 飞马座不规则星系
Pegasus Ridge 飞马脊
Peilisheng 裴丽生（小行星 5273 号）
Peking 北京（小行星 2045 号）
Pele 比利山（木卫一）
Pelican Nebula 鹈鹕星云
PEM （=Photoelastic modulator）光弹调制器
pencil 光锥 又见：*light cone*
pencil beam 笔束
pencil-beam antenna 笔束天线
pencil beam survey 笔束巡天
pencil-beam survey 笔束巡天
pencil rocket 小型火箭
pendulum 摆
pendulum clock 摆钟
pendulum gravimeter 摆式重力仪
pendulum level 摆水准
pendulum mass 摆锤
pendulum tiltmeter 摆式倾斜仪
pendulum vibration 摆动
pen error 笔尖误差（记时仪）

penetrability 贯穿本领 又见: *penetrating power*
penetrating coefficient 贯穿系数
penetrating encounter 穿透交会
penetrating power 贯穿本领 又见: *penetrability*
penetration 贯穿
penetration depth 贯穿深度 又见: *depth of penetration*
penetration height 贯穿高度
penetrative convection 贯穿对流 即: *convective overshooting*
Penghuanwu 彭桓武（小行星48798号）
Penrose diagram 彭罗斯图
Penrose process 彭罗斯过程｜潘羅斯過程〔台〕
Penrose singularity theorem 彭罗斯奇点定理
pentaprism test 五棱镜检验
pentet 五重线 又见: *quintet*
Pentland crater 彭特兰环形山（月球）
penumbra 半影
penumbral eclipse 半影食
penumbral filament 半影纤维
penumbral lunar eclipse 半影月食 又见: *lunar appulse*; *appulse*
penumbral magnitude of eclipse 半影食分
penumbral region 半影区
penumbral wave 半影波（黑子）
pep reaction (=proton-electron-proton reaction) 质子－电子－质子反应，pep反应
Per (=Perseus) 英仙座
peragration 月球公转恒星周
percolation 渗流
Perdita 天卫二十五
Perepelkin crater 佩列佩尔金环形山（月球）
peret 冬季（埃及古历）
perfect cosmological principle 完全宇宙学原理｜完美宇宙論原則〔台〕
perfect fluid 理想流体
perfect gas 理想气体 又见: *ideal gas*
perfectly diffusing sphere 全散射球
perfectly diffusion 全散射
perfect prolate spheroid 理想长等轴椭球体
perfect radiator 理想辐射体｜完全輻射體〔台〕
periapse 近质心点
periapse distance 近质心点距
periapsis 近点，近拱点｜近拱點〔台〕
periapsis distance 近拱点距
periareon 近火点 又见: *perimartian*
periastron 近星点

periastron advance 近星点进动｜近星點前移〔台〕 又见: *advance of the periastron*; *periastron precession*
periastron distance 近星点距
periastron effect 近星点效应（密近双星）
periastron precession 近星点进动｜近星點前移〔台〕 又见: *advance of the periastron*; *periastron advance*
pericenter 近心点
pericenter distance 近心距
pericynthion 近月点 又见: *periselene*; *perilune*
peri-distance 近心点距
peridotite 橄榄岩
perifocus 近主焦点
perigalactic distance 近银心点距
perigalacticon 近银心点 又见: *perigalacticum*; *galactic pericenter*
perigalactic passage 过近银心点
perigalacticum 近银心点 又见: *perigalacticon*; *galactic pericenter*
perigeal tide 近地点潮 又见: *perigean tide*
perigean tide 近地点潮 又见: *perigeal tide*
perigee 近地点
perigee distance 近地点距
perigee kick 近地点启动
perihelic conjunction 近日点合
perihelic opposition 近日点冲
perihelion 近日点
perihelion distance 近日[点]距｜近日距〔台〕
perihelion effect 近日点效应
perihelion motion 近日点进动｜近日點前移〔台〕 又见: *advance of the perihelion*; *perihelion precession*; *precession of perihelion*
perihelion passage 过近日点
perihelion precession 近日点进动｜近日點前移〔台〕 又见: *advance of the perihelion*; *perihelion motion*; *precession of perihelion*
perihelion precession of Mercury 水星近日点进动 又见: *advance of Mercury's perihelion*
perihelion time 过近日点时刻 又见: *time of perihelion passage*
perijove 近木点
perilune 近月点 又见: *periselene*; *pericynthion*
perimartian 近火点 又见: *periareon*
perimercurian 近水点
period-age relation 周期－年龄关系
period-amplitude relation 周期－变幅关系
period change 周期变化 又见: *period variation*; *periodic variation*
period-color relation 周期－颜色关系

period-density relation 周期－密度关系
period determination 周期测定
period discontinuity 周期突变
period-eccentricity 周期－偏心率关系
period gap 周期性空隙
periodic comet 周期彗星
periodic error 周期误差
periodic function 周期函数
periodic inequality 周期差
periodic instability 周期性不稳定
periodicity 周期性 又见：*cyclicity*
periodic motion 周期运动
periodic orbit 周期轨道
periodic oscillation 周期振荡
periodic pattern 周期性方向图
periodic perturbation 周期摄动
periodic point 周期点
periodic process 周期性过程
periodic stream 周期性流星群 | 周期[性]流星雨〔台〕
periodic surface 周期曲面
periodic term 周期项
periodic variable 周期变星
periodic variation 周期变化 又见：*period change; period variation*
period-luminosity-color relation （＝P-L-C relation）周光色关系
period-luminosity curve 周光曲线
period-luminosity relation （＝P-L relation）周光关系
period-mass relation 周期－质量关系
period-mean density relation 周期－平均密度关系
period of light variation 光变周期 又见：*light period*
period of moon's node 月球交点周期
period of revolution 公转周期
periodogram 周期图
periodograph 周期图仪
periodometer 周期计
period-radius relation 周径关系
period resolution 周期分辨率
period-spectrum relation 周谱关系
period variation 周期变化 又见：*period change; periodic variation*
periplutonian 近冥王点 又见：*periplutonium*
periplutonium 近冥王点 又见：*periplutonian*

periposeidon 近海王点
perisaturnian 近土点 又见：*perisaturnium*
perisaturnium 近土点 又见：*perisaturnian*
periscian region 日影环绕区
periselene 近月点 又见：*perilune; pericynthion*
perisphere 外围层
periuranian 近天王点 又见：*periuranium*
periuranium 近天王点 又见：*periuranian*
Perkin crater 珀金环形山（月球）
Perkins Observatory 珀金斯天文台
permafrost layer 永冻层
permanent bench mark 固定水准点
permanent calendar 万年历 又见：*perpetual calendar*
permanent magnet 永磁体
permanent magnetism 永磁性
permeability 磁导率 又见：*magnetic permeability*
permeance 磁导
permissible error 容许误差 又见：*admissible error; allowable error; tolerance*
permitted line 容许谱线
permitted transition 容许跃迁 又见：*allowable transition; allowed transition*
permittivity 电容率
Perot-Fabry interferometer 珀罗－法布里干涉仪
Perot-Fabry spectrometer 珀罗－法布里分光计
perpendicular 垂线，正交
perpendicular collision 垂直碰撞
perpendicular component 垂直分量
perpendicular hydromagnetic shockwave 正交磁流激波
perpetual calendar 万年历 又见：*permanent calendar*
perpetual day 永昼
perpetual night 长夜
Perrine crater 珀赖因环形山（月球）
Perseid meteor shower 英仙流星雨
Perseids 英仙流星群 | 英仙[座]流星雨，八月流星雨〔台〕
Perseus （＝Per）英仙座
Perseus A 英仙射电源 A
Perseus Arm 英仙臂
Perseus association 英仙星协
Perseus chimney 英仙通道
Perseus cluster 英仙星团
Perseus cluster of galaxies 英仙星系团 | 英仙[座]星系團〔台〕

Perseus OB association 英仙 OB 星协
Perseus-Pisces supercluster 英仙－双鱼超星系团
persistence of vision 视觉驻留
persistent train 久现流星余迹
personal difference 人差 | 人[為]差，個人誤差〔台〕 又见：*personal equation; individual error; personal error*
personal equation 人差 | 人[為]差，個人誤差〔台〕 又见：*personal difference; individual error; personal error*
personal error 人差 | 人[為]差，個人誤差〔台〕 又见：*personal difference; personal equation; individual error*
perspective 透视
Perth Observatory 佩斯天文台
perturbation 摄动，微扰 又见：*disturbance*
perturbation amplitude 扰动幅度
perturbation equation 摄动方程
perturbation function 摄动函数 又见：*disturbing function*
perturbation growth 扰动增长
perturbation method 摄动方法
perturbation of dark matter 暗物质扰动
perturbation of entropy 熵扰动
perturbation power spectrum 扰动功率谱
perturbations on sub-Planckian scale 亚普朗克尺度扰动
perturbation spectrum 扰动谱
perturbation theory 摄动理论 | 攝動說〔台〕
perturbative force 摄动力 又见：*disturbing force*
perturbed body 受摄体 又见：*disturbed body; perturbed object*
perturbed coordinate 受摄坐标 又见：*disturbed coordinate*
perturbed element 受摄根数 | 受攝要素，受攝根數〔台〕 又见：*disturbed element*
perturbed motion 受摄运动 又见：*disturbed motion*
perturbed object 受摄体 又见：*disturbed body; perturbed body*
perturbed orbit 受摄轨道 又见：*disturbed orbit*
perturbed state 受摄态
perturbing body 摄动体 又见：*disturbing body; perturbing object*
perturbing object 摄动体 又见：*disturbing body; perturbing body*
perturbing term 摄动项
Petermann crater 彼得曼环形山（月球）
Peterpau 鲍德熹（小行星 34420 号）
petrology 岩石学
Petropavlovskiy crater 彼得罗帕夫洛夫斯基环形山（月球）
Petzval surface 珀兹伐表面
pezograph 气印（陨石） 又见：*regmaglypt*
Pfaffian 普法夫[多项]式
Pfund continuum 普丰德连续区
Pfund lines 普丰德谱线
Pfund series 普丰德线系
PGC （=Preliminary General Catalogue）博斯总星表初编 | 博斯星表初編〔台〕
PG survey （=Palomar Green survey）帕洛玛－格林巡天
PHA （=potentially hazardous asteroid）潜在威胁小行星
Phacd 天玑，北斗三（大熊座 γ） 又见：*Phachd; Phad; Phecda*
Phachd 天玑，北斗三（大熊座 γ） 又见：*Phacd; Phad; Phecda*
Phack 丈人一（天鸽座 α） 又见：*Phact; Phad; Phaet; Phakt*
Phact 丈人一（天鸽座 α） 又见：*Phack; Phad; Phaet; Phakt*
Phad ①天玑，北斗三（大熊座 γ） 又见：*Phacd; Phachd; Phecda* ②丈人一（天鸽座 α） 又见：*Phack; Phact; Phaet; Phakt*
Phaenomena 《物象》（希腊诗人阿拉托斯所著长诗）
Phaet 丈人一（天鸽座 α） 又见：*Phack; Phact; Phad; Phakt*
Phaethon 法厄松（小行星 3200 号）
Phakt 丈人一（天鸽座 α） 又见：*Phack; Phact; Phad; Phaet*
phantom dark energy 幽灵暗能量
phantom field 幽灵场
phantom telescope 仿真望远镜
phase angle 相位角 | 位相角〔台〕
phase center 相位中心
phase change 相变 又见：*phase transformation; phase transition*
phase closure 相位闭合
phase coefficient 相位系数 | 位相係數〔台〕（小行星）
phase collision 相位碰撞
phase contrast wavefront sensor 相位对比波前传感器
phase correlation 相位相关
phase curve 相位曲线 | 位相曲線〔台〕
phased array 相控阵

phased array feed 相控阵馈源
phase decay 相位延迟
phase defect 相位短缺
phase demodulation 相位解调
phase detector 鉴相器
phase diagram 相位图
phase difference 相位差 | [位] 相差〔台〕
phase displacement 相移　又见：*phase shift*
phase effect 相位效应
phase error 相位误差
phase function 相位函数
phase instability 相位不稳定性
phase integral 相位积分 | 相積分〔台〕
phase jitter 相位抖动
phase jump 相位突变 | 位相突變〔台〕
phase lag 相位滞后
phase law 相位律
phase lead 相位超前
phaseless 无相位
phaseless aperture synthesis 无相位孔径综合
phase locking 锁相
phase mask 相位掩模
phase mask coronagraph 相位掩模星冕仪
phase-measuring interferometer 测相干涉仪
phase-measuring interferometry 测相干涉测量
phase mixing 相点混合
phase modulation 调相
phase noise 相位噪声
phase of an eclipse 食相　又见：*phase of eclipse*
phase of eclipse 食相　又见：*phase of an eclipse*
phase of planets 行星相
phase of the moon 月相　又见：*aspect of the moon; lunar aspect; lunar phase; moon's phase*
phase path 相程
phase plane 相平面
phase prediction 相位预测
phase quadrature 90°相
Phase-Referenced Imaging and Microarcsecond Astrometry （＝Prima）相参成像和微角秒天体测量
phase referencing 相位参考法
phase reversal 反相　又见：*antiphase*
phase rotation 相位旋转
phase rotator 相位旋转器
phase screen model 相屏模型
phase sensitive detector 相敏检波器
phase shift 相移　又见：*phase displacement*
phase shifter 相移器
phase-shifting compensator 相移补偿器
phase space 相空间，相宇 | 相空間〔台〕
phase space density 相空间密度
phase-space density 相空间密度
phase-space distribution function 相空间分布函数
phase-space volume 相空间体积
phase spectrum 相位谱
phase stability 相位稳定性
phase-swept interferometer 扫相干涉仪
phase-swept interferometry 扫相干涉测量
phase switching 相位开关
phase-switching interferometer 相位开关干涉仪
phase-switching interferometry 相位开关干涉测量
phase tracking 相位跟踪
phase transformation 相变　又见：*phase transition; phase change*
phase transition 相变　又见：*phase transformation; phase change*
phase velocity 相速度
phasing 定相
phasing antenna 定相天线
phasing camera 相位相机
phasing method 调相方法
phasing sensor 相位传感器
phasing technique 调相技术
Phe （＝Phoenix）凤凰座
Phecda 天玑，北斗三（大熊座γ）　又见：*Phacd; Phachd; Phad*
phenomenological theory 唯象理论
Pherkad 太子，北极一（小熊座γ）
Pherkad Minor 勾陈增五（小熊座λ）
Philae lander 菲莱号着陆器（Churyumov-Gerasimenko 彗星）
Phillips band 菲利普斯谱带
Phillips crater 菲利普斯环形山（月球）
Phillips relation 菲利普斯关系
Philolaus crater ① 菲洛劳斯环形山（月球）② 菲洛劳斯环形山（月球）
Philosophiæ Naturalis Principia Mathematica （＝Principia）【拉】《自然哲学的数学原理》（牛顿著作）
PHL （＝Palomar-Haro-Luyten Catalogue of Blue Stellar Objects）PHL 蓝星体表
Phobos 火卫一
Phocaea 福后星（小行星25号）
Phocaea group 福后星群

Phoebe 土卫九
Phoenicid meteor shower 凤凰流星雨
Phoenicids 凤凰流星群 | 鳳凰座流星雨〔台〕
Phoenix （＝Phe）凤凰座
Phoenix Irregular Galaxy 凤凰座不规则星系
Phoenix Universe 凤凰宇宙模型
Pholus 福鲁斯（小行星 5145 号）
phonon 声子
Phospher 启明星　又见：*Lucifer; Phospherus*
Phospherus 启明星　又见：*Lucifer; Phospher*
phosphorescence 磷光
phosphorescent scintillation 磷光闪烁
phosphorus （＝P）磷（15 号元素）
phoswitch 层叠闪烁体
photino 光微子
photocathode 光阴极
photocell 光电管　又见：*photoelectric tube*
photocenter 光心　又见：*optical center; center of light*
photocentric orbit 光心轨道
photochemistry 光化学
photochronograph 照相记时仪
photoconductive cell 光导管 | 光電導管〔台〕
photoconductive detector 光导检测器
photoconductive effect 光导效应
photocurrent 光电流　又见：*photoelectric current*
photodetachment 光致分离 | 光致脫離〔台〕
photodetector 光电检测器
photo-digitizing system （＝PDS）图像数字仪 | 相片數位化系統〔台〕
photo-disintegration 光致蜕变
photo-dissociation 光致离解 | 光解離〔台〕
photodissociation 光致离解
photo-dissociation region （＝PDR）光致离解区
Photoelastic modulator （＝PEM）光弹调制器
photoelectric absolute magnitude 光电绝对星等
photoelectric absorption 光电吸收
photoelectric absorption coefficient 光电吸收系数
photoelectric astrolabe 光电等高仪
photoelectric astrometry 光电天体测量
photoelectric cathode 光电阴极
photoelectric cell 光电管
photoelectric colorimeter 光电比色计
photoelectric current 光电流　又见：*photocurrent*
photoelectric effect 光电效应
photoelectric guider 光电导星镜
　　又见：*photoelectric guiding telescope*
photoelectric guiding 光电导星
photoelectric guiding telescope 光电导星镜
　　又见：*photoelectric guider*
photoelectric image tube 光电像管
photoelectric instrument 光电仪器
photoelectric ionization 光致电离 | 光致游離〔台〕
　　又见：*photoionization*
photoelectric light curve 光电光变曲线
photoelectric magnitude 光电星等
photoelectric meridian circle 光电子午环
photoelectric microphotometer 光电测微光度计
photoelectric photometer 光电光度计
　　又见：*electrophotometer*
photoelectric photometry 光电测光 | 光電測光術〔台〕　又见：*electrophotometry*
photoelectric polarimeter 光电偏振计
photoelectric polarimetry 光电偏振测量
photoelectric radial-velocityspectrometer 光电视向速度仪　又见：*photoelectric speedometer*
photoelectric receptor 光电接收器
photoelectric solar magnetograph 太阳光电磁像仪
photoelectric spectrophotometer 光电分光光度计
photoelectric spectrophotometry 光电分光光度测量　又见：*electric spectrophotometry*
photoelectric speedometer 光电视向速度仪
　　又见：*photoelectric radial-velocityspectrometer*
photoelectric star-photometer 恒星光电光度计
photoelectric Stokes polarimeter 光电斯托克斯偏振计
photoelectric temperature 光电温度
photoelectric tracking 光电跟踪
photoelectric transit instrument 光电中星仪
photoelectric tube 光电管　又见：*photocell*
photoelectron 光电子
photoelectronic device 光电器件
photoelectronic imaging 光电成像
　　又见：*electrophotonic imaging*
photoelement 光电元件 | 光電元件，光電管〔台〕
photoemission 光电发射
photoemissive cell 光电发射管
photoemitter 光电发射体
photo-emulsion 照相乳胶　又见：*photographic emulsion*
photo-etched structure 光刻结构
photoevaporation 光致蒸发
photo-event 光子事件
photo-excitation 光致激发 | 光電激發〔台〕
Photo Flo 除水渍液（柯达公司产品）

photogrammeter 照相测量仪
photogrammetric method 照相测量法
photogrammetric refraction 照相测量大气折射
photogrammetry 照相测量
photograph 照片
photographic absolute magnitude 照相绝对星等
photographic absorption coefficient 照相吸收系数
photographic albedo 照相反照率
photographic astrometry 照相天体测量学 | 照相天體測量 [術] 〔台〕
photographic astrophotometry 照相天体测光
photographic astrospectroscopy 照相天体分光
photographic calibration 照相定标
photographic camera 照相机
photographic catalogue 照相星表
　又见: *photographic star catalogue; astrographic catalogue*
photographic chart 照相星图　又见: *astrographic chart; astrographic map; astrographic atlas*
photographic density 照相密度
photographic developer 显影剂　又见: *developer; developing agent*
photographic effect 照相效应
photographic emulsion 照相乳胶
　又见: *photo-emulsion*
photographic ephemeris 照相星历表
photographic film 照相胶片
photographic filter 滤色镜
photographic fog 底片雾
photographic instrument 照相仪器
photographic integration 照相累积
photographic latent image 照相潜像
photographic light curve 照相光变曲线
photographic magnetograph 照相磁像仪
photographic magnitude 照相星等
photographic material 照相材料
photographic meridian circle 照相子午环
　又见: *photographic transit circle*
photographic meteor 照相流星
photographic noise 照相噪声
photographic objective 照相物镜
photographic observation 照相观测
photographic photometer 照相光度计
photographic photometry 照相测光 | 照相測光術〔台〕
photographic plate 照相底片
photographic processing 照相处理
photographic radiant 照相辐射点

photographic refractor 折射天体照相仪
　又见: *astrographic refractor*
photographic resolution 照相分辨率
photographic solar magnetograph 太阳照相磁像仪
photographic spectrophotometer 照相分光光度计
photographic spectrophotometry 照相分光光度测量
photographic spectroscopy 照相分光
photographic speed 照相感光速度 | 照相 [感光] 速度〔台〕
photographic star catalogue 照相星表
　又见: *astrographic catalogue; photographic catalogue*
photographic telescope 照相望远镜
photographic tracking 照相跟踪 | 照相追蹤〔台〕
photographic transit circle 照相子午环
　又见: *photographic meridian circle*
photographic transit instrument 照相中星仪
photographic triangulation 照相三角测量
photographic vertical circle 照相垂直环
photographic zenith tube （=PZT）照相天顶筒
photography 照相
photoheliogram 太阳照片 | 太陽[全色]照片〔台〕
photoheliograph 太阳照相仪　又见: *heliograph*
photoionization 光致电离 | 光致游離〔台〕
　又见: *photoelectric ionization*
photoionization cross-section 光电离截面
photoionization heating 光电离致热
photoionization model 光致电离模型
photoionization transition 光电离跃迁
photoionized plasma 光电离等离子体
photoluminescence 光致发光
photometer 光度计
photometer head 光电头
photometric absolute magnitude 测光绝对星等
photometric accuracy 测光精度　又见: *photometric precision*
photometric bench 光度台
photometric binary 测光双星 | 光度雙星〔台〕
　又见: *photometric double star*
photometric calibration 测光定标 | 光度校準〔台〕
photometric catalogue 测光星表 | 恆星光度表〔台〕
photometric data system （=PDS）测光数据仪 | 測光資料系統〔台〕
photometric diameter 测光直径 | 光度直徑〔台〕
photometric distance 测光距离 | 光度距離〔台〕

photometric double star　测光双星 | 光度雙星〔台〕　又见：*photometric binary*
photometric element　测光根数
photometric ellipticity　测光椭率 | 测光椭圆〔台〕
photometric error　测光误差
photometric index　测光指数
photometric method　测光方法
photometric night　测光夜
photometric orbit　测光轨道
photometric paradox　光度佯谬　又见：*luminosity paradox*
photometric parallax　测光视差 | 光度视差〔台〕
photometric period　测光周期
photometric phase　光度相位
photometric precision　测光精度　又见：*photometric accuracy*
photometric redshift　测光红移
photometric scale　光度标
photometric sequence　测光序
photometric solution　测光解
photometric standard　光度标准 | 光度標準，测光標準〔台〕
photometric standard star　光度标准星 | 测光標準星〔台〕　又见：*standard star for photometry*
photometric system　测光系统 | 测光系，光度系〔台〕
photometric telescope　测光望远镜
photometric wedge　测光楔 | 测光劈片〔台〕
photometry　测光 | 测光術〔台〕
photomicrograph　显微照片
photomicrography　显微照相
photomultiplier　光电倍增管　又见：*photomultiplier tube; phototube*
photomultiplier tube　光电倍增管　又见：*photomultiplier; phototube*
photon　光子
photon counter　光子计数器
photon counting　光子计数
photon-counting detector　光子计数检测器
photon decoupling　光子退耦
photon diffusion　光子扩散
photoneutrino　光生中微子 | 光微中子〔台〕
photoneutrino process　光生中微子过程 | 光激微中子過程〔台〕
photon flux　光子流量
photon gas　光子气
photon-phonon collision　光子－声子碰撞
photon-photon absorption　光子－光子吸收

photon recoil effect　光子反冲效应
photon rocket　光子火箭
photon sieve telescope　光子筛望远镜
photon sphere　光子层
photon trapping　光子俘获
photonuclear absorption　光核吸收
photopic curve　适光曲线
photopic eye　适光眼
photopolarimeter　偏振测光计
photopolarimetry　偏振测光
photo-recombination　光致复合
photoreduction　光致还原
photoresistance　光敏电阻
photosource　光源　又见：*light source*
photosphere　光球 | 光球 [層]〔台〕
photospheric abundance　光球丰度
photospheric activity　光球活动
photospheric eruption　光球爆发
photospheric facula　光球光斑 | 光斑，光球層光斑〔台〕
photospheric line　光球谱线
photospheric model　光球模型
photospheric network　光球网络
photospheric radiation　光球辐射
photospheric spectrum　光球光谱
photospheric surface　光球表面
photospheric telescope　光球望远镜
photospheric temperature　光球温度
phototheodolite　照相经纬仪
phototube　光电倍增管　又见：*photomultiplier tube; photomultiplier*
photovisual absorption coefficient　仿视吸收系数
photovisual magnitude　仿视星等
photovisual objective　仿视照相物镜
photovoltaic cell　光伏管
photovoltaic detector　光伏检测器
photovoltaic effect　光伏效应
physical aperture　物理孔径
physical area　物理面积
physical distance　物理距离
physical double　物理双星　又见：*physical pair*
physical element　物理根数
physical horizon　物理地平
physical libration　物理天平动
physical mechanism　物理机制
physical nature　物理本原
physical pair　物理双星　又见：*physical double*

physical parameter 物理参数
physical pendulum 复摆　　又见：*compound pendulum*
Physical Review D 《物理学评论 D 辑》（美国期刊）
physical universe 物理宇宙
physical variable 物理变星
physical visual binary 物理目视双星
physics of stellar atmosphere 恒星大气物理
physics of stellar structure 恒星结构物理
physics of stellar wind 星风物理
Piazzia 皮亚齐（小行星 1000 号）
Piazzi crater 皮亚齐环形山（月球）
PIB model （=primordial isocurvature baryon model）原初等曲率重子模型
Pic （=Pictor）绘架座
Piccard 皮卡尔（法国太阳卫星）
Piccolomini crater 皮科洛米尼环形山（月球）
Pic du Midi Observatory 日中峰天文台（法国）
Pickering lines 皮克林谱线
Pickering series 皮克林线系 | 皮克林 [譜線] 系〔台〕
picosecond 皮秒 | 塵秒〔台〕
Pictet crater 皮克泰环形山（月球）
Pictor （=Pic）绘架座
Pictun 匹克顿（玛雅历日，2880000 日）
　　又见：*Piktun*
picture element 像素 | 像元〔台〕　　又见：*pixel*
picture-point 像点
piecewise parabolic method （=PPM）逐段抛物线法
Pierre Auger Observatory 皮埃尔·俄歇天文台（阿根廷）
piezo effect 压电效应　　又见：*piezoelectric effect*
piezoelectric effect 压电效应　　又见：*piezo effect*
pigeonite 易变辉石
piggybacking photography 背负式摄影
Piktun 匹克顿（玛雅历日，2880000 日）
　　又见：*Pictun*
Pilâtre crater 皮拉特尔环形山（月球）
pile-up boundary 堆积区边界
pile-up effect 堆积效应
pile-up region 堆积区
pillar 基礎
Pillars of Creation 创生之柱（鹰状星云局部）
pillbox antenna 抛物柱盒天线
pinch 箍缩
pinch effect 箍缩效应 | 捏縮效應〔台〕

pinch instability 箍缩不稳定性
pinch-off effect 箍断效应
pinch plasma 箍缩等离子体
pincushion distortion 正畸变
Pingré crater 潘格雷环形山（月球）
pin-hole imaging 针孔成像
pink noise 粉噪声　　即：*1/f noise*
pint-size star 微星
Pinwheel Galaxy 风车星系（M101）
pion π 介子 | 派子〔台〕　　又见：*π-meson*
Pioneer 先驱者 | 先鋒號 [太空船]〔台〕（美国行星际探测器）
Pioneer-Venus 先驱者－金星号（美国金星探测器）
pip 小峰 | 小峰，小突起〔台〕
Pipe Nebula 烟斗星云（Barnard 59, 65, 66, 67, 78）
Pirquet crater 皮尔凯环形山（月球）
Pisces ①（=Psc）双鱼座 ②（=Psc）双鱼宫，訾，亥宫
Pisces-Perseus supercluster 双鱼－英仙超星系团 | 雙魚－英仙 [座] 超星系團〔台〕
Piscid meteor shower 双鱼流星雨
Piscis Australid meteor shower 南鱼流星雨
Piscis Australids 南鱼流星群 | 南魚座流星雨〔台〕
Piscis Austrinus （=PsA）南鱼座
Pistol Nebula 手枪星云
Pistol Star 手枪星云星
pitch angle 旋臂倾角，螺距角 | 旋臂倾角〔台〕
pitch-angle distribution 螺距角分布
pitch-angle redistribution 螺距角再分布
pivotal axis 枢轴
pixel 像素 | 像元〔台〕　　又见：*picture element*
PKS （=Parkes Catalogue of Radio Sources）帕克斯射电源表 | 帕克斯電波源表〔台〕
plage 谱斑　　又见：*flocculus*
plage area 谱斑面积
plage corridor 谱斑区走廊
plage flare 谱斑区耀斑 | 譜斑狀閃焰〔台〕
plage radiation 谱斑辐射
plagioclase 斜长石
Planck blackbody 普朗克黑体
Planck blackbody formula 普朗克黑体公式
Planck blackbody function 普朗克黑体函数
Planck constant 普朗克常数 | 卜朗克常數〔台〕
Planck crater 普朗克环形山（月球）
Planck density 普朗克密度
Planck distribution 普朗克分布 | 卜朗克分佈〔台〕

Planck energy　普朗克能量
Planck era　普朗克期
Planck function　普朗克函数
Planck length　普朗克长度 | 卜朗克長度〔台〕（约为 1.6×10^{-35} 米。）
Planck mass　普朗克质量
Planck satellite　普朗克卫星
Planck's law　普朗克定律 | 卜朗克定律〔台〕
Planck spectrum　普朗克谱
Planck's radiation law　普朗克辐射定律
Planck temperature　普朗克温度
Planck time　普朗克时间 | 卜朗克時間〔台〕（约为 5.4×10^{-44} 秒）
Planck unit　普朗克单位
plane component　扁平子系
plane-concave lens　平凹透镜　又见：*plano-concave lens*
plane-convex lens　平凸透镜　又见：*plano-convex lens*
plane grating　平面光栅
plane mirror　平面镜
planemo　行星质量天体　又见：*planetary-mass object*
plane of emergence　出射面
plane of incidence　入射面
plane of oscillation　振动面
plane of polarization　偏振面　又见：*polarization plane*
plane of reflection　反射面
plane of refraction　折射面
plane of the sky　天球切面
plane-parallel atmosphere　平面平行大气
plane polarization　面偏振
plane-polarized radiation　面偏振辐射
plane-polarized wave　面偏振波
plane subsystem　扁平次系
plane sundial　平面日晷
planet　行星
planetaria　天象仪，行星仪 | 天象儀〔台〕
　　又见：*planetarium*
planetarium　①天文馆 | 天象館〔台〕②天象仪，行星仪 | 天象儀〔台〕　又见：*planetaria*
planetary aberration　行星光行差
planetary alignmentm　行星联珠
Planetary and Space Science　《行星与空间科学》（英国期刊）
planetary assembly　行星[齿轮]组件
planetary astronomy　行星天文学
planetary atmosphere　行星大气

planetary biology　行星生物学
planetary companion　似行星伴天体 | 類行星伴星〔台〕
planetary configuration　行星动态
planetary construction zone　行星组建区
planetary corona　行星冕
planetary cosmogony　行星演化学
planetary dynamo　行星发电机
planetary embryo　行星胎　又见：*embryo of a planet*
planetary ephemeris　行星历表　又见：*planetary table*
planetary evolution　行星演化
planetary geology　行星地质学
planetary ionosphere　行星电离层
planetary magnetic field　行星磁场
planetary magnetosphere　行星磁层
planetary-mass companion　行星质量伴星　又见：*planet-mass companion*
planetary-mass object　行星质量天体　又见：*planemo*
planetary meteorology　行星气象 | 行星氣象學〔台〕
planetary motion　行星运动
planetary nebula　（=PN）行星状星云
planetary nomenclature　行星命名法
planetary occultation　行星掩星
planetary orbit　行星轨道
planetary perturbation　行星摄动
planetary physics　行星物理 | 行星物理學〔台〕
planetary plasmasphere　行星等离子层 | 行星電漿層〔台〕
planetary precession　行星岁差
planetary probe　行星探测器
planetary radio astronomy　行星射电天文学
planetary raw material　原生行星物质
planetary ring　行星环
planetary satellite　行星卫星
planetary seismology　行星地震学
planetary space　行星空间 | 近行星空間〔台〕
Planetary Space Science　（=PSS）《行星空间科学》（欧洲期刊）
planetary stream　行星流星群 | 行星流星雨〔台〕
planetary system　行星系 | 行星系[統]〔台〕
planetary table　行星历表　又见：*planetary ephemeris*
planetary telescope　行星望远镜
planetary temperature　行星温度
planetary theory　行星运动理论

PLAnetary Transits and Oscillations of stars （=PLATO）柏拉图探测器
planetary wind 行星风
planetesimal 星子 | 微行星〔台〕
planetesimal hypothesis 星子假说 | 微行星假說〔台〕（太阳系起源）
planetesimal theory 星子理论 | 微行星理論〔台〕
planet-fall 行星着陆 又见：*planet-landing*
planet formation 行星形成
planet-landing 行星着陆 又见：*planet-fall*
planet-like body 类行星天体
planet-mass companion 行星质量伴星 又见：*planetary-mass companion*
planet-mass object 行星质量天体
planet migration 行星迁移
planetocentric celestial sphere 行星心天球
planetocentric coordinate 行星心坐标
planetocentric latitude 行星心纬度
planetocentric longitude 行星心经度
planetographic celestial sphere 行星面天球
planetographic coordinate 行星面坐标
planetographic latitude 行星面纬度
planetographic longitude 行星面经度
planetography 行星表面学
planetoid 微型行星 | 小行星〔台〕
planetologist 行星学家
planetology 行星学
Planet X X 行星
plane wave 平面波
planisphere 平面天球
planitia 平原低地（行星地貌）
Planitia Descensus 【拉】德森萨斯平原（月球）
plano-concave lens 平凹透镜 又见：*plane-concave lens*
plano-convex lens 平凸透镜 又见：*plane-convex lens*
planum 平原高地 | 大高原〔台〕（行星地貌）
Planum Australe 南部高原（火星）
Planum Boreum 北部高原（火星）
Plaskett crater 普拉斯基特环形山（月球）
Plaskett's star 普拉斯基特星（HD 47129）
plasma 等离子体 | 離子體，電漿〔台〕
plasma astrophysics 等离子天体物理 | 電漿天文物理學〔台〕
plasma-boiler 等离子体加热器
plasma cloud 等离子体云 | 電漿雲〔台〕
plasma diagnostics 等离子体诊断学
plasma diffusion 等离子体扩散
plasma disturbance 等离子体扰动
plasma drift 等离子体漂移
plasma dynamics 等离子体动力学
plasma ejection 等离子体抛射 | 離子體拋射〔台〕
plasma era 等离子时期
plasma frequency 等离子体频率 | 離子體頻率〔台〕
plasma instability 等离子体不稳定性
plasma ion oscillation 等离子振荡 | 離子體振蕩〔台〕 又见：*plasma oscillation*
plasma jet 等离子体喷射结构 | 電漿噴流〔台〕
plasma loop 等离子体环状结构 | 電漿環〔台〕
plasma oscillation 等离子振荡 | 離子體振蕩〔台〕 又见：*plasma ion oscillation*
plasmapause 等离子体层顶 | 電漿層頂〔台〕
plasma physics 等离子体物理
plasma radiation 等离子体辐射
plasma rocket 等离子体火箭
plasma sheet 等离子片
plasmasphere 等离子层 | 電漿層〔台〕
plasma squeezing 等离子体挤压 | 離子體擠壓〔台〕
plasma tail 离子彗尾 又见：*ion tail*
plasma turbulence 等离子体湍动
plasma wave 等离子波 | 離子體波〔台〕
plasmoid 等离子粒团 | 離子粒團〔台〕
plasmon 等离子激元 | 離體子〔台〕
plasmon decay 等离子激元衰变
plasmon model 等离子激元模型
plastic scintillator 塑料闪烁体
plate 底片
Plateau de Bure Interferometer （=PdBI）比尔高原干涉仪（法国）
plateau phase 平台阶段
plateau value 坪值，平台值
plate center 底片中心
plate complexes 复合底片
plate constant 底片常数
plate constant variance 底片常数方差
plate distortion 底片变形 又见：*film distortion*
plate factor 底片因子
plate holder 底片盒
plate library 底片库
plate measuring machine 底片量度仪
plate motion 板块运动 又见：*plate movement*
plate movement 板块运动 又见：*plate motion*
plate overlap method 底片重叠法

plate overlapping 底片重叠	
plate scale 底片比例尺｜底片尺度〔台〕	
plate tectonics 板块构造｜地體構造〔台〕	
platinum （＝Pt）铂（78 号元素）	
PLATO （＝PLAnetary Transits and Oscillations of stars）柏拉图探测器	
Plato crater 柏拉图环形山（月球）	
Platonic year 柏拉图年　即：*great year*	
P-L-C relation （＝period-luminosity-color relation）周光色关系	
PLE （＝pure luminosity evolution）纯光度演化	
Pleiad 昴团星	
Pleiades 昴星团｜昴宿星團〔台〕（NGC 1432）	
Pleiades Nebula 昴星团星云	
Pleione 昴宿增十二（金牛座 28）	
plenilune 满月时期	
plerion 实心超新星遗迹｜實心超新星殘骸〔台〕　又见：*plerionic supernova remnant; plerionic remnant*	
plerionic remnant 实心超新星遗迹｜實心超新星殘骸〔台〕　又见：*plerionic supernova remnant; plerion*	
plerionic SNR （＝plerionic supernova remnant）实心超新星遗迹｜實心超新星殘骸〔台〕	
plerionic supernova remnant （＝plerionic SNR）实心超新星遗迹｜實心超新星殘骸〔台〕　又见：*plerionic remnant; plerion*	
plessite 合纹石	
Ploessl eyepiece 普勒斯尔目镜	
Plössl eyepiece 普洛目镜	
Plough 北斗［七星］（星组）　又见：*Triones; Big Dipper; Northern Dipper; Charles' Wain; Wain; Plow*	
Plow 北斗［七星］（星组）　又见：*Triones; Big Dipper; Northern Dipper; Charles' Wain; Wain; Plough*	
P-L relation （＝period-luminosity relation）周光关系	
plumb 铅垂｜鉛錘〔台〕	
plumbicon 氧化铅光导摄像管　又见：*lead oxide vidicon*	
plumb line 铅垂线｜鉛錘線〔台〕	
plumb-line deviation 垂线偏差　又见：*deviation of the vertical; deflection of vertical*	
Plummer crater 普卢默环形山（月球）	
Plummer's law 普卢默规律	
Plummer's model 普卢默模型	
Plummer sphere 普卢默球	
Plutarch crater 普卢塔克环形山（月球）	
plutino 冥族小天体｜冥族［小］天體〔台〕	
Pluto 冥王星（矮行星）	
plutoid 类冥天体｜冥族小天體〔台〕	
Pluto-Kuiper Belt flyby 冥王星－柯伊伯带飞掠｜冥王星－古柏帶飛掠〔台〕	
Pluto-Kuiper Express 冥王星－柯伊伯带快车	
Plutonian satellite 冥卫　又见：*satellite of Pluto*	
plutonium （＝Pu）钚｜鈽〔台〕（94 号元素）	
P.M. （＝proper motion）自行	
PM （＝post meridiem）【拉】下午	
Pm （＝promethium）钷｜鉕〔台〕（61 号元素）	
p.m. （＝post meridiem）【拉】下午	
P magnitude P 星等	
PM algorithm （＝particle-mesh algorithm）质点－网格算法	
PMC （＝polar mesospheric cloud）极地中层云	
PM code （＝Particle Mesh code）质点－网格程序	
p-mode p 模，压力模｜p 模式〔台〕	
PMS （＝pre-main sequence）主序前｜主星序前〔台〕	
PMS object （＝pre-main sequence object）主序前天体	
PMS star （＝pre-main sequence star）主序前星	
PN （＝planetary nebula）行星状星云	
pnCCD （＝fully depleted pn-junction CCD）全耗尽 pn 结 CCD	
pNGB （＝psuedo Nambu-Goldstone boson）赝南部－戈德斯通玻色子	
Po （＝polonium）钋（84 号元素）	
Pockels cell 泡克耳斯盒	
Pockels cell modulator 泡克耳斯盒调制器	
Pockels effect 泡克耳斯效应	
pocket sextant 袖珍六分仪	
Poczobutt crater 波乔布特环形山（月球）	
POGO （＝Polar Orbiting Geophysical Observatory）极轨地球物理台	
Pogson magnitude scale 波格森星等标　又见：*Pogson scale*	
Pogson ratio 波格森比值｜普格遜比率〔台〕	
Pogson scale 波格森星等标　又见：*Pogson magnitude scale*	
Pogson step method 波格森阶梯法	
Poincaré crater 庞加莱环形山（月球）	
Poincare element 庞加莱根数	
Poincare group 庞加莱群	
Poincare invariant 庞加莱不变量	
Poincare invariant theorem 庞加莱不变量定理	
Poincare spheroid 庞加莱椭球体	
Poincare's surface of section 庞加莱截面	
Poincare theorem 庞加莱定理｜潘卡瑞定理〔台〕

Poincare variable 庞加莱变量
point charge 点电荷
point convective model 点对流模型
pointer 指针　又见：*clock hand*
Pointers 指极星（星组）
pointing 指向，对准
pointing accuracy 指向精度
pointing calibration 指向校准
pointing control system 指向控制系统
pointing correction 指向改正
pointing error 指向误差　又见：*error in pointing*
pointing jitter 指向抖动
point mass 点质量
point of contact 切点
point particle 点粒子
point source 点源
Point Source Catalog Redshift Survey （＝PSCz survey）点源星表红移巡天
point-source model 点源模型
point spread function （＝PSF）点扩散函数
point transformation 点变换
Poisson distribution 泊松分布 | 帕松分布〔台〕
Poisson equation 泊松方程
Poisson error bar 泊松误差棒
Poisson gauge 泊松规范
Poisson noise 泊松噪声
Poisson ratio 泊松比值
Poisson sampling 泊松采样
Poisson spectrum 泊松谱
Polana 波兰（小行星142号）
polar 高偏振星　即：*AM Her binary*
polar angle 极角　又见：*polar cusp*
polar axis 极轴
polar cap 极冠，极盖 | 极冠〔台〕
polar cap absorption （＝PCA）极冠吸收
polar circle 极圈　又见：*arctic circle*
polar coordinate 极坐标
polar cusp 极角　又见：*polar angle*
polar day 极昼
polar diagram 极区图
polar dial 极向日晷 | 极日规〔台〕
polar diameter 极直径
polar displacement 极移　又见：*motion of earth pole; polar motion; polar shift; polar wandering; wandering of the pole*
polar distance 极距

polar distance degrees 余赤纬，极距，去极度 | 去极度〔台〕　又见：*codeclination*
polar effect 磁极效应
polar equatorial coordinate 极向赤道坐标
polar flattening 极扁率　又见：*polar oblateness*
polar gap 极冠隙
polar glow 极光　又见：*aurora; polar light*
polar heliostat 极向定日镜
polarimeter 偏振计
polarimetric radial-velocity-meter 偏振视向速度仪　又见：*polarimetric speedometer*
polarimetric speedometer 偏振视向速度仪　又见：*polarimetric radial-velocity-meter*
polarimetric standard 偏振标准　又见：*polarization standard*
polarimetric standard star 偏振标准星　又见：*polarization standard star*
polarimetry 偏振测量　又见：*polarization measurement*
POLARIS （＝Polar Motion Analysis by Radio Interferometric Surveying）射电干涉测量极移计划
Polaris 勾陈一，北极星（小熊座α）　又见：*North Star*
Polaris almanac 北极星历
Polaris Australis 南极星（南极座σ）
polariscope 偏振光镜
Polaris monitor 北极星检测仪
polarissima 近极星
polarity 极性
polarity reversal 极性反转　又见：*reversal of polarity*
polarity reversal line 极性变换线
polarizability 偏振性
polarization 偏振
polarization angle 偏振角　又见：*angle of polarization*
polarization cloud 偏振云
polarization curve 偏振曲线
polarization effect 偏振效应
polarization interference filter 干涉偏振滤光器 | 偏振干涉滤光器〔台〕
polarization interferometer 偏振干涉仪
polarization interferometry 偏振干涉测量
polarization measurement 偏振测量　又见：*polarimetry*
polarization of background 背景偏振
polarization of light 光偏振　又见：*light polarization*
polarization parameter 偏振参数
polarization photometer 偏振光度计

polarization plane 偏振面　又见：*plane of polarization*
polarization receiver 偏振接收机
polarization sensitive bolometer 偏振敏感测辐射热计
polarization standard 偏振标准　又见：*polarimetric standard*
polarization standard star 偏振标准星
　又见：*polarimetric standard star*
polarization state 偏振态　又见：*state of polarization*
polarization switch 偏振开关
polarization switching 偏振开关
polarization tensor 极化张量
polarization vector 极化矢量
polarization wave 偏振波
polarized antenna 偏振天线
polarized beam 偏振射束　又见：*polarizing beam*
polarized burst 偏振暴
polarized component 偏振分量
polarized light 偏振光
polarized radiation 偏振辐射
polarizer 起偏振器｜偏振器〔台〕
polarizing angle 起偏振角
polarizing beam 偏振射束　又见：*polarized beam*
polarizing beam splitter 偏振射束分离器
polarizing eyepiece 起偏振目镜
polarizing monochromator 起偏振单色器
polarizing plate 起偏振片
polarizing prism 起偏振棱镜
polar latitude 极黄纬
polar light 极光　又见：*aurora; polar glow*
polar longitude 极黄经
polar mesospheric cloud （=PMC）极地中层云
polar motion 极移　又见：*motion of earth pole; polar displacement; polar shift; polar wandering; wandering of the pole*
Polar Motion Analysis by Radio Interferometric Surveying （=POLARIS）射电干涉测量极移计划
polar mounting 极式装置
polar night 极夜
polar oblateness 极向扁率　又见：*polar flattening*
polaroid sheet 偏振片
polar orbit 极轨道
Polar Orbiting Geophysical Observatory （=POGO）极轨地球物理台
polar path 地极轨线｜[地] 極軌線〔台〕
polar period 极性周期
polar plane 极平面

polar plume 极羽（日冕）
polar prominence 极区日珥
polar radius 极半径　又见：*polar semi-diameter*
polar ray 极辐射线｜極射線〔台〕（日冕）
　又见：*polar streamer*
polar region 极区
polar-ring galaxy 极环星系
polar satellite 极轨卫星
polar scope 极轴镜　又见：*polar telescope*
polar semi-diameter 极半径　又见：*polar radius*
polar sequence 近极星序
polar shift 极移　又见：*motion of earth pole; polar motion; polar displacement; polar wandering; wandering of the pole*
polar solar wind 极区太阳风　又见：*solar polar wind*
polar spot 极斑
polar star 极星　又见：*pole star*
polar streamer 极辐射线｜極射線〔台〕（日冕）
　又见：*polar ray*
polar sway 地极颤动
polar telescope 极轴镜　又见：*polar scope*
polar tide 极潮
polar triangle 极三角形｜極三角 [形]〔台〕
polar tube 北极管｜極軸筒〔台〕
polar wandering 极移　又见：*motion of earth pole; polar motion; polar displacement; polar shift; wandering of the pole*
polar wind 极风
polar wobble 地极摆动
polar year 极年
pole-observing Instrument 候极仪
pole of angular momentum 角动量极
pole of figure 形状极
pole of orbit 轨道极
pole of rotation 自转极　又见：*rotation pole*
pole of the ecliptic 黄极　又见：*ecliptic pole*
pole of the equator 赤极
pole-on 极向
pole-on object 极向天体
pole-on star 极向恒星｜極向 [恆] 星〔台〕
pole orbit 地极轨道｜[地] 極軌道〔台〕
pole star 极星　又见：*polar star*
Polet 飞行（苏联科学卫星）
Poleungkuk 保良局（小行星 4562 号）
poleward migration 极向迁移
polhody 地极轨迹
polishing 抛光

pollution 污染
Pollux ① 北河三（双子座 β） 又见：Hercules ② 波吕克斯（法国科学卫星）
polody origin 极原点
poloidal [magnetic] field 极向[磁]场
polonium （=Po）钋（84 号元素）
polyad 多重群
polychromator 多色器
polycyclic aromatic hydrocarbon （=PAH）多环芳烃
Polydeuces 土卫三十四
Polyhymnia 司瑟星（小行星 33 号）
polynomial inflation 多项式暴胀
polyspectrum 多谱
polytrope 多方球 | 多方[次]模型〔台〕
polytropic core 多方球心
polytropic equation of state 多方状态方程
polytropic equilibrium 多方平衡
polytropic gas 多方气体
polytropic gas sphere 多方气体球
polytropic index 多方指数
polytropic model 多方模型
Polzunov crater 波尔祖诺夫环形山（月球）
Pomona 果神星（小行星 32 号）
Poncelet crater 彭赛列环形山（月球）
Pontanus crater 蓬塔诺环形山（月球）
Pontécoulant crater 蓬泰库朗环形山（月球）
poorly directive antenna 弱定向天线
Popigai crater 波皮盖陨星坑（地球）
POP mirror 弹出镜
Popov crater 波波夫环形山（月球）
popular astronomy 大众天文学
population 星族 又见：population of stars; stellar population
population classification 星族分类
population density 粒子数密度
population I 星族 I | 第一星族〔台〕
population II 星族 II | 第二星族〔台〕
population II Cepheid 星族 II 造父变星 即：W Vir type variable; W Vir variable
population III 星族 III | 第三星族〔台〕
Population II main sequence 星族 II 主序
population II object 星族 II 天体
population II star 星族 II 恒星
population inversion 粒子数反转 | 居量反转〔台〕
population I object 星族 I 天体
population I star 星族 I 恒星

population of stars 星族 又见：population; stellar population
population synthesis 星族合成
population type 星族类型
Porcupine 十二面体号（美国卫星） 又见：Dodecapole; Dodecahedron
pore 微黑子 | 小黑點〔台〕 又见：microspot
Porrima 东上相，太微左垣二（室女座 γ）
Porro Prism 波罗棱镜
portable meridian circle 便携式子午环
portable transit instrument 便携式中星仪
Porter crater 波特环形山（月球）
Portia 天卫十二 | 天[王]卫十二〔台〕
positional astronomy 方位天文学
positional error 定位误差
position ambiguity 位置模糊度
position angle （=PA）位置角 | 方位角〔台〕
position-angle effect 位置角效应
position circle 位置圈 | 方位圈〔台〕 又见：circle of position
position encoder 位置编码器
position line 位置线 | 方位線〔台〕 又见：line of position
position micrometer 方位测微器 | 方位測微計〔台〕
Positions and Proper Motions Star Catalogue （=PPM Star Catalogue）PPM 恒星位置和自行表
position sensitive proportional counter 位置灵敏正比计数器
position vector 位置向量
positive 正片
positive correlation 正相关
positive curvature 正曲率
positive electron 正电子 又见：antielectron; positron
positive energy 正能 | 正能量〔台〕
positive energy theorem 正能定理
positive eyepiece 正像目镜 又见：terrestrial eyepiece
positive image 正像
positive ion 正离子 又见：kation
positive leap second 正闰秒
positive lens 正透镜
positron 正电子 又见：antielectron; positive electron
positron-electron annihilation 正电子－电子湮灭
positronium 正原子
POSS （=Palomar Observatory Sky Survey）帕洛玛天图 | 帕洛瑪星圖〔台〕
possibility 可能性
possible error 可能误差

post AGB star　后 AGB 星 | AGB 後星〔台〕
postburst　暴后 | 爆後〔台〕
post-collision merger　碰撞后并合天体
post-common-envelope binary　共包层后双星
post-core-collapse cluster　核坍缩后星团
post-coronal region　冕外区
postdetection bandwidth　检波后带宽
postdetection dedispersing　检波后消色散
posterior probability　后验概率
post-flare loop　耀斑后环 | 閃焰後環〔台〕
post-Galilean transformation　后伽利略变换
post-giant　巨星后天体
post main sequence　（=post MS）主序后
post-main-sequence evolution　主序后演化
post-main-sequence object　主序后天体
post-main-sequence stage　主序后阶段
post-main-sequence star　主序后星　又见：*evolved star*
post-maximum evolution　极大后演化
post-maximum spectrum　极大后光谱
post meridiem　（=p.m.; PM）【拉】下午
post MS　（=post main sequence）主序后
post-Newtonian approximation　后牛顿近似
post-Newtonian celestial mechanics　后牛顿天体力学
post-Newtonian effect　后牛顿效应
postnova　爆后新星　又见：*faded nova; ex-nova*
post-recombination Universe　复合后宇宙
post red-supergiant　红超巨后星
post starburst galaxy　星暴后星系 | 星遽增後星系〔台〕　又见：*poststarburst galaxy*
poststarburst galaxy　星暴后星系 | 星遽增後星系〔台〕　又见：*post starburst galaxy*
post-supernova　爆后超新星　又见：*ex-supernova*
post T-Tauri star　金牛 T 阶段后恒星
potassium　（=K）钾（19 号元素）
potassium-argon dating　钾氩计年 | 鉀氬定年〔台〕
potassium-argon method　钾氩纪年法
potassium-calcium dating　钾钙计年 | 鉀鈣定年〔台〕
potassium-calcium method　钾钙纪年法
Potassium Dihydrogen Phosphate crystal　（=KDP crystal）KDP 晶体
Potato　马铃薯（小行星 88705 号）
potential barrier　势垒 | 勢堰〔台〕
potential-density pair　势－密度对
potential difference　势差 | 勢差，位差〔台〕
potential energy　势能 | 勢能，位能〔台〕
potential energy tensor　势能张量
potential field　势场
potential function　势函数
potentially hazardous asteroid　（=PHA）潜在威胁小行星
potential motion　有势运动
potential perturbation　势扰动
potential well　势阱
Potsdam Astrophysical Observatory　波茨坦天体物理台
Potsdamer Durchmusterung　（=PD）波茨坦巡天星表
Potsdamer Photometrische Durchmusterung　（=PPD）波茨坦测光星表
Potsdamer Spektral-Durchmusterung　（=PSD）波茨坦恒星光谱表
power amplification　功率放大
power amplifier　功率放大器
powered flight　动力飞行
power gain　功率增益
power-law distribution　幂律分布
power-law index　幂律指数
power-law inflation　幂律暴胀
power-law spectrum　幂律谱
power-line noise　电源线噪声
power-line time-signal transfer　电源线时号传递
power pattern　功率方向图
power series　幂级数
power spectral density　功率谱密度
power spectral function　功率谱函数
power spectral index　功率谱指数
power spectrum　功率谱
power spectrum estimation　功率谱估计
Poynting crater　坡印廷环形山（月球）
Poynting-Robertson effect　坡印廷－罗伯逊效应
PP　（=Pulkovo pulsar）普尔科沃脉冲星
PP algorithm　（=particle-particle algorithm）质点－质点算法
pp chain　（=proton-proton chain）质子－质子链
pp cycle　（=proton-proton cycle）质子－质子循环
PPD　（=Potsdamer Photometrische Durchmusterung）波茨坦测光星表
PPM　（=piecewise parabolic method）逐段抛物线法
PPM Star Catalogue　（=Positions and Proper Motions Star Catalogue）PPM 恒星位置和自行表

PPN formalism (＝parameterized post-Newtonian formalism) 参数化后牛顿形式
pp reaction (＝proton-proton reaction) 质子－质子反应
p-process 质子过程 | p－過程〔台〕
Pr (＝praseodymium) 镨（59号元素）
practical astronomy 实用天文学 | 實用天文[學]〔台〕
practical astrophysics 实用天体物理 | 實用天文物理學〔台〕
Praecipua 势四（小狮座46）
Praesepe ①积尸增三（巨蟹座ε）②鬼星团 | 鬼宿星團〔台〕（M44）
Prager crater 普拉格环形山（月球）
Prancing Horse Nebula 跃马星云
Prandtl crater 普朗特环形山（月球）
Prandtl number 普朗特数 | 卜然托數〔台〕
praseodymium (＝Pr) 镨（59号元素）
Praxidike 木卫二十七
pre-amplification 前置放大
pre-amplifier 前置放大器
pre-cataclysmic binary 激变前双星
pre-cataclysmic variable 激变前变星
preceding limb 西边缘，前导边缘　又见：*leading edge; leading limb*
preceding spot 前导黑子，p黑子 | 先導黑子，前導黑子〔台〕　又见：*leading sunspot; preceding sunspot; leading spot*
preceding sunspot 前导黑子，p黑子 | 先導黑子，前導黑子〔台〕　又见：*leading sunspot; preceding spot; leading spot*
precessing-disk model 进动盘模型
precession ①岁差 ②进动　又见：*precessional motion*
precessional constant 岁差常数　又见：*precession constant; constant of precession*
precessional motion 进动　又见：*precession*
precession circle 岁差圈
precession coefficient 岁差系数
precession cone 岁差锥
precession constant 岁差常数　又见：*precessional constant; constant of precession*
precession globe 岁差仪
precession in declination 赤纬岁差
precession in latitude 黄纬岁差
precession in longitude 黄经岁差
precession in right ascension 赤经岁差
precession of orbit 轨道进动

precession of perihelion 近日点进动 | 近日點前移〔台〕　又见：*advance of the perihelion; perihelion precession; perihelion motion*
precession of the earth 地球进动
precession of the equinox 岁差，分点岁差 | 分點歲差〔台〕
precession period 岁差周期
precession rate 进动速率
precise time and time interval (＝PTTI) 精密时刻和时间间隔
precision 精度 | 精[密]度〔台〕
precision analog photon address (＝PAPA) 精密模拟光子定位
precision catalogue 精密星表
precision ephemeris 精密星历表
precision measurement 精密测量
precision radial-velocity spectrometer 精确视向速度光谱仪
precomputed altitude 预算高度
precomputed brightness 预算亮度
precursor effect 前兆效应
precursor object 前身天体 | 前身〔台〕　又见：*progenitor; progenitor object*
precursor pulse 前兆脉冲
precursor star 前身星 | 前身〔台〕　又见：*progenitor; progenitor star*
predetection dedispersing 检波前消色散
predictability 可预报性
predicted asteroid 已预测小行星
predicted comet 已预测彗星 | 預測彗星〔台〕
prediction of solar activity 太阳活动预报　又见：*solar activity prediction*
predissociation 预离解
pre-enrichment 预增丰
prefilter 前置滤波器
preflash 预照光
pre-galactic cloud 前星系云
pre-galactic nucleosynthesis 星系前核合成
pre-galactic star 星系前恒星
pre-Galaxy 前银河系
pre-galaxy 前星系
preionization 预电离
preliminary calculation 初算
preliminary designation 初步命名，初步编号 | 初步命名〔台〕
Preliminary General Catalogue (＝PGC) 博斯总星表初编 | 博斯星表初編〔台〕

preliminary orbit 初始轨道，初轨 | 初 [始] 軌 [道]〔台〕 又见：*primitive orbit; initial orbit*
preliminary position 初定方位
pre-main sequence （＝PMS）主序前 | 主星序前〔台〕
pre-main sequence binary 主序前双星
pre-main sequence object （＝PMS object）主序前天体
pre-main sequence star （＝PMS star）主序前星
pre-maximum brightness 极大前亮度
pre-maximum halt 极大前息（新星光变）
pre-maximum spectrum 极大前光谱
pre-monochromatization 预单色化
pre-monochromator 前置单色器
Pre-Nectarian System 前酒海结构系统
prenova 爆前新星
pre-nova object 新星爆前天体
pre-nova spectrum 新星爆前光谱
preplanetary disk 前行星盘
pre-planetary object 行星状星云前天体
pre-planetary star 行星状星云前恒星
preprocessor 预处理机
presolar matter 太阳前物质
presolar nebula 太阳前星云
presolar object 太阳前天体
Presqu'île crater 半岛陨星坑（地球）
Press-Schechter approach 普雷斯－谢克特方法 又见：*Press-Schechter method*
Press-Schechter formalism 普雷斯－谢克特公式
Press-Schechter mass function （＝PS mass function）普雷斯－谢克特质量函数
Press-Schechter method 普雷斯－谢克特方法 又见：*Press-Schechter approach*
Press-Schechter theory 普雷斯－谢克特理论
pressure balance 压力平衡 又见：*pressure equilibrium*
pressure broadening 压力致宽
pressure effect 压力效应
pressure equilibrium 压力平衡 又见：*pressure balance*
pressure-induced band 压力感生谱带
pressure-induced spectrum 压力感生光谱
pressure ionization 压致电离
pressure scale height 压力标高 | 壓力尺度高〔台〕
pressurized proportional counter 高气压式正比计数器
pre-stellar body 星前体
pre-stellar cloud 星前云
pre-stellar evolution 星前演化
pre-stellar matter 星前物质
pre-stellar object 星前天体
pre-supernova 爆前超新星
pre-supernova binary 超新星爆前双星
pre-supernova object 超新星爆前天体
pre-telescope astronomy 望远镜前天文学
pre-white dwarf 白矮前身星
Priamus 普丽阿姆斯（小行星 884 号）
Priestley crater 普里斯特利环形山（月球）
Prijipati 八谷一（御夫座 δ）
Prima （＝Phase-Referenced Imaging and Microarcsecond Astrometry）相参成像和微角秒天体测量
Prima Giedi 牛宿增六（摩羯座 α） 又见：*Algedi Prima*
Prima Hyadum 毕宿四（金牛座 γ） 又见：*Hyadum I*
primary antenna 初级天线，原天线
Primary Atmosphere hypothesis 原气说 又见：*Original Gas hypothesis*
primary beam 初级波束，原波束
primary body 主天体
primary circle 基本圆
primary clock ① 主钟 ② 母钟 又见：*master clock*
primary component 主星 又见：*primary star*
primary constant 初始常数
primary cosmic radiation 原宇宙线 又见：*primary cosmic rays*
primary cosmic rays 原宇宙线 又见：*primary cosmic radiation*
primary crater 主陨击坑 | 初级陨击坑〔台〕
primary distance indicator 一级示距天体
primary eclipse 主食
primary feed 主馈
primary image 主像 又见：*main image*
primary lens 主透镜
primary maximum 主极大 又见：*principal maximum; main maximum*
primary minimum 主极小 又见：*principal minimum; main minimum*
primary mirror 主镜
primary mirror-cell 主镜室
primary optical axis 主光轴
primary particle 原粒子
primary pattern 初级方向图
primary period 主周期
primary source 本源

primary spectrum 主谱，第一级光谱 | 主光谱〔台〕
primary standard 基准
primary star 主星 又见: *primary component*
primary surface 主表面 | 原表面〔台〕
primary telescope 主望远镜 又见: *principal telescope*
primary triangulation 一等三角测量
primary wave （＝P wave）P 波，初波，地震纵波
prime focus 主焦点 又见: *principal focus*
prime-focus astrophotography 主焦点天体摄影
prime focus cage 主焦观测室
prime focus corrector 主焦改正镜
prime-focus spectrograph 主焦摄谱仪
prime meridian 本初子午线，本初子午圈 | 本初子午線〔台〕 又见: *first meridian*
prime plane 卯酉面
primeval atom 原初原子 | 原始原子〔台〕
primeval fireball 原初火球 | 原始火球〔台〕 又见: *primordial fireball*
primeval galaxy 原初星系
primeval nebula 原始星云 | 原雲〔台〕 又见: *primordial nebula*
prime vertical 卯酉圈
prime vertical dial 卯酉面日晷
prime vertical transit 卯酉面中星仪
primitive nebula 原星云 | 原[星]雲〔台〕 又见: *proto-nebula*
primitive orbit 初始轨道，初轨 | 初[始]軌[道]〔台〕 又见: *preliminary orbit; initial orbit*
primordial abundance 原始丰度 | 原初豐度〔台〕
primordial binary 原始双星 又见: *primordial binary stars*
primordial binary stars 原始双星 又见: *primordial binary*
primordial black hole 太初黑洞 | 原初黑洞〔台〕
primordial element abundance 原初元素丰度
primordial field 原发场
primordial fireball 原初火球 | 原始火球〔台〕 又见: *primeval fireball*
primordial galaxy 原始星系 | 原初星系〔台〕
primordial gas cloud 原始气云
primordial helium 原始氦 | 原初氦〔台〕
primordial hydrogen 原始氢 | 原初氢〔台〕
primordial isocurvature baryon model （＝PIB model）原初等曲率重子模型

primordial nebula 原始星云 | 原雲〔台〕 又见: *primeval nebula*
primordial nucleosynthesis 原初核合成
primordial spectrum 原初功率谱
primordial star 原恒星 又见: *protostar; original star*
primoridal neutrino 原初中微子
primum mobile 宗动天（托勒玫体系）
principal axis 主轴
principal axis of inertia 惯性主轴
principal component analysis （＝PCA）主成分分析
principal elliptic term 主椭圆项
principal focus 主焦点 又见: *prime focus*
principal maximum 主极大 又见: *primary maximum; main maximum*
principal minimum 主极小 又见: *primary minimum; main minimum*
principal nutation 主章动
principal plane 主平面 | 主[平]面〔台〕
principal plane pattern 主平面方向图
principal planet 大行星 又见: *major planet*
principal series 主线系
principal solution 主解
principal spectrum 主光谱
principal telescope 主望远镜 又见: *primary telescope*
principal vertical circle 天球子午圈 又见: *celestial meridian*
Principia （＝Philosophiæ Naturalis Principia Mathematica）【拉】《自然哲学的数学原理》（牛顿著作）
principle of detailed balancing 细致平衡原理
principle of equivalence 等效原理 又见: *equivalence principle; equivalent principle*
principle of least action 最小作用原理
principle of mediocrity 平庸原理
principle of minimal coupling 最小耦合原理
principle of optical equivalence 光学等效原理
principle of superposition 叠加原理
principle of virtual work 虚功原理
printing chronograph 打印记时仪
printing theodolite 打印经纬仪
prior baseline 预知基线
prior estimate 先验估计
prior probability 先验概率
prism astrolabe 棱镜等高仪 又见: *prismatic astrolabe*

prismatic astrolabe　棱镜等高仪　又见：*prism astrolabe*
prismatic binoculars　棱镜双目望远镜
prismatic camera　棱镜照相机 | 棱鏡照相儀〔台〕
prismatic sextant　棱镜六分仪
prismatic spectrograph　棱镜摄谱仪　又见：*prism spectrograph*
prismatic spectroscope　棱镜分光镜　又见：*prism spectroscope*
prismatic spectrum　棱镜光谱
prismatic transit instrument　棱镜中星仪 | 折軸中星儀〔台〕
prism spectrograph　棱镜摄谱仪　又见：*prismatic spectrograph*
prism spectroscope　棱镜分光镜　又见：*prismatic spectroscope*
probability curve　概率曲线
probability density　概率密度
probability density function　概率密度函数
probability distribution　概率分布
probability distribution function　(=PDF) 概率分布函数
probable error　(=p.e.) 概差, 可几误差 | 概差〔台〕
probable value　概然值
problem of many bodies　多体问题　又见：*many-body problem; multiple-body problem*
problem of n bodies　n 体问题　又见：*N-body problem*
problem of three bodies　三体问题　又见：*3-body problem; three-body problem*
problem of two bodies　二体问题　又见：*2-body problem; two-body problem*
procaryote　原核生物
Proceedings of the Astronomical Society of Australia　(=PASA)《澳大利亚天文学会会刊》
Prochion　南河三（小犬座 α）　又见：*Procion; Procyon*
Procion　南河三（小犬座 α）　又见：*Prochion; Procyon*
Proctor crater　普罗克特环形山（月球）
Procyon　南河三（小犬座 α）　又见：*Prochion; Procion*
production rate　产星率, 产生率
progenitor　① 前身天体 | 前身〔台〕　又见：*progenitor object; precursor object* ② 前身星 | 前身〔台〕　又见：*progenitor star; precursor star*
progenitor bias　前身天体偏袒
progenitor nuclei　原粒子核

progenitor object　前身天体 | 前身〔台〕　又见：*progenitor; precursor object*
progenitor star　前身星 | 前身〔台〕　又见：*progenitor; precursor star*
Prognostication Classic of the Kaiyuan Era　《开元占经》　又见：*Treatise on Astrology in the Kaiyuan Reign*
Prognoz　预报号（苏联科学卫星）
prograde　顺行　又见：*direct motion; prograde motion*
prograde merger　顺行并合
prograde motion　顺行　又见：*direct motion; prograde*
prograde orbit　顺行轨道　又见：*direct orbit*
program object　待测天体
program star　待测星, 纲要星 | 綱要星〔台〕
Progress　进步号（俄罗斯太空飞行器）
progressive error　行差, 齿隙差　又见：*progressive inequality*
progressive inequality　行差, 齿隙差　又见：*progressive error*
progressive wave　前进波　又见：*advancing wave*
projected baseline　投影基线
projected correlation function　投影相关函数
projected density　投影密度
projected density profile　投影密度轮廓
projected semi-major axis　投影半长轴
projected velocity　投影速度
projection angle　投影角
projection effect　投影效应
projection operator　投影算子
Project Mercury　水星计划 | 水星計畫〔台〕（美国载人航天计划）
prolate spheroid　长球体 | 長球, 長球面〔台〕
prolate spheroidal coordinate　长等轴椭球坐标
Promethei Terra　普罗米修斯高地（火星）
Prometheus　土卫十六
promethium　(=Pm) 钷 | 鉕〔台〕（61号元素）
prominence　日珥　又见：*solar prominence*
prominence flare　日珥耀斑 | 日珥閃焰〔台〕
prominence knot　日珥结
prominence phenomenology　日珥现象学
prominence spectrometer　日珥分光仪
prominence spectroscope　日珥分光镜
prominence streamer　日珥射流
promontorium　岬 | 峽〔台〕
Promontorium Agarum　【拉】阿格鲁姆海角（月球）
Promontorium Agassiz　【拉】阿加西海角（月球）

Promontorium Archerusia	【拉】阿切鲁西亚海角（月球）
Promontorium Deville	【拉】德维尔海角（月球）
Promontorium Fresnel	【拉】菲涅尔海角（月球）
Promontorium Heraclides	【拉】赫拉克利德海角（月球）
Promontorium Kelvin	【拉】开尔文海角（月球）
Promontorium Laplace	【拉】拉普拉斯岬（月球）
Promontorium Taenarium	【拉】泰纳里厄姆海角（月球）

propagation 传播
propagation constant 传播常数
propagation delay 传播迟滞，传播时延 | 傳播遲滯〔台〕
propagation medium 传播媒质
propagation time 传播时间 又见: *travel time*
propagation velocity 传播速度
propagator 传播子，传播函数
propane 丙烷（C_3H_8）
propellant 推进剂 又见: *propellent*
propellent 推进剂 又见: *propellant*
proper direction 本征方向
proper distance 固有距离
proper element 本征根数
proper function 本征函数 又见: *eigenfunction*
proper length 静长度
proper mass 静质量 又见: *rest mass*
proper motion （＝P.M.）自行
proper motion group 自行群
proper motion in declination 赤纬自行
proper motion in right ascension 赤经自行
proper motion member 自行成员星 | 自行成员〔台〕 又见: *proper motion membership*
proper motion membership 自行成员星 | 自行成員〔台〕 又见: *proper motion member*
proper motion star 大自行星 | 大自行 [恆] 星〔台〕
proper plane 本征平面
proper reference frame 固有参考架 | 固有参考坐標〔台〕
proper reference system 固有参考系
proper star 基准星
proper tetrad 本征基，本征四维标架 | 本徵基〔台〕
proper time 原时，固有时 | 原時〔台〕
proper value 本征值 又见: *eigenvalue*
proper volume 固有体积
proplyd 原行星盘 又见: *protoplanetary disk*

proportional counter 正比计数器
proportional counting 正比计数
propulsion 推进
Propus 钺（双子座 η）
propyne 丙炔，甲基乙炔（C_3H_4） 又见: *methyl-acetylene*
Proserpina 冥后星（小行星 26 号）
Prospero 天卫十八
protactinium （＝Pa）镤（91 号元素）
Proteus 海卫八
proto-binary 原双星 又见: *binary protostar*
proto-brown-dwarf 原褐矮星
proto-cluster 原星团
proto-cluster of galaxies 原星系团
proto-disk 原盘
proto-earth 原地球
protogalactic cloud 原银河云
protogalactic object 原银河系天体
proto-Galaxy 原银河系
protogalaxy 原星系
proto-globular cluster 原球狀星团
proto-Jupiter 原木星
Proton 质子号（苏联运载火箭）
proton 质子
proton decay 质子衰变
proto-nebula 原星云 | 原 [星] 雲〔台〕 又见: *primitive nebula*
proton-electron-proton reaction （＝pep reaction）质子－电子－质子反应，pep 反应
proton event 质子事件
proton flare 质子耀斑 | 質子閃焰〔台〕
protonosphere 质子层
proton-proton chain （＝pp chain）质子－质子链
proton-proton cycle （＝pp cycle）质子－质子循环
proton-proton reaction （＝pp reaction）质子－质子反应
proton reaction 质子反应
proton splash 质子激散
proton stream 质子射流
proto-planet 原行星 又见: *original planet*
protoplanetary cloud 原行星云 又见: *protoplanetary nebula*
protoplanetary disk 原行星盘 又见: *proplyd*
proto-planetary nebula 原行星状星云
protoplanetary nebula 原行星云 又见: *protoplanetary cloud*

protoplanetary system 原行星系
protoprism 原棱镜
proto-shell star 原气壳星
protosolar cloud 原太阳云
protosolar collapse 原太阳坍缩
protosolar contraction 原太阳收缩
protosolar nebula 原太阳星云
protostar 原恒星 又见：*original star; primordial star*
protostellar association 原星协
protostellar cloud 原恒星云
protostellar collapse 原恒星坍缩
protostellar core 原恒星核
protostellar disk 原恒星盘
protostellar jet 原恒星喷流
protostellar matter 原恒星物质
protostellar outflow 原恒星外向流
proto-subnebula 原亚星云
protosun 原太阳 又见：*original sun*
protubérance 【法】日珥
provisional designation 暂定名，临时编号
provisional value 暂定值 | 暂定值，初值〔台〕
provisional weight 暂定权
Proxima Centauri [半人马] 比邻星 | [半人馬座] 比鄰星〔台〕（半人马座 α 星 C）
proximity effect 邻近效应
Príbram meteorite 普日布拉姆陨星
PsA （=Piscis Austrinus）南鱼座
Psamathe 海卫十
Psc ①（=Pisces）双鱼座 ②（=Pisces）双鱼宫，訾，亥宫
PSCz survey （=Point Source Catalog Redshift Survey）点源星表红移巡天
PSD （=Potsdamer Spektral-Durchmusterung）波茨坦恒星光谱表
pseudo body-fixed system 准地固坐标系
pseudobulge 赝球核
pseudo-Cepheid 赝造父变星 | 假造父變星〔台〕
pseudonucleus 假核
pseudo-potential 赝势
pseudotensor 赝张量
pseudo-variable 赝变星 | 假變星〔台〕
PSF （=point spread function）点扩散函数
PS mass function （=Press-Schechter mass function）普雷斯－谢克特质量函数
p-spot （=leading sunspot）前导黑子，p 黑子 | 先導黑子，前導黑子〔台〕
PSR （=pulsar）脉冲星 | 脈衝星，波霎〔台〕

PSS ①（=Planetary Space Science）《行星空间科学》（欧洲期刊）②（=Palomar Sky Survey）帕洛玛天图 | 帕洛瑪星圖〔台〕 即：*POSS*
PST （=Pacific standard time）太平洋标准时
pseudo Nambu-Goldstone boson （=pNGB）赝南部－戈德斯通玻色子
p-sunspot （=leading sunspot）前导黑子，p 黑子 | 先導黑子，前導黑子〔台〕
Psyche 灵神星（小行星 16 号）
Pt （=platinum）铂（78 号元素）
Ptolemaeus crater 托勒玫环形山 | 托勒米環形山〔台〕（月球）
Ptolemaic constellation 托勒玫星座
Ptolemaic system 托勒玫体系 | 托勒米 [宇宙] 體系〔台〕
Ptolemaism 托勒玫学说 | 托勒米學說〔台〕
ptolemaist 托勒玫学人
PTTI （=precise time and time interval）精密时刻和时间间隔
P-type asteroid P 型小行星
Pu （=plutonium）钚 | 鈽〔台〕（94 号元素）
Publications of the American Astronomical Society （=PAAS）《美国天文学会会刊》
Publications of The Astronomical Society of Australia 《澳大利亚天文学会会刊》
Publications of The Astronomical Society of Japan 《日本天文学会会刊》
Publications of the Astronomical Society of Japan （=PASJ）《日本天文学会会刊》
Publications of the Astronomical Society of the Pacific （=PASP）《太平洋天文学会会刊》
Puchezh-Katunki crater 普切日－卡通基陨星坑（地球）
Puck 天卫十五 | 天 [王] 衛十五〔台〕
puff 喷焰（太阳）
Puiching 培正（小行星 77138 号）
Pulcherrima 梗河一（牧夫座 ε） 又见：*Izar; Mirach; Mirak; Mirrak*
Pulkovo Main Astronomical Observatory 普尔科沃天文主台
Pulkovo pulsar （=PP）普尔科沃脉冲星
pulsar （=PSR）脉冲星 | 脈衝星，波霎〔台〕
pulsar back end 脉冲星后端
pulsar evolution 脉冲星演化
pulsar signal 脉冲星信号
pulsar synchronizer 脉冲星同步器
pulsar time scale 脉冲星时标
pulsar time transfer 脉冲星时间传递
pulsar wind 脉冲星风

pulsatance 角频率 又见: *angular frequency*
pulsating active nucleus 脉动活动核
pulsating activity 脉动活动
pulsating binary 脉动双星
pulsating radio source 脉冲射电源 | 脈衝電波源〔台〕
pulsating star 脉动星 | 脈動[變]星〔台〕
　又见: *pulsator*
pulsating universe 振荡宇宙 又见: *oscillating universe*
pulsating variable 脉动变星 又见: *pulsation variable*
pulsating X-ray source 脉冲X射线源
pulsation 脉动
pulsation acceleration 脉动加速
pulsation axis 脉动对称轴
pulsation constant 脉动常数
pulsation equation 脉动方程
pulsation frequency 脉动频率
pulsation instability 脉动不稳定性 | 脈動不穩定[性]〔台〕
pulsation instability strip 脉动不稳定带
pulsation mass 脉动质量
pulsation mode 脉动模式
pulsation phase 脉动相位
pulsation pole 脉动极
pulsation rate 脉冲率 又见: *pulse rate*
pulsation theory 脉动理论 | 脈動說〔台〕
pulsation variable 脉动变星 又见: *pulsating variable*
pulsator 脉动星 | 脈動[變]星〔台〕
　又见: *pulsating star*
pulse 脉冲 又见: *impulse*
pulse broadening 脉冲展宽
pulse counter 脉冲计数器 又见: *impulse counter*
pulse counting 脉冲计数
pulse-counting photometer 脉冲计数光度计
pulse delay 脉冲延迟
pulse duration 脉冲时间
pulse energy 脉冲能量
pulse intensity 脉冲强度
pulse light curve 脉冲光变曲线
pulse nulling 脉冲消失
pulse nulling phenomenon 脉冲消失现象
　又见: *nulling phenomenon*
pulse period 脉冲周期
pulse profile 脉冲轮廓
pulse rate 脉冲率 又见: *pulsation rate*
pulse signal 脉冲信号 | 脈衝訊號〔台〕

pulse smearing 脉冲拖影
pulse structure 脉冲结构
pulse-to-pulse correlation 脉冲间相关性
pulse width 脉冲宽度 又见: *duration of pulse*
pulse window 脉冲窗
pump frequency 抽运频率
pumping 抽运 | 起能〔台〕
pumping mechanism 抽运机制
pumping source 抽运源
punch-card machine 洞卡计算机
puncture counter 击穿计数器
Pup （=Puppis）船尾座
pupil 光瞳
pupil densifier 光瞳密实器
pupil-plane beam combiner 瞳面合束器
pupil-plane interferometer 瞳面干涉仪
Puppid-Velid meteor shower 船尾－船帆流星雨
Puppis （=Pup）船尾座
Puppis A 船尾射电源A
Purbach crater 普尔巴赫环形山（月球）
Pure Brightness 清明（节气） 又见: *Fresh Green; Clear and Bright*
pure coordinated time 纯粹协调时
pure density evolution （=PDE）纯密度演化
pure electron event 纯电子事件
pure local time 纯地方时 | 純粹地方時〔台〕
pure luminosity evolution （=PLE）纯光度演化
purely scattering 纯散射
purely scattering atmosphere 纯散射大气
pure temperature radiation 纯温度辐射
pure Trojan group 纯特洛伊群
Purkinje effect 浦尔金耶效应 | 普肯頁效應〔台〕
Purple Forbidden Enclosure 紫微垣
Purple Mountain 紫金山（小行星3494号）
Purple Mountain Observatory 紫金山天文台
Purple Palace 紫宫
PVD （=pairwise velocity dispersion）成对速度弥散
PV Telescopii star 望远镜PV型星
P wave （=primary wave）P波, 初波, 地震纵波
Pwyll crater 浦伊尔环形山（木卫二）
pycnonuclear reaction 超密态核反应
pygmy star 特矮星
pyramidal antenna 角锥天线
pyramid wavefront sensor 四棱锥波前传感器
pyranometer 日射强度计

pyrex 硼硅酸玻璃，派勒克斯玻璃 | 硼矽酸玻璃，耐熱玻璃〔台〕
pyrheliometer 太阳热量计 | 日射強度計〔台〕
pyrheliometry 太阳热量测量
pyrogenetic rock 火成岩　又见：*igneous rock*
pyrolite 上幔岩

pyrometry 高温测量
pyroxene 辉石
Pythagoras crater 毕达哥拉斯环形山（月球）
Pyx （＝Pyxis）罗盘座
Pyxis （＝Pyx）罗盘座
PZT （＝photographic zenith tube）照相天顶筒

Q

Q-ball Q-球
QBO （=quasi-biennial oscillation）准双年振荡
Q-branch Q 谱带 | Q-支〔台〕
QCD （=quantum chromodynamics）量子色动力学
QED （=quantum electrodynamics）量子电动力学
Qian 钱钟书（小行星 189347 号）
Qiansanqiang 钱三强（小行星 25240 号）
Qianxuesen 钱学森（小行星 3763 号）
Qinghai 青海（小行星 2255 号）
Q magnitude Q 星等
QPO （=quasi-periodic oscillation）准周期振荡
QSG （=quasi-stellar galaxy）类星星系
QSO （=quasi-stellar object）类星体
QSRS （=quasi-stellar source）类星射电源 | 類星電波源〔台〕
QSS （=quasi-stellar source）类星射电源 | 類星電波源〔台〕
Q-type asteroid Q 型小行星
Quadrans Muralis 象限仪座（现已不用）
quadrant ① 象限 ② 象限仪 又见：sector
quadrant altazimuth 地平象限仪 又见：azimuth quadrant
quadrantal triangle 象限三角形
quadrant detector 象限探测器
Quadrantid meteor shower 象限仪流星雨
Quadrantids 象限仪流星群 | 象限儀 [座] 流星雨〔台〕
quadratic estimator 二次方估计
quadratic magnitude term 星等二次项
quadratic Stark effect 斯塔克二次效应 | 二次史塔克效應〔台〕
quadratic sum 平方和
quadrature 方照
quadraxial mounting 四轴装置 又见：4-axis mounting; four-axis mounting

quadraxial tracking frame 四轴跟踪架
quadraxial tracking system 四轴跟踪系统
quadric 二次曲面
quadruple star 四合星
quadrupole 四极
quadrupole effect 四极效应
quadrupole formula 四极公式
quadrupole matrix 四极矩阵
quadrupole moment 四极矩
quadrupole radiation 四极辐射
quadrupole term 四极项
qualitative analysis 定性分析
qualitative method 定性方法
quality factor 品质因数
quantitative analysis 定量分析
quantitative spectral classification 定量光谱分类
quantization 量子化
quantum amplifier 量子放大器
quantum chromodynamics （=QCD）量子色动力学
quantum cosmology 量子宇宙学 | 量子宇宙論〔台〕
quantum defect 量子数亏损 | 量子欠缺〔台〕
quantum effect 量子效应
quantum efficiency 量子效率
quantum electrodynamics （=QED）量子电动力学
quantum field 量子场
quantum fluctuation 量子涨落
quantum gravitation 量子引力
quantum gravity 量子引力
quantum of flux 磁通量量子
quantum solid 量子固体
quantum state 量子态
quantum statistics 量子统计 | 量子統計 [學]〔台〕
quantum transition 量子跃迁
quantum universe 量子宇宙

quantum well infrared detector （＝QWIP）量子阱红外探测器
quantum yield 量子产额 | 量子產率〔台〕
Quanzhou 泉州（小行星 3335 号）
Quaoar 夸奥尔（小行星 50000 号）
quardruple-lensed quasar 引力透镜效应四像类星体
quark 夸克 | 夸克，夸子〔台〕
Quark era 夸克时期
quark-gluon plasma 夸克胶子等离子体
quark-hadron phase 夸克－强子相
quark-hadron phase transition 夸克－强子相跃迁
quark star 夸克星　又见：*quark-star*
quark-star 夸克星　又见：*quark star*
quarter-wave plate 四分之一波片 | 四分之一波[晶]片〔台〕
quartet ① 四重线 | 四重[谱]线，四合系统〔台〕 ② 四重态
quartz clock 石英钟　又见：*quartz crystal clock*
quartz crystal 石英晶体
quartz crystal clock 石英钟　又见：*quartz clock*
quartz crystal resonator 石英共振器
quartz extensometer 石英伸缩仪
quartz glass 石英玻璃
quartz-polaroid monochromator 石英偏振片单色仪 | 石英偏振片單色器〔台〕
quartz prism 石英棱镜
quartz spectrograph 石英摄谱仪 | 石英[棱镜]攝譜儀〔台〕
quasar 类星体　又见：*quasi-stellar object*
quasar astronomy 类星体天文学
quasar core 类星体核
quasar-galaxy pair 类星体－星系对
quasar halo 类星体晕
quasar infrared source 类星体红外源
quasar jet 类星体喷流
quasar radio source 类星体射电源
quasi-absolute method 准绝对方法
quasi-adiabatic convection 准绝热对流
quasi-biennial oscillation （＝QBO）准双年振荡
quasi-diurnal variation 准周日变化
quasi-elastic scattering 准弹性散射
quasi-equilibrium process 准平衡过程
quasi-equilibrium state 准平衡态
quasi-ergodic system 准各态历经系统
quasi-eruptive prominence 准爆发日珥
quasi-isothermal layer 准等热层

quasi-linear theory 准线性理论
quasilinear transition 准线性转变
quasi-longitutdinal oscillation 准纵向振荡
quasi-longitutdinal propagation 准纵向传播
quasi-neutrality 准中性
quasi-optical feed 准光学馈源
quasi-optical wave 准光波
quasi-optic technique 准光学技术
quasi-particle 准粒子
quasi-periodic function 准周期函数
quasi-periodic orbit 拟周期轨道，准周期轨道 | 準周期軌道〔台〕
quasi-periodic oscillation （＝QPO）准周期振荡
quasi-periodic variable 准周期变星
quasi-periodic variation 准周期变化
quasi-point radio source 类点射电源
quasi-polymer structure 似聚合物结构 | 似聚合體結構〔台〕
quasi-sidereal time 准恒星时
quasi-stable state 准稳态　又见：*quasi-stationary state*
quasi-static approximation 准静态近似
quasi-static process 准静态过程 | 似穩過程〔台〕
quasi-stationary density wave 准稳密度波
quasi-stationary level 准稳级
quasi-stationary spectrum 准稳定谱
quasi-stationary spiral structure 准稳旋涡结构
quasi-stationary state 准稳态　又见：*quasi-stable state*
quasi-stationary wave 准稳波
quasi-steady state 准稳态
quasi-stellar blue galaxy 蓝类星星系
quasi-stellar galaxy （＝QSG）类星星系
quasi-stellar object （＝QSO）类星体　又见：*quasar*
quasi-stellar radio source 类星射电源 | 類星電波源〔台〕　又见：*quasi-stellar source*
quasi-stellar red galaxy 红类星星系
quasi-stellar source （＝QSRS; QSS）类星射电源 | 類星電波源〔台〕　又见：*quasi-stellar radio source*
quasi-thermal flare 准热耀斑
quasi-transverse oscillation 准横向振荡
quasi-transverse propagation 准横向传播
quasi-uniform time 准均匀时
Q/U decomposition Q/U 分解
Quetelet crater 凯特尔环形山（月球）
quiescence 宁静态
quiescent brightness 宁静态亮度

quiescent galaxy 宁静星系
quiescent luminosity 宁静态光度
quiescent prominence 宁静日珥
quiescent radiation 宁静辐射　又见: *quiet radiation*
quiescent spectrum 宁静光谱
quiescent star formation 宁静恒星形成
quiet component 宁静分量
quiet corona 宁静日冕
quiet day 宁静日　又见: *calm day*
quiet optical sun 宁静光学太阳
quiet radiation 宁静辐射　又见: *quiescent radiation*
quiet radio emission 宁静射电 | 寧靜電波〔台〕
　　又见: *quiet radio radiation*
quiet radio radiation 宁静射电 | 寧靜電波〔台〕
　　又见: *quiet radio emission*
quiet radio sun 宁静射电太阳 | 寧靜電波太陽〔台〕
quiet solar radiation 宁静太阳辐射

quiet solar radio radiation 宁静太阳射电 | 寧靜太陽電波〔台〕
quiet sun 宁静太阳
quiet sun noise 宁静太阳噪声 | 寧靜太陽雜訊〔台〕
quiet thermal emission 宁静热发射
quiet thermal radiation 宁静热辐射
quintessence 精质
quintessential inflation 精质化暴胀
quintet ① 五重线　又见: *pentet* ② 五重态
quintom dark energy 精灵暗能量
quintuple star 五合星
Quintuplet Cluster 五合星团 | 五胞胎星團〔台〕
　　又见: *Quintuplet Star Cluster*
Quintuplet Star Cluster 五合星团 | 五胞胎星團〔台〕　又见: *Quintuplet Cluster*
Ququinyue 曲钦岳（小行星 3513 号）
Q value Q 值
QWIP （=quantum well infrared detector）量子阱红外探测器

R

R.A. （＝right ascension）赤经
RA （＝right ascension）赤经
Ra （＝radium）镭（88号元素）
RAA （＝Research in Astronomy and Astrophysics）《天文和天体物理学研究》（中国期刊）
Rabbi Levi crater 拉比列维环形山（月球）
radar astronomy 雷达天文学
radar echo 雷达回波
radar mapper 雷达测绘器
radar mapping 雷达测绘
radar meteor 雷达流星
radar observation 雷达观测
radar ranging 雷达测距
radar survey 雷达巡天
radar telescope 雷达望远镜
Radcliffe Observatory 拉德克利夫天文台
radial acceleration ①径向加速度 ②视向加速度
radial action 径向作用量
radial arc 径向弧
radial coordinate 径向坐标
radial defocusing 径向偏焦
radial distortion 径向畸变
radial filament 径向暗条
radial grating 径向光栅
radial inversion 径向反演
radially streaming 径向股流
radial momentum 径向动量
radial motion ①径向运动 ②视向运动
radial-orbit instability 径向轨道不稳定性
radial oscillation 径向振荡
radial period 径向周期
radial pulsation 径向脉动
radial pulsator 径向脉动星
radial velocity ①径向速度 ②视向速度
　　又见：*line-of-sight velocity*
radial-velocity curve 视向速度曲线
RAdial Velocity Experiment （＝RAVE）视向速度实验
radial-velocity orbit 分光解　又见：*spectroscopic orbit*
radial-velocity reference star 视向速度参考星
radial-velocity scanner 视向速度仪
　　又见：*radial-velocity spectrometer*
radial-velocity scanning 视向速度扫描
radial-velocity spectrometer ①（＝RVS）视向速度仪　又见：*radial-velocity scanner* ②恒星视向速度仪　又见：*stellar speedometer*
radial-velocity standard star 视向速度标准星
　　又见：*standard star for radial-velocity; standard-velocity star*
radial-velocity survey 视向速度巡天
radial-velocity trace 视向速度描迹
　　又见：*radial-velocity tracing*
radial-velocity tracing 视向速度描迹
　　又见：*radial-velocity trace*
radial wavelength 径向波长
radial wavenumber 径向波数
radian 弧度
radiance 面辐射强度 | 辐射率〔台〕
　　又见：*radiancy*
radiancy 面辐射强度 | 辐射率〔台〕
　　又见：*radiance*
radiant 辐射点　又见：*radiant point*
radiant density 辐射密度　又见：*radiation density*
radiant energy 辐射能　又见：*radiation energy*
radiant flux 辐射流量 | 辐射通[量]〔台〕
　　又见：*flux of radiation; radiation flux*
radiant intensity 辐射强度
radiant of comet 彗星辐射点
radiant of meteor 流星辐射点
radiant of moving cluster 移动星团辐射点
radiant point 辐射点　又见：*radiant*
radiating ridge 辐射纹
radiation 辐射
radiation background 辐射背景

radiation belt	辐射带
radiation belt of the Earth	地球辐射带
radiation coefficient	辐射系数
radiation compression	辐射压缩
radiation constant	辐射常数
radiation damage	辐射损伤
radiation damping	辐射阻尼
radiation density	辐射密度　又见: *radiant density*
radiation-density constant	辐射密度常数
radiation dominated era	辐射主导期, 辐射占优期｜輻射主導期〔台〕
radiation-dominated fluid	辐射主导流体
radiation dominated gas	辐射主导气体, 辐射占优气体
radiation dominated universe	辐射主导宇宙, 辐射占优宇宙｜輻射主導宇宙〔台〕
radiation dose	辐射剂量
radiation drag	辐射拖曳
radiation effect	辐射效应
radiation efficiency	辐射效率
radiation energy	辐射能　又见: *radiant energy*
radiation-energy density	辐射能密度
radiation entropy	辐射熵
radiation era	辐射期｜輻射時代〔台〕
radiation field	辐射场
radiation flow	辐射流｜輻射流 [量]〔台〕
radiation flux	辐射流量｜輻射通 [量]〔台〕 又见: *flux of radiation; radiant flux*
radiation length	辐射长度
radiationless transition	无辐射跃迁
radiation lifetime	辐射寿命
radiation mechanism	辐射机制
radiation pattern	辐射方向图
radiation plasma turbulence	辐射等离子体湍流
radiation pressure	辐射压｜輻射壓 [力]〔台〕
radiation receiver	辐射接收器
radiation region	辐射区
radiation resistance	辐射电阻
radiation scattering	辐射散射
radiation spectrum	辐射谱
radiation temperature	辐射温度｜電波溫度〔台〕 又见: *radiometric temperature; radiative temperature*
radiation transport	辐射输运　又见: *radiative transport*
radiation zone	辐射带｜輻射層〔台〕
radiative braking	辐射制动
radiative capture	辐射俘获　又见: *trapping of radiation*
radiative cooling	辐射致冷
radiative decay model	辐射衰变模型
radiative dissociation	辐射离解
radiative envelope	辐射包层｜輻射殼〔台〕
radiative equilibrium	辐射平衡
radiative feedback	辐射反馈
radiative fluid	辐射流体
radiative heating	辐射致热
radiatively inefficient accretion flow	（＝RIAF）辐射低效吸积流
radiative recombination	辐射复合
radiative relaxation	辐射弛豫
radiative temperature	辐射温度｜電波溫度〔台〕 又见: *radiometric temperature; radiation temperature*
radiative temperature gradient	辐射温度梯度
radiative transfer	辐射转移
radiative transport	辐射输运　又见: *radiation transport*
radiative viscosity	辐射黏度
radiator	辐射体, 辐射器｜輻射體〔台〕
radical	基
radio absorption	射电吸收
radioactive age dating	放射性计年
radioactive atom	放射性原子
radioactive dating	放射性计年
radioactive decay	放射性衰变
radioactive element	放射性元素
radioactive nucleus	放射性核
radio active star	射电活跃恒星｜電波活躍恆星〔台〕
radioactive substance	放射性物质
radio active sun	射电活跃太阳｜電波活躍太陽〔台〕
radioactivity	放射性｜放射性, 放射作用, 放射現象〔台〕
radio afterglow	射电余辉
radio antenna	射电天线
radio arm	射电臂｜電波臂〔台〕
radio astrometry	射电天体测量学｜電波天體測量學〔台〕
RadioAstron	射电天文号（俄罗斯卫星） 又见: *Spektr-R*
radio astronomical observatory	射电天文台｜電波天文台〔台〕　又见: *radio observatory*
radio astronomical receiver	射电天文接收机
radio astronomical station	射电天文站

Radio Astronomical Telescope of the Academy of Sciences 600　（＝RATAN-600）RATAN600 米射电望远镜
radio astronomy　射电天文学 | 電波天文 [學]〔台〕
Radio Astronomy Explorer　（＝RAE）射电天文探测器
radio astronomy satellite　射电天文卫星
radio astrophysics　射电天体物理 | 電波天文物理學〔台〕
radio background　射电背景辐射 | 電波背景輻射〔台〕又见：*radio background radiation*
radio background radiation　射电背景辐射 | 電波背景輻射〔台〕又见：*radio background*
radio binary　射电双星 | 電波雙星〔台〕
radio bridge　射电桥 | 電波橋〔台〕
radio brightness　射电亮度 | 電波亮度〔台〕
radio brightness distribution　射电亮度分布
radio brightness temperature　射电亮度温度 即：*radio temperature*
radio brown dwarf　射电褐矮星
radio burst　射电暴 | 電波爆發〔台〕
radiocarbon age　放射性碳龄
radiocarbon dating　放射性碳计年 | 放射性碳定年〔台〕
radio channel　射电波道
radiochemical neutrino detector　放射化学式中微子探测器
radio continuum　射电连续辐射
radio contour　射电轮廓图 | 電波輪廓圖〔台〕
radio core　射电核
radio corona　射电冕 | 電波冕〔台〕
radio counterpart　射电对应体 | 電波對應體〔台〕
radio depth　射电深度
radio diameter　射电直径 | 電波直徑〔台〕
radio direction finding　无线电定向
radio dish　射电圆面天线 | 電波碟形天線〔台〕 又见：*radio disk*
radio disk　射电圆面天线 | 電波碟形天線〔台〕 又见：*radio dish*
radio disk temperature　射电圆面天线温度
radio Doppler effect　射电多普勒效应
radio double source　射电双源
radio early-type emission-line star　（＝radio ETELS）射电早型发射线星
radio echo　无线电回波
radio echo method　无线电回波法
radio eclipse　射电食 | 電波食〔台〕
radio emission　射电辐射 | 電波發射〔台〕

radio ETELS　（＝radio early-type emission-line star）射电早型发射线星
radio flare　射电耀发 | 電波閃焰〔台〕
radio flare star　射电耀星 | 電波 [閃] 焰星〔台〕
radio flux　射电流量 | 電波流量〔台〕
radio frequency　（＝RF）射频
radio-frequency amplifier　射频放大器
radio-frequency band　射频段
radio-frequency disk　射频圆面天线
radio-frequency interference　射频干涉
radio-frequency line　射频谱线
radio-frequency mass spectrometer　射频质谱仪
radio-frequency radiation　射频辐射
radio-frequency signal　射频信号
radio-frequency spectrum　射频谱 | 電波頻譜〔台〕
radio galaxy　射电星系 | 電波星系〔台〕
radio-galaxy Hubble diagram　射电星系哈勃图谱
radiogenic heat　放射热
radiograph　射电图 | 電波圖〔台〕又见：*radio map; radio picture*
radio halo　射电晕 | 電波暈〔台〕
radio heliograph　射电日像仪 | 電波日像儀〔台〕
radio holography　射电全息法
radio image　射电像 | 電波像〔台〕
radio index　射电指数 | 電波指數〔台〕
radio interference　射电干涉
radio interferometer　射电干涉仪 | 電波干涉儀〔台〕
radio interferometry　射电干涉测量 | 電波干涉測量〔台〕
radio-intermediate quasar　中介射电类星体
radio-isophote　射电等强线 | 電波等強線〔台〕
radioisotope thermoelectric generator　（＝RTG）放射性同位素热电发生机
radio jet　射电喷流 | 電波噴流〔台〕
radio light curve　射电变光曲线
radio line　射电谱线 | 電波譜線〔台〕又见：*radio spectral line*
radio-link interferometer　无线电中继干涉仪 | 電波連接干涉儀〔台〕
radio-link interferometry　无线电中继干涉测量
radio lobe　射电瓣
radiolocational astronomy　无线电定位天文学 | 電波定位天文 [學]〔台〕
radio loud quasar　射电类星体 | 電波類星體〔台〕又见：*radio quasar*
radioloud quasar　强射电类星体 | 電波強類星體〔台〕

radioloud star　强射电星 | 電波強星〔台〕
radioloud sun　强射电太阳
radio luminosity　射电光度 | 電波光度〔台〕
radio magnitude　射电星等 | 電波星等〔台〕
radio map　射电图 | 電波圖〔台〕
　　又见：*radiograph; radio picture*
radio meteor　射电流星 | 電波流星〔台〕
radiometer　辐射计
radiometric dating　放射性同位素计年
radiometric magnitude　辐射星等
radiometric temperature　辐射温度 | 電波溫度〔台〕　又见：*radiation temperature; radiative temperature*
radiometry　辐射测量　又见：*actinometry*
radio multi-source　多重射电源 | 多重電波源〔台〕
　　又见：*multiple radio source*
radio navigation　无线电导航
radio nebula　射电星云 | 電波星雲〔台〕
radio noise　射电噪声 | 電波雜訊〔台〕
radio noise-burst　射电噪暴 | 電波雜訊爆〔台〕
radio nonthermal brightness temperature　非热射电亮度温度
radio nova　射电新星 | 電波新星〔台〕
radio nova remnant　射电新星遗迹
radionuclide　放射性核素
radio observation　射电观测
radio observatory　射电天文台 | 電波天文台〔台〕
　　又见：*radio astronomical observatory*
radio occultation　掩射电源
radiophotography　无线电照相
radio photon　射电光子
radio photosphere　射电光球
radio picture　射电图 | 電波圖〔台〕
　　又见：*radiograph; radio map*
radio plage　射电谱斑 | 電波譜斑〔台〕
radio plage component　射电谱斑分量
radio polarimeter　射电偏振计 | 電波偏振計〔台〕
radio polarimetry　射电偏振测量
radio pollution　射电污染
radio pulsar　射电脉冲星 | 電波脈衝星〔台〕
radio quasar　射电类星体 | 電波類星體〔台〕
　　又见：*radio loud quasar*
radio quiet quasar　射电宁静类星体 | 電波寧靜類星體〔台〕
radio quiet sun　射电宁静太阳 | 電波寧靜太陽〔台〕
radio radiation　射电 | 電波輻射〔台〕
radio radiation spectrum　射电频谱 | 電波頻譜〔台〕　又见：*radio spectrum*

radio recombination　射电复合
radio recombination line　射电复合谱线 | 電波復合[譜]線〔台〕
radio scintillation　射电闪烁 | 電波閃爍〔台〕
radio sextant　射电六分仪 | 電波六分儀〔台〕
radio sky　射电天空 | 電波天空〔台〕
radio sky map　射电天图 | 電波星圖〔台〕
radio sky temperature　天空射电温度
radiosonde　无线电探空仪 | 無線電探空儀，雷送〔台〕
radio source　射电源 | 電波源〔台〕
radio-source catalogue　射电源表 | 電波源表〔台〕
radio source count　射电源计数 | 電波源計數〔台〕
　　又见：*radio-source counting*
radio-source counting　射电源计数 | 電波源計數〔台〕　又见：*radio source count*
radio-source distribution　射电源分布
radio source reference system　射电源参考系 | 電波源參考系〔台〕
radio-source spectrum　射电源频谱
radio-source structure　射电源结构
radio spectral index　射电谱指数 | 電波譜指數〔台〕
radio spectral line　射电谱线 | 電波譜線〔台〕
　　又见：*radio line*
radio spectrograph　射电频谱仪 | 電波頻譜儀〔台〕
radio spectroheliogram　太阳射电频谱图 | 電波太陽單色圖〔台〕
radio spectroheliograph　太阳射电频谱仪 | 電波太陽單色儀〔台〕
radio spectrometer　射电频谱计
radio spectroscope　射电频谱镜
radio spectroscopy　射电频谱测量
radio spectrum　射电频谱 | 電波頻譜〔台〕
　　又见：*radio radiation spectrum*
radio spot　射电斑
radio star　射电星 | 電波星〔台〕
radio storm　射电暴 | 電波爆〔台〕
radio structure　射电结构
radio subsolar temperature　日下点射电温度
radio sun　射电太阳 | 電波太陽〔台〕
radio sunspot　射电太阳黑子
radio supernova　射电超新星 | 電波超新星〔台〕
radio supernova remnant　射电超新星遗迹
radio survey　射电巡天 | 電波巡天〔台〕
radio-tail galaxy　带尾射电星系 | 電波尾星系〔台〕
　　又见：*tail radio galaxy*
radio-tail object　带尾射电天体 | 電波尾天體〔台〕

radio telescope 射电望远镜 | 電波望遠鏡〔台〕
radio telescope array 射电望远镜阵 | 電波望遠鏡陣〔台〕
radio temperature 射电温度 | 電波溫度〔台〕
radio thickness 射电厚度
radio time code 无线电时码
radio time signal 无线电时号 | 無線電時間記錄〔台〕 又见: wireless time signal
radio tracking 无线电跟踪
radio trail 射电拖曳
radio-trail source 拖曳射电源
radio wave 射电波 | [無線]電波〔台〕
radio waveguide 射电波导
radio wavelength band 射电波段
radio window 射电窗口 | 電波窗〔台〕
radium （＝Ra）镭（88 号元素）
radius-luminosity relation 半径－光度关系 | [半]徑光[度]關係〔台〕
radius-mass relation 半径－质量关系 | [半]徑質[量]關係〔台〕
radius of curvature 曲率半径
radius of gyration 回旋半径 又见: gyroradius
radius of influence 影响半径
radius of shadow 影半径
radius vector 矢径 | 矢徑，向徑〔台〕
radome 天线罩
radon （＝Rn）氡（86 号元素）
RAE （＝Radio Astronomy Explorer）射电天文探测器
Rahu 罗睺（黄白道升交点）
Raimond crater 雷蒙环形山（月球）
rainband 雨带
Rain Water 雨水（节气）
raking-up effect 集积效应
RAM （＝random access memory）随机存取存储器
Raman effect 拉曼效应
Raman scattering 拉曼散射
Raman spectroscopy 拉曼光谱学
Raman spectrum 拉曼光谱
ramjet 冲压喷气
rampart 环壁
rampart crater 壁垒陨击坑
ram pressure 冲压
ram-pressure confinement 冲压约束
ram-pressure stripping 冲压剥离
Ramsauer effect 冉绍尔效应

Ramsay crater 拉姆齐环形山（月球）
Ramsden circle 冉斯登环
Ramsden disk 冉斯登圆面
Ramsden eyepiece 冉斯登目镜
Rana 天苑三（波江座 δ） 又见: Theemini
random access memory （＝RAM）随机存取存储器
random background object 无序背景天体
random deviation 随机偏差
random displacement 随机位移
random distribution 随机分布
random error 随机误差
random event 随机事件
random fluctuation 随机起伏，无规起伏
random motion 随机运动，无规运动 | 隨機運動〔台〕
randomness 随机性
random noise 随机噪声
random peculiar motion 随机本动
random polarization 随机偏振
random sample 随机样本
random sampling 随机取样
random trajectory 随机轨迹
random variable 随机变量 又见: stochastic variable
random variation 随机变化
random velocity 随机速度
random walk 随机游动 | 無規行走〔台〕
range counter 程长计数器
range data 变幅数据
range finder 测距仪 又见: distance gauge; telemeter
range of light-variation 光变幅
Ranger 徘徊者号 | 遊騎兵號〔台〕（美国月球探测器系列）
range-range navigation 测距－测距导航
range rate 程长速率
ranging 测距
Rankine-Hugoniot equation 兰金－于戈尼奥方程 | 藍欽－雨果方程〔台〕
Rankine-Hugoniot jump condition 兰金－于戈尼奥跳变条件
Rankine-Hugoniot relation 兰金－于戈尼奥关系
Rankine temperature scale 兰氏温标
rank of commensurability 通约秩
RAO （＝Robotic Autonomous Observatory）程控自主天文台
Raoult's law 拉乌尔定律 | 拉午耳定律〔台〕
rapid burst 快暴

rapid burster 快暴源 | 快爆源〔台〕
rapid development 快速显影
rapidly oscillating Ap star 快速振荡 Ap 星
rapidly varying component 快变分量
rapid neutron capture process （＝r-process）快过程，r 过程 | 中子快捕獲過程，r 過程〔台〕
rapid nova 快新星　又见：*fast nova*
rapid rotator 快转星，快转天体
rapid variable 快变星 | 迅速變星〔台〕
rare-earth element 稀土元素
rarefied plasma 稀薄等离子体
rare gas 稀有气体　又见：*noble gas*
RAS ①（＝Royal Astronomical Society）英国皇家天文学会 ②（＝Russian Academy of Sciences）俄罗斯科学院
Rasalague 候（蛇夫座 α）　又见：*Rasalhague*
Rasalas 轩辕十（狮子座 μ）
Ras Algethi 帝座（武仙座 α）　又见：*Rasalgethi*
Rasalgethi 帝座（武仙座 α）　又见：*Ras Algethi*
Rasalhague 候（蛇夫座 α）　又见：*Rasalague*
RASC （＝Royal Astronomical Society of Canada）加拿大皇家天文学会
Ras Elased Australis 轩辕九（狮子座 ε）　又见：*Algenubi; Asad Australis*
R association R 星协
Rastaban 天棓三（天龙座 β）　又见：*Alwaid*
raster 光栅 | 光域〔台〕　即：*grating*
raster scan mapping 逐行扫描成图法
RATAN-600 （＝Radio Astronomical Telescope of the Academy of Sciences 600）RATAN600 米射电望远镜
rate coefficient 比率系数
rated value 额定值
rate of clock 钟速　又见：*clock rate*
rate of diffusion 扩散率　又见：*diffusivity*
rate of discovery 发现率
rate of escape 逃逸率
rate of formation 诞生率
rate of occurrence 掩食率
rate of stellar extinction 恒星消亡率
rational horizon 地心地平　又见：*geocentric horizon*
RAVE （＝RAdial Velocity Experiment）视向速度实验
raw value 原始值 | 原始數值，草值〔台〕
ray and horn 芒角
Raychudhuri equation 瑞楚德胡瑞方程
ray crater 辐射状陨击坑
Rayleigh crater 瑞利环形山（月球）

Rayleigh criterion 瑞利判据
Rayleigh distribution 瑞利分布
Rayleigh guide star 瑞利引导星
Rayleigh-Jeans approximation 瑞利－金斯近似
Rayleigh-Jeans formula 瑞利－金斯公式
Rayleigh-Jeans law 瑞利－金斯定律
Rayleigh-Jeans region 瑞利－金斯区域
Rayleigh-Jeans side 瑞利－金斯侧边
Rayleigh limit 瑞利极限
Rayleigh number 瑞利数
Rayleigh resolution 瑞利分辨率
Rayleigh scattering 瑞利散射
Rayleigh-Taylor instability 瑞利－泰勒不稳定性
ray path 线程
rays of the moon 月面辐射纹 | [月面] 輻射紋〔台〕　又见：*lunar rays*
Razin effect 拉津效应
Razin-Tsytovich effect 拉津－楚托维奇效应
razor-thin disk 无限薄盘
Razumov crater 拉祖莫夫环形山（月球）
Rb （＝rubidium）铷（37 号元素）
R-branch R 谱带 | R－支〔台〕
R Canis Majoris star 大犬 R 型星
RCBG （＝Reference Catalogue of Bright Galaxies）亮星系表
R Class asteroid R 型小行星
R CMa binary 大犬 R 型双星　即：*R CMa star*
R CMa star 大犬 R 型星
R CMa variable 大犬 R 型变星　即：*R CMa star*
RC optics （＝Ritchey-Chretien optics）RC 光学系统
R Coronae Borealis star 北冕 R 型星
R CrA Dark Cloud 南冕 R 暗云
R CrB star 北冕 R 型星
R CrB variable 北冕 R 型变星　即：*R CrB star*
RCS （＝reaction control system）反冲控制系统
RC system （＝Ritchey-Chretien system）RC 系统 | RC 系統，里奇－克萊琴系統〔台〕
RC telescope （＝Ritchey-Chretien telescope）RC 望远镜 | RC 望遠鏡，里奇－克萊琴望遠鏡〔台〕
Re （＝rhenium）铼（75 号元素）
re-absorption 再吸收
reactance amplifier 电抗耦合放大器
reacting force 反作用力
reaction chain 反应链
reaction control system （＝RCS）反冲控制系统
reaction cross-section 反作用截面

reading 读数
reading error 读数误差
reading micrometer 读数测微计
reading microscope 读数测微镜|讀數顯微鏡〔台〕
read only memory （＝ROM）只读存储器
readout 读出
readout noise 读出噪声
real anomaly 真近点角 又见：*true anomaly*
real axis 实轴
real brightness 真亮度
real ellipticity 真椭率
real flattening 真扁率
real focus 实焦点
real image 实像
real singularity 实奇点
real space 实空间
real-space clustering 实空间成团性
real sun 真太阳 又见：*true sun*
real time （＝RT）实时
Real Time Data Service （＝RTDS）实时资料服务
real-time detection 实时检测
real-time determination 实时测定
real-time display 实时显示
real-time interferometry 实时干涉测量|即时干涉测量〔台〕
real-time measurement 实时测量
real-time observation 实时观测
real-time synchronization 实时同步|即時同步〔台〕
re-arrangement 重排列
rear surface 后表面（陨星） 又见：*back surface*
Reaumur thermometric scale 列氏温标
re-capture 再俘获
receding galaxy 退行星系
receding object 退行天体
receding star 退行星
receiver noise 接收机噪声
receiving antenna 接收天线
receiving area 接收面积 又见：*collecting area*
receiving system 接收系统
recession 退行 又见：*regression*
recession velocity 退行速度 又见：*velocity of recession; regression velocity*
reciprocal mass 质量倒数
reciprocal medium 倒易介质
reciprocity 倒易性

reciprocity curve 倒易性曲线
reciprocity failure 倒易性失效
reciprocity law 倒易律
re-circulating correlator 循环相关器
recoil 反冲 又见：*bounce-back*
recoil electron 反冲电子
recoil particle 反冲粒子
re-collapse 再坍缩
recombination 复合
recombination coefficient 复合系数
recombination continuum 复合连续区
recombination epoch 复合期|復合紀元〔台〕
recombination era 复合时期
recombination line 复合线
recombination line emission 复合线发射
recombination radiation 复合辐射
recombination spectrum 复合谱
recombination time 复合时间
reconnaissance satellite 侦察卫星
re-connection 重连
RECONS （＝Research Consortium on Nearby Stars）近星研究会
reconstruction of density field 密度场重构
recorder 记录仪
recording Doppler comparator 多普勒自记比较仪
recording equipment 记录设备
recording instrument 记录仪器
recording micrometer 自记测微计 又见：*self-recording micrometer*
recording microphotometer 自记测微光度计
record time 记录时刻
recovery package 回收装置
rectangular aperture 矩形孔径
rectangular coordinate 直角坐标
rectangular horn 矩形喇叭
rectangular source 矩形源
rectascension 【德】赤经 即：*RA*
rectifiable model 可矫正模型
rectification 矫频|矯正，糾正法〔台〕
rectilinear jet 直线喷流
rectilinear solution 直线解
recurrence formula 递推公式 又见：*recurrent formula*
recurrence frequency 重复频率
recurrence theorem 回归定理
recurrent burst 复现暴|再發爆發〔台〕
recurrent disturbance 复现扰动

recurrent eruption 复现爆发　又见：*recurrent explosion; recurrent outburst*
recurrent explosion 复现爆发　又见：*recurrent eruption; recurrent outburst*
recurrent flare 再现耀斑 | 再發閃焰〔台〕
recurrent formula 递推公式　又见：*recurrence formula*
recurrent nova 再发新星　又见：*repeated nova; repeating nova*
recurrent outburst 复现爆发　又见：*recurrent eruption; recurrent explosion*
recursion formula 递归公式
recursive filter 循环滤波器
recycled pulsar 再生脉冲星
re-cycling 再循环
Red Bird 朱鸟（四象）　又见：*Vermilion Bird*
red branch 红支
red coronal line 日冕红线
reddened galaxy 红化星系
reddened object 红化天体
reddened quasar 红化类星体
reddened star 红化星
reddening 红化
reddening correction 红化改正
reddening law 红化定律
reddening line 红化线 | 紅化曲線〔台〕
reddening ratio 红化比
red dot finder 红点寻星镜
red dwarf 红矮星　又见：*red dwarf star*
red dwarf flare star 红矮耀星
red dwarf star 红矮星　又见：*red dwarf*
red giant 红巨星　又见：*red giant star*
red giant branch （＝RGB）红巨星支
red giant star 红巨星　又见：*red giant*
red giant tip 红巨星支上端 | 紅巨星支尖〔台〕　又见：*red giant up*
red giant up 红巨星支上端 | 紅巨星支尖〔台〕　又见：*red giant tip*
red horizontal-branch （＝RHB）红水平支
re-distribution 再分布
red magnitude 红星等
red nebulous object （＝RNO）红色云状体
Red Oval 小红斑（木星）
Red Rectangle Nebula 红矩形星云 | 紅矩星雲〔台〕（AFCRL 618-1343）
red-sensitive plate 红敏底片
redshift 红移

redshift-angular diameter relation 红移－角径关系
redshift-apparent magnitude diagram 红移－视星等图
redshift controversy 红移论争
redshift correction 红移改正 | 紅移修正〔台〕
redshift cutoff 红移截断
redshift-distance relation 红移－距离关系
redshift distortion 红移畸变
redshift effect 红移效应
redshift evolution 红移演化
redshift-magnitude relation 红移－星等关系
redshift parameter 红移参数
redshift space 红移空间
redshift space distortion 红移空间畸变
redshift survey 红移巡天
Red Spot 大红斑 | 紅斑〔台〕（木星）
Red Spot hollow 大红斑穴（木星）
red straggler 红离散星 | 紅脫序星〔台〕
red supergiant 红超巨星
red supergiant stars 红超巨星
reduced latitude 归化纬度，归心纬度　又见：*parametric latitude*
reduced mass 折合质量 | 折合質量；約化質量〔台〕
reduced particle 约化粒子
reduced potential 约化势
reduced proper motion 归化自行
reduced radiation 减弱辐射
reduced surface density 约化面密度
reduced width 约化宽度
reduction factor 归算因子
reduction method 归算法
reduction parameter 归算参数
reduction principle 约化原理
reduction to sea-level 海平面订正　又见：*sea-level correction*
reduction to the equator 赤道订正
reduction to the meridian 子午订正
reduction to the sun 太阳订正
red variable 红变星
red white dwarf 红白矮星
re-emission 再发射 | 再發射，再輻射〔台〕
reentry module 返回舱，再入舱
reentry trajectory 再入轨道
reentry velocity 再入速度 | 重入[大氣]速度〔台〕
Rees-Sciama effect 里斯－夏默效应

Reference Catalogue of Bright Galaxies （＝RCBG）亮星系表
reference circle 基本圈 | 基本大圓〔台〕
  又见：fundamental circle
reference ellipsoid 参考椭球 | 參考橢圓體〔台〕
reference frame 参考架，参考系 又见：frame of reference
reference great circle 参考大圆
reference orbit 参考轨道
reference source 参考源 | 參考源，定標源〔台〕
reference star 参考星
reference system 参考系
reference time scale 参考时标 | 時間參考尺度〔台〕
reflectance 反射 又见：reflection; reflexion
reflectance spectrum 反射光谱 又见：reflection spectrum
reflecting power 反射本领 又见：reflective power
reflecting Schmidt telescope 反射式施密特望远镜
reflecting telescope 反射望远镜 又见：catoptric telescope; mirror telescope; reflector
reflection 反射 又见：reflectance; reflexion
reflection amplifier 反射放大器
reflection angle 反射角 又见：angle of reflection
reflection at critical angle 临界角反射
reflection coefficient 反射系数 又见：coefficient of reflection
reflection effect 反射效应
reflection filament 反射暗条
reflection grating 反射光栅
reflection nebula 反射星云
reflection spectrum 反射光谱 又见：reflectance spectrum
reflection-type maser 反射型微波激射器
reflection variable 反射变星
reflective power 反射本领 又见：reflecting power
reflectivity 反射率 | 反射本領，反射率〔台〕
reflector ①反射镜 又见：mirror ②反射望远镜
  又见：reflecting telescope; catoptric telescope; mirror telescope
reflector-antenna 反射天线
reflector-corrector 折反射望远镜
  又见：catadioptric telescope
reflex feed system 反射式馈源系统
reflexion 反射 又见：reflectance; reflection
reflex motion 反应运动 | 反折運動〔台〕
refracting angle 折射角 又见：angle of refraction

refracting telescope 折射望远镜 又见：refractor; dioptric telescope
refraction 折射
refraction anomaly 折射反常
refraction coefficient 折射系数 又见：coefficient of refraction
refraction constant 折射常数 又见：constant of refraction
refraction correction 折射改正 | [大氣] 折射改正〔台〕
refraction halo 折射晕
refraction law 折射定律
refraction table 折射表
refractive index 折射率 又见：index of refraction
refractive power 折射本领
refractivity 折射率 又见：refringence
refractor 折射望远镜 又见：refracting telescope; dioptric telescope
refractory 耐熔质
refringence 折射率 又见：refractivity
regeneration time 再生时标
Regge pole 雷吉极点
Regge trajectory 雷吉轨道
Regge-Wheeler equation 雷吉－惠勒方程
Regiomontanus crater 雷乔蒙塔努斯环形山（月球）
Regional Warning Center （＝RWC）太阳活动区域警报中心
region of corotation 公转区
regmaglypt 气印（陨石） 又见：pezograph
regolith 浮土 | 表岩屑〔台〕
Regor 天社一（船帆座γ） 又见：Al Suhail al Muhlif
regression 退行 又见：recession
regression analysis 回归分析
regression coefficient 回归系数
regression curve 回归曲线
regression equation 回归方程
regression line 回归线 又见：tropic
regression of galaxy 星系退行
regression of nodes 交点退行 又见：regression of the node; nodal regression
regression of the node 交点退行 又见：regression of nodes; nodal regression
regression period 回归周期
regression velocity 退行速度 又见：velocity of recession; recession velocity
regular cluster 规则星团
regular cluster of galaxies 规则星系团

regular galaxy 规则星系
regularity 正则性
regularity condition ① 正则条件 ② 正规条件
regularization transformation 正规化变换
regular nebula 规则星云
regular observation 常规观测　又见: *routine observation*
regular orbit 有序轨道，规则轨道 | 有序軌道〔台〕　又见: *ordered orbit*
regular perturbation system 正则摄动系统
regular point 正则点
regular satellite 规则卫星
regular variable 规则变星
regulator 调整器
Regulus 轩辕十四（狮子座 α）　又见: *Cor Leonis*
reheating 再热
reheating from inflation 暴胀再热
reheating redshift 再加热红移
Reichenbach crater 雷亨巴赫环形山（月球）
reionization 再电离
reionization era 再电离时期
reionization phase 再电离阶段
reionization redshift 再电离红移
Reiskessel crater 赖斯克塞尔陨星坑（地球）
Reissner-Nordstrom metric 赖斯内尔－诺德斯特洛姆度规
relative abundance 相对丰度
relative acceleration 相对加速度
relative aperture ① 相对口径 ② 相对孔径
relative catalogue 相对星表
relative determination 相对测定
relative energy 相对能量
relative energy distribution 相对能量分布
relative error 相对误差
relative focal length 相对焦距
relative gradient 相对梯度
relative intensity 相对强度
relative ionospheric opacity meter 电离层相对不透明度计 | 游離層相對不透明度計〔台〕　又见: *riometer*
relative luminosity factor 相对光度因子
relative measurement 相对测量
relative motion 相对运动　又见: *relative movement*
relative movement 相对运动　又见: *relative motion*
relative number 相对数
relative number of spots 黑子相对数　又见: *sunspot relative number*
relative number of sunspots 太阳黑子相对数
relative orbit 相对轨道
relative parallax 相对视差
relative photometry 相对测光 | 相對光度測量，相對測光〔台〕
relative position 相对位置
relative potential 相对势
relative proper motion 相对自行
relative radial velocity 相对视向速度
relative radio emitting power 相对射电发射功率
relative stability 相对稳定度
relative sunspot number 相对黑子数　即: *relative number of spots; sunspot relative number*
relative velocity 相对速度
relativistically expanding source 相对论性膨胀源
relativistic astrophysics 相对论天体物理学 | 相對論[性]天文物理學〔台〕
relativistic beaming 相对论性射束
relativistic bremsstrahlung 相对论性韧致辐射 | 相對論[性]制動輻射〔台〕
relativistic correction 相对论改正 | 相對論修正〔台〕
relativistic cosmology 相对论宇宙学 | 相對論[性]宇宙論〔台〕
relativistic deflection of light 相对论性光线偏折
relativistic degeneracy 相对论简并性
relativistic disk 相对论性圆面
relativistic Doppler shift 相对论性多普勒位移
relativistic Doppler term 相对论性多普勒项
relativistic effect 相对论效应
relativistic electron 相对论性电子
relativistic enthalpy 相对论性焓
relativistic expansion 相对论性膨胀
relativistic field equation 相对论场方程
relativistic fluid dynamics 相对论流体动力学
relativistic gas 相对论性气体
relativistic jet 相对论性喷射 | 相對論性噴流〔台〕
relativistic Maxwell distribution 相对论性麦克斯韦分布
relativistic mechanics 相对论力学
relativistic particle 相对论性粒子
relativistic periastron advance 相对论性近星点进动　又见: *relativistic periastron precession*
relativistic periastron precession 相对论性近星点进动　又见: *relativistic periastron advance*
relativistic perihelion advance 相对论性近日点进动　又见: *relativistic perihelion precession*

relativistic perihelion precession　相对论性近日点进动　又见：*relativistic perihelion advance*
relativistic plasma　相对论性等离子体
relativistic shift　相对论性移动
relativistic star　相对论性星
relativistic velocity　相对论性速度
relativistic zone　相对论性区
relativity principle　相对性原理
relativity shift　相对论移动 | 相對論位移〔台〕
relaxation　弛豫
relaxation effect　弛豫效应 | 鬆弛效應〔台〕
　　又见：*effect of relaxation*
relaxation rate　弛豫速率
relaxation time　弛豫时间 | 鬆弛時間〔台〕
　　又见：*time of relaxation*
relaxed cluster　弛豫星团
relay optics　中继光学系统
relevant parameter　相关参数
reliability　置信度，可靠度 | 可信度，可靠度〔台〕
relic magnetic field　遗迹磁场
relic of supernova　超新星遗迹 | 超新星殘骸〔台〕
　　又见：*supernova remnant; remnant of supernova*
relics of the Big Bang　大爆炸遗迹
reluctance　磁阻
reluctivity　磁阻率
remainder function　余函数
remaining star　残留星
remnant of nova　新星遗迹 | 新星殘骸〔台〕
remnant of supernova　超新星遗迹 | 超新星殘骸〔台〕　又见：*supernova remnant; relic of supernova*
remote clock　远距钟
remote control　遥控　又见：*distance control; telecontrol*
remote control system　遥控系统
remote equipment　遥控装置
remote operation　遥控运作
remote satellite　远距卫星
remote sensing　遥感
remote star　远距星
remote telescope　遥控望远镜
Remus　林卫二（小行星 Sylvia 的卫星）
rendezvous　会合
Renoir region　雷诺阿地区（水星）
renormalization　重正化
renormalization group　重正化群
re-orientation　重取向
repeatability　重复性

repeated nova　再发新星　又见：*recurrent nova; repeating nova*
repeating nova　再发新星　又见：*recurrent nova; repeated nova*
replenishment rate　补充率
replica of grating　复制光栅
representative point　代表点
reproducibility　复现性，复制性
repulsive force　斥力
repulsive potential　推斥势
repulsive state　推斥态
re-radiation　再辐射 | 再發射，再輻射〔台〕
Research Associate Observatoire de Besancon　【法】贝桑松联合天文台
Research Consortium on Nearby Stars　（＝RECONS）近星研究会
Research in Astronomy and Astrophysics　（＝RAA）《天文和天体物理学研究》（中国期刊）
reset　清零，复位
residual aberration　残余像差
residual calculation　残差计算
residual delay　剩余延迟
residual error　残差
residual fringe frequency　剩余条纹频率
residual fringe phase　剩余条纹相位
residual gauge mode　残留规范模
residual intensity　剩余强度 | 殘[餘]強度〔台〕
residual radial velocity　剩余视向速度 | 殘[餘]視向速度〔台〕
residual strong nuclear force　残余强核力
residual velocity　剩余速度 | 殘餘速度〔台〕
resisting medium　阻尼介质
resistive instability　电阻不稳定性
resistivity　电阻率
resolution　分辨率 | 鑑別率，解像力〔台〕
resolution box　分辨率框
resolution circle　分辨率圆
resolution ellipse　分辨率椭圆
resolution filter　分辨滤波器
resolution limit　分辨极限
resolution-limited observation　限定分辨率观测
resolution of velocity　速度分解
resolvable source　可分辨源
resolving power　分辨本领 | 鑑別本領〔台〕
resolving time　分辨时间
resonance　共振

resonance absorption 共振吸收
resonance capture 共振俘获 | 共振捕獲〔台〕
resonance condition 共振条件
resonance coupling 共振耦合
resonance cross-section 共振截面
resonance domain 共振域
resonance gap 共振空区
resonance line 共振谱线 | 共振[譜]線〔台〕
resonance orbit 共振轨道 又见：*resonant orbit*
resonance overlap theory 共振重叠理论
resonance radiation 共振辐射
resonance satellite 共振卫星
resonance spectrum 共振谱
resonance state 共振态
resonance transition 共振跃迁
resonant absorption 共振吸收
resonant cavity 共振腔
resonant frequency 共振频率
resonant gravitational wave detector 共振型引力波探测器
resonant mode 共振模
resonant orbit 共振轨道 又见：*resonance orbit*
resonant particle 共振粒子
resonant reaction 共振反应
resonant scattering spectrometer 共振散射光谱仪
resonant structure 共振结构
resonator 共振器
response curve 响应曲线 | 反應曲線〔台〕
response time 响应时间 | 反應時間〔台〕
responsiveness 响应性
responsive plane 响应平面
responsivity 响应度
rest energy 静能
rest frame 静止参考架 | 靜止坐標系〔台〕
rest mass 静质量 又见：*proper mass*
restoration technique 复原技术
restored image 复原图像
restored map 复原图
restricted cosmological principle 限制性宇宙学原理 | 狹義宇宙論原則〔台〕
restricted problem 限制性问题 | 設限問題〔台〕
restricted problem of three bodies 限制性三体问题 | 設限三體問題〔台〕 又见：*restricted three-body problem*
restricted theory 限制性理论

restricted three-body problem 限制性三体问题 | 設限三體問題〔台〕 又见：*restricted problem of three bodies*
resultant velocity 合速度
re-synchronization 再同步
Ret （＝Reticulum）网罟座
retardation 减速
retarded baseline 推迟基线
retarded Green function 推迟格林函数
retarded motion 减速运动
retarded offset 推迟偏置
retarded potential 推迟势
reticle 十字丝 又见：*reticule*
Reticon 雷地康管
Reticon array 雷地康管阵
reticule 十字丝 又见：*reticle*
Reticulum （＝Ret）网罟座
Retina Nebula 视网膜星云（IC 4406）
retractable enclosure 伸缩式围罩
retrograde merger 逆行并合
retrograde motion 逆行 又见：*retrogression; antecedence; backward movement*
retrograde orbit 逆行轨道
retrograde stationary 逆留
retrogression 逆行 又见：*retrograde motion; antecedence; backward movement*
retroreflector 后向反射器
retro-rocket 制动火箭
return-beam tube 回束摄像管
return wave 回波，反向波 又见：*echo wave; back wave*
Reull Vallis 鲁尔谷（火星）
Reuven Ramaty High Energy Solar Spectroscopic Imager （＝RHESSI）太阳高能光谱成像探测器
reverberation mapping 反响映射
reversal 反变，自食 | 反變〔台〕
reversal of polarity 极性反转 又见：*polarity reversal*
reverse azimuth 反方位角 又见：*inverse azimuth*
reverse shock 反向激波
reverse sign 反号
reversibility 可逆性
reversible film 反转软片 | 可能轉換片，兩面用轉片〔台〕
reversible level 回转水准
reversible mounting 回转装置
reversible pendulum 可倒摆
reversible process 可逆过程

reversible transit circle 回转子午环
reversing layer 反变层 又见: inversion layer
reversing prism 倒像棱镜 又见: inverting prism
Revised Harvard Photometry （＝RHP）哈佛测光星表修订版 | 哈佛恆星測光表修訂版〔台〕 又见: Harvard Revised Photometry
Revised New General Catalogue of Nonstellar Astronomical Objects （＝RNGC）非星天体新总表修订版
Revised Shapley-Ames Catalog （＝RSA Catalog）沙普利－艾姆斯星表修订版
Revista Mexicana De Astronomia Y Astrofisica 【西】《墨西哥天文学与天体物理学杂志》
revival group 重现群
revival sunspot ①重现黑子 ②重现太阳黑子
revival sunspot group 重现黑子群
revolution 公转, 绕转 | 公轉〔台〕
revolution of micrometer screw 测微计螺旋周值
Reynold's number 雷诺数
RF （＝radio frequency）射频
RFT （＝richest-field telescope）特广视场望远镜
R galaxy R 星系
RGB （＝red giant branch）红巨星支
RGO （＝Royal Greenwich Observatory）格林尼治皇家天文台 | 格林 [威治] 皇家天文台〔台〕
RGU color system RGU 颜色系统
RGU photometry RGU 测光
Rh （＝rhodium）铑（45 号元素）
RHB （＝red horizontal-branch）红水平支
Rhea 土卫五
Rhea Mons 瑞亚山（金星）
Rheasilvia basin 雷亚希尔维亚盆地（灶神星）
rhenium （＝Re）铼（75 号元素）
rheonomic constraint 不稳定约束
rheostat 变阻器 又见: varistor
RHESSI （＝Reuven Ramaty High Energy Solar Spectroscopic Imager）太阳高能光谱成像探测器
rhodium （＝Rh）铑（45 号元素）
rhombic antenna 菱形天线 又见: diamond antenna
RHP （＝Revised Harvard Photometry）哈佛测光星表修订版 | 哈佛恆星測光表修訂版〔台〕
rhumb line 恒向线 又见: loxodrome
rhythmic time signal 科学式时号 | 節奏時號〔台〕
RIAF （＝radiatively inefficient accretion flow）辐射低效吸积流
Riccioli crater 里乔利环形山（月球）
Ricci scalar 里奇标量
Ricci tensor 里奇张量
Riccius crater 利玛窦环形山（月球）
Ricco crater 里科环形山（月球）
rice grain 米粒组织（太阳） 又见: granulation
rice-grain effect 米粒效应
Richardson crater 里查孙环形山（月球）
Richardson-Lucy algorithm 理查森－卢西算法
rich cluster ①富星系团 又见: rich cluster of galaxies; rich galaxy cluster ②富星团 又见: rich star cluster; rich cluster of stars
rich cluster of galaxies 富星系团 又见: rich galaxy cluster; rich cluster
rich cluster of stars 富星团 又见: rich star cluster; rich cluster
RICH detector （＝ring imaging Cherenkov detector）环形成像切伦科夫探测器
richest-field telescope （＝RFT）特广视场望远镜
rich-field adapter 广角转接器
rich galaxy cluster 富星系团 又见: rich cluster of galaxies; rich cluster
richness class 富度级
richness index 富度指数
richness parameter 富度参数
rich star cluster 富星团 又见: rich cluster of stars; rich cluster
ridge 山脊
ridge horn 脊形喇叭
ridge square horn 皱纹方形喇叭
Riemann-Christoffel curvature tensor 黎曼－克里斯托弗尔曲率张量
Riemann crater 黎曼环形山（月球）
Riemann ellipsoid 黎曼椭球体
Riemann space 黎曼空间 | 瑞曼空間〔台〕
Riemann space curvature 黎曼空间曲率
Riemann tensor 黎曼张量
Rigel 参宿七（猎户座 β）
right angle prism 直角棱镜
right ascension （＝R.A.; RA）赤经
right ascension circle 赤经度盘 又见: right ascension setting circle
right ascension setting circle 赤经度盘 又见: right ascension circle
right-hand circular polarization 右旋圆偏振
right-handed coordinate system 右手坐标系
right-handed polarization 右旋偏振
right-handed reference frame 右手参考架
right-hand rule 右手定则
right sphere 正交天球 | 垂直球〔台〕
rigid body rotation 刚体自转

rigid crust　刚性壳
rigid earth　刚性地球
rigid halo　刚体晕
rigidity　刚性
Rigil Kent　南门二（半人马座α）　又见：*Rigil Kentaurus*
Rigil Kentaurus　南门二（半人马座α）　又见：*Rigil Kent*
Rigl al Awwa　亢宿增七（室女座μ）
Riley crater　赖利环形山（金星）
rill　沟纹，溪 | 细沟〔台〕　又见：*rille*
rille　沟纹，溪 | 细沟〔台〕　又见：*rill*
rille crater　沟纹环形山 | 细沟環形山〔台〕
Rima Agatharchides　【拉】阿格瑟奇德斯溪（月球）
Rima Agricola　【拉】阿格里科拉溪（月球）
Rima Archytas　【拉】阿契塔溪（月球）
Rima Ariadaeus　【拉】阿里亚代乌斯溪 | 阿麗阿黛月溪〔台〕（月球）
Rima Artsimovich　【拉】阿尔齐莫维奇溪（月球）
Rima Billy　【拉】比伊溪（月球）
Rima Birt　【拉】伯特溪（月球）
Rima Brayley　【拉】布雷利溪（月球）
Rima Cauchy　【拉】柯西溪（月球）
Rima Delisle　【拉】德利尔溪（月球）
Rima Diophantus　【拉】丢番图溪（月球）
Rima Draper　【拉】德雷伯溪（月球）
Rimae Alphonsus　【拉】阿方索溪（月球）
Rimae Apollonius　【拉】阿波罗尼奥斯溪（月球）
Rimae Archimedes　【拉】阿基米德溪（月球）
Rimae Aristarchus　【拉】阿利斯塔克溪（月球）
Rimae Arzachel　【拉】阿尔扎赫尔溪（月球）
Rimae Atlas　【拉】阿特拉斯溪（月球）
Rimae Bode　【拉】波得溪（月球）
Rimae Bürg　【拉】比格溪（月球）
Rimae Daniell　【拉】丹聂耳溪（月球）
Rimae Darwin　【拉】达尔文溪（月球）
Rimae Doppelmayer　【拉】多佩尔迈尔溪（月球）
Rimae Fresnel　【拉】菲涅尔溪（月球）
Rimae Gassendi　【拉】伽桑狄溪（月球）
Rimae Grimaldi　【拉】格里马尔迪溪（月球）
Rimae Gutenberg　【拉】谷登堡溪（月球）
Rimae Hase　【拉】哈泽溪（月球）
Rimae Hevelius　【拉】赫维留溪（月球）
Rimae Hypatia　【拉】希帕蒂娅溪（月球）
Rimae Janssen　【拉】让桑溪（月球）
Rimae Littrow　【拉】利特罗夫溪（月球）
Rimae Maclear　【拉】麦克利尔溪（月球）
Rimae Maestlin　【拉】梅斯特林溪（月球）
Rimae Maupertuis　【拉】莫佩尔蒂溪（月球）
Rimae Menelaus　【拉】米尼劳斯溪（月球）
Rimae Mersenius　【拉】梅森溪（月球）
Rimae Parry　【拉】帕里溪（月球）
Rimae Pettit　【拉】佩蒂特溪（月球）
Rimae Plato　【拉】柏拉图溪（月球）
Rimae Plinius　【拉】普利纽斯溪（月球）
Rimae Prinz　【拉】普林茨溪（月球）
Rimae Ramsden　【拉】拉姆斯登溪（月球）
Rimae Riccioli　【拉】里乔利溪（月球）
Rimae Ritter　【拉】里特尔溪（月球）
Rimae Römer　【拉】罗默溪（月球）
Rimae Sosigenes　【拉】索西琴尼溪（月球）
Rimae Sulpicius Gallus　【拉】加卢斯溪（月球）
Rimae Theaetetus　【拉】特埃特图斯溪（月球）
Rima Euler　【拉】欧拉溪（月球）
Rimae Vasco da Gama　【拉】达·伽马溪（月球）
Rimae Zupus　【拉】祖皮溪（月球）
Rima Flammarion　【拉】弗拉马里翁溪（月球）
Rima Galilaei　【拉】伽利略溪（月球）
Rima G.Bond　【拉】乔·邦德溪（月球）
Rima Hadley　【拉】哈德利溪（月球）
Rima Hesiodus　【拉】赫西奥德斯溪（月球）
Rima Hyginus　【拉】希吉努斯溪 | 海金努斯月溪〔台〕（月球）
Rima Marius　【拉】马里乌斯溪（月球）
Rima Messier　【拉】梅西叶溪（月球）
Rima Oppolzer　【拉】奥波尔策溪（月球）
Rima Planck　【拉】普朗克溪 | 卜朗克月溪〔台〕（月球）
Rima Schroedinger　【拉】薛定谔溪 | 薛丁格月溪〔台〕（月球）
Rima Sharp　【拉】夏普溪（月球）
Rima Sheepshanks　【拉】希普尚克斯溪（月球）
Rima Sirsalis　【拉】希萨尔斯溪 | 希薩利斯月溪〔台〕（月球）
Rima Suess　【拉】休斯溪（月球）
Rima T.Mayer　【拉】托·迈耶溪（月球）
rima（复数：rimae）　【拉】沟纹，溪 | 裂缝〔台〕　即：*rill; rille*
Rindler space　林德勒空间
ring antenna　环状天线
ring aperture　环状孔径　又见：*annular aperture*
ring arc　环弧

ring array　环状天线阵
ring current　环电流
ring division　环缝
ringed barred galaxy　有环棒旋星系　又见：*ringed barred spiral galaxy*
ringed barred spiral galaxy　有环棒旋星系　又见：*ringed barred galaxy*
ringed planet　有环行星
ring galaxy　环状星系
ring imaging Cherenkov detector　（＝RICH detector）环形成像切伦科夫探测器
ring micrometer　环状测微计 | 環形測微計〔台〕
ring-moon　行星环卫星　又见：*moom*
ring mountain　环形山 | 環形山，[陨石] 坑洞，火山口〔台〕　又见：*crater*
Ring Nebula　指环星云（M57）
ring nebula　环状星云　又见：*annular nebula*
ring of Jupiter　木星环　又见：*Jupiter's ring; Jovian ring*
ring of Neptune　海王星环　又见：*Neptune's ring; Neptunian ring*
ring of Saturn　土星环 | 土星 [光] 環〔台〕　又见：*Saturnian ring*
ring of Uranus　天王星环　又见：*Uranian ring; Uranus' ring*
ring plain　环形平原
ring-plane crossing　环面穿越
ring radio telescope　带形射电望远镜
ring system　光环系
ring transformation　环变换
riometer　电离层相对不透明度计 | 游離層相對不透明度計〔台〕　又见：*relative ionospheric opacity meter*
rise phase　上升阶段
rise time　上升时间
rising limit　出限
rising point　上升点
rising prominence　上升日珥
Ritchey-Chretien optics　（＝RC optics）RC 光学系统
Ritchey-Chretien system　（＝RC system）RC 系统 | RC 系統，里奇－克萊琴系統〔台〕
Ritchey-Chretien telescope　（＝RC telescope）RC 望远镜 | RC 望遠鏡，里奇－克萊琴望遠鏡〔台〕
Ritz combination principle　里茨组合原则 | 瑞茲加成原則〔台〕
Ritz crater　里茨环形山（月球）
rizalite　玻璃陨石

R Lep　（＝Hind's Crimson star）欣德深红星
RM　（＝rotation measure）旋转量
R magnitude　R 星等
RMC　（＝Rosette Molecular Cloud）玫瑰分子云
R Monocerotis　麒麟 R 星云 | 麒麟 [座]R 星雲〔台〕（NGC 2261）又见：*R Monocerotis nebula*
R Monocerotis nebula　麒麟 R 星云 | 麒麟 [座]R 星雲〔台〕（NGC 2261）又见：*R Monocerotis*
rms　（＝root mean square）均方根
rms deviation　（＝root mean square deviation）均方差 | 均方 [根] 差〔台〕
rmse　（＝mean error）中误差
rms value　（＝root mean square value）均方根值
rms velocity　（＝root mean square velocity）均方根速度
Rn　（＝radon）氡（86 号元素）
RNGC　（＝Revised New General Catalogue of Nonstellar Astronomical Objects）非星天体新总表修订版
RNO　（＝red nebulous object）红色云状体
Roberts crater　罗伯茨环形山（月球）
Robertson crater　罗伯逊环形山（月球）
Robertson-Walker metric　罗伯逊－沃克规度 | 勞勃遜－厄克規度〔台〕
Robotic Autonomous Observatory　（＝RAO）程控自主天文台
robotic observatory　程控天文台
robotic telescope　程控望远镜 | 自動望遠鏡〔台〕
robust estimation　稳健估计
Rocard scattering　罗卡散射
Rocca crater　罗卡环形山（月球）
Rochechouart crater　罗什舒阿尔陨星坑（地球）
Roche crater　洛希环形山（月球）
Roche Division　洛希环缝（土星）
Roche limit　洛希极限
Roche lobe　洛希瓣
Roche-lobe overflow　洛希瓣超流
Roche model　洛希模型
Roche radius　洛希半径
Roche surface　洛希面
Roche zone　洛希带
rocket astronomy　火箭天文学
rocket-borne instrument　箭载仪器
rocket-borne telescope　箭载望远镜
rocket infrared astronomy　火箭红外天文学
rocket mechanism　火箭机制
rocket train　多级火箭　又见：*multi-stage rocket; step rocket*

rocket ultraviolet astronomy 火箭紫外天文学
rocket X-ray astronomy 火箭 X 射线天文学
rocket γ-ray astronomy 火箭 γ 射线天文学
rocking mirror 摆动反光镜 | 擺動面鏡〔台〕
rocking telescope 摆动望远镜
rock magnetism 岩石磁性
rockoon 气球发射高空探测火箭
rocky core 岩核
rocky dwarf 岩质矮行星
rocky planet 岩质行星
ROE （=Royal Observatory, Edinburgh）爱丁堡皇家天文台
rolling hangar 机库
ROM （=read only memory）只读存储器
Roman calendar 罗马历
Roman era 罗马纪元
Roman indiction 罗马小纪
Roman Republican Calendar 罗马共和历
Rome Observatory 罗马天文台
Romulus 林卫一（小行星 Sylvia 的卫星）
Romulus and Remus crater 罗慕路斯与雷穆斯环形山（土卫四）
Ronchi test 戎奇检验
Röntgen crater 伦琴环形山（月球）
Röntgen Satellite （=ROSAT）伦琴 X 射线天文台 | 倫琴 X 光天文台〔台〕（德国天文卫星）
Rood-Sastry classification 路德－沙斯特里分类
Rood-Sastry system 路德－沙斯特里分类系统
Rood-Sastry type 鲁德－萨斯特瑞型
roof prism 屋脊棱镜
Rooftop 危宿（二十八宿）
Room 房宿（二十八宿） 又见: Chamber
Root 氐宿（二十八宿）
root mean square （=rms）均方根
root mean square deviation （=rms deviation）均方差 | 均方 [根] 差〔台〕
root mean square error 中误差 又见: mean error
root mean square value （=rms value）均方根值
root mean square velocity （=rms velocity）均方根速度
root of unit 单位根
Roque de los Muchachos Observatory 穆查丘斯罗克天文台（西班牙）
Rosalind 天卫十三 | 天[王]衛十三〔台〕
ROSAT （=Röntgen Satellite）伦琴 X 射线天文台 | 倫琴 X 光天文台〔台〕（德国天文卫星）
Rosemary Hill Observatory 罗斯玛丽山天文台

Rosenberger crater 罗森贝格尔环形山（月球）
Rosenbluth potential 罗森布卢特势
Rosetta （=Rosetta spacecraft）罗塞塔号探测器
Rosetta spacecraft （=Rosetta）罗塞塔号探测器
rosette 玫瑰花结（日芒）
rosette method 系列法
Rosette Molecular Cloud （=RMC）玫瑰分子云
Rosette Nebula 玫瑰星云 | 薔薇星雲，玫瑰星雲〔台〕（NGC 2237）
Rossby number 罗斯贝数
Rossby wave 罗斯贝波 | 洛士贝波〔台〕
Ross camera 罗斯照相机
Rosseland crater 罗斯兰环形山（月球）
Rosseland mean 罗斯兰平均
Rosseland mean absorption coefficient 罗斯兰平均吸收系数
Rosseland mean opacity 罗斯兰平均不透明度
Rosseland opacity 罗斯兰不透明度
Rosseland theorem 罗斯兰定理
Rossiter effect 罗西托效应 | 洛西特效應〔台〕
Rossi X-ray Timing Explorer （=RXTE）罗西 X 射线时变探测器 | 羅西 X 光時變探測器〔台〕（美国）
Ross objective 罗斯物镜
Ross telescope 罗斯望远镜
Rotanev 瓠瓜四（海豚座 β）
rotating axis 自转轴 又见: axis of rotation; rotation axis
rotating body 旋转体
rotating chopper 旋转斩波器
rotating disk 旋转圆面，旋转盘
rotating-lobe interferometer 旋转瓣干涉仪
rotating magnetic dipole 旋转磁偶极子
rotating object 旋转天体
rotating polarizer 旋转起偏器
rotating radio transient （=RRAT）自转型暂现射电源
rotating sector 旋转遮光板
rotating shell 旋转气壳
rotating shutter 旋转快门 | 轉動快門〔台〕
rotating star 自转星
rotating variable 自转变星 又见: rotational variable
rotating wave-plate 旋转波片
rotation 自转 又见: axial rotation
rotational aperture synthesis 自转孔径综合
rotational axis precession 自转轴进动 又见: spin precession

rotational broadening　自转致宽
rotational constant　转动常数
rotational contour　转致轮廓　又见：rotational profile
rotational energy　转动能　又见：energy of rotation
rotational evolution　自转演化
rotational inclination　自转轴倾角
rotational inertia　转动惯量　又见：rotation inertia
rotational instability　自转不稳定性
rotational line　转动谱线
rotational modulation　自转调制
rotational modulation collimator　旋转调制准直器
rotational momentum　自转角动量
rotational parallax　自转视差
rotational period　自转周期　又见：rotation period; axial period
rotational phase　自转相位　又见：rotation phase
rotational profile　转致轮廓　又见：rotational contour
rotational state　转动态
rotational structure　转动结构
rotational transition　转动跃迁
rotational variable　自转变星　又见：rotating variable
rotational velocity　自转速度
rotational velocity-space velocity relation　自转速度－空间速度关系
rotational velocity-spectral type relation　自转速度－光谱型关系
rotation axis　自转轴　又见：rotating axis; axis of rotation
rotation center　旋转中心
rotation curve　自转曲线
rotation-curve decomposition　旋转曲线分解
rotation effect　自转效应
rotation frequency　自转频率
rotation inertia　转动惯量　又见：rotational inertia
rotation measure　（＝RM）旋转量
rotation number　旋转数
rotation of binary　双星绕转
rotation of ecliptic　黄道转动
rotation of galaxy　星系自转　又见：galactic rotation
rotation of galaxy cluster　星系团旋转
rotation of the earth　地球自转　又见：earth rotation
rotation of the Galaxy　银河系自转　又见：Galactic rotation
rotation of the line of apsides　拱线转动
rotation period　自转周期　又见：rotational period; axial period
rotation phase　自转相位　又见：rotational phase
rotation pole　自转极　又见：pole of rotation
rotation rate　自转速率
rotation spectrum　转动光谱
rotation synthesis　自转综合孔径｜自轉合成〔台〕
rotation temperature　转动温度｜自轉溫度〔台〕
rotation-vibration band　转动－振动谱带｜轉振譜帶〔台〕
rotation-vibration spectrum　转动－振动光谱
rotation-vibration state　转动－振动态
rotator　自转天体，转子
roton　旋子
Rotten Egg Nebula　臭蛋星云（OH 231.84 +4.22）
rounded value　舍入值｜約整值〔台〕
rounding error　舍入误差｜約整誤差〔台〕　又见：roundoff error
roundoff　舍入
roundoff error　舍入误差｜約整誤差〔台〕　又见：rounding error
Routh limit　劳斯极限
routine observation　常规观测　又见：regular observation
row error　行误差
Rowland-circle spectrometer　罗兰圆光谱仪
Rowland crater　罗兰环形山（月球）
Rowland ghost line　罗兰鬼线
Rowland grating　罗兰光栅
Royal Astronomical Society　（＝RAS）英国皇家天文学会
Royal Astronomical Society of Canada　（＝RASC）加拿大皇家天文学会
Royal Greenwich Observatory　（＝RGO）格林尼治皇家天文台｜格林[威治] 皇家天文台〔台〕
Royal Observatory, Edinburgh　（＝ROE）爱丁堡皇家天文台
Rozhdestvenskiy crater　罗日杰斯特文斯基环形山（月球）
R-parity conservation　R 宇称守恒
r-process　（＝rapid neutron capture process）快过程，r 过程｜中子快捕獲過程，r 過程〔台〕
RQPNMLK color system　RQPNMLK 颜色系统
RRab Lyrae star　天琴 RRab 型星｜天琴[座]RRab 型星〔台〕
RRa Lyrae star　天琴 RRa 型星｜天琴[座]RRa 型星〔台〕
RRAT　（＝rotating radio transient）自转型暂现射电源
RRb Lyrae star　天琴 RRb 型星｜天琴[座]RRb 型星〔台〕

RRc Lyrae star　天琴 RRc 型星 | 天琴 [座]RRc 型星〔台〕

R-region　R 区　又见：*R zone*

RR Lyrae star　天琴 RR 型变星，天琴 RR 型星 | 天琴 [座]RR[型] 變星〔台〕　又见：*RR Lyrae variable*

RR Lyrae variable　天琴 RR 型变星，天琴 RR 型星 | 天琴 [座]RR[型] 變星〔台〕　又见：*RR Lyrae star*

RRs Lyrae star　天琴 RRs 型星 | 天琴 [座]RRs 型星〔台〕

RRs variable　RRs 型变星

RR Telescopii star　望远镜 RR 型星 | 望遠鏡 [座]RR 型星〔台〕

RR Telescopii variable　望远镜 RR 型变星　即：*RR Telescopii star*

RSA Catalog　（＝Revised Shapley-Ames Catalog）沙普利－艾姆斯星表修订版

RS Canum Venaticorum binary　（＝RS CVn binary）猎犬 RS 型双星 | 獵犬 [座]RS 型雙星〔台〕

RS Canum Venaticorum star　（＝RS CVn star）猎犬 RS 型星 | 獵犬 [座]RS 型星〔台〕

RS Canum Venaticorum variable　（＝RS CVn variable）猎犬 RS 型变星

RS Coupling　（＝Russell-Saunders coupling）罗素－桑德斯耦合

RS CVn binary　（＝RS Canum Venaticorum binary）猎犬 RS 型双星 | 獵犬 [座]RS 型雙星〔台〕

RS CVn star　（＝RS Canum Venaticorum star）猎犬 RS 型星 | 獵犬 [座]RS 型星〔台〕

RS CVn variable　（＝RS Canum Venaticorum variable）猎犬 RS 型变星

R star　R 型星　又见：*R-type star*

RT　（＝real time）实时

RTDS　（＝Real Time Data Service）实时资料服务

RTG　（＝radioisotope thermoelectric generator）放射性同位素热电发生机

RT Ser star　巨蛇 RT 型星 | 巨蛇 [座]RT 型星〔台〕

RT Ser variable　巨蛇 RT 型变星　即：*RT Ser star*

R-type star　R 型星　又见：*R star*

Ru　（＝ruthenium）钌（44 号元素）

rubber second　负闰秒　又见：*negative leap second*

rubble pile structure　砾石堆结构

rubidium　（＝Rb）铷（37 号元素）

rubidium clock　铷钟

rubidium gas cell　铷气泡 | 銣氣胞〔台〕

rubidium maser　铷微波激射器 | 銣邁射〔台〕

rubidium-strontium dating　铷－锶纪年

Rubin-Ford effect　鲁宾－福特效应

ruby　红宝石

Ruchba　天津增三十七（天鹅座 o²）

Ruchbah　阁道三（仙后座 δ）

Rudolphine table　鲁道夫星表（开普勒编制）

Ruianzhongxue　瑞安中学（小行星 4073 号）

Rukbat　天渊三（人马座 α）　又见：*Alrami*

Rukh　天津二（天鹅座 δ）　又见：*Urakhga*

ruled grating　刻划光栅

ruled-surface map　直纹曲面图

Rumford crater　拉姆福德环形山（月球）

runaway coalescence　逃逸结合

runaway electron　逃逸电子

runaway growth　失控式生长

runaway star　速逃星

run error　行差

Runge-Kutta method　龙格－库塔法

Running Chicken Nebula　半人马 λ 星云（IC 2944）

running mean　移动平均

running penumbral wave　半影行波

Runrun Shaw　邵逸夫（小行星 2899 号）

rupes　【拉】峭壁，陡岩 | 峭壁〔台〕

Rupes Altai　阿尔泰峭壁 | 阿勒泰峭壁〔台〕（月球）

Rupes Boris　【拉】鲍里斯峭壁（月球）

Rupes Cauchy　【拉】柯西峭壁（月球）

Rupes Kelvin　【拉】开尔文峭壁（月球）

Rupes Liebig　【拉】李比希峭壁（月球）

Rupes Mercator　【拉】墨卡托峭壁（月球）

Rupes Recta　【拉】直壁（月球）

Rupes Toscanelli　【拉】托斯卡内利峭壁（月球）

Russell crater　罗素环形山（月球）

Russell diagram　罗素图 | 赫羅圖〔台〕

Russell Gap　罗素环缝（土星）

Russell mixture　罗素混合物

Russell model　罗素模型

Russell-Saunders coupling　（＝RS Coupling）罗素－桑德斯耦合

Russell-Vogt theorem　罗素－福格特定理 | 羅素－沃克定理〔台〕

Russia　俄罗斯（小行星 232 号）

Russian Academy of Sciences　（＝RAS）俄罗斯科学院

Russian Virtual Observatory　（＝RVO）俄罗斯虚拟天文台

ruthenium　（＝Ru）钌（44 号元素）

Rutherford crater 卢瑟福环形山（月球）
Rutilicus 河中，天市右垣一（武仙座 β）
　又见：*Korneforos; Kornephoros*
RVO （=Russian Virtual Observatory）俄罗斯虚拟天文台
RVS （=radial-velocity spectrometer）视向速度仪
RV Tauri star 金牛 RV 型星 | 金牛 [座]RV 型 [变] 星〔台〕
RV Tauri variable 金牛 RV 型变星　即：*RV Tauri star*
RW Aurigae star （=RW Aur star）御夫 RW 型星 | 御夫 [座]RW 型星〔台〕
RW Aurigae variable （=RW Aur variable）御夫 RW 型变星　即：*RW Aur star*
RW Aur star （=RW Aurigae star）御夫 RW 型星 | 御夫 [座]RW 型星〔台〕
RW Aur variable （=RW Aurigae variable）御夫 RW 型变星　即：*RW Aur star*
RWC （=Regional Warning Center）太阳活动区域警报中心
RXTE （=Rossi X-ray Timing Explorer）罗西 X 射线时变探测器 | 羅西 X 光時變探測器〔台〕（美国）
Rydberg constant 里德伯常数 | 芮得柏常數〔台〕
Rydberg correction 里德伯改正
Rydberg frequency 里德伯频率
Ryle Telescope 瑞尔射电望远镜
Rynin crater 雷宁环形山（月球）
R zone R 区　又见：*R-region*

S

S (＝sulfur/sulphur）硫（16 号元素）
S0 galaxy　S0 星系 | S0 型 [旋涡] 星系〔台〕
　　即：lenticular galaxy
S&T （＝Sky and Telescope）《天空与望远镜》（美国期刊）
SA （＝spherical aberration）球差 | 球 [面像] 差〔台〕
SAAO （＝South African Astronomical Observatory）南非天文台
Sabik　宋，天市左垣十一（蛇夫座 η）
Sacajawea Patera　萨卡贾维亚托边火山（金星）
Sachs Patera　萨克斯托边火山（金星）
Sachs-Wolfe effect　萨克斯－沃尔夫效应
Sacramento Peak Observatory （＝SPO）萨克拉门托峰天文台
Sacrobosco crater　萨克罗博斯科环形山（月球）
Sadachbia　坟墓二（宝瓶座 γ）　又见：Sadalachbia
Sadalachbia　坟墓二（宝瓶座 γ）　又见：Sadachbia
Sadalbari　离宫二（飞马座 μ）
Sadalmelik　危宿一（宝瓶座 α）
saddle point　鞍点
Sadir　天津一（天鹅座 γ）　又见：Sador; Sadr
Sador　天津一（天鹅座 γ）　又见：Sadir; Sadr
Sadr　天津一（天鹅座 γ）　又见：Sadir; Sador
Saenger crater　森格尔环形山（月球）
SAF （＝Societe Astronomique de France）【法】法国天文学会
safety screw　安全螺旋
SAFIR （＝Single Aperture Far-Infrared Observatory）单孔径远红外天文台
Safronov number　萨夫罗诺夫数
Sa galaxy　Sa 型星系
SagDEG （＝Sagittarius Dwarf Elliptical Galaxy）人马矮椭圆星系
SagDIG （＝Sagittarius Dwarf Irregular Galaxy）人马不规则矮星系
Sagitta （＝Sge）天箭座

sagittal　弧矢
sagittal plane　矢面
Sagittarious Arm　人马臂（银河）
Sagittarius　①（＝Sgr）人马座②（＝Sgr）人马宫，析木，寅宫
Sagittarius Arm　人马臂
Sagittarius cloud　人马星云
Sagittarius Dwarf Elliptical Galaxy （＝SagDEG）人马矮椭圆星系
Sagittarius Dwarf Galaxy　人马矮星系 | 人馬 [座] 矮星系〔台〕
Sagittarius Dwarf Irregular Galaxy （＝SagDIG）人马不规则矮星系
Sagittarius star cloud　人马恒星云
Sagittarius Window Eclipsing Extrasolar Planet Search （＝SWEEPS）人马窗口凌星系外行星搜索
Saha crater　萨哈环形山（月球）
Saha equation　萨哈方程 | 沙哈方程〔台〕
Saha formula　萨哈公式 | 沙哈公式〔台〕
Saha ionization equation　萨哈电离方程 | 沙哈游離方程〔台〕
Saha ionization formula　萨哈电离公式 | 沙哈游離公式〔台〕
SAI （＝Societa Astronomica Italiana）【意】意大利天文学会
sailing star　蓬星
Saint Martin crater　圣马丁陨星坑（地球）
Saiph　参宿六（猎户座 κ）
Sakata model　坂田模型
Sakharov condition　萨哈罗夫条件
Sakharov oscillation　萨哈罗夫振荡
Sakigake　先驱号空间探测器
Salm　离宫五（飞马座 τ）
Salpeter function　萨尔皮特函数
Salpeter mass function　萨尔皮特质量函数
Salpeter process　萨尔皮特过程 | 索彼得過程〔台〕

Salpeter time　萨尔皮特时间

SALT　（＝Southern African Large Telescope）南非大型望远镜 | 南非大型遠鏡〔台〕

Salyut　礼炮号（苏联空间站）

samarium　（＝Sm）钐（62 号元素）

SAMBO　（＝simultaneous auroral multi-balloon observation）多气球同步极光观测

SAMPEX　（＝Solar, Anomalous, and Magnetospheric Particle Explorer）太阳异常性/磁层粒子探索者

sample averaging　取样平均

sampler　取样器，采样器

sample size　样本量

sample time　取样时间

sampling clock　取样钟

sampling frequency　取样频率

sampling rate　取样率，采样率

sampling theorem　取样定理

sampling up-the-ramp　斜升采样

Sandage-Loeb test　桑德奇－勒布检验

sand-bank model　沙堆模型

sand clock　沙漏 | 沙漏 [鐘]〔台〕　又见：*sand glass; hour glass; sandy clock*

sand dune　沙丘

sand glass　沙漏 | 沙漏 [鐘]〔台〕　又见：*sand clock; hour glass; sandy clock*

S Andromedae　仙女 S 超新星 | 仙女 [座]S 超新星〔台〕

sandy clock　沙漏 | 沙漏 [鐘]〔台〕　又见：*sand clock; sand glass; hour glass*

Sanford crater　桑福德环形山（月球）

Sanshui　三水（小行星 3509 号）

SAO　（＝Smithsonian Astrophysical Observatory）史密松天体物理台 | 史密松天文物理台〔台〕

Sao　海卫十一

SAO Catalog　SAO 星表

SAO RAS　（＝Special Astrophysical Observatory）特设天体物理台

SAO Star Catalogue　（＝Smithsonian Astrophysical Observatory Star Catalog）SAO 星表

Sapas Mons　萨帕斯山（金星）

Sappho　赋神星（小行星 80 号）

SAR　（＝synthetic-aperture radar）综合孔径雷达 | 合成孔徑雷達〔台〕

Sargas　尾宿五（天蝎座 θ）

Sarin　魏，天市左垣一（武仙座 δ）

Saros　沙罗周期　又见：*Saros cycle*

Saros cycle　沙罗周期　又见：*Saros*

Saros series　沙罗食周

Sarton crater　萨尔顿环形山（月球）

SAS　（＝Small Astronomical Satellite）SAS 小天文卫星

SAS-A　乌呼鲁 X 射线卫星 | 自由號 [X 光天文衛星]〔台〕　又见：*Uhuru; SAS I*

SAS I　乌呼鲁 X 射线卫星 | 自由號 [X 光天文衛星]〔台〕　又见：*Uhuru; SAS-A*

Saskia crater　萨斯基亚环形山（金星）

SAT　（＝stepped atomic time）跳跃原子时 | 步進原子時〔台〕

satellite　卫星

satellite altimetry　卫星测高

satellite antenna　人卫天线

satellite astronomy　卫星天文学

satellite borne instrument　人卫运载仪器

satellite-borne telescope　星载望远镜

satellite camera　人卫照相机

satellite camera with quadraxial mounting　四轴式人卫照相机

satellite camera with quadraxial tracking frame　四轴式人卫跟踪照相机

satellite coronagraphy　人卫日冕照相术

satellite Doppler navigation system　人卫多普勒导航系统

satellite Doppler tracking　卫星多普勒测量 | 衛星都卜勒測量〔台〕

satellite eclipse　卫星食　又见：*eclipse of satellite*

satellite galaxy　伴星系　又见：*companion galaxy*

satellite geodesy　卫星大地测量学

satellite laser ranging　（＝SLR）卫星激光测距 | 衛星雷射測距〔台〕

satellite life　人卫寿命

satellite line　伴线

satellite navigation　卫星导航

satellite nebula　伴星云

satellite of Jupiter　木卫 | 木 [星] 衛 [星]〔台〕　又见：*Jovian satellite; Jupiter's satellite*

satellite of Mars　火卫　又见：*Martian satellite; Martian moon*

satellite of Neptune　海卫 | 海 [王] 衛〔台〕　又见：*Neptunian satellite; Neptune's satellite*

satellite of Pluto　冥卫　又见：*Plutonian satellite*

satellite of Saturn　土卫　又见：*Saturnian satellite; Saturn's satellite*

satellite of Uranus　天卫 | 天 [王] 衛〔台〕　又见：*Uranian satellite*

Satellite per Astronomia X　（＝SAX; BeppoSAX）【意】贝波 X 射线天文卫星

satellite ranging　人卫测距
satellite-sensed event　卫星可测事件
satellite sunspot　卫星黑子
satellite system　卫星系 | 衛星系〔统〕〔台〕
satellite-to-satellite tracking　（＝SST）人卫－人卫跟踪
satellite tracking　人卫跟踪 | 衛星追蹤〔台〕
satellite tracking camera　人卫跟踪照相机
satellite trail　卫星痕迹
satellite transit　卫星凌行星
satellite triangulation　卫星三角测量
satelloid　卫星体
satelloon　球载卫星　又见：balloon satellite
saturated absorption　饱和吸收
saturated absorption line　饱和吸收线
saturated line　饱和谱线 | 飽和 [譜] 線〔台〕
saturation effect　饱和效应
saturation factor　饱和因子
saturation function　饱和函数
saturation pump　饱和抽运
Saturn　土星，镇星
Saturn-crossing asteroid　越土小行星
Saturnian ring　土星环 | 土星 [光] 環〔台〕
　　又见：ring of Saturn
Saturnian ringlet　土星细环
Saturnian satellite　土卫　又见：Saturn's satellite; satellite of Saturn
saturnicentric coordinate　土心坐标
saturnicentric latitude　土心纬度
saturnicentric longitude　土心经度
saturnicentric ring latitude　土心环纬度
saturnicentric ring longitude　土心环经度
saturnigraphic coordinate　土面坐标
saturnigraphic latitude　土面纬度
saturnigraphic longitude　土面经度
saturnigraphy　土面学
Saturn Nebula　土星状星云（NGC 7009）
Saturn shine　土星反照
Saturn's satellite　土卫　又见：Saturnian satellite; satellite of Saturn
saucer crater　碟形环形山
sausage-type instability　腊肠型不稳定性
Saussure crater　索绪尔环形山（月球）
Sawiskera　萨维斯克拉（小行星 Teharonhiawako 的卫星）
SAX　（＝Satellite per Astronomia X）【意】贝波 X 射线天文卫星

SB　（＝spectroscopic binary）分光双星
Sb　（＝antimony）锑（51 号元素）
SB 1　（＝single-lined spectroscopic binary）单谱分光双星，单谱双星 | 單線 [分光] 雙星〔台〕
SB 2　（＝double-lined [spectroscopic] binary）双谱 [分光] 双星 | 復綫 [分光] 雙星〔台〕
SBa galaxy　SBa 型星系
S band　S 波段
SBb galaxy　SBb 型星系
SBc galaxy　SBc 型星系
SB galaxy　（＝barred spiral galaxy）棒旋星系，SB 型星系
SBI　① （＝short baseline interferometer）短基线干涉仪　② （＝short baseline interferometry）短基线干涉测量
Sc　（＝scandium）钪（21 号元素）
scalar antenna　标量天线
scalar CMB anisotropy　标量微波背景各向异性
scalar curvature　标量曲率
scalar curvature of space-time　时空标量曲率
scalar density　标量密度
scalar field　标量场 | 純量場〔台〕
scalar fluctuation　标量涨落
scalar horn　梯形喇叭
scalar mode　标量模
scalar perturbation　标量扰动
scalar potential　标量势 | 純量勢〔台〕
scalar-tensor gravity　标量－张量引力
scalar-tensor theory　标量－张量理论 | 純量－張量理論〔台〕
scalar virial theorem　标量位力定理
scale division　刻度　又见：graduation
scale factor　① 标度因子　② 比例因子
scale factor of the universe　① 宇宙标度因子　又见：cosmic scale factor　② 宇宙标度因子
scale-free solution　尺度无关解
Scale-free spectrum　尺度无关功率谱
scale height　标高 | 尺度高〔台〕
scale-invariance　标度不变性
scale-invariant perturbation　尺度不变扰动
scale-invariant spectrum　尺度不变功率谱
scale length　标长
scale pair　标度星对
scale value　标度值
Scaliger crater　斯卡利杰环形山（月球）
scaling　定标
scaling factor　定标因子

scaling law　定标定律
scaling of string networks　弦网络标度化
scaling relation　标度关系
scandium　（＝Sc）钪（21号元素）
scanner　扫描器
scanning great circle　扫描大圆
scanning modulation collimator　扫描调制准直器
scanning spectrometer　扫描分光仪 | 掃描分光計〔台〕
scaphe　仰仪
scattered disk object　（＝SDO）SDO型天体，散盘型天体
scattered light　散射光
scattered radiation　漫辐射，散射辐射
scattered reflection　漫反射　又见：*diffuse reflection*
scattering　散射
scattering angle　散射角
scattering atmosphere　散射大气
scattering by interstellar media　（＝ISS）星际散射
scattering coefficient　散射系数
scattering cross-section　散射截面
scattering diffusion　散射扩散
scattering effect　散射效应
scattering kernel　散射核
scattering power　散射本领
SCD　（＝swept charge device）扫式电荷器件
Scd galaxy　Scd型星系
sCDM model　（＝standard Cold Dark Matter model）标准冷暗物质模型
Sceptrum　九斿增四（波江座53）
Sc galaxy　Sc型星系
Schaeberle crater　舍贝勒环形山（月球）
Schatzman mechanism　沙兹曼机制
Scheat　室宿二（飞马座β）　又见：*Menkib*
Schechter function　谢克特函数
Schechter's law　谢克特定律
Schedar　王良四（仙后座α）　又见：*Schedir*
Schedir　王良四（仙后座α）　又见：*Schedar*
Scheiner crater　沙伊纳环形山（月球）
schematic diagram　示意图
Scheuer-Readhead hypothesis　朔伊尔－里德黑德假说
Schiaparelli crater　斯基亚帕雷利环形山（火星）
schiefspiegler telescope　席夫施皮格勒望远镜
Schiller crater　席勒环形山（月球）
Schilt photometer　西尔特光度计
Schlesinger crater　施莱辛格环形山（月球）
Schliemann crater　谢里曼环形山（月球）
Schlüter crater　施吕特环形山（月球）
Schmidt aplanatic camera　施密特齐明球透镜照相机
Schmidt camera　施密特照相机 | 施密特[式]照相機〔台〕
Schmidt-Cassegrain configuration　施密特－卡塞格林光路系统
Schmidt-Cassegrain telescope　施密特－卡塞格林望远镜 | 施密特－卡塞格林[式]望遠鏡〔台〕
Schmidt correcting plate　施密特改正镜 | 施密特改正板〔台〕　又见：*Schmidt corrector; Schmidt plate*
Schmidt corrector　施密特改正镜 | 施密特改正板〔台〕　又见：*Schmidt correcting plate; Schmidt plate*
Schmidt galactic model　施密特银河系模型
Schmidt-Holsapple scaling law　施密特－霍尔萨普尔定标定律
Schmidt law　施密特定律
Schmidt model　施密特模型
Schmidt optics　施密特光学系统
Schmidt plate　施密特改正镜 | 施密特改正板〔台〕　又见：*Schmidt correcting plate; Schmidt corrector*
Schmidt telescope　施密特望远镜 | 施密特[式]望遠鏡〔台〕
Schneller crater　施内勒尔环形山（月球）
Schonberg-Chandrasekhar limit　申贝格－钱德拉塞卡极限　即：*Chandrasekhar-Schoenberg limit*
Schottky barrier mixer　肖特基势垒混频管
Schrödinger crater　薛定谔环形山（月球）
Schröter effect　施洛特效应
Schubert crater　舒伯特环形山（月球）
Schumacher crater　舒马赫环形山（月球）
Schumann-Runge continuum　舒曼－龙格连续谱
Schuster crater　舒斯特环形山（月球）
Schuster mechanism　舒斯特机制
Schuster problem　舒斯特问题
Schuster-Schwarzschild atmosphere　舒斯特－施瓦西大气
Schwabe cycle　施瓦贝循环
Schwarz inequality　施瓦茨不等式
Schwarzschild black hole　施瓦西黑洞 | 史瓦西黑洞〔台〕
Schwarzschild coordinate　施瓦西坐标
Schwarzschild-Couder Cherenkov telescope　施瓦西－库德型切伦科夫望远镜
Schwarzschild crater　施瓦西环形山（月球）
Schwarzschild criterion　施瓦西判据

Schwarzschild distribution function 施瓦西分布函数
Schwarzschild filling factor 施瓦西填充因子
Schwarzschild geometry 施瓦西几何
Schwarzschild index 施瓦西指数 | 史瓦西指數〔台〕
Schwarzschild lens 施瓦西透镜
Schwarzschild metric 施瓦西度规
Schwarzschild-Milne equation 施瓦西－米尔恩方程
Schwarzschild model 施瓦西模型
Schwarzschild radius 施瓦西半径 | 史瓦西半徑〔台〕
Schwarzschild singularity 施瓦西奇点 | 史瓦西奇異點〔台〕
Schwarzschild sphere 施瓦西球 | 史瓦西球〔台〕
Schwarzschild telescope 施瓦西望远镜
Schwarzschild time 施瓦西时间
Schwarzschild velocity ellipsoid 施瓦西速度椭球
Schwassmann-Wachmann 1 施瓦斯曼－瓦赫曼 1 号彗星
Schwassmann-Wachmann 2 施瓦斯曼－瓦赫曼 2 号彗星
Schwassmann-Wachmann 3 施瓦斯曼－瓦赫曼 3 号彗星
sciagraphy 星影计时法
sciametry 日月食理论
Science and Technology SATellite （＝STSAT）科学技术卫星（韩国）
scintillation 闪烁 又见：*twinkling*
scintillation counter 闪烁计数器 又见：*scintillometer*
scintillation effect 闪烁效应
scintillation spectrum 闪烁谱
scintillation telescope 闪烁望远镜
scintillator neutrino detector 闪烁体式中微子探测器
scintillometer 闪烁计数器 又见：*scintillation counter*
Scl （＝Sculptor）玉夫座
scleronomic constraint 稳定约束
Scl-For galaxy （＝Sculptor-Fornax galaxy）玉夫－天炉星系
SCNA （＝sudden cosmic noise absorption）宇宙噪声突降 | 宇宙雜訊突然吸收，宇宙雜訊突減〔台〕
Sco ① （＝Scorpius）天蝎座 ② （＝Scorpius）天蝎宫，大火，卯宫
Sco-Cen association 天蝎－半人马星协 又见：*Scorpius-Centaurus association*

S-component 缓变分量 又见：*slowly varying component*
scopulus 杂相陡岩 | 不規則斷崖〔台〕
Scoresby crater 斯科斯比环形山（月球）
Scorpius ① （＝Sco）天蝎座 ② （＝Sco）天蝎宫，大火，卯宫
Scorpius-Centaurus Association 天蝎－半人马星协
Scorpius-Centaurus association 天蝎－半人马星协 又见：*Sco-Cen association*
Scorpius OB 1 天蝎 OB 1 星协
scotopic curve 适暗曲线
scotopic eye 适暗眼
scotopic luminosity curve 暗光光度曲线
scotopic vision 晨昏视觉 | 暗黑視覺，桿體視覺〔台〕
Scott crater 斯科特环形山（月球）
Scott effect 斯科特效应 | 斯卡特效應〔台〕
SCP （＝Supernova Cosmology Project）超新星宇宙学计划
scratch dial 刻线日晷 又见：*scratch sundial*
scratch sundial 刻线日晷 又见：*scratch dial*
screen brightness 屏亮度
screening effect 屏蔽效应 又见：*shielding effect*
screw micrometer 螺旋测微器 | 螺旋測微計〔台〕
screw microscope 螺旋测微镜
screw value 螺旋周值
SC star SC 型星
Sct （＝Scutum）盾牌座
SCUBA （＝Submillimetre Common User Bolometer Array）亚毫米波普通用户辐射热计阵列
Sculptor （＝Scl）玉夫座
Sculptor Dwarf Galaxy 玉夫座矮星系
Sculptor-Fornax galaxy （＝Scl-For galaxy）玉夫－天炉星系
Sculptor galaxy 玉夫星系 又见：*Sculptor system*
Sculptor group 玉夫星系群 | 玉夫[座]星系群〔台〕
Sculptorids 玉夫流星群 | 玉夫[座]流星雨〔台〕
Sculptor supercluster 玉夫超星系团
Sculptor system 玉夫星系 又见：*Sculptor galaxy*
Sculptor void 玉夫巨洞 | 玉夫[座]巨洞〔台〕
Scutulum 海石二（船底座ι） 又见：*Aspidiske; Turyeish; Turais*
Scutum （＝Sct）盾牌座
Scutum Star Cloud 盾牌座恒星云
SD （＝semi-diameter）半径

s.d. （＝standard deviation）标准偏差 | 標準差〔台〕
SDC （＝Stellar Data Center）恒星资料中心
SDD （＝silicon drift detector）硅漂移探测器
Sd galaxy Sd 型星系
SDM （＝Solar Diameter Monitor）太阳直径监测器
SDO ① （＝Solar Dynamics Observatory）太阳动力学观测台 ② （＝scattered disk object）SDO 型天体，散盘型天体
S Dor star 剑鱼 S 型星
SDSS （＝Sloan Digital Sky Survey）斯隆数字化巡天 | 斯隆數位化巡天〔台〕
S-duality S 对偶
SE （＝standard earth）标准地球
Se （＝selenium）硒（34 号元素）
s.e. （＝standard error）标准误差
SEA （＝sudden enhancement of atmospherics）天电突增
Seagull Nebula 海鸥星云（IC 2177）
sea interferometer 海岸干涉仪 又见：*cliff interferometer*
sea level 海平面
sea-level correction 海平面订正 又见：*reduction to sea-level*
Search for Extraterrestrial Intelligence （＝SETI）地外文明探索 | 地 [球] 外智慧生物搜寻〔台〕
search for extraterrestrial life 地外生命搜寻 | 地 [球] 外生命搜寻〔台〕
searching ephemeris 寻星历表 | 搜尋星曆表〔台〕
Seares crater 西尔斯环形山（月球）
Seares' formulae 西尔斯公式
Seashell Galaxy 扇贝星系（PGC 48894）
seasonal fluctuation 季节性起伏
seasonal phenomenon 季节性现象
seasonal variation 季节性变化
SEB （＝south equatorial belt）南赤道带（木星）
Secchi classification 塞奇分类
Sechenov crater 谢切诺夫环形山（月球）
second 秒
secondary axis 副轴
secondary blockage 副镜遮拦
secondary circle 副圈
secondary clock 子钟 又见：*slave clock*
secondary component 次星 | 伴星〔台〕 又见：*secondary star*

secondary cosmic radiation 次级宇宙线 | 衍生宇宙線〔台〕 又见：*secondary cosmic rays*
secondary cosmic rays 次级宇宙线 | 衍生宇宙線〔台〕 又见：*secondary cosmic radiation*
secondary crater 次级陨击坑
secondary distance indicator 次级示距天体
secondary eclipse 次食
secondary electron 次级电子
secondary element 次级元素
secondary emission 次级发射
secondary fluctuation 次起伏
secondary flux of atomspheric γ-rays 大气 γ 射线次级流量
secondary flux of cosmic rays 宇宙线次级流量
secondary focus 副焦点
secondary image 副像
secondary maximum 次极大
secondary minimum 次极小
secondary mirror 副镜
secondary-mirror modulation 副镜调制
secondary-mirror modulator 副镜调制器
secondary mirror support tower 副镜支承塔
secondary mirror tripod 副镜三脚架
secondary optical axis 副光轴
secondary particle 次级粒子
secondary pattern 次级方向图
secondary period 副周期 | 次周期〔台〕
secondary prominence 次生日珥
secondary radiant 次辐射点
secondary reflector 副反射面 又见：*subreflector*
secondary resonance 次共振 | 次级共振〔台〕
secondary source 次源
secondary spectrum 二级光谱 又见：*second order spectrum*
secondary standard 二级标准
secondary star 次星 | 伴星〔台〕 又见：*secondary component*
secondary surface 副表面，辅助面
secondary tail 副彗尾
secondary wave 次波
second beat 秒拍
second contact 食既，第二切
second control 秒控制
second cosmic velocity 第二宇宙速度
second difference 二次差
Second Index Catalogue of Nebulae and Clusters of Stars （＝IC II）星云星团新总表续编 II

second of arc	（＝arcsec）角秒　又见：arc second
second of time	时秒
second order day number	二阶日数
second order phase transition	二级相变
second order spectrum	二级光谱　又见：secondary spectrum
second order term	二阶项
second pendulum	秒摆
second quantization	二次量子化
second summation	二次和
second transit	第二次中天
second type surface	副表面
sector	象限仪　又见：quadrant
sector boundary	扇形边界
sectorial harmonics	扇谐函数
sector shutter	扇形快门
sector structure	扇形结构　又见：fan-like structure
secular aberration	长期光行差
secular acceleration	长期加速度 | 長期加速〔台〕
secular equation	长期差　又见：secular inequality
secular evolution	长期演化
secular inequality	长期差　又见：secular equation
secular instability	长期不稳定性
secular motion	长期运动
secular parallax	长期视差
secular perturbation	长期摄动　又见：long-term perturbation
secular polar motion	长期极移 | 長年極移〔台〕
secular precession	长期进动
secular resonance	长期共振
secular stability	长期稳定性 | 長期穩定[度]〔台〕　又见：long-term stability
secular tendency	长期趋势　又见：secular trend
secular term	长期项
secular trend	长期趋势　又见：secular tendency
secular variable	长期变星　又见：secular variation
secular variation	长期变星　又见：secular variable
Secunda Giedi	牛宿二（摩羯座 α^2）　又见：Algiedi Secunda; Algedi Secunda
SED	（＝spectral energy distribution）光谱能量分布，能谱分布 | 光譜能量分佈〔台〕
Sedan Crater	色当弹坑（美国）
sedimentary rock	沉积岩，水成岩
Sedna	赛德娜（小行星 90377 号）
Sednoid	类赛德娜天体
Sedov phase	谢多夫阶段
seed-field model	种子场模型
seed nucleus	籽核 | 基核〔台〕
seeing	视宁度 | 視相，大氣寧靜度〔台〕
seeing disk	视宁圆面，视影圆面 | 視影盤面〔台〕
seeing disturbance	视宁度扰动
seeing image	视影，视宁像 | 視影〔台〕
seeing management	视宁度控管
seeing monitor	视宁度监测器
Seeliger paradox	西利格佯谬
seesaw mechanism	跷跷板机制
Segin	阁道二（仙后座 ε）　又见：Navi
Seginus	招摇（牧夫座 γ）　又见：Haris
segmented mirror	拼合镜面
segmented mirror telescope	（＝SMT）拼合镜面望远镜
segmented optics	拼合镜面光学系统
segregation	分层
SEGUE	（＝Sloan Extension for Galactic Understanding and Exploration）SEGUE 巡天（SDSS 巡天子项目）
Seidel crater	赛德尔环形山（月球）
seism	地震　又见：earthquake
seismic belt	地震带
seismic focus	震源　又见：seismic origin; seismic source
seismicity	地震活动
seismic origin	震源　又见：seismic focus; seismic source
seismic source	震源　又见：seismic focus; seismic origin
seismic surface wave	地震面波
seismic wave	地震波
seismic zone	地震区
seismo-astronomy	地震天文
seismogram	震波图
seismograph	地震仪
seismology	地震学
seismometer	地震计
seismophysics	地震物理
seismoscope	地震示波仪
Sekhmet	狮女星（小行星 5381 号）
selection effect	选择效应　又见：selective effect
selection function	选择函数
selection rule	选择定则
selective absorption	选择吸收 | 選擇[性]吸收〔台〕
selective absorption coefficient	选择吸收系数　又见：coefficient of selective absorption
selective effect	选择效应　又见：selection effect
selective emission	选择发射
selective emission coefficient	选择发射系数

selective light pressure　选择光压 | 選擇 [性] 光壓〔台〕
selective radiation　选择辐射
selective radiation process　选择辐射过程
selective scattering　选择散射 | 選擇 [性] 散射〔台〕
selective scattering coefficient　选择散射系数 | 選擇 [性] 輻射係數〔台〕
selectivity　选择性
SELENE　（＝Selenological and Engineering Explorer）月球学与工程探测器，月亮女神号（日本月球探测器）
selenium　（＝Se）硒（34 号元素）
selenium cell　硒光电池
selenium photometer　硒光度计
selenocentric coordinate　月心坐标
selenodesy　月面测量 | 月面測量學〔台〕
selenofault　月面断层
selenogony　月球起源说
selenograph　月面图　又见：*lunar map; selenographic chart; selenographic map*
selenographic chart　月面图　又见：*selenograph; lunar map; selenographic map*
selenographic coordinate　月面坐标
selenographic latitude　月面纬度
selenographic longitude　月面经度
selenographic map　月面图　又见：*selenograph; lunar map; selenographic chart*
selenography　月面学
selenoid　月球卫星　又见：*lunar satellite*
Selenological and Engineering Explorer　（＝SELENE）月球学与工程探测器，月亮女神号（日本月球探测器）　又见：*KAGUYA*
selenologist　月球学家
selenology　月球学 | 月面學〔台〕
selenomorphology　月面形态学
selenophysical phenomenon　月球物理现象
selenophysics　月球物理 | 月球物理學〔台〕
selenothermy　月热学
self-absorbed radiation　自吸收辐射
self-absorbed synchrotron　自吸收同步辐射
self-absorption　自吸收
self-adjoint　自伴
self-calibration method　自定标方法
self-coherent camera　自相干相机
self-consistency　自洽性
self-consistent field　自洽场
self-consistent solution　自洽解
self-containment　自持

self-correction　自校正
self-correlation　自相关　又见：*autocorrelation*
self-energy　自能量
self-field　原场 | 自身場〔台〕
self-gravitating body　自引力天体
self-gravitation　自引力 | 自吸引，自身重力〔台〕
self-inductance　自感
self-induction　自感应 | 自感 [應]〔台〕
self-luminance　自发光
self-luminous body　自发光体
self-luminous train　自发光余迹 | 自發光 [流星] 餘跡〔台〕
self-magnetic field　自磁场
self-noise　原噪声
self-propagating star formation　自传播恒星形成
self-recording micrometer　自记测微计　又见：*recording micrometer*
self-regulation　自调节
self-reproducing universe　自增殖宇宙
self-scanning　自扫描
self-shielding　自屏蔽
self-similar collapse　自相似坍缩
self-similar evolution　自相似演化
self-similar growth　自相似增长
self-similarity　自相似性
self-similar model　自相似模型
self-stimulated emission　自受激发射
self-supporting grating　自持光栅
Selove receiver　塞洛夫接收机
Selqet　蝎女星（小行星 136818 号）
selsyn　自整角机　又见：*autosyn*
Semele　化女星（小行星 86 号）
semi-amplitude　半变幅，半振幅 | 半變幅〔台〕
semi-analytical galaxy formation　半解析星系形成
semi-analytical model　半解析模型
semi-annual variation　半年变化
semi-barium star　半钡星
semiclassical quantum gravity　半经典量子引力
semiconductor X-ray detector　半导体 X 射线探测器
semiconductor γ-ray detector　半导体 γ 射线探测器
semi-convection　部分对流
semidefinite time　半确定时
semi-detached binary　半接双星 | 半分離雙星〔台〕　又见：*semidetached binary*

semidetached binary 半接双星 | 半分離雙星〔台〕
 又见: *semi-detached binary*
semi-diameter （＝SD）半径
semi-diurnal arc 半日弧 | 半日周弧〔台〕
semi-diurnal tide 半日潮
semi-duration 半期
semi-forbidden line 半禁线
semi-infinite atmosphere 半无限大气
semi-latus rectum 半通径
semi-lunar period 半月周期
semi-lunar variation 半月变化
semi-major axis 半长轴，半长径 | 半長軸〔台〕
semi-meridian 半子午线
semi-minor axis 半短轴，半短径 | 半短徑〔台〕
semi-quadrate 半象限差（45°）
semi-regular variable （＝SR variable）半规则变星
semi-revolution 半周
sense 指向 | 旋向〔台〕
sense of revolution 公转方向
sense of rotation 自转方向
sensibility 灵敏度   又见: *sensitivity*
sensing device 传感器   又见: *sensor*
sensitivity 灵敏度   又见: *sensibility*
sensitivity constant 灵敏度常数
sensitivity curve 灵敏度曲线
sensitivity distribution 灵敏度分布
sensitivity equation 灵敏度方程
sensitization 敏化 | 增感，敏化〔台〕
 又见: *hypersensitization; hypersensitizing*
sensitizer 敏化剂
sensitogram 感光图 | 感光度圖〔台〕
sensitometer 感光计 | 露光計〔台〕
 又见: *actinometer*
sensitometry 感光测量
sensor 传感器   又见: *sensing device*
SEP （＝solar energetic particle）太阳高能粒子
separated source 分离源
separate-nucleus hypothesis 分离核心假说
separatrice 分界
Sepedet 天狗周（埃及古历）
septet ① 七重线   又见: *heptet* ② 七重态
sequence of giants 巨星序
sequence of stellar spectra 恒星光谱序
sequence of sub-dwarfs 亚矮星序 | 次矮星序〔台〕
 又见: *subdwarf sequence*
sequence of sub-giants 亚巨星序

sequence of supergiants 超巨星序
sequence of white dwarfs 白矮星序
sequential estimation 序贯估计
sequential processing 序贯处理
Ser （＝Serpens）巨蛇座
serendipitous X-ray source 偶遇 X 射线源 | 偶遇 X 光源〔台〕
serendipitous γ-ray source 偶遇 γ 射线源
Serenitatis Basin 澄盆地
series limit 系限
series spectrum 线系谱
Serpens （＝Ser）巨蛇座
Serpens Caput 巨蛇头
Serpens Cauda 巨蛇尾
Serpens Dark Cloud 巨蛇暗云
Serpens molecular cloud 巨蛇分子云
serpentine 蛇纹石
Serrurier truss 赛路里桁架
Sersic index 塞西克指数
Sersic profile 塞西克轮廓
service module （＝SM）服务舱
Serving-maid 须女   又见: *Lowly concubine of emperor; Waiting-maid*
servo control receiver 伺服接收机
servo-drive 伺服驱动器
servo-system 伺服系统
sesqui-quadrate 一个半象限差（135°）
SEST （＝Sweden-ESO Submillimetre Telescope）瑞典－欧南台亚毫米波望远镜，SEST 亚毫米波望远镜
Se star Se 星，S 型发射线星
Setebos 天卫十九
SETI （＝Search for Extraterrestrial Intelligence）地外文明探索 | 地[球]外智慧生物搜尋〔台〕
Set of Identifications, Measurements, and Bibliography for Astronomical Data （＝SIMBAD）辛巴达天文数据库
setting circle 定位度盘
setting limit 落限
setting point 下落点
set-up error 安装误差   又见: *mounting error*
Seven Luminaries 七曜，七政
Seven Sisters 七姊妹星   即: *Pleiades*
Seven Stars 星宿（二十八宿）
Severe Cold 大寒（节气）   又见: *Greater Cold*
Sex （＝Sextans）六分仪座
sexagesimal cycle 六十干支周

sexagesimal system 六十分制
Sextans （＝Sex）六分仪座
Sextans Dwarf Galaxy 六分仪座矮星系
Sextans Ridge 六分仪脊
sextant ① 六分仪 ② 纪限仪
sextet ① 六重线 ② 六重态
sextile 六十度角距
sextile aspect 六分方位
sextuple star 六合星
Seyfert crater 赛弗特环形山（月球）
Seyfert galaxy 赛弗特星系 | 西佛星系〔台〕
Seyfert nucleus 赛弗特星系核
Seyfert's Sextet 赛弗特六重星系 | 西佛六重星系〔台〕
SFA （＝sudden field anomaly）场强突异
SFD （＝sudden frequency drift）频率突漂
SFE （＝star-formation efficiency）恒星形成效率
sferics 天电 又见: *atmospherics; stray*
sfermion 标量费米子
SFR （＝star-formation rate）恒星形成率
SFTS （＝standard frequency and time signal）标准时频 | 標準時號〔台〕
SFU （＝solar flux unit）太阳流量单位 | 太陽通量單位〔台〕
S galaxy （＝spiral galaxy）旋涡星系 | 螺旋[狀]星系〔台〕
SGB （＝subgiant branch）亚巨星支 | 次巨星序〔台〕
Sge （＝Sagitta）天箭座
SGMC （＝supergiant molecular cloud）超巨分子云
SGR （＝soft gamma repeater）软 γ 射线复现源 | 軟 γ[射線]重覆爆發源〔台〕
Sgr ① （＝Sagittarius）人马座 ② （＝Sagittarius）人马宫, 析木, 寅宫
SGT （＝soft γ-ray transient）软 γ 射线暂现源 | 軟 γ 射線瞬變源〔台〕
SH （＝southern hemisphere）南半球
Shaanxi 陕西（小行星 2263 号）
Shack-Hartmann wavefront sensor 沙克－哈特曼波前传感器
shadow band 影带 又见: *shadow belt*
shadow band structure 影带结构
shadow belt 影带 又见: *shadow band*
shadow cone 影锥
shadow definer 影符, 景符 | 景符〔台〕
shadow eclipse 影食
shadow matter 虚影物质

Shadow-Planet 太岁，岁阴 又见: *counter-Jupiter*
shadow transit 卫影凌行星 又见: *transit of shadow*
Shakespeare region 莎士比亚地区（水星）
Shakura-Sunyaev disc 沙库拉－苏尼阿耶夫盘
Sham 左旗一（天箭座 α） 又见: *Alsahm*
Shandong 山东（小行星 2510 号）
Shandongdaxue 山东大学（小行星 29467 号）
Shane Telescope 沙因望远镜
Shanghai 上海（小行星 2197 号）
Shantou 汕头（小行星 3139 号）
Shao 邵正元（小行星 1881 号）
shaped Cassegrain antenna 整形卡塞格林天线
shaped dual reflector 异形双反射面
shaped-pupil coronagraph 光瞳整形星冕仪
shaped reflector 异形反射面
shape factor 形状因子
shape function 形状函数
shape relation 形状关系
Shapley-Ames catalogue 沙普利－艾姆斯星系表
Shapley concentration 沙普利聚集度
Shapley supercluster 沙普利超星系团
Sharatan 娄宿一（白羊座 β） 又见: *Sheratan*
Sharonov crater 沙罗诺夫环形山（月球）
sharp image 锐像
Sharpless catalogue 沙普利斯亮星云表
sharp line 锐谱线
sharpness 清晰度
sharp series 锐线系
shatter cone 碎裂锥岩
Shaula 尾宿八（天蝎座 λ）
Shayn crater 沙因环形山（月球）
shear distortion 剪切畸变
shear-dominated encounter 剪切主导的交会
shear field 剪切场
shearing 剪切
shearing effect 剪切效应
shearing interferometer wavefront sensor 剪切干涉仪波前传感器
shear instability 剪切不稳定性
shear layer 剪切层
shear stabilization 剪切致稳
shear tensor 剪切张量
shear viscosity 剪切黏性
shear viscosity coefficient 剪切黏性系数
sheath 鞘
Sheliak 渐台二（天琴座 β） 又见: *Shelyak; Shiliak*

shell burning	壳层燃烧
shell crossing	壳层交叉，壳层穿越
shell galaxy	壳状星系
shell remnant	壳状遗迹
shell source	壳源
shell-source model	壳源模型
shell star	气壳星　又见：*envelope star*
shell structure	壳层结构
shell supernova remnant	壳状超新星遗迹
Shelyak	渐台二（天琴座 β）　又见：*Sheliak; Shiliak*
shemu	埃及古历夏季
Shenchunshan	沈君山（小行星 202605 号）
Shen Guo	沈括（小行星 2027 号）
Shenzhen	深圳（小行星 2425 号）
Shenzhou	神舟（小行星 8256 号）
shepherding satellite	牧羊犬卫星　又见：*shepherd satellite; shepherd moon*
shepherd moon	牧羊犬卫星　又见：*shepherd satellite; shepherding satellite*
shepherd satellite	牧羊犬卫星　又见：*shepherding satellite; shepherd moon*
Sheratan	娄宿一（白羊座 β）　又见：*Sharatan*
shergottite	辉熔长石无球粒陨石
Shergotty	（＝SNC meteorite）SNC 陨星　又见：*Nakhla and Chassigny meteorite*
Shergotty meteorite	谢尔戈蒂陨星
Sherrington crater	谢灵顿环形山（月球）
SHG	（＝spectroheliography）太阳单色光照相术
shielding effect	屏蔽效应　又见：*screening effect*
shielding variable	屏蔽变星　又见：*obscured variable*
shield volcano	盾状火山
shift theorem	移位定理
shift vector	位移矢量
Shihwingching	施永青（小行星 64289 号）
Shiliak	渐台二（天琴座 β）　又见：*Sheliak; Shelyak*
shimmer	闪光　又见：*flash; flashing light; lightening flash; lightening*
Shinsei	新星太阳卫星
Shirakatsi crater	希拉卡齐环形山（月球）
Shi Shen crater	石申环形山（月球）
Shizu-Mao	始祖懋（小行星 132825 号）
SHM	（＝simple harmonic motion）简谐运动
shock front	激波波前　又见：*shock wave-front*
shock heating	激波加热
shock temperature	激波温度
shock wave	激波｜衝擊波，震波〔台〕
shock wave-front	激波波前　又见：*shock front*
shock-wave ionization	激波电离
Shoemaker crater	①舒梅克环形山（月球）②舒梅克陨星坑（地球）
shooting star	流星　又见：*meteor; falling star*
Shorrtt clock	雪特钟　又见：*Shortt clock*
short-arc method	短弧法
short-axis tube orbit	短轴管形轨道
short baseline interferometer	（＝SBI）短基线干涉仪
short baseline interferometry	（＝SBI）短基线干涉测量
short burst	短期爆发　又见：*short duration burst*
short count	短期积日
Short crater	肖特环形山（月球）
short distance scale	短距离尺度
short duration burst	短期爆发　又见：*short burst*
short-lived star	短寿命星
short-period Cepheid	短周期造父变星｜短周期造父[型]變星〔台〕
short-period comet	短周期彗星
short-period perturbation	短周期摄动
short-period term	短周期项
short-period variable	短周期变星
short-range field	短程场
short-range force	短程力
Shortt clock	雪特钟　又见：*Shorrtt clock*
short-term stability	短期稳定度
short wave branch	短波支
short-wave fade-out	（＝SWF）短波衰退
short-wavelength radiation	短波辐射
Shorty crater	肖蒂环形山（月球）
shot noise	散粒噪声｜散粒雜音〔台〕
shower meteor	属群流星
shower radiant	流星雨辐射点
shut-off	断路
shutter blade	快门叶片
shutter correction	快门改正
shutter sweep correction	快门扫掠改正
shuttle	航天飞机　又见：*space shuttle*
Shuttle Pointed Autonomous Research Tool for Astronomy	（＝SPARTAN）斯巴达人号
SI	（＝international system of units）国际单位制
Si	（＝silicon）硅（14 号元素）
Si:As infrared array	硅掺砷红外面阵
Si:Ga infrared array	硅掺镓红外面阵
Siarnaq	土卫二十九
Sichuan	四川（小行星 2215 号）

Sickle of Leo 狮子座镰刀（星组）
SiC mirror 碳化硅镜
SID （＝sudden ionospheric disturbance）电离层突扰 | 游離層突發性擾動〔台〕
siddhanta 《印度历数书》
sidelit view 侧照视像
sidelobe 旁瓣
sidelobe level 旁瓣电平
sidelobe subtraction 减旁瓣法
sidereal astronomy 恒星天文
sidereal chronometer 恒星时计 | 恆星 [時] 時計〔台〕
sidereal clock 恒星钟
sidereal day 恒星日　又见：*equinoctial day*
sidereal drive 转仪钟　又见：*drive clock; driving clock; clock drive; sidereal driving clock*
sidereal driving clock 转仪钟　又见：*drive clock; driving clock; clock drive; sidereal drive*
sidereal hour 恒星小时
sidereal hour-angle 恒星时角
sidereal midnight 恒星时子夜
sidereal minute 恒星时分
sidereal month 恒星月
sidereal noon 恒星时正午
sidereal period 恒星周期
sidereal rate 恒星周天转速
sidereal revolution 恒星周
sidereal second 恒星时秒
sidereal time （＝ST）恒星时
sidereal universe 恒星世界
sidereal year 恒星年
siderite 铁陨星，铁陨石，陨铁 | 隕鐵，鐵 [質] 隕石〔台〕　又见：*aerosiderite; iron meteorite; meteoric iron*
siderograph 恒星时记时器
siderolite 石铁陨星 | 石隕鐵〔台〕　又见：*lithosiderite; stony-iron meteorite*
siderophyre 古铜鳞英石铁陨石
siderostat 定星镜
Siding Spring Observatory （＝SSO）赛丁泉天文台（澳大利亚）
Siedentopf crater 西登托普夫环形山（月球）
Sierpinski crater 谢尔平斯基环形山（月球）
Sif Mons 西夫山（金星）
sighting 照准
sighting telescope 瞄准望远镜
sighting-tube 窥管，望筒　又见：*dioptra*

sighting-tube ring 四游仪　又见：*movable sighting set; component of the four displacements*
sight-line velocity 视向速度
sign 宫　又见：*house*
signal 信号，讯号
signal analysis 信号分析
signal band 信号频带 | 信號 [頻] 帶〔台〕
signal covariance matrix 信号协方差矩阵
signal detection 信号检测
signal enhancement 信号增强
signal lag 信号滞后
signal-noise ratio 信噪比 | 信 [號] 噪比〔台〕　又见：*signal/noise ratio; signal to noise ratio*
signal/noise ratio （＝SNR; S/N）信噪比 | 信 [號] 噪比〔台〕　又见：*signal-noise ratio; signal to noise ratio*
signal processing 信号处理
signal source 信号源
signal strength 信号强度
signal to noise ratio 信噪比 | 信 [號] 噪比〔台〕　又见：*signal/noise ratio; signal-noise ratio*
SIGNE 3 （＝Solar Interplanetary Gamma-Neutron Experiment 3）信使 3 号（法国天文卫星）
significant figure 有效数字
signs of zodiac 黄道十二宫　又见：*zodiacal signs*
SII （＝stellar intensity interferometer）恒星强度干涉仪
Sikhote-Alin meteorite 希霍特－阿林陨星
Sikorsky crater 西科尔斯基环形山（月球）
silica 硅土 | 二氧化矽〔台〕
silicaceous asteroid 硅质小行星
silicate 硅酸盐
silicon （＝Si）硅（14 号元素）
silicon burning 硅燃烧
silicon diode array 硅二极管阵
silicon drift detector （＝SDD）硅漂移探测器
silicon intensifier target （＝SIT）硅增强靶　又见：*silicon intensifying target*
silicon intensifying target 硅增强靶　又见：*silicon intensifier target*
silicon monosulfide 一硫化硅（SiS）
silicon monoxide 一氧化硅（SiO）
silicon pore optics （＝SPO）硅孔光学系统
silicon star 硅星
silicon strip detector （＝SSD）硅条探测器
Siljan Ring crater 锡利扬湖陨星坑（地球）
Silk damping 西尔克阻尼
silver （＝Ag）银（47 号元素）
silver coating 镀银　又见：*silvering*

silvering 镀银　又见：*silver coating*
Simaqian 司马迁（小行星 12620 号）
SIMBAD （＝Set of Identifications, Measurements, and Bibliography for Astronomical Data）辛巴达天文数据库
similarity 相似性
similarity model 相似模型
similarity transformation 相似变换
simple burst 简单暴
simple ecliptic armillary 黄道经纬仪　又见：*ecliptic armillary sphere*
simple equatorial armillary 赤道经纬仪　又见：*equatorial armillary sphere*
simple harmonic motion （＝SHM）简谐运动
simple interferometer 相加干涉仪　又见：*adding interferometer*
simple pendulum 单摆
simple stellar population （＝SSP）简单恒星星族
simultaneity 同时性
simultaneous altitude 同时高度，同时地平纬度
simultaneous auroral multi-balloon observation （＝SAMBO）多气球同步极光观测
simultaneous collision 同时碰撞
simultaneous differential imager 同步差分成像仪
simultaneous observation 同时观测，同步观测 | 同時觀測〔台〕
sine parallax 正弦视差
sine wave 正弦波　又见：*sinusoidal wave*
single antenna 单天线
Single Aperture Far-Infrared Observatory （＝SAFIR）单孔径远红外天文台
single-aperture radio telescope 单孔径射电望远镜
single-arc method 单弧法
single-beam polarimeter 单光束偏振计
single-cavity maser amplifier 单腔量子放大器
single-channel photometer 单通道光度计
single-channel tunable spectrometer 单通道调谐频谱仪
single-dish radio telescope 单碟射电望远镜
single-lined binary 单谱分光双星，单谱双星 | 單線 [分光] 雙星〔台〕　又见：*single-lined spectroscopic binary; single-spectrum binary*
single-lined spectroscopic binary （＝SB 1）单谱分光双星，单谱双星 | 單線 [分光] 雙星〔台〕　又见：*single-lined binary; single-spectrum binary*
single-prism spectrograph 单棱镜摄谱仪
single-prism spectrometer 单棱镜分光仪

single-prism spectroscope 单棱镜分光镜
single-sideband receiver 单边带接收机
single-spectrum binary 单谱分光双星，单谱双星 | 單線 [分光] 雙星〔台〕　又见：*single-lined spectroscopic binary; single-lined binary*
single-station method 单站法
single-station observation 单站观测
single stellar population 单一星族
singlet 单谱线，单一态，单透镜 | 單 [譜] 線，單透鏡〔台〕
singular coordinate 奇异坐标
singular isothermal sphere 奇异等温球
singularity ① 奇点 | 奇異點〔台〕　又见：*singular point* ② 奇异性
singularity of the universe 宇宙奇点　又见：*cosmic singularity*
singular periodic solution 奇周期解
singular perturbation system 奇异摄动系统
singular point 奇点 | 奇異點〔台〕　又见：*singularity*
singular solution 奇解，奇异解
singular state 奇态　又见：*odd state*
singular value decomposition （＝SVD）奇异值分解
Sinope 木卫九
sinuous rille 弯曲沟纹
sinus 【拉】湾
Sinus Aestuum 【拉】浪湾（月球）
Sinus Amoris 【拉】爱湾（月球）
Sinus Asperitatis 【拉】狂暴湾（月球）
Sinus Concordiae 【拉】和谐湾（月球）
Sinus Fidei 【拉】信赖湾（月球）
Sinus Honoris 【拉】荣誉湾（月球）
Sinus Iridum 【拉】虹湾（月球）
Sinus Lunicus 【拉】眉月湾（月球）
Sinus Medii 【拉】中央湾（月球）
sinusoid 正弦曲线
sinusoidal projection 正弦投影
sinusoidal wave 正弦波　又见：*sine wave*
Sinus Roris 【拉】露湾（月球）
Sinus Successus 【拉】成功湾（月球）
SiO maser 一氧化硅微波激射
Siphon 渴乌（中国古代虹吸管）
Sippar Sulcus 西帕沟（木卫三）
Sirian companion 天狼伴星
Sirius 天狼 [星]（大犬座 α）　又见：*Canicula; Aschere; Dog Star*
Sirona 细女星（小行星 116 号）

Sirrah 壁宿二（仙女座 α） 又见：*Alpheratz*
SIRTF （＝Space Infrared Telescope Facility）空间红外望远镜设备 即：*SST*
SIS mixer （＝superconductor-insulator-superconductor mixer）超导 SIS 混频器
SIT （＝silicon intensifier target）硅增强靶
site protection 台址保护
site selection 选址 又见：*site testing*
site telescope 选址望远镜 又见：*test telescope*
site testing 选址 又见：*site selection*
Situla 虚梁三（宝瓶座 κ）
six-color photoelectric photometry 六色光电测光
six-color photometry 六色测光 又见：*6-color photometry*
six-color system ①六色测光系统 ②六色测光系统
Six-degree Field Galaxy Survey （＝6dFGS）6 度视场星系巡天 又见：*6dF Galaxy Survey*
Sixth Cambridge Survey of radio sources （＝6C survey）第 6 剑桥射电源巡天
SKA （＝Square Kilometer Array）平方千米 [射电望远镜] 阵 | 平方公里 [電波望遠鏡] 陣〔台〕
Skalnate Pleso Atlas of the Heavens 《捷克天图》
Skat 羽林军二十六（宝瓶座 δ）
Skathi 土卫二十七
skewness 歪斜度 | 偏斜度〔台〕
skin effect 趋肤效应, 集肤效应 | 表面效應〔台〕
Sklodowska crater 斯克洛多夫斯卡环形山（月球）
Skoll 土卫四十七
Sky and Telescope （＝S&T）《天空与望远镜》（美国期刊）
sky atlas 天图 又见：*celestial chart; sky map; atlas*
sky background 天空背景
sky background noise 天空背景噪声 | 天空背景雜訊〔台〕
sky background radiation 天空背景辐射
sky brightness 天空亮度
sky brightness temperature 天空亮度温度
sky clock 天钟
sky compass 天罗盘
sky factor 天光因子
sky fixed component 跟踪天空分量
sky flat 天光平场
sky fog 天雾
sky-horn switching 天空喇叭开关
Skylab 天空实验室（美国空间站）
Skylark 云雀号（英国探空火箭）

sky-light 天光
sky map 天图 又见：*sky atlas; celestial chart; atlas*
sky measuring scale 量天尺 又见：*cosmic yardstick*
Skynet 天网（英国卫星计划）
sky noise 天空噪声
sky patrol 巡天观测 又见：*sky survey*
sky phenomena 天象
sky radiation 星空辐射
sky spectrum 天光光谱
sky survey 巡天观测 又见：*sky patrol*
sky-wave 天波
SL9 （＝Comet Shoemaker-Levy 9）舒梅克－列维 9 号彗星 | 舒梅克－李維 9 號彗星〔台〕
slab structure 板块结构
slant range 斜距
slat collimator 条形准直器
Slate Islands crater 斯莱特群岛陨星坑（地球）
slave clock 子钟 又见：*secondary clock*
slepton 标轻子
Slight Cold 小寒（节气） 又见：*Lesser Cold*
Slight Heat 小暑（节气） 又见：*Lesser Heat*
Slight Snow 小雪（节气） 又见：*Light Snow*
slim disc 扁盘
Slipher crater 斯里弗环形山（月球）
slipping 失次
slit 光缝 | 狭缝〔台〕
slit image 光缝像 | 狭缝像〔台〕
slit jaw 光缝面
slit-jaw observation 光缝面观测
slitless spectrogram 无缝光谱 又见：*slitless spectrum*
slitless spectrograph 无缝摄谱仪
slitless spectroscopy 无缝分光
slitless spectrum 无缝光谱 又见：*slitless spectrogram*
slit spectrogram 有缝光谱 又见：*slit spectrum*
slit spectrograph 有缝摄谱仪 | 狹縫攝譜儀〔台〕
slit spectroscopy 有缝分光
slit spectrum 有缝光谱 又见：*slit spectrogram*
Sloan Digital Sky Survey （＝SDSS）斯隆数字化巡天 | 斯隆數位化巡天〔台〕
Sloan Extension for Galactic Understanding and Exploration （＝SEGUE）SEGUE 巡天（SDSS 巡天子项目）
slow angular variable 慢角变量
slow development 慢速显影
slow-drift burst 慢漂暴 | 慢漂爆發〔台〕
slowing-down time 减速时间

slowing factor 慢变因子
slowly varying component （=SVC）缓变分量
  又见：S-component
slowly varying potential 缓变势
slow magnetosonic-wave speed （=SMS）慢磁声波速
slow-mode wave 慢波 又见：slow wave
slow motion 慢动，迟行 | 慢動〔台〕
slow neutron capture process （=s-process）慢过程，s 过程 | 中子慢捕獲過程，s 過程〔台〕
slow nova 慢新星
slow plate 低灵敏度底片
slow pulsar 慢脉冲星
slow pulsator 慢脉动星
slow-roll approximation 慢滚近似
  又见：slow-rolling approximation
slow-rolling approximation 慢滚近似
  又见：slow-roll approximation
slow-rolling phase 慢滚阶段
slow rotator 慢自转星
slow-spinning black hole 慢自旋黑洞
slow supernova 慢超新星
slow wave 慢波 又见：slow-mode wave
SLR （=satellite laser ranging）卫星激光测距 | 衛星雷射測距〔台〕
SM （=service module）服务舱
Sm （=samarium）钐（62 号元素）
SMA （=Sub-Millimeter Array）SMA 亚毫米波射电望远镜阵 | 次毫米波[電波望遠鏡]陣〔台〕
small-angle anisotropy 小角度各向异性
Small Astronomical Satellite （=SAS）SAS 小天文卫星
small circle 小圆
small cycle 小纪（15 年） 又见：indiction
small-diameter source 小角径源
Small Explorer （=SMEX）小型探测器项目（美国）
Small Magellanic Cloud （=SMC）小麦云，小麦哲伦云
Small Missions for Advanced Research and Technology （=SMART-1）斯玛特－1 月球探测器
small scale anisotropy 小尺度各向异性
small-scale damping 小尺度衰减
small scale inhomogeneity 小尺度不均匀性
Small Scientific Satellite （=SSS）SSS 小科学卫星
small Solar System body 太阳系小天体

SMART-1 （=Small Missions for Advanced Research and Technology）斯玛特－1 月球探测器
SMBH （=supermassive black hole）特大质量黑洞 | 超大質量黑洞〔台〕
SMC （=Small Magellanic Cloud）小麦云，小麦哲伦云
SMD （=sudden magnetic disturbance）急始磁扰
smearing effect 涂污效应
smearing function 涂污函数
smearing-out image 亏损星像
SMEX （=Small Explorer）小型探测器项目（美国）
SMF （=stellar mass function）恒星质量函数
Smithsonian Astrophysical Observatory （=SAO）史密松天体物理台 | 史密松天文物理台〔台〕
Smithsonian Astrophysical Observatory Star Catalog （=SAO Star Catalogue）SAO 星表
SMM （=Solar Maximum Mission）太阳极大[年]使者 | 太陽極大期任務衛星〔台〕（美国）
smoked glass 熏烟玻璃 | 薰煙玻璃〔台〕
smoke train 冒烟余迹
Smoluchowski crater 斯莫卢霍夫斯基环形山（月球）
smoothed distribution 修匀分布
smoothed-particle hydrodynamics （=SPH）平滑质点流体动力学
smoothing 平滑，修匀 | 修匀〔台〕
smoothing effect 平滑效应
smoothing kernel 平滑核
SMS （=slow magnetosonic-wave speed）慢磁声波速
SMT ①（=Submillimeter Telescope）SMT 亚毫米波望远镜 ②（=segmented mirror telescope）拼合镜面望远镜
SMTO （=Submillimeter Telescope Observatory）亚毫米波天文台
SMY （=solar maximum year）太阳峰年
S/N （=signal/noise ratio）信噪比 | 信[號]噪比〔台〕
Sn （=tin）锡（50 号元素）
Snail Nebula 蜗牛星云（NGC 6543）
Snake Nebula 蛇形暗云（Barnard 72）
  又见：Dark S Nebula
snapshot 快照
SNC meteorite （=Shergotty）SNC 陨星
Snellius crater 斯涅尔环形山（月球）
Snell's law 斯涅尔定律 | 司乃耳定律〔台〕

sneutrino 标量中微子
SNLS （＝Supernova Legacy Survey）超新星遗珍巡天
snooper satellite 间谍卫星　又见：*spy satellite*
Snowman crater 雪人环形山（灶神星）
snow-plow model 雪耙模型
snowplow phase 雪耙相
SNR ①（＝signal/noise ratio）信噪比 | 信[號]噪比〔台〕②（＝supernova remnant）超新星遗迹 | 超新星殘骸〔台〕
SN type I I型超新星　又见：*type I supernova*
SN type II II型超新星　又见：*type II supernova*
SNU （＝solar neutrino unit）太阳中微子单位 | 太陽微中子單位〔台〕
SN（复数：*supernovae*）（＝supernova）超新星
SOAR （＝Southern Astrophysical Research Telescope）南方天体物理研究望远镜
Societa Astronomica Italiana （＝SAI）【意】意大利天文学会
Societe Astronomique de France （＝SAF）【法】法国天文学会
sodium （＝Na）钠（11号元素）
sodium guide star 钠引导星
sodium-iodide detector （＝NaI detector）碘化钠探测器 | 碘化鈉偵測器〔台〕
sodium vapour lamp 钠蒸气灯
SOFIA （＝Stratospheric Observatory for Infrared Astronomy）索菲亚平流层红外天文台 | 索菲雅[平流層紅外天文台]〔台〕
soft binary 软双星
softened point-mass potential 软化点质量势
soften gravity 软化引力
softening length 软化长度
softening parameter 软化参数
soft gamma repeater （＝SGR）软γ射线复现源 | 軟γ[射線]重覆爆發源〔台〕　又见：*soft γ-ray repeater*
soft phase 缓变阶段 | 緩變相〔台〕
soft supersymmetry breaking 软超对称破缺
soft ware 软件
software radio telescope 软件式射电望远镜
soft X-ray background 软X射线背景
soft X-ray object 软X射线天体
soft X-ray repeater （＝SXR）软X射线再现源 | 軟X光重覆爆發源〔台〕
soft X-ray source 软X射线源 | 軟X光源〔台〕
soft X-ray transient （＝SXT）软X射线暂现源 | 軟X光瞬變源〔台〕

soft γ-ray burst repeater 软γ射线暴再现源
soft γ-ray repeater 软γ射线复现源 | 軟γ[射線]重覆爆發源〔台〕　又见：*soft gamma repeater*
soft γ-ray source 软γ射线源
soft γ-ray transient （＝SGT）软γ射线暂现源 | 軟γ射線瞬變源〔台〕
SOHO （＝Solar and Heliospheric Observatory）索贺号，太阳和日球层探测器 | SOHO太陽觀測衛星，太陽與太陽圈觀測衛星〔台〕（欧美）
Sojourner 旅居者号火星车
sol 火星日
Solar, Anomalous, and Magnetospheric Particle Explorer （＝SAMPEX）太阳异常性/磁层粒子探索者
Solar-A 阳光号（日本太阳观测卫星）　又见：*Yohkoh*
solar active region 太阳活动区
solar activity 太阳活动
solar activity cycle 太阳活动周 | 太陽[活動]周期〔台〕　又见：*solar cycle*
solar activity prediction 太阳活动预报　又见：*prediction of solar activity*
solar air mass 太阳大气质量
Solar and Heliospheric Observatory （＝SOHO）索贺号，太阳和日球层探测器 | SOHO太陽觀測衛星，太陽與太陽圈觀測衛星〔台〕（欧美）
solar annual tide 太阳周年潮
solar antapex 太阳背点
solar apex 太阳向点
solar atlas 太阳图
solar atmosphere 太阳大气
Solar-B 日出号（日本太阳卫星）　又见：*Hinode*
solar battery 太阳能电池　又见：*solar cell*
solar-blind detector 日盲探测器
solar breeze 微太阳风
solar burst 太阳暴 | 太陽爆發〔台〕
solar calendar 阳历
solar cell 太阳能电池　又见：*solar battery*
solar cell paddle 太阳能电池叶片　又见：*solar paddle*
solar chromosphere 太阳色球 | 太陽色球[層]〔台〕
solar constant 太阳常数
solar continuum emission 太阳连续谱发射
solar corona 日冕
solar coronal hole 冕洞 | [日]冕洞〔台〕　又见：*coronal hole*
solar coronal loop 冕环　又见：*coronal loop*
solar cosmic rays 太阳宇宙线

solar cycle 太阳活动周 | 太陽 [活動] 周期〔台〕	**solar irradiance** 太阳辐照度
又见: *solar activity cycle*	**solar irradiation** 太阳辐照
solar day 太阳日	**solarization** 负感现象
solar diagonal 太阳棱镜观测镜	**solarized image** 光致负感像
Solar Diameter Monitor （＝SDM）太阳直径监测器	**solar latitude** 日面纬度 又见: *heliolatitude; heliographic latitude; sun's latitude*
solar disk 太阳圆面，日面 | 太陽圓面，日面，日輪〔台〕	**solar-like oscillation** 类太阳振荡 又见: *Sun-like oscillation*
solar distance 太阳距离	**solar-like star** 类太阳恒星 又见: *Sun-like star*
solar disturbance 太阳扰动	**solar limb** 日面边缘
Solar Dynamics Observatory （＝SDO）太阳动力学观测台	**solar local radio source** 太阳局部射电源
solar dynamo 太阳发电机	**solar longitude** 太阳经度 | 日面經度〔台〕
solar eclipse 日食 又见: *eclipse of the sun*	**solar loop** 太阳环状物
solar eclipse expedition 日食观测队	**solar luminosity** 太阳光度
solar eclipse limit 日食限 又见: *solar ecliptic limit*	**solar magnetic circle** 太阳磁圈
solar ecliptic limit 日食限 又见: *solar eclipse limit*	**solar magnetic cycle** 太阳磁周
solar electron event 太阳电子事件	**solar magnetic field** 太阳磁场
solar energetic particle （＝SEP）太阳高能粒子	**solar magnetism** 太阳磁性，太阳磁学
solar energy 太阳能	**solar magnetograph** 太阳磁像仪 | 太陽磁 [場] 強 [度] 計〔台〕 又见: *magnetoheliograph*
solar energy panel 太阳能板	**solar magnetogrph** 太阳磁象仪
solar equation 日躔 | 太陽差，日躔〔台〕	**solar mass** 太阳质量
solar eruption 太阳爆发	**solar-mass star** 太阳质量恒星
solar extreme-ultraviolet burst 太阳极紫外暴	**Solar Maximum Mission** （＝SMM）太阳极大 [年] 使者 | 太陽極大期任務衛星〔台〕（美国）
solar eyepiece 太阳目镜	**solar maximum year** （＝SMY）太阳峰年
solar facula 太阳光斑	**solar microwave burst** 太阳微波暴 | 太陽微波爆發〔台〕
solar filament 太阳暗条	**solar modulation** 太阳调制
solar filtergram 太阳单色像 | 太陽單色光照片〔台〕 又见: *spectroheliogram*	**solar month** 太阳月
solar flare 太阳耀斑 | 太陽閃焰〔台〕	**solar motion** 太阳运动
solar flare effect 耀斑效应	**solar mottle** 日芒 | 日芒，日斑〔台〕 又见: *mottle*
solar flocculus 太阳谱斑	**solar nebula** 太阳星云 | 太陽 [星] 雲〔台〕
solar flux 太阳流量 | 太陽通量〔台〕	**solar neighbourhood** 太阳附近空间
solar flux unit （＝SFU）太阳流量单位 | 太陽通量單位〔台〕	**solar neutrino** 太阳中微子 | 太陽微中子〔台〕
solar gale 太阳风暴 又见: *solar storm*	**solar neutrino deficit** 太阳中微子亏缺 | 太陽微中子虧缺〔台〕
solar global parameter 太阳全球参数	**solar neutrino flux** 太阳中微子流量
solar granulation 太阳米粒组织	**solar neutrino problem** 太阳中微子问题
solar hard X-ray burst 太阳硬 X 射线暴	**solar neutrino unit** （＝SNU）太阳中微子单位 | 太陽微中子單位〔台〕
solar hour 太阳小时	**solar noise** 太阳噪声
solar impulsive X-ray burst 太阳脉冲式 X 射线暴	**solar nutation** 太阳章动
solar infrared radiation 太阳红外辐射 又见: *infrared solar radiation*	**solar oblateness** 太阳扁度 | 太陽扁率，太陽扁度〔台〕
solar interior 太阳内部	**solar observation** 太阳观测
Solar Interplanetary Gamma-Neutron Experiment 3 （＝SIGNE 3）信使 3 号（法国天文卫星）	**solar optical telescope** 太阳光学望远镜
solar-interplanetary magnetic loop 太阳－行星际磁环	**solar orbit telescope** 环日轨道望远镜

solar oscillation	太阳振荡
solar outburst	太阳大爆发
solar paddle	太阳能电池叶片　又见：solar cell paddle
solar panel	太阳能电池板
solar parallax	太阳视差
solar patrol	太阳巡视
solar phase angle	日地张角｜太陽相角〔台〕
solar photosphere	太阳光球｜太陽光球[層]〔台〕
Solar Physics	《太阳物理》（荷兰期刊）
solar physics	太阳物理学｜太陽物理[學]〔台〕
solar polar wind	极区太阳风　又见：polar solar wind
solar probe	太阳探测器
solar prominence	日珥　又见：prominence
solar proton event	太阳质子事件
solar proton flux	太阳质子流量
solar pulsation	太阳脉动
solar radiation	太阳辐射
Solar Radiation and Climate Experiment	（＝SORCE）太阳辐射与大气实验卫星
Solar Radiation and Thermospheric Satellite	（＝SRATS）太阳号（日本高层大气和太阳观测卫星）　又见：Taiyo
Solar Radiation Monitoring Satellite	（＝SOLRAD）索拉德号，太阳辐射监测卫星｜太陽輻射監測衛星〔台〕（美国）
solar radiation pressure	太阳辐射压
solar radio astronomy	太阳射电天文｜太陽電波天文學〔台〕
solar radio burst	太阳射电暴｜太陽電波爆發〔台〕
solar radio dynamic spectrograph	太阳射电动态频谱仪
solar radio radiation	太阳射电
solar radius	太阳半径
solar ring	环式日晷｜環[式]日規〔台〕
solar rotation	太阳自转
solar seeing monitor	太阳视宁度监测仪
solar semi-diurnal tide	太阳半日潮
solar service	太阳服务｜太陽聯合觀測，太陽服務〔台〕
solar soft X-ray burst	太阳软X射线暴
solar spectrograph	太阳摄谱仪
solar spectrum	太阳光谱
solar spicule	太阳针状体
solar spot	太阳黑子｜[太陽]黑子〔台〕　又见：sunspot
solar spot prominence	太阳黑子日珥
solar star	太阳型恒星｜太陽型星〔台〕　又见：solar-type star
solar storm	太阳风暴　又见：solar gale
solar system	太阳系
solar system object	太阳系天体
Solar System Research	《太阳系研究》（俄罗斯期刊）
solar telescope	太阳望远镜
solar term	节气
Solar-Terrestrial Energy Programme	（＝STEP）日地能源计划
solar-terrestrial environment	日地环境
solar-terrestrial phenomenon	日地现象
solar-terrestrial physics	日地物理｜日地物理學〔台〕
solar-terrestrial relation	日地关系　又见：solar-terrestrial relationship
Solar Terrestrial Relation Observatory	（＝STEREO）日地关系观测台（美国）
solar-terrestrial relationship	日地关系　又见：solar-terrestrial relation
solar-terrestrial space	日地空间
solar-terrestrial space physics	日地空间物理
solar tidal wave	日潮波
solar tide	日潮
solar time	太阳时
solar tower	太阳塔｜太陽觀測塔〔台〕
solar transition region	太阳过渡区
solar-type star	太阳型恒星｜太陽型星〔台〕　又见：solar star
solar ultraviolet imaging telescope	太阳紫外成像望远镜
solar ultraviolet radiation	太阳紫外辐射
solar unit	太阳单位
solar wind	太阳风
solar wind boundary	太阳风边界
solar X-ray burst	太阳X射线暴
solar X-ray emission	太阳X射线发射
solar X-ray imaging	太阳X射线成像
solar X-ray imaging telescope	太阳X射线成像望远镜
solar X-ray irradiance	太阳X射线辐照度
solar X-ray radiation	太阳X射线辐射｜太陽X光輻射〔台〕
solar year	太阳年　即：natural year; tropical year
solar γ-ray burst	太阳γ射线暴
solar γ-ray emission	太阳γ射线发射
solar γ-ray radiation	太阳γ射线辐射

Soleil compensator 索莱伊补偿器 | 索累補償器〔台〕即：*Babinet compensator*
solid angle 立体角 又见：*spatial angle*
solid body 固体
solid-body rotation 固体自转
solid earth 固体地球
solid-state detector 固体探测器
solid-state diode array 固体二极管阵
solid-state imaging detector 固体成像探测器
solid-state laser 固体激光器
solid-state X-ray detector 固体 X 射线探测器
solid-state γ-ray detector 固体 γ 射线探测器
solid tide 固体潮 | [物] 體潮〔台〕又见：*body tide*
Solis Planum 索利斯高原（火星）
solitary 孤子性
solitary kinetic Alfvén wave 动理学阿尔文孤波
solitary wave 孤子波
soliton 孤立子
soliton star 孤子星
SOLRAD （＝Solar Radiation Monitoring Satellite）索拉德号，太阳辐射监测卫星 | 太陽輻射監測衛星〔台〕（美国）
solstices 二至点，二至日 | 二至點〔台〕又见：*solstitial points*
solstitial colure ① 二至圈 ② 二至圈
solstitial points 二至点，二至日 | 二至點〔台〕又见：*solstices*
solve-for parameter 待估参数
Solwind 太阳风号（美国天文卫星）
Sombrero galaxy 草帽星系（NGC 4594）又见：*Sombrero Hat galaxy*
Sombrero Hat galaxy 草帽星系（NGC 4594）又见：*Sombrero galaxy*
Sommerfeld crater 索末菲环形山（月球）
Song of Pacing the Heavens 《步天歌》
Songyuan 松原（小行星 23686 号）
sonic radius 声半径
sonic surface 声表面
Sonneberg Observatory 宗涅贝格天文台
Sonneberg variable 宗涅贝格天文台变星
Soochow astronomical chart 苏州石刻天文图 又见：*Suzhou Planisphere*; *Soochow planisphere*
Soochow planisphere 苏州石刻天文图 又见：*Suzhou Planisphere*; *Soochow astronomical chart*
SORCE （＝Solar Radiation and Climate Experiment）太阳辐射与大气实验卫星
Sothic cycle 天狼周
Sothic year 天狼年

sound absorption 声吸收
sound-emitting fireball 发声火流星 又见：*detonating fireball*
sound horizon 声波视界
sounding balloon 探空气球
sounding rocket 探空火箭
sound velocity 声速 又见：*acoustic velocity*; *speed of sound*
sound wave 声波 又见：*acoustic wave*
source brightness 源亮度
source brightness distribution 源亮度分布
source count 源计数 又见：*source counting*
source counting 源计数 又见：*source count*
source function 源函数
source identification 源证认
source model fitting 源模型拟合
source of energy 能源 又见：*energy source*
source peeling 剥源
source place 源位置 又见：*source position*
source position 源位置 又见：*source place*
source restoration 复源法
source shape 源形状
source-size effect 源大小效应
source structure 源结构
source subtraction 减源法
source survey 源巡天
south 南
South African Astronomical Observatory （＝SAAO）南非天文台
South Atlantic Anomaly 南大西洋近点角
south-bound node 降交点 又见：*descending node*
south celestial pole 南天极
South crater 索思环形山（月球）
south ecliptic pole 南黄极
south equatorial belt （＝SEB）南赤道带（木星）
Southern African Large Telescope （＝SALT）南非大型望远镜 | 南非大望遠鏡〔台〕
Southern Astrophysical Research Telescope （＝SOAR）南方天体物理研究望远镜
southern celestial hemisphere 南天 [区] 又见：*southern sky*
Southern Coalsack ① 煤袋星云 又见：*Coalsack Dark Nebula*; *Coalsack Nebula* ② 南煤袋
Southern Dipper 南斗（星组）又见：*Milk Dipper*
southern elongation 南大距
southern galaxy 南天星系
southern hemisphere （＝SH）南半球

southern latitude 南纬
southern light 南极光　又见: *aurora australis*
southern limit 南界限
southern limit of eclipse 南食限
Southern Palace Red Bird 南宫朱鸟
Southern Pinwheel 南风车星系（M83）
Southern Pleiades 南天七姐妹星团（IC 2602）
southern polarity 南极性 | 南 [磁] 極性〔台〕
southern reference star catalogue （=SRS catalogue）南天参考星表
southern sky 南天 [区]　又见: *southern celestial hemisphere*
Southern Sky Survey 南天天图
southern star 南天恒星
Southern tropic 南回归线　又见: *Tropic of Capricorn*
south following star 东南星
south frigid zone 南寒带
South Galactic Cap 南银冠 | 南银 [極] 冠〔台〕
South Galactic Pole 南银极
south-north asymmetry 南北不对称性　又见: *north-south asymmetry*
south point 南点
south polar distance 南极距
South Polar Group 南极星系群　即: *Sculptor group*
south pole 南极
South Pole-Aitken Basin 南极－艾特肯盆地（月球）
South Pole Telescope （=SPT）SPT 南极点望远镜（美国）
south preceding star 西南星
south south temperate belt （=SSTB）南南温带（木星）
south temperate belt （=STB）南温带（木星）
south temperate zone 南温带（地球）
south tropical zone 南热带（地球）
Soviet Mountains 苏维埃山脉（月球）
Soviet variable star （=SVS）苏联变星
Soyuz 联盟号 | 聯合號 [太空船]〔台〕（苏联载人飞船）
SPA （=sudden phase anomaly）位相突异 | 相位突異〔台〕
space absorption 空间吸收
space aeronautics 航天学
space age 空间时代，太空时代
space-air vehicle 空天飞机
space astrometry 空间天体测量学 | 太空天體測量學〔台〕
space astronomy 空间天文学 | 太空天文學〔台〕

space astrophysics 空间天体物理 | 太空天文物理學〔台〕
space-based interferometer 空基干涉仪
space-based observation 空基观测
space-based telescope 空基望远镜
space biology 空间生物学
space charge 空间电荷
space charge wave 空间电荷波
space concentration 空间聚度
spacecraft 太空飞行器，宇宙飞船，航天器 | 太空船，宇宙飛船〔台〕
spacecraft astronomy 宇宙飞船天文学
spaced antennas 间隔天线阵　又见: *spaced array*
spaced array 间隔天线阵　又见: *spaced antennas*
space debris 太空垃圾
space density 空间密度
space density profile 空间密度轮廓
space distribution 空间分布　又见: *distribution in space; spatial distribution*
space dynamics 空间动力学
space engineering 太空工程
space environment 空间环境
space exploration 空间探测 | 太空探測〔台〕
space-fixed coordinate system 空固坐标系
space flight 航天，太空飞行 | 太空飛行〔台〕
space geodesy 空间大地测量 | 太空大地測量〔台〕
space geodynamics 空间地球动力学
Space Infrared Telescope Facility （=SIRTF）空间红外望远镜设备　即: *SST*
Space Infrared Telescope for Cosmology and Astrophysics （=SPICA）宇宙学和天体物理空间红外望远镜，天门号
space instrumentation 航天设备
space inversion 空间反演
Spacelab 空间实验室 | 太空實驗室〔台〕（美国航天飞机舱室）
spacelike infinity 类空无限远
spacelike interval 类空间隔
spacelike path 类空路径
spacelike separation 类空分隔
space mission 航天任务
space motion 空间运动 | 太空運動〔台〕
space observation 空间观测
space observatory 空间天文台 | 太空天文台〔台〕
space of constant curvature 常曲率空间
space particle 空间粒子，太空粒子
space photography 太空照相

space pollution　太空污染
space probe　空间探测器 | 太空 [探测] 船〔台〕
space probe instrument　空间飞船运载仪器
space quantization　空间量子化
space radio astrometry　空间射电天体测量 | 太空電波天體測量〔台〕
space radio astronomy　空间射电天文 | 太空電波天文學〔台〕
space reddening　空间红化
space research　空间研究 | 太空研究〔台〕
space science　空间科学 | 太空科學〔台〕
Space Science Reviews　《空间科学评论》（荷兰期刊）
space ship　太空船
space shuttle　航天飞机　又见: *shuttle*
Space Solar Telescope　（=SST）空间太阳望远镜（中国）
space station　空间站 | 太空站，宇宙站〔台〕
space suit　太空服
space technology　空间技术
space telescope　空间望远镜 | 太空望遠鏡〔台〕
　又见: *free-flying telescope*
Space Telescope Science Data Analysis System　（=STSDAS）空间望远镜科学数据分析系统
Space Telescope Science Institute　（=STScI）空间望远镜研究所 | 太空望遠鏡 [科學] 研究所〔台〕（美国）
spacetime　时空
spacetime continuum　时空连续统
space-time curvature　时空曲率　又见: *curvature of space-time*
spacetime metric　时空度规
spacetime singularity　时空奇点
spacetime wave　时空波
space transportation system　航天运输系统
space transporter　航天运载工具
Space Variable Objects Monitor　（=SVOM）空间变源监视器（中法高能天文卫星）
space vehicle　航天器 | 太空船〔台〕
space velocity　空间速度
spacewalk　太空行走
Space Weather　《空间天气》（美国期刊）
space weather　空间天气 | 太空氣候〔台〕
Space Weather Prediction Center　空间天气预报中心
spallation　散裂
spallation cross-section　散裂截面

Spanish Virtual Observatory　（=SVO）西班牙虚拟天文台
spar　多筒望远镜，组合太阳望远镜 | 多路望遠鏡裝置〔台〕
spark chamber　火花室 | 電花室〔台〕
spark line　电花谱线
spark spectrum　电花光谱
sparse aperture　稀疏孔径　又见: *dilute aperture*
sparse aperture masking　稀疏孔径掩模法
SPARTAN　（=Shuttle Pointed Autonomous Research Tool for Astronomy）斯巴达人号
spatial angle　立体角　又见: *solid angle*
spatial brightness　空间亮度
spatial coherence　空间相干性
spatial correlation function　空间相关函数
spatial dispersion　空间弥散
spatial distribution　空间分布　又见: *distribution in space*; *space distribution*
spatial domain　空间域
spatial filter　空间滤波器
spatial filtering　空间滤波
spatial frequency　空间频率
spatial frequency characteristic　空间频率特征
spatially closed metric　空间闭合度规
spatially open metric　空间开放度规
spatial mass segregation　空间质量分层
spatial motion　空间运动
spatial resolution　空间分辨率
spatial spectrum　空间谱 | 空間 [光] 譜〔台〕
spatial strain　空间应变
spatial structure　空间结构
spatial velocity　空间速度
spatiography　空间测量学
Special Astrophysical Observatory　（=SAO RAS）特设天体物理台
special color-index　特殊色指数
special coordinate　特殊坐标
special maximum resolution emulsion　特高分辨率乳胶
special perturbation　特殊摄动
special relativity　狭义相对论
specific enthalpy　比焓
specific flux　流量比
specific gravity　比重　又见: *specific weight*
specific heat　比热
specific impulse　比冲量
specific intensity　比强度

specific intensity of radiation　比辐射强度
specific internal energy　比内能
specific moment　比矩
specific pressure　比压
specific star-formation rate　比恒星形成率
specific thrust　比推力
specific weight　比重　又见: *specific gravity*
speckle　斑点
speckle camera　斑点照相机
speckle imaging　斑点成像法
speckle interferometer　斑点干涉仪 | 散斑干涉仪〔台〕
speckle interferometry　斑点干涉测量 | 散斑干涉法〔台〕
speckle masking　斑点掩模
speckle pattern　斑点图样
speckle photometry　斑点测光
speckle spectroscopy　斑点分光
speckle technique　斑点技术
specklogram　斑点干涉像
spectracon　光谱摄像管
spectral albedo　分光反照率
spectral analysis　光谱分析,波谱分析,谱分析　又见: *spectrum analysis*
spectral analyzer　频谱分析仪
spectral band　谱带
spectral calibration lamp　光谱定标灯
spectral catalogue　光谱表
spectral class　光谱型　又见: *spectral type*
spectral classification　光谱分类
spectral comparator　比长仪　又见: *comparator; spectrocomparator*
spectral continuum　连续谱区
spectral cutoff　谱截断
spectral density　谱密度
spectral-density model　谱密度模型
spectral distribution　谱分布　又见: *spectrum distribution*
spectral duplicity　光谱成双性
spectral energy distribution　（=SED）光谱能量分布, 能谱分布 | 光譜能量分佈〔台〕
spectral estimate　频谱估计
spectral flux density　谱流量密度
spectral folding　谱折叠
spectral function　谱函数
spectral index　谱指数 | 光譜指數〔台〕　又见: *spectrum index*

spectral intensity　谱强度
spectral irradiation　谱照度
spectral line　谱线　又见: *spectrum line*
spectral line absorption　谱线吸收
spectral line broadening　谱线展宽
spectral line correlation　谱线相关
spectral line emission　谱线发射
spectral line formation　谱线形成　又见: *line formation*
spectral line narrowing　谱线变窄
spectral line receiver　谱线接收机　又见: *line receiver*
spectral line shift　谱线位移　又见: *line displacement; line shift*
spectral line source　谱线发射源
spectral line splitting　谱线分裂　又见: *line splitting*
spectral line width　谱线宽度　又见: *breadth of spectral line*
spectral moment　谱矩
spectral parameter　谱参数
spectral power　谱功率
spectral power spectrum　谱功率谱
spectral purity　光谱纯度
spectral range　光谱范围
spectral region　① 光谱区 ② 频谱区
spectral resolution　谱分辨率
spectral resolving power　谱分辨本领
spectral response　① 光谱响应 ② 频谱响应
spectral selectivity　① 光谱选择性 ② 频谱选择性
spectral sensitivity　分光敏度
spectral sensitometry　分光敏度测量
spectral sequence　光谱序
spectral series　光谱线系 | [光]譜線系〔台〕
spectral synthesis modeling　光谱合成模型
spectral term　光谱项
spectral tilt　光谱倾斜
spectral transmission　分光透射
spectral type　光谱型　又见: *spectral class*
spectral window　频谱窗　又见: *spectrum window*
spectra-spectroheliograph　太阳光谱单色光照相仪
spectra-spectroheliography　太阳光谱单色光照相术
spectrobologram　分光测热图
spectrobolometer　分光测热计
spectrocomparator　比长仪　又见: *spectral comparator; comparator*

spectro-enregistreur des vitesses　分光速度记录仪
　　又见: *enregistreur des vitesses*
spectrogram　① 光谱图 | 光譜〔台〕② 频谱图
spectrograph　① 摄谱仪 ② 频谱仪
　　又见: *spectrometer*
spectrographic orbit　摄谱轨道，摄谱解 | 攝譜軌道，分光軌道〔台〕
spectrography　摄谱
spectrohelio-cinematograph　太阳分色电影仪
spectroheliogram　太阳单色像 | 太陽單色光照片〔台〕　又见: *solar filtergram*
spectroheliograph　太阳单色光照相仪
spectroheliography　（＝SHG）太阳单色光照相术
spectrohelioimaging　太阳分光成像
spectrohelioscope　太阳单色光观测镜
spectrohelioscopy　太阳分光
spectrometer　① 频谱仪　又见: *spectrograph* ② 分光仪 | 分光計，光譜儀〔台〕
spectrometry　分光测量
spectrophotograph　分光光度照相仪
spectrophotography　分光光度照相
spectrophotometer　分光光度仪 | 分光光度計〔台〕
spectrophotometric gradient　分光光度梯度
spectrophotometric standard star　分光光度标准星　又见: *standard star for spectrophotometry*
spectrophotometric temperature　分光光度温度
spectrophotometry　分光光度测量
spectropolarimeter　分光偏振计，偏振频谱仪 | 分光偏振譜儀〔台〕
spectropolarimetry　分光偏振测量
spectroprojector　光谱投影仪 | 分光映射儀〔台〕
spectropyrheliometer　太阳分光测热计
spectropyrheliometry　太阳分光测热法
spectroradiometer　分光辐射计
spectroradiometry　分光辐射测量
spectroscope　分光镜
spectroscopic analysis　分光分析
spectroscopic binary　（＝SB）分光双星
spectroscopic distance　分光距离
spectroscopic element　分光根数 | 测谱要素，分光要素〔台〕
spectroscopic luminosity　分光光度
spectroscopic orbit　① 分光轨道 | 测谱轨道，分光軌道〔台〕② 分光解　又见: *radial-velocity orbit*
spectroscopic parallax　分光视差
spectroscopic period　分光周期
spectroscopic redshift　光谱红移

Spectroscopic Survey Telescope　（＝SST）分光巡天望远镜
spectroscopy　① 光谱学 | 光譜學，分光學〔台〕② 频谱学
spectrosensitometer　分光感光计
spectrosensitometry　分光感光测量
spectrum　① 光谱　又见: *optical spectrum* ② 频谱
spectrum analysis　① 频谱分析 ② 光谱分析，波谱分析，谱分析　又见: *spectral analysis*
spectrum analyzer　频谱分析器
spectrum binary　光谱双星
spectrum degradation　能谱软化
spectrum-density diagram　光谱－密度图
spectrum distribution　谱分布　又见: *spectral distribution*
spectrum expander　频谱展宽器
spectrum field　谱场
spectrum index　谱指数 | 光譜指數〔台〕
　　又见: *spectral index*
spectrum intensity　频谱强度
spectrum line　谱线　又见: *spectral line*
spectrum-luminosity diagram　光谱－光度图
　　即: *HR diagram*
spectrum of turbulence　湍谱 | 湍[流]譜〔台〕
spectrum-rate of variation relation　光谱－光变率关系
spectrum renormalization　频谱重正化
Spectrum-Roentgen-Gamma　（＝SRG）X射线－γ射线谱天文卫星
spectrum scanner　光谱扫描仪
spectrum scanning　光谱扫描
spectrum variable　光谱变星
spectrum window　频谱窗　又见: *spectral window*
specular density　定向密度
specular reflection　单向反射，镜反射 | 鏡反射，單向反射〔台〕
specular reflector　单向反射器
speculum　镜齐，镜用合金
speculum metal　镜用金属
speed of light　光速　又见: *velocity of light*
speed of light circle　光速圈
speed of photographic plate　照相底片感光速度
speed of rotation of the ecliptic　黄道回转速度
speed of sound　声速　又见: *acoustic velocity; sound velocity*
Spektr-R　射电天文号（俄罗斯卫星）
　　又见: *RadioAstron*
Spencer Jones crater　斯潘塞·琼斯环形山（月球）

SPH （＝smoothed-particle hydrodynamics）平滑质点流体动力学
sphere of action 作用范围　又见：*sphere of activity; sphere of influence*
sphere of activity 作用范围　又见：*sphere of action; sphere of influence*
sphere of Edoxus 埃多克斯球
sphere of gravity 重力范围　又见：*gravisphere*
sphere of influence 作用范围　又见：*sphere of action; sphere of activity*
spherical aberration （＝SA）球差 | 球 [面像] 差〔台〕
spherical accretion 球对称吸积
spherical albedo 球面反照率
spherical angle 球面角
spherical antenna 球面天线 | 球天線〔台〕
spherical astronomy 球面天文学 | 球天文 [學]〔台〕
spherical collapse 球形坍缩
spherical component 球状子系
spherical coordinate 球坐标 | 球 [面] 坐標〔台〕
spherical distribution 球面分布
spherical excess 球面角超
spherical function 球函数
spherical galaxy 球状星系　又见：*globular galaxy*
spherical harmonic expansion 球谐展开
spherical harmonic oscillator 球谐振荡
spherical harmonics 球谐函数
spherical indicatrix of scattering 球面散射指示量
spherical lens 球面透镜
spherical meniscus 球面弯月镜 | 球面彎月 [形透] 鏡〔台〕
spherical mirror 球面镜
spherical model 球状模型
spherical nebula 球狀星雲
spherical reflector 球反射面
spherical shell 球壳
spherical space 球空间 | 球面空間〔台〕
spherical subsystem 球狀次系
spherical triangle 球面三角形 | 球面三角 [形]〔台〕
spherical trigonometry 球面三角学 | 球面三角 [學]〔台〕
spherical wave 球面波
spheroidal coordinate 椭球坐标　又见：*ellipsoidal coordinate*
spheroidal dwarf galaxy 矮椭球星系　又见：*dwarf spheroidal galaxy*
spheroidal galaxy 椭球星系
spheroid of revolution 旋转椭球　又见：*ellipsoid of gyration*
spherometer 球径计，球面曲率计
spherule 小球体
SPICA （＝Space Infrared Telescope for Cosmology and Astrophysics）宇宙学和天体物理空间红外望远镜，天门号
Spica 角宿一（室女座 α）　又见：*Azimech; Epi*
spicule 针状体，针状物 | 針狀體〔台〕
spider 网支架
spider diffraction 网架衍射 | 十字繞射〔台〕
spike burst 尖峰暴 | 尖峰爆發〔台〕
spillover 漏失
spillover radiation 漏失辐射 | 溢流輻射〔台〕
spin 自旋，旋转 | 自旋〔台〕
spinar 磁旋体　又见：*magnetoid*
spin axis 自旋轴，自转轴
spin casting 自旋印模
Spindle galaxy 纺锤星系 | 紡錘狀星系〔台〕（NGC 3115）
spindle nebula 纺锤星云 | 紡錘狀星雲〔台〕
spindle-shaped meteor 纺锤状流星
spin-down 自旋减慢
spinel 尖晶石　又见：*spinelle*
spinelle 尖晶石　又见：*spinel*
spin evolution 自旋演化
spin-flip collision 自旋翻转碰撞
spin-flip transition 自旋翻转跃迁
spinodal decomposition 旋节线分解
spinor 旋量
spin-orbit coupling 轨旋耦合
spin-orbit resonance 轨旋共振　又见：*orbit-rotation resonance*
spin-orbit transition 轨旋跃迁
spin phase 自旋相位
spin precession 自转轴进动　又见：*rotational axis precession*
spin-spin coupling 旋旋耦合
spin temperature 自旋温度
spin-up 自旋加快
spiral arm 旋臂
spiral arm tracer 示臂天体
spiral feature 旋涡构形
spiral galaxy （＝S galaxy）旋涡星系 | 螺旋 [狀] 星系〔台〕
spiral jet 螺旋喷流

spiral nebula 旋涡星云 | 螺旋星雲〔台〕	**sporadic solar γ-ray emission** 偶现太阳 γ 射线发射
spiral orbit 螺旋轨道　又见：*helical orbit*	**sporadic X-ray source** 偶现 X 射线源 | 偶现 X 光源〔台〕
spiral pattern 旋涡图像	**sporadic γ-ray source** 偶现 γ 射线源
spiral potential 旋涡势	**Spörer minimum** 斯波勒极小期
spiral structure 旋涡结构 | 螺旋狀結構〔台〕	**Spörer's law** 斯波勒定律
spiral tracer 旋涡示踪天体	**spot activity** 黑子活动
spirit level 酒精水准器	**spot area** 黑子区
Spirit Mars Exploration Rover 勇气号火星探测车	**spot group** 黑子群 | [太陽] 黑子群〔台〕　又见：*sunspot group*
Spirograph Nebula 滚筒仪星云（IC 418）	**spot number** 黑子数　又见：*sunspot number*
Spitzer-Oort hypothesis 斯皮策－奥尔特假说 | 史匹哲－歐特假說〔台〕	**spot penumbra** 黑子半影　又见：*sunspot penumbra*
Spitzer Space Telescope （＝SST）斯皮策空间望远镜 | 斯皮策 [紅外] 太空望遠鏡〔台〕	**spotted area** 黑子覆盖区 | 黑子覆蓋面積〔台〕
	spotted star 富黑子恒星
splash-down 溅落	**spot umbra** 黑子本影　又见：*sunspot umbra*
splendid star 景星　又见：*great star*	**spray** 日喷
spline fit 样条拟合	**spray prominence** 喷射日珥 | 噴散日珥〔台〕
Splinter Galaxy 木刺星系（NGC 5907）	**spread function** 扩展函数
split detector 分离探测器	**Spring Equinox** 春分（节气）　又见：*Vernal Equinox*
SPO ①（＝Sacramento Peak Observatory）萨克拉门托峰天文台 ②（＝silicon pore optics）硅孔光学系统	**spring equinox** 春分点　又见：*vernal equinox; vernal point; first point of Aries*
Sponde 木卫三十六	**Spring Festival** 春节
sponge topology 海绵状拓扑	**Springfield mount** 斯普林菲尔德装置　又见：*Springfield mounting*
spontaneous emission 自发发射	
spontaneous emission coefficient 自发发射系数	**Springfield mounting** 斯普林菲尔德装置　又见：*Springfield mount*
spontaneous recombination 自发复合	
spontaneous scattering 自发散射	**spring tide** 大潮　又见：*high water*
spontaneous scattering coefficient 自发散射系数	**SPRINT-A** 火崎号（日本极紫外天文卫星）　又见：*Hisaki*
spontaneous star formation 自发恒星形成	**s-process** （＝slow neutron capture process）慢过程，s 过程 | 中子慢捕獲過程，s 過程〔台〕
spontaneous supersymmetry breaking 自发超对称破缺	**Sproul Observatory** 斯普罗尔天文台
spontaneous symmetry breaking 自发对称性破缺	**SPT** （＝South Pole Telescope）SPT 南极点望远镜（美国）
spontaneous transition 自发跃迁	**spurious disk** 虚圆面 | 虛盤面〔台〕
sporadic burst 偶现暴	**spurious radiation** 伪辐射
sporadic coronal condensation 偶现日冕凝块	**sputnik** 【俄】人造卫星
sporadic meteor 偶现流星	**sputtering** 溅射
sporadic radio burst 偶现射电暴 | 偶現電波爆發〔台〕	**spy satellite** 间谍卫星　又见：*snooper satellite*
sporadic radio emission 偶现射电	**square aperture** 方孔径
sporadic radio source 偶现射电源 | 偶現電波源〔台〕	**square degree** 平方度
	Square Kilometer Array （＝SKA）平方千米 [射电望远镜] 阵 | 平方公里 [電波望遠鏡] 陣〔台〕
sporadic solar radio burst 偶现太阳射电暴	**square-law detection** 平方律检波
sporadic solar radio emission 偶现太阳射电	**square-law detector** 平方律检波器
sporadic solar X-ray burst 偶现太阳 X 射线暴	**square-shaped array** 矩形天线阵
sporadic solar X-ray emission 偶现太阳 X 射线发射	**square signal** 方信号
sporadic solar γ-ray burst 偶现太阳 γ 射线暴	

square wave　方波
square-wave modulation　方波调制
squark　标夸克
SQUID　（=superconducting quantum interference device）超导量子干涉器件
Sr　（=strontium）锶（38 号元素）
sr　（=steradian）球面度
SRATS　（=Solar Radiation and Thermospheric Satellite）太阳号（日本高层大气和太阳观测卫星）
SRG　（=Spectrum-Roentgen-Gamma）X 射线－γ 射线谱天文卫星
SRS catalogue　（=southern reference star catalogue）南天参考星表
SR variable　（=semi-regular variable）半规则变星
SSA　（=synchrotron self-absorption）同步加速自吸收
SS coupling　SS 耦合
SS Cygni star　天鹅 SS 型星 | 天鵝 [座]SS 型星〔台〕
SSD　（=silicon strip detector）硅条探测器
SSM　（=standard solar model）标准太阳模型
SSO　（=Siding Spring Observatory）赛丁泉天文台（澳大利亚）
SSP　（=simple stellar population）简单恒星星族
SSS　（=Small Scientific Satellite）SSS 小科学卫星
SST　①（=Space Solar Telescope）空间太阳望远镜（中国）②（=Spectroscopic Survey Telescope）分光巡天望远镜 ③（=Spitzer Space Telescope）斯皮策空间望远镜 | 斯皮策 [紅外] 太空望遠鏡〔台〕④（=satellite-to-satellite tracking）人卫－人卫跟踪
S star　S 型星　又见：*S-type star*
SSTB　（=south south temperate belt）南南温带（木星）
SSWF　（=sudden short-wave fade-out）短波突衰　即：*Mögel-Dellinger effect*
ST　（=sidereal time）恒星时
stability　稳定性 | 穩定性，穩定度〔台〕
stability criterion　稳定性判据
stability domian　稳定区域　又见：*stable region*
stability probability　稳定性概率
stabilization　稳定化
stable clustering　稳定成团性
stable coordinate　稳定坐标
stable equilibrium　稳定平衡
stable orbit　稳定轨道
stable prominence　稳定日珥
stable region　稳定区域　又见：*stability domian*
stable star　稳定星　又见：*stationary star*
stable state　稳定态
Stackel potential　斯塔克尔势
Stadius crater　斯塔迪乌斯环形山（月球）
stage of ionization　电离级　又见：*ionization level*
staircase function　阶梯函数
standard antenna　标准天线
standard atmosphere　标准大气
standard candle　标准烛光
standard Cold Dark Matter model　（=sCDM model）标准冷暗物质模型
standard coordinate　标准坐标
standard cosmological model　标准宇宙模型
standard deviation　（=s.d.）标准偏差 | 標準差〔台〕
standard earth　（=SE）标准地球
standard epoch　标准历元
standard error　（=s.e.）标准误差
standard frequency　标准频率　又见：*etalon frequency*
standard frequency and time signal　（=SFTS）标准时频 | 標準時號〔台〕
standard inflation　标准暴胀
standardised candle method　标准化烛光法
standardization　标准化，正则化 | 標準化，正規化〔台〕
standard meridian　标准子午线，标准子午面 | 標準子午圈，標準子午線〔台〕
standard model　标准模型
standard model of cosmology　标准宇宙学模型
standard model of particle physics　标准粒子物理模型
standard noon　标准时正午
standard one-zone model　标准单区模型
standard rotational-velocity star　自转速度标准星
standard ruler　标准尺
standard solar model　（=SSM）标准太阳模型
standard source　校准源，标准噪声发声器
standard spheroid　标准旋转椭球
standard star　标准星
standard star for photometry　光度标准星 | 測光標準星〔台〕　又见：*photometric standard star*
standard star for radial-velocity　视向速度标准星　又见：*radial-velocity standard star; standard-velocity star*
standard star for spectral classification　光谱分类标准星

standard star for spectrophotometry 分光光度标准星 又见：*spectrophotometric standard star*
standard system 标准系统
standard time 标准时 又见：*etalon time*
standard time-signal 标准时号
standard time-zone 标准时区
standard-velocity star 视向速度标准星
又见：*radial-velocity standard star; standard star for radial-velocity*
standing wave 驻波 又见：*stationary wave*
standing wave pattern 驻波图案
standstill 停变期
star 恒星
star association 星协 又见：*stellar association*
star atlas 星图集 | 星圖〔台〕 又见：*stellar atlas*
star background 恒星背景
starbirth activity 恒星诞生活动
starburst 星暴 | 星邉增〔台〕
starburst galaxy 星暴星系 | 星邉增星系〔台〕
starburst nucleus 星暴核
star catalogue 星表 又见：*catalogue of stars*
star chart 星图册 又见：*stellar chart*
star cloud 恒星云
star cluster 星团 又见：*stellar cluster; cluster of stars; cluster*
star color 星色
star complex 恒星复合体 又见：*stellar complex*
star count 恒星计数 又见：*star counting; star gauge*
star counting 恒星计数 又见：*star count; star gauge*
star density 恒星密度
star dial 星晷
star-dial time-determining instrument 星晷定时仪
star distribution 恒星分布
star drift 星流 又见：*star streaming; drift of stars*
Stardust 星尘号（美国彗星探测器）
star evolution 恒星演化
star finder 寻星镜 又见：*finderscope; viewfinder; finder*
star formation 恒星形成 又见：*formation of stars; stellar formation*
star-formation activity 恒星形成活动
star-formation burst 产星暴
star-formation efficiency （＝SFE）恒星形成效率
star-formation process 恒星形成过程
star-formation rate （＝SFR）恒星形成率
star-formation region 恒星形成区
又见：*star-forming region; star-producing region*
star-formation threshold 恒星形成阈值
star-forming galaxy 产星星系
star-forming nebula 产星星云
star-forming phase 产星阶段
star-forming region 恒星形成区
又见：*star-producing region; star-formation region*
star gauge 恒星计数 又见：*star count; star counting*
stargazer 天文爱好者 | 業餘天文學家，天文愛好者〔台〕 又见：*amateur astronomer; astrophile*
star ghost 恒星鬼像
star group 恒星群 又见：*group of stars*
starhopping 星桥法
star identification 恒星证认 | 恆星識別〔台〕
star identifier 认星器
star image 星像 又见：*stellar image*
star imaging 恒星成像
Stark broadening 斯塔克展宽 | 史塔克致寬〔台〕
Stark effect 斯塔克效应 | 史塔克效應〔台〕
Starlette 小星号（法国测地卫星）
starlight 星光
star-like infrared object 类星红外天体
star-like nucleus 类星星系核
star-like object 类星天体 | 星狀天體〔台〕
star-like optical object 类星光学天体
star-like radio object 类星射电天体
star-like remnant 星状遗迹
Star Manual of the Masters Gan and Shi 《甘石星经》
star map 星图 又见：*stellar map; atlas*
star-mass object 恒星质量天体
star motion 恒星运动 又见：*stellar motion*
star name 星名
star navigation 恒星导航 又见：*stellar guidance*
star nomenclature 恒星命名 又见：*stellar nomenclature*
star number 星号，星数 | 星數〔台〕
star observation platform 观星台
star origin 恒星起源
star-pair method 星对法
starpatch 星斑 又见：*starspot*
star photometer 恒星光度计 又见：*stellar photometer*
star photometry 恒星测光 又见：*stellar photometry*
star place 恒星位置
star poor region 贫星区
star portent 星象 又见：*aspect astrology*

star-producing region　恒星形成区
　　又见：*star-forming region; star-formation region*
starquake　星震
starry sky　星空
star sensor　星敏感器｜恆星傳感器〔台〕
star sight　恒星目测
starspot　星斑　又见：*starpatch*
starspot cycle　星斑周期
star streaming　星流　又见：*star drift; drift of stars*
star swarm　恒星群
star system　恒星系统　又见：*stellar system*
starting value　初值　又见：*initial value*
star trail　星像迹线
state function　态函数
state of excitation　激发态
state of ionization　电离态　又见：*ionic state*
state of matter　物态
state of polarization　偏振态　又见：*polarization state*
static electric field　静电场　又见：*electrostatic field*
static limit　静态极限｜靜止極限，無位移極限〔台〕　又见：*stationary limit*
static magnetic field　静磁场
static model　静态模型｜靜止[宇宙]模型〔台〕
static property　静态性质
static spherical potential　静态球对称势
static universe　静态宇宙
stationarity　稳态
stationary　留
stationary camera　固定式照相机
stationary limit　静态极限｜靜止極限，無位移極限〔台〕　又见：*static limit*
stationary line　无位移谱线
stationary meteor　驻留流星
stationary mode　固定式
stationary model　稳态模型｜静[宇宙]模型〔台〕
stationary phase method　稳相位法
stationary point　留点
stationary potential　稳定势｜穩定勢，穩定位，穩定電勢，穩定電位〔台〕
stationary prominence flare　稳定日珥型耀斑｜穩定日珥型閃焰〔台〕
stationary radiant　固定辐射点
stationary random process　平稳随机过程
stationary satellite　静止卫星
stationary satellite-camera　固定式人卫照相机
stationary shell　稳定气壳
stationary star　稳定星　又见：*stable star*

stationary state　定态
stationary time series　平稳时间序列
stationary wave　驻波　又见：*standing wave*
station beam　站集波束
station coordinate　测站坐标
station correction　测站改正
station drift　测站漂移
station error　测站误差
statistical analysis　统计分析
statistical astronomy　统计天文学
statistical betatron process　统计感应加速过程
statistical broadening　统计展宽｜統計致寬〔台〕
statistical correlation　统计相关　又见：*statistical dependence*
statistical dependence　统计相关　又见：*statistical correlation*
statistical distribution　统计分布
statistical ensemble　统计系综
statistical equilibrium　统计平衡
statistical error　统计误差
statistical estimation　统计性估计
statistical Fermi process　统计费米过程
statistical fluctuation　统计起伏
statistical orbit-determination　统计定轨
statistical parallax　平均视差，统计视差
　　又见：*mean parallax*
statistical weight　统计权｜統計權[重]〔台〕
STB　（＝south temperate belt）南温带（木星）
steadiness　稳定度
steady field　稳定场，规则场
steady motion　稳定运动，定态运动
steady spiral pattern　稳态旋涡图案
steady state　稳恒态｜穩態〔台〕
steady-state cosmology　稳恒态宇宙学｜穩態宇宙論〔台〕
steady-state model　稳恒态模型｜穩態模型〔台〕
steady-state theory　稳恒态学说｜穩態學說〔台〕
steady sunspot　稳态黑子
steady velocity　稳定速度
Stebbins crater　斯特宾斯环形山（月球）
Stebbins-Whitford photometry　斯特宾斯－惠特福德测光
Stebbins-Whitford system　斯特宾斯－惠特福德测光系统
steel-yard clepsydra　秤漏
Steen River crater　斯廷河陨星坑（地球）
steepest descent method　最速下降法

steep spectrum　陡谱
steep-spectrum quasar　陡谱类星体
steep-spectrum radio quasar　陡谱射电类星体
steep-spectrum radio source　陡谱射电源
steep-spectrum source　陡谱源
steerability　可动性
steerable antenna　可动天线
steerable radio telescope　可动射电望远镜
Stefan-Boltzmann constant　斯特藩－玻尔兹曼常数｜史特凡－波兹曼常数〔台〕
Stefan-Boltzmann law　斯特藩－玻尔兹曼定律
Stefan crater　斯特藩环形山（月球）
Stefan's constant　斯特藩常数
Stefan's law　斯特藩定律｜史特凡定律〔台〕
Stein Crater Field crater　斯坦环形山群（金星）
Steins　斯坦因斯（2867号小行星）
Stella Mira　刍藁增二（鲸鱼座 o）　又见：*Mira*
stellar aberration　恒星光行差
stellar abundance　恒星元素丰度
stellar activity　恒星活动
stellar age　恒星年龄
stellar aggregate　星集　又见：*stellar aggregation*
stellar aggregation　星集　又见：*stellar aggregate*
stellar angle　非旋转原点时角，恒星角
stellar association　星协　又见：*star association*
stellar astronomy　恒星天文学
stellar astrophysics　恒星天体物理｜恆星天文物理學〔台〕
stellar atlas　星图集｜星圖〔台〕　又见：*star atlas*
stellar atmosphere　恒星大气
stellarator　仿星器
stellar birth　恒星诞生
stellar birthrate　恒星产生率
stellar black hole　恒星级黑洞
stellar brightness　恒星亮度
stellar cannibalism　恒星吞食
stellar cataclysm　恒星激变
stellar catastrophe　恒星灾变
stellar chain　星链
stellar chart　星图册　又见：*star chart*
stellar chromosphere　恒星色球
stellar classification　恒星分类
stellar cluster　星团　又见：*star cluster; cluster of stars; cluster*
stellar collapse　恒星坍缩｜恆星塌縮〔台〕
stellar collision　恒星碰撞
stellar complex　恒星复合体　又见：*star complex*

stellar component　恒星成分
stellar constitution　恒星结构　又见：*stellar structure*
stellar corona　星冕
stellar cosmogony　恒星演化学　又见：*astrogony*
Stellar Data Center　（＝SDC）恒星资料中心
stellar diameter　恒星直径
stellar disk　恒星圆面
stellar dynamics　恒星动力学
stellar dynamo　恒星发电机
stellar eclipse　星食
stellar embryo　恒星胎　又见：*embryo of a star*
stellar encounter　恒星交会
stellar envelope　恒星包层
stellar environment　恒星环境
stellar evolution　恒星演化
stellar evolution chronometer　恒星演化时计
stellar field　星场
stellar flare　恒星耀斑｜恆星閃焰〔台〕
stellar flat field　恒星平场
stellar formation　恒星形成　又见：*star formation; formation of stars*
stellar group　恒星群
stellar guidance　恒星导航　又见：*star navigation*
stellar halo　恒星晕
stellar image　星像　又见：*star image*
stellar inertial guidance　恒星惯性导航
stellar infrared radiation　恒星红外辐射　又见：*infrared stellar radiation*
stellar intensity interferometer　（＝SII）恒星强度干涉仪
stellar interferometer　恒星干涉仪
stellar interferometry　恒星干涉测量
stellar interior　恒星内部
stellar interior structure　恒星内部结构
stellar jet　恒星喷流
stellar kinematics　恒星运动学
stellar luminosity　恒星光度
stellar luminosity class　恒星光度级
stellar magnetic field　恒星磁场
stellar magnitude　星等　又见：*magnitude*
stellar map　星图　又见：*star map; atlas*
stellar mass　恒星质量
stellar-mass black hole　恒星质量黑洞
stellar mass function　（＝SMF）恒星质量函数
stellar metallicity　恒星金属丰度
stellar model　恒星模型
stellar motion　恒星运动　又见：*star motion*

stellar nomenclature 恒星命名 又见：*star nomenclature*
stellar occultation 星掩源
stellar orbit 恒星轨道
stellar oscillation 恒星振荡
stellar parallax 恒星视差
stellar photometer 恒星光度计 又见：*star photometer*
stellar photometry 恒星测光 又见：*star photometry*
stellar photosphere 恒星光球
stellar physics 恒星物理学
stellar polarimeter 恒星偏振计
stellar polarimetry 恒星偏振测量
stellar polarization 恒星偏振
stellar population 星族 又见：*population of stars; population*
stellar radiation 恒星辐射
stellar radio astronomy 恒星射电天文学
stellar radio radiation 恒星射电
stellar radio source 恒星射电源
stellar reddening 恒星红化
stellar reference system 恒星参考系
stellar remnant 恒星遗迹
stellar ring 恒星环
stellar rotation 恒星自转
stellar seismology 星震学 又见：*asteroseismology; astroseismology*
stellar spectrograph 恒星摄谱仪
stellar spectroscope 恒星分光仪
stellar spectroscopy 恒星光谱学
stellar spectrum 恒星光谱
stellar speedometer 恒星视向速度仪 又见：*radial-velocity spectrometer*
stellar stability 恒星稳定性
stellar statistics 恒星统计 | 恒星統計 [學]〔台〕
stellar structure 恒星结构 又见：*stellar constitution*
stellar surface 恒星表面
stellar system 恒星系统 又见：*star system*
stellar temperature 恒星温度
stellar ultraviolet radiation 恒星紫外辐射
stellar universe 恒星宇宙
stellar wind 星风 | 恆星風〔台〕
stellar X-ray radiation 恒星X射线辐射
stellar γ-ray radiation 恒星γ射线辐射
STEP （＝Solar-Terrestrial Energy Programme）日地能源计划
step by step method 逐步法

Stephano 天卫二十
Stephan's Quintet 斯蒂芬五重星系 | 崐史特凡五重星系〔台〕
step length 步长 又见：*step size*
step method 光阶法 即：*Argelander method*
stepped atomic time （＝SAT）跳跃原子时 | 步進原子時〔台〕
stepped filter 阶梯滤光片 | 階梯濾光板〔台〕
stepped sector 阶梯遮光板
stepped slit 阶梯狭缝
stepped weakener 阶梯减光板
step rocket 多级火箭 又见：*multi-stage rocket; rocket train*
step size 步长 又见：*step length*
stepwise regression 逐步回归
steradian （＝sr）球面度
STEREO （＝Solar Terrestrial Relation Observatory）日地关系观测台（美国）
STEREO A 前导空间观测台
STEREO B 后随空间观测台
stereo-comparator 体视比较仪
stereogram 球极投影图
stereographic grid 球极投影格网 | 球極平面投影網〔台〕
stereographic mapping 球极绘制
stereographic net 球极投影网
stereographic projection 球极投影 | 球極平面投影〔台〕
stereoscope 体视镜 | 立體鏡〔台〕
stereoscopic observation 体视观测
stereoscopic photography 体视照相 | 立體照相〔台〕
stereotelescope 体视望远镜 又见：*tele-stereoscope*
stereotheodolite 体视经纬仪 | 立體經緯儀〔台〕
sterile neutrino 惰性中微子
Stern-Gerlach experiment 斯特恩－盖拉赫实验
Stern molecular beam technique 斯特恩分子束技术
Sterope 昴宿三（金牛座21）又见：*Asterope*
Stetson crater 斯特森环形山（月球）
Steward Observatory 斯图尔德天文台
sticking coefficient 黏附系数
stigmatic camera 消像散照相机
stigmatic optics 消像散光学
stigmatic spectrograph 消像散摄谱仪
still image 静止像
stimulated emission 受激发射
stimulated radiation 受激辐射

Stingray Nebula 刺魟星云（Hen-1357）
STJ （＝superconducting tunnel junction）超导隧道结
St. John crater 圣约翰环形山（月球）
stochastic acceleration 随机加速
stochastic bias 随机偏袒
stochastic dependence 随机相依
stochastic inflation 随机暴胀
stochastic orbit 随机轨道
stochastic plasma heating 随机等离子体加热
stochastic process 随机过程
stochastic variable 随机变量　又见：*random variable*
Stokes crater 斯托克斯环形山（月球）
Stokes meter 斯托克斯参数测量仪
Stokes' number 斯托克斯参数｜史托克士參數〔台〕　又见：*Stokes' parameter*
Stokes' parameter 斯托克斯参数｜史托克士參數〔台〕　又见：*Stokes' number*
Stokes parameters 斯托克斯参数
Stokes polarimetry 斯托克斯偏振测量
Stomach 胃宿（二十八宿）
Stonehenge 巨石阵
stony-iron meteorite 石铁陨星｜石隕鐵〔台〕
　　又见：*lithosiderite; siderolite*
stony meteorite 石陨星，石陨石｜石質隕石〔台〕
　　又见：*asiderite; aerolith; aerolite*
stop band 不透射带
Stopping-place 昴宿（二十八宿）　又见：*Ball of wool*
stopping point 爆炸点
stop-watch 停表｜停錶，馬錶〔台〕
storage beam tube 束管　又见：*beam tube*
storage bulb 储存泡
storage capacity 存储容量
stored energy 储能
storm burst 噪暴｜雜訊爆〔台〕　又见：*noise storm*
storm radiation 噪暴辐射
Strabo crater 斯特拉博环形山（月球）
straight angle 平角
straight fan 直扇状流
straight spectrum 直线谱
straight tail 直彗尾
straight wall 竖墙｜地塹〔台〕
strain 应变
strainmeter 应变计
strain tensor 应变张量
strange attractor 奇引源

strangeness 奇异性
strangeness number 奇异数
strange particle 奇异粒子　又见：*exotic particle*
strange star 奇异星　又见：*exotic star*
Strangways crater 斯特兰韦斯陨星坑（地球）
stratification 层化
stratified rock 层状岩
stratigraphy 地层学
straton 层子
stratopause 平流层顶
stratosphere 平流层
Stratospheric Observatory for Infrared Astronomy （＝SOFIA）索菲亚平流层红外天文台｜索菲雅[平流層紅外天文台]〔台〕
stratospheric telescope 平流层望远镜
Stratton crater 斯特拉顿环形山（月球）
stratus cloud 层云
stray 天电　又见：*atmospherics; sferics*
stray factor 杂散因子
stray field 杂散场
stray light 杂散光
streak photography 条层照相
streaming flow 流注流
streaming motion 流注运动
streamline 冕流，流线｜流線〔台〕
Street crater 斯特里特环形山（月球）
strehl raio 斯特列尔比
strength 强度　又见：*intensity*
strengthening 强化
stress 应力
stress distribution 应力分布
stressed Ge:Ga infrared array 应力型锗掺镓红外面阵
stress strain relation 应力应变关系
stress tensor 应力张量
stretching force 张力　又见：*tensile force*
strewn field 散播场
Stride 奎宿（二十八宿）　又见：*Legs*
striding level 跨水准｜跨水準儀〔台〕
string cosmology 弦宇宙学
string network 弦网络
string scale 弦标度
string spacetime 弦时空
string theory 弦论
strip brightness 带亮度
strip chart 带状图
strip-chart recorder 纸带记录仪

stripped nucleus 裸核 又见: *bare nucleus*
stripped plasma 全电离等离子体
stripscan 带状扫描
stroboscope 闪光仪
Ströemberg asymmetrical drift 斯特龙贝格不对称漂移
Ströemberg color system 斯特龙贝格颜色系统
Ströemberg diagram 斯特龙贝格图
Ströemgren four-color index 斯特龙根四色指数 | 史壯格倫四色指數〔台〕
Ströemgren four-color photometry 斯特龙根四色测光
Ströemgren photometry 斯特龙根测光
Ströemgren radius 斯特龙根半径 | 史壯格倫半徑〔台〕
Ströemgren sphere 斯特龙根球区 | 史壯格倫球〔台〕
Ströemgren system 斯特龙根系统
Strömgren crater 斯特龙根环形山（月球）
Strong Anthropic Principle 强人择原理
strong burst 强暴 又见: *intense burst*
strong correlation 强相关
strong encounter 强交会
strong energy condition 强能量条件
strong equivalence principle 强等效原理 | 強等效原則〔台〕
strong force field 强力场
strong gravitational lensing 强引力透镜
strong interaction 强相互作用 | 強交互作用〔台〕
strong Jeans theorem 强金斯定理
strong nuclear interaction 强核相互作用
strong source 强源 又见: *intense source*
strong stability 强稳定性
strong turbulence 强湍动
strontium （＝Sr）锶（38号元素）
strontium age-dating 锶测年
strontium method 锶测年法
strontium star 锶星
structural geology 构造地质学 | 構造地質[學]〔台〕
structure formation 结构形成
structure formation problem 结构形成问题
Struve crater 斯特鲁维环形山（月球）
STSAT （＝Science and Technology SATellite）科学技术卫星（韩国）
STScI （＝Space Telescope Science Institute）空间望远镜研究所 | 太空望遠鏡[科學]研究所〔台〕（美国）
STSDAS （＝Space Telescope Science Data Analysis System）空间望远镜科学数据分析系统
S-type asteroid ① S 型小行星 ② S 型小行星
S-type star S 型星 又见: *S star*
Styx 冥卫五
Sualocin 瓠瓜一（海豚座 α）
subaperture 子孔径
sub-arcsecond radio astronomy 亚角秒射电天文学
Subaru Telescope 昴星团望远镜 | 昴望遠鏡〔台〕
subastral point 星下点 又见: *substellar point*
subatomic particle 亚原子粒子
subaurural latitude 极光下点纬度
subband 次能带
Subbotin crater 苏博京环形山（月球）
subcategory 子范畴
subcenter 副中心, 子中心
subclass 次型
subcluster 次团
subclustering 次成团
subcondensation 子凝聚物
sub-damped Lyα system （＝sub-DLAs）亚阻尼莱曼 α 系统
sub-DLAs （＝sub-damped Lyα system）亚阻尼莱曼 α 系统
subdwarf 亚矮星 | 次矮星〔台〕
subdwarf B star B 型亚矮星
subdwarf O star O 型亚矮星
subdwarf sequence 亚矮星序 | 次矮星序〔台〕 又见: *sequence of sub-dwarfs*
subflare 亚耀斑 | 次閃焰〔台〕
subflare kernel 亚耀斑核
subgiant 亚巨星 | 次巨星〔台〕
subgiant branch （＝SGB）亚巨星支 | 次巨星序〔台〕
subgiant spiral galaxy 亚巨旋涡星系
subgroup 子群
subhalo 子晕
subhalo mass function 子晕质量函数
subharmonic 副谐波
sub-horizon evolution 亚视界演化
subjective catalog 主观星表
sublimation 升华
subluminous star 亚光度恒星 | 次光度恆星〔台〕
sublunar point 月下点
submanifold 子流形
submarine relief 海底起伏

submarine ridge 海脊
submarine structure 海底结构
submarine trench 海沟
submarine valley 海谷
submarine volcano 海底火山
submicroscopic particle 亚微观粒子
submilli-arc-second astrometry 亚毫角秒天体测量
submilli-arc-second optical astrometry 亚毫角秒光学天体测量
submilli-arc-second radio astrometry 亚毫角秒射电天体测量
Sub-Millimeter Array （＝SMA）SMA 亚毫米波射电望远镜阵 | 次毫米波 [電波望遠鏡] 陣〔台〕
submillimeter astronomy 亚毫米波天文学 | 次毫米波天文學〔台〕 又见：*submillimeter-wave astronomy*
submillimeter blackground 亚毫米波背景
submillimeter observatory 亚毫米波天文台
submillimeter photometry 亚毫米波测光
submillimeter radio astronomy 亚毫米波射电天文学
submillimeter source 亚毫米波源
submillimeter space astronomy 亚毫米波空间天文学
Submillimeter Telescope （＝SMT）SMT 亚毫米波望远镜
Submillimeter Telescope Observatory （＝SMTO）亚毫米波天文台
submillimeter wave 亚毫米波 | 次公釐波，次毫米波〔台〕
submillimeter-wave astronomy 亚毫米波天文学 | 次毫米波天文學〔台〕 又见：*submillimeter astronomy*
Submillimeter Wave Astronomy Satellite （＝SWAS）SWAS 亚毫米波天文卫星 | 次毫米波天文衛星〔台〕
submillimetre astronomy 亚毫米波天文学
Submillimetre Common User Bolometer Array （＝SCUBA）亚毫米波普通用户辐射热计阵列
submillisecond optical pulsar 亚毫秒光学脉冲星
submillisecond pulsar 亚毫秒脉冲星
submodulator 副调制器
suboceanic structure 洋底结构
subordinate line 辅线
subpoint 下点
subpulse 次脉冲
Subra 轩辕十五（狮子座 o）
subradar point 雷达下点

subreflector 副反射面 又见：*secondary reflector*
subroutine 子例程
subsatellite 子卫星
subsatellite point 卫星下点
subscript 下标
subsolar point 日下点
subsolar point altitude 日下点高度
subsolar point temperature 日下点温度
subspace 子空间
substellar companion 亚恒星伴星
substellar object 亚恒星天体 | 次恆星天體〔台〕
substellar point 星下点 又见：*subastral point*
substellar system 亚恒星系
substile 日晷指针方向线 又见：*substyle*
substorm 亚暴 | 次爆〔台〕
substructure 子结构
substyle 日晷指针方向线 又见：*substile*
subsynchronism 亚同步
subsynchronous rotation 亚同步自转
subsystem 次系 | 次 [星] 系〔台〕
subtracting double-pass spectrograph 相减式双通摄谱仪
subtype 次型 | 次 [光谱] 型〔台〕
successive approximation 逐次近似法 | 逐次逼近法，逐次近似法〔台〕
successive consequent 后继点系列
Sudbury crater 萨德伯里陨星坑（地球）
sudden cosmic noise absorption （＝SCNA）宇宙噪声突降 | 宇宙雜訊突然吸收，宇宙雜訊突減〔台〕
sudden disappearance of filament 暗条突逝 又见：*filament sudden disappearance*
sudden enhancement of atmospherics （＝SEA）天电突增
sudden field anomaly （＝SFA）场强突异
sudden frequency drift （＝SFD）频率突漂
sudden ionospheric disturbance （＝SID）电离层突扰 | 游離層突發性擾動〔台〕
sudden magnetic disturbance （＝SMD）急始磁扰
sudden phase anomaly （＝SPA）位相突异 | 相位突異〔台〕
sudden production 突然产生
sudden short-wave fade-out （＝SSWF）短波突衰 即：*Mögel-Dellinger effect*
Sudingqiang 苏定强（小行星 19366 号）
Sudongpo 苏东坡（小行星 145588 号）
Suhail 老人星（船底座 α） 又见：*Canopus*

Suiqizhong　广州七中（小行星145546号）
Suisei　彗星号探测器（日本）
Sulafat　渐台三（天琴座 γ）　又见：*Sulaphat*
Sulaphat　渐台三（天琴座 γ）　又见：*Sulafat*
sulcus（复数：sulci）　皱沟
sulfur dioxide　二氧化硫（SO_2）
sulfur monoxide　一氧化硫（SO）
sulfur/sulphur　（＝S）硫（16号元素）
sulphuric acid　硫酸
summation convention　求和约定
summer　夏｜夏[季]〔台〕
Summer Solstice　夏至（节气）　又见：*June Solstice*
summer solstice　夏至点　又见：*June solstice; first point of Cancer*
summer time　夏令时｜日光節約時間〔台〕
　又见：*daylight saving time*
Summer Triangle　夏夜大三角（星组）
Sumner crater　萨姆纳环形山（月球）
Sumner line　位置线，萨姆纳线
sun　太阳
sundial　日晷｜日晷，日规〔台〕　又见：*dial*
Sundman theorem　宋德曼定理
sundog　幻日　又见：*mock sun; parhelion*
Sunflower Galaxy　葵花星系（M63）
sungrazer　掠日彗星，掠日天体
sungrazing comet　掠日彗星
sunlight　日光　又见：*day light*
Sun-like activity　类太阳活动
Sun-like oscillation　类太阳振荡　又见：*solar-like oscillation*
Sun-like star　类太阳恒星　又见：*solar-like star*
sunlit aurora　日耀极光
sunlit surface　日耀表面
sunpillar　日柱
sunrise　日出
sunrise terminator　日出明暗界
sunseeker　寻日器
sunset　日没
sunset terminator　日没明暗界
sunshine　日照
sun's latitude　日面纬度　又见：*heliolatitude; heliographic latitude; solar latitude*
sun's longitude　日面经度　又见：*heliolongitude; heliographic longitude*
sunspot　太阳黑子｜[太阳]黑子〔台〕　又见：*solar spot*
sunspot component　黑子分量

sunspot cycle　黑子周｜黑子周期〔台〕
sunspot flare　黑子耀斑｜黑子闪焰〔台〕
sunspot group　黑子群｜[太阳]黑子群〔台〕
　又见：*spot group*
sunspot maximum　黑子极大期
sunspot minimum　黑子极小期
sunspot number　黑子数　又见：*spot number*
sunspot penumbra　黑子半影　又见：*spot penumbra*
sunspot polarity　黑子极性
sunspot prominence　黑子日珥
sunspot radiation　黑子辐射｜黑子辐射，日斑辐射〔台〕
sunspot relative number　黑子相对数
　又见：*relative number of spots*
sunspot spectrum　黑子光谱
sunspottedness　有黑子态
sunspot umbra　黑子本影　又见：*spot umbra*
sunspot zone　黑子带｜[太阳]黑子带〔台〕
sunward tail　向日彗尾
sun-weather relationship　太阳－天气关系
Sunyaev-Zel'dovich cooling　（＝S-Z cooling）苏尼阿耶夫－泽尔多维奇冷化，SZ 冷化
Sunyaev-Zel'dovich distortion　苏尼阿耶夫－泽尔多维奇畸变
Sunyaev-Zel'dovich effect　（＝S-Z effect）苏尼阿耶夫－泽尔多维奇效应，SZ 效应｜蘇尼阿耶夫－澤爾多維奇效應〔台〕
super-achromatic wave-plate　超消色散波片
superadiabatic convection　超绝热对流
superaerodynamics　高空空气动力学｜高空熱氣動力學〔台〕
superassociation　超星协
superbolide　超火流星
super-Chandrasekhar model　超钱德拉塞卡模型
super-civilization　超级文明世界
supercluster　超星系团　又见：*galaxy supercluster*
superconducting hot electron bolometer mixer　（＝HEB mixer）超导 HEB 混频器
superconducting quantum interference device　（＝SQUID）超导量子干涉器件
superconducting tunnel junction　（＝STJ）超导隧道结
superconductive cosmic string　超导宇宙弦
superconductivity　超导性
superconductor-insulator-superconductor mixer　（＝SIS mixer）超导 SIS 混频器
supercooling　过冷却
supercorona　超冕，超日冕｜超日冕〔台〕

Supercosmos 超级多功能底片测量仪
superdense star 超密星 | 超密 [恒] 星〔台〕
super-Eddington accretion 超爱丁顿吸积
superelastic collision 超弹性碰撞
super energetic stellar explosion 超大能量恒星爆发
superexcitation 超激发
superfluid 超流体
superfluid core 超流体核 | 超流體星核〔台〕
superfluid gyroscope 超流回转仪
superfluidity 超流性
supergalactic coordinate 超星系坐标
supergalactic plane 超星系道面
supergalactic streamer 超星系流状结构
supergalaxy 超星系 又见: *hypergalaxy*
supergiant 超巨星 又见: *supergiant star*
supergiant elliptical 超巨椭圆星系
  又见: *supergiant elliptical galaxy*
supergiant elliptical galaxy 超巨椭圆星系
  又见: *supergiant elliptical*
supergiant galaxy 超巨星系
supergiant molecular cloud （=SGMC）超巨分子云
supergiant radio galaxy 超巨射电星系
supergiant star 超巨星 又见: *supergiant*
supergranular cell 超米粒元胞 | 超米粒組織泡〔台〕
supergranulation 超米粒组织
  又见: *hypergranulation*
supergranule 超米粒 | 超米粒組織〔台〕
  又见: *hypergranule*
supergravity 超引力
superheavy boson 超重波色子
superheterodyne receiver 超外差接受机
super-high density 超高密度
super-horizon perturbation 超视界扰动
super-horizon solution 超视界解
superhump 长驼峰
superhumper 长驼峰星
superhump period 长驼峰周期
super-inflation 超暴胀
superionization 超电离
superior conjunction 上合
superior eclipse limit 上食限
superior epoch 上元 又见: *grand origin*
superior planet 外行星 | 地外行星〔台〕
superior transit 上中天 又见: *upper culmination; upper transit*

Super-K （=Super Kamiokande）超级神冈（日本中微子探测器）
Super Kamiokande （=Super-K）超级神冈（日本中微子探测器）
superlight source 超光速源 又见: *superluminal source*
superluminal illusion 超光速幻觉
superluminal jet 超光速喷流
superluminal motion 超光速运动
superluminal source 超光速源 又见: *superlight source*
superluminous star 超高光度星 | 高光度恆星〔台〕
supermagnetosonic turbulent 超磁性声速湍流
supermagnetosonic wave 超磁性声波
supermassive black hole （=SMBH）特大质量黑洞 | 超大質量黑洞〔台〕
supermassive rotator 特大质量旋体 | 超大質量天體〔台〕
supermassive star 特大质量星 | 超大質量恆星〔台〕 又见: *heavy weight star*
supermaximum 长极大 | 超極大〔台〕
super-metal-poor star 特贫金属星
  又见: *ultrametal-poor star*
super-metal-rich star 特富金属星 | 超量金屬星〔台〕
supermultiplet 超多重线 | 超多重 [谱] 線〔台〕
supernova classification 超新星分类
Supernova Cosmology Project （=SCP）超新星宇宙学计划
supernova ejecta 超新星抛射物
supernova envelope 超新星包层
supernova explosion 超新星爆发 又见: *supernova outburst*
supernova feedback 超新星反馈
supernova Hubble diagram 超新星哈勃图
Supernova Legacy Survey （=SNLS）超新星遗珍巡天
supernova neutrino 超新星中微子
Supernova of 1054 1054 年超新星 | 1054 超新星〔台〕（CM Tau）
supernova outburst 超新星爆发 又见: *supernova explosion*
supernova radio souce 超新星射电源
supernova rate 超新星频数
supernova remnant （=SNR）超新星遗迹 | 超新星殘骸〔台〕 又见: *relic of supernova; remnant of supernova*
supernova shell 超新星壳层
supernova shock 超新星激波

supernova survey　超新星巡天
supernova-γ-ray burst connection　超新星－γ暴关联
supernova（复数：supernovae）　（＝SN）超新星
superoutburst　长爆发
superpartner　超对称伙伴
superpenumbra　超半影
superpenumbral region　超半影区
superplanet　超大行星
superposition　叠置
superpotential　超势
super-relativistic effect　极端相对论性效应
super resonance region　超级共振区
super-rotation　特快自转
supersaturation　过饱和
super-Schmidt　超施密特
super-Schmidt camera　超施密特照相机
super-Schmidt telescope　超施密特望远镜
super-short period Cepheid　超短周期造父变星
　　又见：*ultra-short-period Cepheid*
supersoft X-ray source　超软 X 射线源 | 超軟 X 光源〔台〕
supersonic accretion　超声速吸积
supersonic aerodynamics　超声速气体动力学
supersonic flow　超声流
supersonic speed　超声速　又见：*supersonic velocity*
supersonic velocity　超声速　又见：*supersonic speed*
super star cluster　超星团
superstring　超弦
superstring theory　超弦理论
supersymmetric particle　超对称粒子
supersymmetry　超对称
supersynthesis aperture　超综合孔径
supervoid　超巨洞
superwind galaxy　超级风星系
supplementary angle　补角
supplementary condition　附加条件
supporting system　支承系统　又见：*support system*
support mechanism　支承机制
support system　支承系统　又见：*supporting system*
supposed value　假设值 | 假設值，可能值〔台〕
suppressed-lobe interferometer　抑瓣干涉仪 | 折瓣干涉儀〔台〕
suprathermal bremsstrahlung　超热韧致辐射
suprathermal electron　超热电子

supreme epoch　太极上元　又见：*supreme ultimate grand origin*
Supreme Subtlety Enclosure　太微垣
supreme ultimate grand origin　太极上元
　　又见：*supreme epoch*
surface acoustic wave　表面声波
surface activity　表面活动
surface brightness　面亮度 | 表面亮度〔台〕
surface brightness fluctuation　面亮度起伏
surface-brightness fluctuation　面亮度起伏，表面亮度涨落
surface-brightness fluctuation technique　面亮度起伏方法
surface brightness method　面亮度法
surface brightness profile　面亮度轮廓
surface brightness temperature　面亮度温度
surface channel CCD　表面型 CCD
surface charge　面电荷
surface density　面密度
surface element　面元
surface error　表面公差
surface flux　表面流量
surface gravity　表面重力
surface of constant geopotential　等重力势面
surface of discontinuity　间断面，突变面
surface of section　截面　又见：*cross section*
surface of zero velocity　零速度面　又见：*zero velocity surface*
surface photometer　面源光度计
surface photometry　面源测光　又见：*area photometry*
surface temperature　表面温度
surface transformation　表面变换
surface wave　表面波
surge　日浪 | 涌浪日珥〔台〕
surge prominence　冲浪日珥 | 涌浪日珥〔台〕
Surtur　土卫四十八
surveillance network　监测台网
survey　巡天　又见：*patrol survey; all-sky patrol*
survey catalogue　巡天星表
surveying astronomy　巡天天文
Surveyor　探测者号（美国月球探测器）
survey telescope　巡天望远镜
survivability　存活概率
SUSI　（＝Sydney University Stellar Interferometer）悉尼大学恒星干涉仪
suspected nova　疑似新星

suspected supernova 疑似超新星
suspected variable 疑似变星
suspended level 悬水准　又见：*hanging level*
suspended standard instrument 悬正仪
Suttungr 土卫二十三
SU UMa star 大熊SU型星
Suzaku 朱雀号（日本X射线天文卫星）
　又见：*Astro-EII*
Suzhou 苏州（小行星2719号）
Suzhou Planisphere 苏州石刻天文图
　又见：*Soochow planisphere; Soochow astronomical chart*
Suzhousanzhong 苏州三中（小行星5013号）
SVC （＝slowly varying component）缓变分量
SVD （＝singular value decomposition）奇异值分解
SVO （＝Spanish Virtual Observatory）西班牙虚拟天文台
SVOM （＝Space Variable Objects Monitor）空间变源监视器（中法高能天文卫星）
SVS （＝Soviet variable star）苏联变星
S Vul star 狐狸S型星
Swan band 斯旺谱带 | 史萬譜帶〔台〕
swan-like structure 天鹅云状结构
Swan Nebula 天鹅星云（M17）　即：*Omega Nebula; ω Nebula*
SWAS （＝Submillimeter Wave Astronomy Satellite）SWAS亚毫米波天文卫星 | 次毫米波天文衛星〔台〕
S wave S波
Sweden-ESO Submillimetre Telescope （＝SEST）瑞典－欧南台亚毫米波望远镜，SEST亚毫米波望远镜
sweep-frequency receiver 扫频接收机
　又见：*panoramic receiver*
sweeping star 扫帚星
sweep rate 扫描速率
SWEEPS （＝Sagittarius Window Eclipsing Extrasolar Planet Search）人马窗口凌星系外行星搜索
swept charge device （＝SCD）扫式电荷器件
swept-frequency interferometer 扫频干涉仪
swept-lobe interferometer 扫瓣干涉仪　又见：*lobe sweeping interferometer*
SWF （＝short-wave fade-out）短波衰退
Swift （＝Swift Gamma-ray Burst Explorer）雨燕γ射线暴探测器（美国）
Swift Gamma-ray Burst Explorer （＝Swift）雨燕γ射线暴探测器（美国）
Swift-Tuttle Comet 斯威夫特－塔特尔彗星

Swift γ-Ray Burst Mission 雨燕γ射线暴任务
Swift γ-ray observatory 雨燕γ射线天文台（美国卫星）　即：*Swift*
swing amplification 摆动放大
swing-by trajectory 借力飞行弹道 | 借力飛行軌道〔台〕
swinging arc 摆弧
swinging axis 摆动轴
swirl 涡流 | 旋涡〔台〕　又见：*eddy; vortex flow*
switched receiver 开关接收机
switching time 开关时间
switch-signal generator 开关信号发生器
SW Sex star 六分仪SW型星 | 六分儀[座]SW型星〔台〕
SX Ari star 白羊SX型星
SX Phe star 凤凰SX型星 | 鳳凰[座]SX型星〔台〕
SXR （＝soft X-ray repeater）软X射线再现源 | 軟X光重覆爆發源〔台〕
SXT （＝soft X-ray transient）软X射线暂现源 | 軟X光瞬變源〔台〕
Sycorax 天卫十七
Sydney list 悉尼射电星表
Sydney University Stellar Interferometer （＝SUSI）悉尼大学恒星干涉仪
Sylvester crater 西尔维斯特环形山（月球）
Sylvia 林神星（小行星87号）
symbiotic binary 共生双星
symbiotic Mira 共生刍藁
symbiotic nova 共生新星
symbiotic recurrent nova 共生再发新星
symbiotic star 共生星
symbiotic variable 共生变星　又见：*combination variable*
symmetrical mounting 对称型装置
symmetric top 对称陀螺
symmetry 对称
sympathetic flare 感生耀斑 | 和應閃焰〔台〕
sympathetic radio burst 相应射电暴 | 和應電波爆發〔台〕
sympathetic X-ray burst 相应X射线暴
sympathetic γ-ray burst 相应γ射线暴
Symphonie 交响曲号（法德美通信卫星）
symplectic group 辛群
synchro-Compton radiation 同步康普顿辐射
synchro-cyclotron radiation 同步回旋加速辐射
synchrone 等时线　又见：*isochronal line*
synchronism 同步性 | 同步〔台〕
synchronization 同步化

synchronous detection 同步检测
synchronous detector 同步检测器
synchronous emission 同步发射
synchronous gauge 同步规范
synchronous observation 同步观测
synchronous orbit 同步轨道
synchronous rotation 同步绕转
synchronous satellite 同步卫星
synchrotron damping 同步加速阻尼
synchrotron damping rate 同步加速阻尼速率
synchrotron emission 同步加速发射
synchrotron instability 同步加速不稳定性
synchrotron loss 同步加速耗损
synchrotron mechanism 同步辐射机制
synchrotron radiation 同步加速辐射 | 同步加速輻射，磁阻尼輻射〔台〕
synchrotron radiation source 同步加速辐射源
   又见：*synchrotron source*
synchrotron self-absorption （＝SSA）同步加速自吸收
synchrotron source 同步加速辐射源
   又见：*synchrotron radiation source*
synchrotron spectrum 同步加速辐射谱
sync pulse 同步脉冲
syndyname 等力线 又见：*isodynamic line*
synodic month 朔望月，太阴月 又见：*lunar month; moon month; month of the phases; lunation*
synodic motion 会合运动
synodic period 会合周期
synodic revolution 会合周
synodic rotation 会合自转 | 會合自轉周〔台〕
synodic rotation number 会合自转周数
synodic rotation period 会合自转周期
synodic year 会合年
synoptic chart 全景图
synoptic map of the sun 太阳综合图
synthesis aperture 综合孔径 | 孔徑合成〔台〕
   又见：*synthetic aperture; aperture synthesis; synthesized aperture*
synthesis imaging 综合成像
synthesis map 综合图 又见：*synthesizer*
synthesis radio telescope 综合孔径射电望远镜 | 孔徑合成電波望遠鏡〔台〕 又见：*aperture synthesis radiotelescope*
synthesis telescope 综合孔径望远镜 | 孔徑合成望遠鏡〔台〕

synthesized aperture 综合孔径 | 孔徑合成〔台〕
   又见：*synthetic aperture; aperture synthesis; synthesis aperture*
synthesized beam 综合波束
synthesized pattern 综合方向图
synthesizer 综合图 又见：*synthesis map*
synthetic aperture 综合孔径 | 孔徑合成〔台〕
   又见：*aperture synthesis; synthesis aperture; synthesized aperture*
synthetic-aperture radar （＝SAR）综合孔径雷达 | 合成孔徑雷達〔台〕
synthetic focal array 焦面综合阵
synthetic light-curve 综合光变曲线
synthetic observation 内插同步观测
synthetic quartz 人造石英
synthetic spectrum 合成光谱
Syrma 亢宿二（室女座ι）
systematic betatron process 系统感应加速过程
systematic deviation 系统偏差，系统偏离
systematic error 系统误差 | 系統[誤]差〔台〕
   又见：*biased error*
systematic Fermi process 系统费米过程
systematic velocity 系统速度 | 質心速度〔台〕
system III longitude 系统 III 经度
system II longitude 系统 II 经度
system I longitude 系统 I 经度
system noise 系统噪声 | 系統雜訊〔台〕
system noise temperature 系统噪声温度
system of astronomical constants 天文常数系统
   又见：*astronomical constant system*
system of coordinate 坐标系 又见：*coordinate system*
system of fixed ecliptic 固定黄道坐标系
system of particles 质点系
system of rays 辐射纹系统
system temperature 系统温度
sytall 微晶玻璃 又见：*cervit; zerodur; glass-ceramic*
syzygetic tide 朔望潮 又见：*syzygial tide*
syzygial tide 朔望潮 又见：*syzygetic tide*
syzygy 朔望
S-Z cooling （＝Sunyaev-Zel'dovich cooling）苏尼阿耶夫－泽尔多维奇冷化，SZ 冷化
S-Z effect （＝Sunyaev-Zel'dovich effect）苏尼阿耶夫－泽尔多维奇效应，SZ 效应 | 蘇尼阿耶夫－澤爾多維奇效應〔台〕
Szilard crater 齐拉特环形山（月球）

T

Ta （=tantalum）钽（73 号元素）
Tabit 参旗六（猎户座 π）
table of the moon 月球运动表
tabular interval 列表间隔
tachocline 差旋层（太阳）
tachyon 快子 | 超光速粒子〔台〕
tachyonic field 快子场
Tacubaya National Observatory 塔库巴亚国立天文台（墨西哥）
tadpole-shaped curve of zero velocity 蝌蚪形零速度线
taenite 白沸石 | 鎳紋石〔台〕
Tagish Lake meteorite 塔吉什湖陨石
TAI （=Temp Atomique Internationale）【法】国际原子时 即：International Atomic Time
taikonaut 太空人，航天员 即：astronaut; cosmonaut
Tail 尾宿（二十八宿）
tail axis 彗尾轴
tail band 彗尾谱带
tail-disconnection event 断尾事件 又见：disconnection event
tail kink 彗尾扭结
tail knot 彗尾结
tail of comet 彗尾 又见：cometary tail; comet tail
tail of the earth 地尾
tail radio galaxy 带尾射电星系 | 電波尾星系〔台〕 又见：radio-tail galaxy
tail ray 彗尾射线
tail streamer 彗尾流束
Taiwan 台湾（小行星 2169 号）
Taiyo 太阳号（日本高层大气和太阳观测卫星） 又见：Solar Radiation and Thermospheric Satellite
Taiyuan 太原（小行星 2514 号）
Talcott level 太尔各特水准 | 泰爾各特水準〔台〕
Talcott method 太尔各特方法 | 泰爾各特法〔台〕
Talita 上台一，三台一（大熊座 ι） 又见：Talitha
Talitha 上台一，三台一（大熊座 ι） 又见：Talita

Talitha Australis 上台二，二台二（大熊座 κ） 又见：Al Kaprah; Alphikra Australis
TAMS （=terminal-age main-sequence）终龄主序
Tamyeunleung 谭远良（小行星 5709 号）
tangency condition 切向运动条件
tangential acceleration 切向加速度
tangential arc 正切弧 | 切弧〔台〕
tangential coordinate 切向坐标
tangential distortion 正切畸变
tangential force 切向力
tangential motion 切向运动
tangential plane 切面
tangential velocity 切向速度
tangent space 切空间
Tangshan 唐山（小行星 2778 号）
Tangtisheng 唐涤生（小行星 64295 号）
Tania Australis 中台二，三台四（大熊座 μ）
Tania Borealis 中台一，三台三（大熊座 λ）
Tanjiazhen 谈家桢（小行星 3542 号）
tantalum （=Ta）钽（73 号元素）
TAO （=Tokyo Astronomical Observatory）东京天文台
tape perforator 纸带穿孔机
tape recorder 磁带记录器
tapered distribution 锥形分布
tapering 锥削法
tapering function 锥形函数
Tarantula 蜘蛛星云（NGC 2070） 又见：Tarantula Nebula
Tarantula Nebula 蜘蛛星云（NGC 2070） 又见：Tarantula
Tarazad 河鼓三（天鹰座 γ） 又见：Tarazed
Tarazed 河鼓三（天鹰座 γ） 又见：Tarazad
Target of Opportunity 机遇目标
target particle 靶粒子
Tarqeq 土卫五十二
T array T 形天线阵

Tarvos 土卫二十一
Tashikuergan 塔什库尔干（小行星 48799 号）
T association T 星协
Ta Tsun 太尊（大熊座 ψ）
Tau ①（＝Taurus）金牛座 ②（＝Taurus）金牛宫，大梁，酉宫
Taurid meteor shower 金牛座流星雨
Taurids 金牛流星群 | 金牛 [座] 流星雨〔台〕
Taurus ①（＝Tau）金牛座 ②（＝Tau）金牛宫，大梁，酉宫
Taurus cluster 金牛星团
Taurus-Littrow Valley 金牛－利特罗峡谷（月球）
Taurus molecular cloud （＝TMC）金牛分子云
Taurus Moving Cluster 金牛移动星团
Taurus stream 金牛星流
taxonomy 分类学
Taygeta 昴宿二（金牛座 19） 又见：Taygete
Taygete ① 昴宿二（金牛座 19） 又见：Taygeta ② 木卫二十
Taylor column 泰勒柱
Taylor instability 泰勒不稳定性
Tb （＝terbium）铽（65 号元素）
TC （＝meridian circle）子午环
Tc （＝technetium）锝 | 鎝〔台〕（43 号元素）
TCB （＝Barycentric Coordinate Time）质心坐标时
TCDM （＝Tilted Cold Dark Matter model）倾斜冷暗物质模型
TCG （＝Geocentric Coordinate Time）地心坐标时
T-class asteroid T 型小行星
T-corona T 日冕
TD-1 TD－1 紫外天文卫星
TDB （＝Barycentric Dynamical Time）质心力学时
TDE （＝tidal disruption event）潮汐瓦解事件
TDI （＝time-delay integration）时延积分
TDRSS （＝tracking and data-relay satellite system）跟踪和数据传输卫星系统
TDT （＝Terrestrial Dynamical Time）地球力学时
T-duality T－对偶
T dwarf T 型矮星
Te （＝tellurium）碲（52 号元素）
tearing mode instability 撕裂模不稳定性
Tebbutt Comet 特巴特彗星（C/1861 J1）
technetium （＝Tc）锝 | 鎝〔台〕（43 号元素）
technetium line 锝谱线
technetium star 锝星

tectogenesis 构造运动
tectonic crustal phenomenon 地壳构造现象
tectonics 大地构造学 又见：geotectonics
tectonism 构造作用
tectosphere 构造层
Tegmen 水位四（巨蟹座 ζ） 又见：Tegmine
Tegmine 水位四（巨蟹座 ζ） 又见：Tegmen
Teharonhiawako 造神星（小行星 88611 号）
Teide Observatory 泰德峰天文台（西班牙）
Teisserenc crater 泰塞朗・德博尔环形山（月球）
Tejat Posterior 井宿一（双子座 μ）
tektite 玻璃陨体 | 似曜石〔台〕
Tel （＝Telescopium）望远镜座
telecommunication satellite 通信卫星
tele-compressor 缩焦镜
telecontrol 遥控 又见：distance control; remote control
tele-extender 增焦镜
telemeter 测距仪 又见：distance gauge; range finder
telemetering 遥测
telemetering monochrometer 遥测单色计
telemetry 遥测技术 | 遙測術〔台〕
telemetry frequency band 遥测频带
teleobjective 望远物镜
telephone time signal 电话时间信号
telerecorder 遥测记录仪
telescope 望远镜
telescope array 望远镜阵
telescope construction 望远镜建造
telescope control system 望远镜控制系统
telescope design 望远镜设计
telescope dome 望远镜圆顶
telescope driving system 望远镜驱动装置
telescope instrumentation 望远镜附属仪器
telescope mirror 望远镜镜面
telescope mounting 望远镜装置
telescope optical system 望远镜光学系统
telescope pier 镜墩
telescope time 望远镜观测时间
telescope tracking system 望远镜跟踪系统
telescope tube 镜筒
telescopic aperture 望远镜孔径
telescopic astronomy 望远镜天文学
telescopic comet 望远镜彗星
telescopic meteor 望远镜流星
telescopic object 望远镜天体
telescopic radiant 望远镜辐射点

telescopic resolution 望远镜分辨率
telescopic star 望远镜恒星
Telescopio Nazionale Galileo （＝TNG）【意】国立伽利略望远镜
Telescopium （＝Tel）望远镜座
telescopy 望远镜学 | 望遠鏡學，望遠鏡製造學〔台〕
teleseism 远震
tele-stereoscope 体视望远镜　又见：*stereotelescope*
Telesto 土卫十三
Television and Infrared Observation Satellite （＝TIROS）泰罗斯号（美国气象卫星）
television camera （＝TV camera）电视照相机
television detector 电视型探测器
　又见：*television-type detector*
television guider （＝TV guider）电视导星镜
television guiding （＝TV guiding）电视导星
television synchronizing pulse 电视同步脉冲
television-type detector 电视型探测器
　又见：*television detector*
television-type sensor 电视型传感器
telluric band 大气谱带 | [地球] 大氣譜帶〔台〕
telluric line 大气谱线 | [地球] 大氣譜線〔台〕
　又见：*atmospheric line*
tellurion 地球仪
tellurium （＝Te）碲（52 号元素）
tellurometer 微波测距仪
Temp Atomique Internationale （＝TAI）【法】国际原子时　即：*International Atomic Time*
Tempel's 1 comet 坦普尔 1 号彗星 | 譚普 1 號彗星〔台〕
Tempel's comet 坦普尔彗星
Tempel-Tuttle comet 坦普尔－塔特尔彗星
temperate zone 温带
temperature coefficient 温度系数
temperature compensation 温度补偿
temperature control 温度控制
temperature control equipment 温度控制装置
temperature distribution 温度分布　又见：*thermal distribution*
temperature effect 温度效应
temperature fluctuation 温度起伏
temperature gradient 温度梯度　又见：*thermal gradient*
temperature inversion 逆温 | 溫度逆增〔台〕
temperature scale 温标　又见：*thermometric scale*
temperature sequence 温度序
temperature-spectrum relation 温度—光谱关系

temperature zone 温度区带
template mismatch 模板错配
tempon 时元（10^{-23} 秒）
temporal coherence 时间相干性
temporal resolution 时间分辨率　又见：*time resolution*
temporary star 暂星 | 新星〔台〕　即：*guest star*
temporary station 临时台站
ten-day period 旬　又见：*10-day period*
ten-day star 旬星　又见：*decans*
Tenma 天马号（日本 X 射线天文卫星）
　又见：*Astro-B*
tensile force 张力　又见：*stretching force*
tensor density 张量密度
tensor fluctuation 张量涨落
tensor mode 张量模
tensor perturbation 张量扰动
tensor virial theorem 张量位力定理
TEO （＝Terrestrial Ephemeris Origin）地球历书零点
terbium （＝Tb）铽（65 号元素）
Terebellum 狗国（星组，古代星官）
Teresateng 邓丽君（小行星 42295 号）
term analysis 光谱项分析
terminal 终端
terminal-age main-sequence （＝TAMS）终龄主序
terminal synchrone 终端等时线
termination principle 终止原则
termination shock 终端激波
termination shock region 终端激波区
terminator 明暗界线 | 明暗 [界] 線，晝夜 [界] 線〔台〕
Terpsichore 司舞星（小行星 81 号）
terra 地，台地
terraced wall 台地墙
terraforming 地球化
terrence 地表
terrestrial atmosphere 地球大气　又见：*earth atmosphere*
terrestrial biosphere 地球生物圈
terrestrial branch 地支
terrestrial coordinate 地球坐标
terrestrial coordinate system 地球坐标系
Terrestrial Dynamical Time （＝TDT）地球力学时
terrestrial ellipsoid 地球椭球体　又见：*earth ellipsoid*

Terrestrial Ephemeris Origin （=TEO）地球历书零点

terrestrial exoplanet 系外类地行星 | [太阳] 系外類地行星〔台〕 又见：*exo-Earth; Earth-like exoplanet*

terrestrial eyepiece 正像目镜 又见：*positive eyepiece*

terrestrial globe 地球 又见：*Earth; globe*

terrestrial gravitation 地球引力 又见：*earth attraction*

terrestrial heat 地热

Terrestrial Intermediate Origin （=TIO）地球中间零点

Terrestrial Intermediate Reference System （=TIRS）地球中间参考系

terrestrial kilometric radiation （=TKR）地球千米波辐射

terrestrial latitude 地面纬度

terrestrial longitude 地面经度

terrestrial magnetic disturbance 地磁扰动 又见：*geomagnetic disturbance*

terrestrial magnetism 地磁 又见：*geomagnetism*

terrestrial-mass planet 地球质量行星

terrestrial meridian 地面子午线

terrestrial parallel 地面纬圈 | 地面緯度圈〔台〕

terrestrial physics 地球物理学

terrestrial planet 类地行星

Terrestrial Planet Finder （=TPF）类地行星搜索者 | 類地行星發現者〔台〕

terrestrial pole 地极 又见：*earth pole*

terrestrial radiation 地球辐射

terrestrial reference system 地球参考系

terrestrial refraction 地球大气折射 | 地面 [大氣] 折射〔台〕

terrestrial space 近地空间 | 地球空間〔台〕 又见：*geospace*

terrestrial telescope 正像望远镜

Terrestrial Time （=TT）地球时

tertiary distance indicator 三级示距天体

tertiary reflector antenna 三级反射面天线

TES （=transition edge sensor）超导转换边界传感器

tesometer 伸缩仪

TESS （=Transiting Exoplanet Survey Telescope）凌星系外行星巡天望远镜

tesselated mirror 镶嵌反射镜 | 嵌合反射鏡〔台〕

tesseral harmonic 田谐函数 | 田形調和函數, 田諧函數〔台〕

Test and Training Satellite （=TTS）TTS 试验和训练卫星

test function 测试函数

test particle 试探粒子

test star 试验星

test telescope 选址望远镜 又见：*site telescope*

Tethys 土卫三

tetrad 四元基，四维标架

Teviot Vallis 蒂维厄特谷（火星）

TFD （=time and frequency dissemination）时频发播

Th （=thorium）钍（90 号元素）

Thabit 参宿增卅六（猎户座 υ）

Thalassa 海卫四

thalassoid 【拉】海，盆地（月面）

Thalia 司剧星（小行星 23 号）

thallium （=Tl）铊（81 号元素）

thallium atomic beam 铊原子束

thallium detector 铊探测器

Tharsis Bulge 塔尔西斯突出部（火星）

Thebe 木卫十四

The Bright Star Catalogue （=BS）亮星星表 又见：*Catalogue of Bright Stars*

The Corrective Optics Space Telescope Axial Replacement （=COSTAR）COSTAR 光学改正系统

Theemim 天园九（波江座 ν）

Theemin 九州殊口四（波江座 ν） 又见：*Beemim*

Theemini 天苑三（波江座 δ） 又见：*Rana*

Theia Mons 忒伊亚山（金星）

Thelxinoe 木卫四十二

THEMIS （=Thermal Emission Imaging System）热辐射成像系统（火星探测相机）

Themis 司理星（小行星 24 号）

Themis family 司理星族

Themisto 木卫十八

theodolite 经纬仪

Theophilus crater 西奥菲勒斯环形山（月球）

theoretical astronomy 理论天文学

theoretical astrophysics 理论天体物理学 | 理論天文物理學〔台〕

theoretical line-profile 谱线理论轮廓

theoretical parallax 理论视差

Theory of Bright Heavens 昕天论 又见：*Discourse on the Diurnal Revolution of the Heavens*

Theory of Canopy-Heavens 盖天说

theory of continental drift 大陆漂移说

theory of cycles 循环理论

theory of everything 万物理论

theory of expounding appearance　宣夜说
又见：*theory of infinite heavens; theory of expounding appearance in the night sky; Infinite Empty Space*

theory of expounding appearance in the night sky　宣夜说　又见：*theory of expounding appearance; theory of infinite heavens; Infinite Empty Space*

theory of infinite heavens　宣夜说　又见：*theory of expounding appearance; theory of expounding appearance in the night sky; Infinite Empty Space*

theory of isostasy　均衡说
theory of nucleosynthesis　核合成理论
theory of probability　概率论
theory of relativity　相对论
Theory of Sphere-Heavens　浑天说
Theory of Stable Heavens　安天论　又见：*Discourse on the Conformation of the Heavens*
theory of star-streaming　星流学说
theory of stellar evolution　恒星演化理论
theory of terpidation　颤动理论
Theory of Vaulting Heavens　穹天论
又见：*Discourse on the Vastness of the Heavens*
thermal background　热背景
thermal background emission　热背景发射
thermal background radiation　热背景辐射
thermal black-body radiation　热黑体辐射
thermal bremsstrahlung　热轫致辐射 | 熱制動輻射〔台〕
thermal broadening　热致宽
thermal component　热分量
thermal conductivity　导热率 | 導熱性，導熱率〔台〕
thermal confinement　热约束
thermal decoupling　热退耦
thermal diffusion　热扩散　又见：*heat diffusion*
thermal diffusion timescale　热扩散时标
thermal distribution　温度分布　又见：*temperature distribution*
thermal Doppler broadening　热多普勒致宽
thermal effect　热效应
thermal emission　热发射
Thermal Emission Imaging System（＝THEMIS）热辐射成像系统（火星探测相机）
thermal energy　热能
thermal energy-density　热能密度
thermal equilibrium　热平衡
thermal evolution　热演化
thermal excitation　热激发
thermal flexure　热[致]弯曲

thermal gradient　温度梯度　又见：*temperature gradient*
thermal history　热史
thermal inertia　热惯性
thermal instability　热不稳定性　又见：*heat instability*
thermal ionization　热致电离 | 熱致游離〔台〕
thermalization　热化
thermal noise　热噪声 | 熱雜訊〔台〕
thermal plume　热柱
thermal pulse　热脉冲
thermal radiation　热辐射 | 熱[致]輻射〔台〕
thermal radiator　热辐射体
thermal radio radiation　热射电
thermal radio wave　热射电波
thermal scattering　热散射
thermal source　热源　又见：*heat source*
thermal spectrum　热谱
thermal stability　热稳定性
thermal structure　热结构
thermal time-scale　热时标
thermal velocity　热速度
thermion　热离子
thermionic emission　热离子发射
thermistor　热敏电阻
thermocouple　热电偶　又见：*thermoelectric couple; thermoelement*
thermodynamic equilibrium　热动平衡 | 熱力平衡〔台〕
thermodynamic function　热力学函数
thermodynamic potential　热力学势
thermodynamic temperature-scale　热力学温标
thermodynamic time-scale　热力学时标
thermoelectric couple　热电偶　又见：*thermocouple; thermoelement*
thermoelectric current　热电流
thermoelectric effect　热电效应
thermoelectric generator　热电发生器
thermoelectric photometer　热电光度计
thermoelectric power　①温差电势率 ②温差电功率
thermoelectromotive force　温差电动势
thermoelement　热电偶　又见：*thermocouple; thermoelectric couple*
thermohaline convection　温盐合成对流 | 熱鹽對流〔台〕
thermoluminescence　热致发光
thermometric scale　温标　又见：*temperature scale*

thermonuclear explosion 热核爆发
thermonuclear flash model 热核闪模型
thermonuclear reaction 热核反应
thermonuclear runaway 热核剧涨
thermopause 热层顶 | 增温層，熱氣層〔台〕
thermosphere 热层 | 熱力層〔台〕
thermostat 恒温器
theta pinch 角箍缩
Thetis 海女星（小行星 17 号）
thick CCD 厚型 CCD
thick disk 厚盘
thick-disk population 厚盘星族 | 厚盤族〔台〕
thick gravitational lense 厚引力透镜
thick-mirror Schmidt camera 厚反射镜施密特照相机
thick-target model 厚靶模型
Thiele-Burrau regularization 蒂利－布劳正规化
Thiessen crater 蒂森环形山（月球）
thin accretion disk 薄吸积盘
thin disk 薄盘
thin-disk population 盘族，薄盘族 | [星系] 盤 [星] 族，薄盤族〔台〕 又见: disk population
thin exponential disk 指数型薄盘
thin foil mirror 薄箔镜
thin homoeoid 薄同形体
thin layer 薄层
thinned CCD 薄型 CCD
thin-screen model 薄屏模型
thin-target model 薄靶模型
thin track 细径迹
thioformaldehyde 硫代甲醛（H_2CS）
Third Cambridge Catalogue of Radio Sources （＝3C catalogue）第 3 剑桥射电源表，3C 射电源表
third contact 生光，第三切 | 生光〔台〕
third cosmic velocity 第三宇宙速度
third integral 第三积分
third light 第三光源
third quarter 下弦 又见: last quarter
third radiation belt 第三辐射带
Thirty Meter Telescope （＝TMT）30 米望远镜 | 三十米望遠鏡〔台〕
Thisbe 尽女星（小行星 88 号）
tholus （复数: tholi） 山丘
Thomas-Fermi theory 托马斯－费米理论 | 托馬士－費米理論〔台〕
Thomson crater 汤姆孙环形山（月球）

Thomson cross-section 汤姆孙截面
Thomson effect 汤姆孙效应
Thomson scattering 汤姆孙散射 | 湯姆遜散射〔台〕
Thomson scattering coefficient 汤姆孙散射系数
Thomson scattering cross-section 汤姆孙散射截面
thorium （＝Th）钍（90 号元素）
Thorne-Zytkow object 索恩－祖特阔夫天体
three-axial ellipsoid 三轴椭球体
three-axis mount 三轴装置 又见: triaxial mount; triaxial mounting; 3-axis mounting; three-axis mounting
three-axis mounting 三轴装置 又见: triaxial mount; triaxial mounting; 3-axis mounting; three-axis mount
three-body collision 三体碰撞 又见: triple collision
three-body problem 三体问题 又见: 3-body problem; problem of three bodies
three-body recombination 三体复合
three-color photometry 三色测光 | 三色光度測量，三色測光〔台〕 又见: 3-color photometry
three-degree Kelvin radiation 3K 辐射 又见: 3K radiation
three-dimension classification 三元光谱分类 | 三維分類法〔台〕 又见: three-dimension spectral classification
three-dimension spectral classification 三元光谱分类 | 三維分類法〔台〕 又见: three-dimension classification
three-drift hypothesis 三流假说
Three Enclosures 三垣
three-halves power law 二分之三幂律
three-hour geomagnetic planetary index 三小时全球地磁指数
three-kiloparsec arm 三千秒差距臂 又见: 3 kpc Arm
three luminary set 三辰仪 又见: component of the three arrangers of time
three-mirror anastigmat （＝TMA）三反 [射镜] 消像散系统
three-phase CCD 三相 CCD
three-point correlation function 三点相关函数
three-prism spectrograph 三重棱镜摄谱仪
three-reflector feed system 三反射面馈源系统
three-satellite constellation 三星星座
Three Sequences Calendar 三统历
Three-stars 参宿（二十八宿） 又见: Triads; Investigator
three-way Doppler 三程多普勒
three-way light time 三程光行时
threshold density 阈密度

threshold energy　阈能 | 底限能〔台〕
threshold of sensitivity　灵敏度阈
threshold value　阈值
throughput　透光率
thruster　推进器
Thrymr　土卫三十
Thuban　右枢，紫微右垣一（天龙座α）
又见：*Adib*
Thule　图勒（小行星279号）
Thule group　图勒群
thulium　（=Tm）铥（69号元素）
Thumbprint Nebula　指纹星云
Thyone　木卫二十九
Thyra　赛娜（小行星115号）
Ti　（=titanium）钛（22号元素）
Tian-guan guest star　天关客星
Tianjin　天津（小行星2209号）
Tianyahaijiao　天涯海角（小行星9668号）
tidal acceleration　潮汐加速度
tidal action　潮汐作用
tidal approximation　潮汐近似
tidal arm　潮汐臂
tidal bulge　潮汐隆起
tidal capture　潮汐俘获
tidal capture rate　潮汐俘获率
tidal capture time　潮汐俘获时标
tidal chart　潮汐图
tidal deformation　潮汐形变 | 潮汐變形〔台〕
tidal disruption　潮汐扰动　又见：*tidal disturbance*
tidal disruption event　（=TDE）潮汐瓦解事件
tidal distortion　潮汐畸变
tidal disturbance　潮汐扰动　又见：*tidal disruption*
tidal dwarf galaxy　潮汐矮星系
tidal effect　潮汐效应
tidal evolution　潮汐演化
tidal factor　潮汐因子
tidal force　引潮力 | 潮汐力〔台〕　又见：*tidal generation force; tide raising force*
tidal friction　潮汐摩擦
tidal generation force　引潮力 | 潮汐力〔台〕
又见：*tidal force; tide raising force*
tidal height　潮高
tidal hypothesis　潮汐假说
tidal indicator　示潮器
tidal instability　潮汐不稳定性
tidal interaction　潮汐相互作用
tidal lag　潮滞

tidal potential　潮汐势 | 潮汐势，潮汐位〔台〕
tidal radius　潮力半径
tidal range　潮差
tidal shock　潮汐激波
tidal station　验潮站
tidal stream　潮汐流
tidal stripping　潮汐剥离
tidal tail　潮尾
tidal tensor　潮汐张量
tidal theory　潮汐理论
tidal torgue　潮汐扭矩
tidal torque　潮汐力矩
tidal wave　潮汐波
tidal yielding　潮汐屈服
tide　潮汐
tide generating potential　引潮势
tide raising force　引潮力 | 潮汐力〔台〕
又见：*tidal force; tidal generation force*
Tienchanglin　田长霖（小行星3643号）
tight-coupling approximation　紧密耦合近似
tight-coupling limit　紧密耦合极限
tightly coupled solution　紧密耦合解
tightly wound arm　紧卷旋臂　又见：*closely coiled arm*
tight-winding approximation　紧卷近似
Tikhomirov crater　季霍米罗夫环形山（月球）
Tikhov crater　季霍夫环形山（月球）
tile beam　片集波束
tiltable antenna　倾角可调天线
tiltable plate radio telescope　带形射电望远镜
又见：*tiltable plate telescope*
tiltable plate telescope　带形射电望远镜
又见：*tiltable plate radio telescope*
tilt angle　倾角　又见：*angle of tilt; obliquity*
Tilted Cold Dark Matter model　（=TCDM）倾斜冷暗物质模型
tiltmeter　倾斜仪
time accuracy　时间精度
time and frequency dissemination　（=TFD）时频发播
time and frequency metrology　时频计量
time average　时间平均　又见：*time averaging*
time averaging　时间平均　又见：*time average*
time axis　时轴
time-base　时基
time-base instability　时基不稳定性
time-base stability　时基稳定性

time chart 时角图
time coarse check 时间粗校
time code 时间码
time comparison 时间比对
time constant 时间常数
time correction 时间改正
time correlation 时间相关
time curve 时间曲线
time delay 时延 | 時間延遲〔台〕
time-delay integration （＝TDI）时延积分
time delay relay 时延继电器
time determination 测时
time diagram 时间图
time difference 时间差
time dilatation 时间变慢
time dilation 时间延缓
time dilation effect 时间延缓效应
time dispersion 时间弥散
time dissemination 时间发播
time-distance helioseismology 时距日震学
time distribution 时间分布
time domain 时域
time duration 持续时间
time element 时间元素
time error 时间误差
time evolution operator 时间演化算子
time increment 时间增量
time interval 时段 | 時間間隔，時段〔台〕
time keeper 记时仪 又见: *chronograph*
time-keeping 守时
time lag 时滞
timelike infinity 类时无限远
timelike path 类时路径
timelike separation 类时分隔
time mark 时标 又见: *time scale; hour index*
time marker 时号器 | 時間記號〔台〕
time measurement 时间计量
time meridian 计时子午圈
time of coincidence （＝TOC）重合时间
time of exposure 曝光时间 | 曝光時間，露光時間〔台〕
time of light 光行时间
time of periastron passage 过近星点时刻
time of perihelion passage 过近日点时刻
  又见: *perihelion time*
time of reception 接收时刻

time of relaxation 弛豫时间 | 鬆弛時間〔台〕
  又见: *relaxation time*
time of transmission 发播时刻 | 發播時間〔台〕
time piece 时计 又见: *chronometer; horologe*
time prediction 时间预测
time profile 时间轮廓
time pulse 时间脉冲
time receiving 收时
time reckoning 计时法
time reference station 时间基准站
time reflection 时间反演 又见: *time reversal*
time resolution 时间分辨率 又见: *temporal resolution*
time resolved spectroscopy 时间分辨分光测量
time response 时间响应
time reversal 时间反演 又见: *time reflection*
time scale 时标 又见: *hour index; time mark*
time scale coordination 时标协调
time service 授时，时间服务 | 授時；時間工作；時間服務〔台〕
time signal 时号 又见: *hour mark*
time signal station 时辰站 | 時號站〔台〕
time standard 时间标准
time star 时星 | [测] 時星〔台〕
time step 时间跳跃 | 時階〔台〕
time synchronism 时间同步 又见: *time synchronization*
time synchronization 时间同步 又见: *time synchronism*
time tick 报时信号
time transfer 时间比对
time variable 时间变量
time voice announcement 语言报时
time zone 时区
timing 计时 | 定時〔台〕
timing age 计时年龄
timing device 计时装置
timing equipment 计时仪器 又见: *timing instrument*
timing error 计时误差
timing instrument 计时仪器 又见: *timing equipment*
timing system 计时系统 | 定時系統〔台〕
Timiryazev crater 季米里亚泽夫环形山（月球）
tin （＝Sn）锡（50 号元素）
tincle 粒迹（陨石）
Tinkaping 田家炳（小行星 2886 号）
TIO （＝Terrestrial Intermediate Origin）地球中间零点

TIO locator　TIO 定位角
TIO meridian　TIO 子午圈
tip of the red giant branch　（=TRGB）红巨星支上端 | 紅巨星支尖〔台〕　即：*red giant tip; red giant up*
tip-tilt correction　倾斜改正
tip-tilt mirror　斜置反射镜
tired light　老化光
tired-light model　光老化宇宙模型
TIROS　（=Television and Infrared Observation Satellite）泰罗斯号（美国气象卫星）
TIRS　（=Terrestrial Intermediate Reference System）地球中间参考系
Tiselius crater　蒂塞利乌斯环形山（月球）
Tisserand's criterion　蒂塞朗判据
Tisserand's relation　蒂塞朗关系
Titan　土卫六 | 土衛六，泰坦〔台〕
Titania　天卫三 | 天[王]衛三〔台〕
titanium　（=Ti）钛（22 号元素）
Titius-Bode law　提丢斯－波得定则 | 波提定律〔台〕　又见：*Titius-Bode's law*
Titius-Bode's law　提丢斯－波得定则 | 波提定律〔台〕　又见：*Titius-Bode law*
Titius crater　提丢斯环形山（月球）
TKR　（=terrestrial kilometric radiation）地球千米波辐射
Tl　（=thallium）铊（81 号元素）
TLP　（=transient lunar phenomenon）月球暂现现象 | 月球瞬變現象〔台〕
Tm　（=thulium）铥（69 号元素）
TMA　（=three-mirror anastigmat）三反[射镜]消像散系统
TMC　（=Taurus molecular cloud）金牛分子云
TMSS　（=Two-Micron Sky Survey）2 微米巡天
TMT　（=Thirty Meter Telescope）30 米望远镜 | 三十米望遠鏡〔台〕
TNG　① （=Galileo National Telescope）国立伽利略望远镜（意大利）② （=Telescopio Nazionale Galileo）【意】国立伽利略望远镜
TNO　（=trans-Neptunian object）海外天体
TOAD　（=tremendous outburst amplitude dwarf nova）巨爆幅矮新星
Toby Jug Nebula　蝴蝶星云（IC 2220）
　　又见：*Butterfly Nebula*
TOC　（=time of coincidence）重合时间
Tohil Mons　托希尔山（木卫一）
Tokamak　托卡马克 | 托卡馬〔台〕（聚变实验装置）

Tokyo Astronomical Observatory　（=TAO）东京天文台
tolerance　容许误差　又见：*admissible error; permissible error; allowable error*
Tolies eyepiece　托勒斯目镜
Tolman-Bondi solution　托尔曼－邦迪解
tomograph　三维结构图
tonalpohuali calendar　托纳波胡利历（阿兹特克历）
Tonantzintla National Astrophysical Observatory　托南钦特拉国立天文台（墨西哥）
toner　调频器，调色剂
Tongling　铜陵（小行星 12418 号）
Tookoonooka crater　图库努卡陨星坑（地球）
Toomre sequence　图姆尔星系并合序
Toomre's model　图姆尔模型
Toomre's stability criterion　图姆尔稳定性判据
top-down galaxy formation　自上而下式星系形成
top-down scenario　自上而下图景
top-hat filter　高帽滤波器
top hat window function　高帽窗函数
top layer　顶层
topocentric aberration　站心光行差
topocentric coordinate　站心坐标 | 地面點坐標〔台〕
topocentric libration　站心天平动，站心秤动
topocentric place　站心位置
topocentric zenith distance　站心天顶距
topography　地形测量学
topological defect　拓扑缺陷
topological space　拓扑空间
topology of universe　宇宙拓扑
Torcularis Septentrionalis　右更四（双鱼座 o）
tornado prominence　龙卷日珥　又见：*waterspout prominence*
toroidal [magnetic] field　环向[磁]场 | 環形磁場〔台〕
toroidal equilibrium　环体平衡
toroidal structure　环状结构
torque　扭矩　又见：*torsional moment; twisting moment*
torque force　扭力　又见：*torsion; torsional force; twisting force*
torquetum　赤基黄道仪
torrid zone　热带　又见：*tropical zone; tropics*
torsion　扭力　又见：*torque force; torsional force; twisting force*
torsional force　扭力　又见：*torque force; torsion; twisting force*

torsional moment 扭矩 　又见：*torque; twisting moment*
torsional oscillation 扭转振荡
torsional pendulum 扭摆
torsion balance 扭秤
torsion tensor 挠率张量
torus（复数：**tori**） 环面
total absorption 总吸收
total absorption coefficient 总吸收系数
total absorption spectrometer 总吸收分光计
total acceleration 总加速度
total amplitude 总变幅，全振幅
total angular momentum 总角动量
total-annular eclipse 全环食　又见：*annular-total eclipse*
total binding energy 总结合能
total brightness 总亮度
total cross-section 总截面
total eclipse 全食　又见：*totality*
total eclipse beginning 食既，全食始　又见：*beginning of totality*
total eclipse end 全食终 | 生光〔台〕　又见：*end of totality*
total eclipse of the moon 月全食　又见：*total lunar eclipse*
total eclipse of the sun 日全食　又见：*total solar eclipse*
total energy 总能量
total flux 总流量 | 總通量〔台〕
total half-width 全半宽（谱线）
total heat 热函　又见：*heat content*
totality 全食　又见：*total eclipse*
total lunar eclipse 月全食　又见：*total eclipse of the moon*
total magnitude 总星等
total number density 总数密度
total-power receiver 总功率接收机
total-power telescope 总功率望远镜
total pressure 总压力
total proper-motion 总自行
total quantum number 总量子数
total radiation 总辐射
total reflection 全反射
total reflection prism 全反射棱镜
total scattering 总散射
total sky brightness 全天亮度
total solar eclipse 日全食　又见：*total eclipse of the sun*
total solar radiation 太阳总辐射
total starlight 全天星光
total visual magnitude 目视总星等
touchdown 着地
Tournesol 向日葵号科学卫星（法国）
Toutatis 托塔蒂斯（小行星 4179 号）
tower telescope 塔式望远镜
TPF （＝Terrestrial Planet Finder）类地行星搜索者 | 類地行星發現者〔台〕
TrA （＝Triangulum Australe）南三角座
TRACE （＝Transition Region and Coronal Explorer）TRACE 太阳探测器 | 太陽過渡區和日冕探測器〔台〕
trace element 示踪元素
tracker 跟踪器
tracker solution 跟踪解
tracking and data-relay satellite system （＝TDRSS）跟踪和数据传输卫星系统
tracking axis 跟踪轴
tracking camera 跟踪照相机
tracking counter 跟踪计数器
tracking error 跟踪误差 | 追蹤誤差〔台〕
tracking mode 跟踪方式
tracking power 跟踪本领
tracking telescope 跟踪望远镜
tracking velocity 跟踪速度
trade wind 信风
traditional astronomy 传统天文学
trailer spot 后随黑子，f 黑子 | 尾隨黑子〔台〕　又见：*following sunspot; following spot; trailer sunspot*
trailer sunspot 后随黑子，f 黑子 | 尾隨黑子〔台〕　又见：*following sunspot; following spot; trailer spot*
trailing 旋闭
trailing arm 曳臂 | 尾隨旋臂〔台〕　又见：*following arm*
trailing edge 后边缘
trailing hemisphere 后随半球
trailing wave 曳波
train of waves 波列　又见：*wave train*
trajectory 轨道，弹道 | 軌跡〔台〕
transducer 转换器
transfer between orbits 轨道转移 | 軌道變換〔台〕　又见：*orbit transfer*
transfer curve 转移曲线
transfer efficiency 传能效率
transfer equation 转移方程 | 轉換方程〔台〕　即：*equation of radiative transfer*
transfer function 转移函数
transfer orbit 转移轨道 | 轉換軌道〔台〕

transformation coefficient 归化系数
transformation of coordinates 坐标变换
    又见: *conversion of coordinates; coordinate transformation*
transform fault 转换断层
transient event 暂现事件 | 瞬變事件〔台〕
transient lunar phenomenon (=TLP)月球暂现现象 | 月球瞬變現象〔台〕
transient perturbation 暂现扰动
transient spiral pattern 暂现旋涡图案
    又见: *transitory spiral pattern*
transient X-ray burst 暂现X射线暴 | 瞬變X光爆發〔台〕
transient X-ray burster 暂现X射线暴源
transient X-ray source 暂现X射线源 | 瞬變X光爆發〔台〕
transient γ-ray burst 暂现γ射线暴 | 瞬變γ射線爆發〔台〕
transient γ-ray burster 暂现γ射线暴源
transient γ-ray source 暂现γ射线源 | 瞬變γ射線源〔台〕
transit 凌
transit circle 子午环   又见: *meridian circle*
transiter 中天记录器
transiting exoplanet 凌星系外行星
Transiting Exoplanet Survey Telescope (=TESS)凌星系外行星巡天望远镜
transiting planet 凌星行星
transit instrument 中星仪   又见: *meridian instrument*
transition coefficient 跃迁系数
transition edge sensor (=TES)超导转换边界传感器
transition frequency 跃迁频率
transition probability 跃迁概率 | 躍遷機率〔台〕
transition radiation 跃迁辐射
transition region 过渡区(太阳)
Transition Region and Coronal Explorer (=TRACE)TRACE太阳探测器 | 太陽過渡區和日冕探測器〔台〕
transition region line 过渡区谱线
transit method 凌星方法
transit of Jovian satellite 木卫凌木   又见: *transit of Jupiter's satellite*
transit of Jupiter's satellite 木卫凌木   又见: *transit of Jovian satellite*
transit of Mercury 水星凌日
transit of shadow 卫影凌行星   又见: *shadow transit*
transit of Venus 金星凌日
transit optical telescope 中天光学望远镜

transitory spiral pattern 暂现旋涡图案
    又见: *transient spiral pattern*
transit radio telescope 中天射电望远镜
transit telescope 中天望远镜
transit time 中天时刻, 渡越时间 | 中天時刻〔台〕
transit time effect 切换时间效应, 渡越时间效应
translation 平移
translational motion 平移运动
translational velocity 平移速度
translocatin 联测定位
translunar space 月外空间 | 月[轨道]外空間〔台〕
translunar orbit 月外轨道
transmission coefficient 透射系数 | 透射係數, 傳輸係數〔台〕
transmission grating 透射光栅
transmission line 传输线
transmission loss 传输损耗
transmission system 传输系统
transmissivity 透射率
transmittance 透射比
transmitting antenna 发射天线
trans-Neptunian object (=TNO)海外天体
trans-Neptunian planet 海外行星
    又见: *ultra-Neptunian planet*
trans-Neptunian space 海外空间
transparency 透明度   又见: *diaphaneity*
transparent 透明
transplanetary space 行星外空间
trans-Pluto 冥外行星(假设)
    又见: *trans-Plutonian planet*
trans-Plutonian planet 冥外行星(假设)
    又见: *trans-Pluto*
transponder 应答器
transport equation 输运方程
transport phenomenon 输运现象
transport process 输运过程
transpose 转置
transversal filter 横向滤波器
transversal wave 横波   又见: *transverse wave*
transverse component 横向分量
transverse conductivity 横向传导率
transverse field 横向磁场, 横场 | 横[向]场〔台〕   又见: *transverse magnetic field*
transverse heat conduction 横向热传导
transverse magnetic field 横向磁场, 横场 | 横[向]场〔台〕   又见: *transverse field*
transverse parallax screw 横向视差螺旋

transverse propagation　横向传播
transverse spherical aberration　横向球面视差 | 横向球面像差〔台〕
transverse velocity　横向速度
transverse wave　横波　又见: *transversal wave*
transverse Zeeman effect　横向塞曼效应
trapeze　经纬度格
Trapezium Cluster　猎户四边形星团
Trapezium of Orion　猎户四边形天体 | 獵戶座四邊形〔台〕
trapped gas　俘获气体
trapped particle　俘获粒子
trapped radiation　俘获辐射
trapped surface　俘获面
trapping of gas　气体俘获
trapping of particles　粒子俘获
trapping of radiation　辐射俘获　又见: *radiative capture*
trapping of waves　波俘获
traveling beam tube　行束管
traveling wave　行波
traveling-wave antenna　行波天线
traveling-wave maser　(＝TWM) 行波微波激射器
traveling-wave tube　(＝TWT) 行波管
traveling wire　动丝　又见: *movable wire*
travel path　传播路径
travel time　传播时间　又见: *propagation time*
Treatise on Astrology in the Kaiyuan Reign　《开元占经》　又见: *Prognostication Classic of the Kaiyuan Era*
treatment of data　数据处理　又见: *data handling; data processing*
tree algorithm　树算法
Tremaine-Gunn limit　特里梅因－冈恩极限
tremendous outburst amplitude　巨爆幅
tremendous outburst amplitude dwarf nova　(＝TOAD) 巨爆幅矮新星
TRGB　(＝tip of the red giant branch) 红巨星支上端 | 紅巨星支尖〔台〕　即: *red giant tip; red giant up*
Tri　(＝Triangulum) 三角座
triad　三元基，三维标架
triadic mean　三点平均
Triads　参宿（二十八宿）　又见: *Investigator; Three-stars*
trial curve　试验曲线
triangular distortion　三角畸变
triangular point　三角点

triangulation　三角测量
Triangulum　(＝Tri) 三角座
Triangulum Australe　(＝TrA) 南三角座
Triangulum Galaxy　三角星系（M33）　又见: *Triangulum Nebula*
Triangulum Nebula　三角星系（M33）　又见: *Triangulum Galaxy*
triaxial distribution　三轴分布
triaxiality parameter　三轴性参数
triaxial mount　三轴装置　又见: *triaxial mounting; 3-axis mounting; three-axis mount; three-axis mounting*
triaxial mounting　三轴装置　又见: *triaxial mount; 3-axis mounting; three-axis mount; three-axis mounting*
triaxial system　三轴系统
tri-dimensional spectroscopy　三维分光
tridymite　鳞石英
Trifid Nebula　三叶星云 | 三裂星雲〔台〕（M20）
trigger mechanism　触发机制
trigonal aspect　三分方位
trigonal point　三分点
trigonometric parallax　三角视差
Trinculo　天卫二十一
Trio in Leo　狮子三重星系　又见: *Leo Triplet*
Triones　北斗 [七星]（星组）　又见: *Big Dipper; Northern Dipper; Charles' Wain; Wain; Plough; Plow*
triple-alpha process　三 α 过程（核反应）　又见: *triple-α process*
triple asteroid　三合小行星
triple collision　三体碰撞　又见: *three-body collision*
triple galaxy　三重星系
triple quasar　三合类星体
triple star　三合星
triplet　① 三合透镜 | 三合 [透] 鏡〔台〕② 三重线 ③ 三重态
triplet lens　三合透镜
triple-α process　三 α 过程（核反应）　又见: *triple-alpha process*
tripod　三脚架
triquetrum　三角仪
trispectrum　三谱
tritium　氚（^{3}H，氢同位素）
tritium decay　氚衰变
Triton　海卫一 | 海 [王] 衛一〔台〕
triton　氚核
trivial anthropic principle　平凡人择原理
troilite　陨硫铁
Troilus　特洛伊鲁斯（小行星 1208 号）
Trojan asteroid　特洛伊型小行星　又见: *Trojans*

Trojan group　特洛伊群
Trojans　特洛伊型小行星　又见：*Trojan asteroid*
tropic　回归线　又见：*regression line*
tropical century　回归世纪
tropical month　分至月，回归月
tropical revolution　分至周
tropical year　① 岁实 ② 回归年　又见：*natural year*
tropical zone　热带　又见：*torrid zone; tropics*
Tropic of Cancer　北回归线　又见：*Northern tropic*
Tropic of Capricorn　南回归线　又见：*Southern tropic*
tropics　热带　又见：*torrid zone; tropical zone*
tropopause　对流层顶
troposphere　对流层
tropospheric attenuation　对流层衰减
tropospheric delay　对流层延迟
tropospheric refraction　对流层折射
tropospheric refraction correction　对流层折射改正
true absorption　真吸收
true absorption coefficient　真吸收系数
true altitude　真高度
true anomaly　真近点角　又见：*real anomaly*
true continuous absorption　真连续吸收
true declination　真赤纬
true diameter　真直径
true distance　真距离
true equator　真赤道
true equinox　真春分点
true error　真误差
true field　有效视场
true horizon　真地平
true latitude　真纬度
true libration　真天平动，真秤动 | 真天平動〔台〕
true longitude　真经度
true mean place　真平位置 | 真平位[置]〔台〕
true meridian　真子午线，真子午圈
true midnight　真子夜
true modulus　真距离模数 | 真[距離]模數〔台〕
true motion　真动
true noon　真正午
true orbit　真轨道
true place　真位置　又见：*true position*
true pole　真天极 | 真極〔台〕
true position　真位置　又见：*true place*
true radiant　真辐射点
true revolution　真公转周

true right ascension　真赤经
true rotation　真自转周
true selective absorption　真选择吸收 | 真選擇性吸收〔台〕
true semidiameter　真半径
true sidereal day　真恒星日
true sidereal time　真恒星时 | 視恆星時〔台〕
　　又见：*apparent sidereal time*
true solar day　真太阳日
true solar time　真太阳时 | 視[太陽]時〔台〕
　　又见：*apparent solar time*
true sun　真太阳　又见：*real sun*
true time　真时
true vertex　真奔赴点
true zenith distance　真天顶距
Trumpler crater　特朗普勒环形山（月球）
Trumpler's classification　特朗普勒分类
Trumpler's star　特朗普勒星 | 庄普勒星〔台〕
truncated cone　截锥　又见：*frustum of a cone*
truncated distribution　截尾分布
truncation error　截断误差　又见：*cut-off error*
Tsai　蔡章献（小行星 2240 号）
Tsander crater　灿德尔环形山（月球）
　　又见：*Zander crater*
Tsanghinchi　曾宪梓（小行星 3388 号）
Tseraskiy crater　采拉斯基环形山（月球）
　　又见：*Ceraski crater*
T-shaped array　T 形望远镜阵 | T 形[望遠鏡]陣〔台〕
Tsih　策（仙后座 γ）
Tsiolkovskiy crater　齐奥尔科夫斯基环形山（月球）
T star　T 型星
Tsuchinshan comet　紫金山彗星
Tsuihark　徐克（小行星 23258 号）
Tsytovich effect　齐托维奇效应
TT　（＝Terrestrial Time）地球时
T Tauri phase　金牛 T 阶段
T Tauri star　金牛 T 型星 | 金牛[座]T[型]變星〔台〕
T Tauri variable　金牛 T 型变星
T Tauri wind　金牛 T 星风
TTS　（＝Test and Training Satellite）TTS 试验和训练卫星
T-type asteroid　T 型小行星
tube of force　力线管 | 力管〔台〕
tube phenomenon　管筒现象
tube photometer　管光度计

Tuc (=Tucana) 杜鹃座
Tucana (=Tuc) 杜鹃座
Tucana Dwarf Galaxy 杜鹃座矮星系
Tully-Fisher method 塔利－费希尔方法
Tully-Fisher relation 塔利－费希尔关系
Tun 顿（玛雅历日，360天）
tunable filter 可调滤光器
tunable reflection maser 调谐反射式微波激射器
tunable traveling-wave maser 可调行波式微波激射器
tungsten (=W) 钨（74号元素）
Tunguska-class NEA 通古斯型近地小行星
Tunguska crater 通古斯陨星坑 | 通古斯隕石坑〔台〕（地球）
Tunguska event 通古斯事件
tuning fork diagram 音叉图（星系）
tunnel effect 隧道效应 | 穿隧效應〔台〕
tunneling 隧[道贯]穿
Tunuunik crater 图努尼克陨星坑（地球）
Turais 海石二（船底座ι） 又见: *Aspidiske; Scutulum; Turyeish*
turbidity 像晕
turbopause 湍流层顶
turbule 湍流元 又见: *turbulence element*
turbulence broadening 湍流致宽
turbulence cell 湍流胞
turbulence element 湍流元 又见: *turbule*
turbulence heating 湍动致热
turbulence spectrum 湍动谱
turbulence theory 湍流理论
turbulent dynamo 湍流发电机
turbulent flow 湍流 | 湍流，涡流〔台〕
turbulent line-width 湍动线宽
turbulent magnetic field 湍动磁场 | 湍流磁場〔台〕
turbulent motion 湍动 | 湍動，渦動〔台〕
turbulent plasma 湍动等离子体
turbulent velocity 湍速 | 湍流速度〔台〕
turnaround radius 回缩半径
Turner method 特纳法
turning moment 转矩
turning point 转向点
turn-off age 折向点年龄，拐点年龄 | 轉離年齡〔台〕
turn-off mass 折向点质量，拐点质量 | 轉離質量〔台〕
turn-off point 折向点，拐点

turn-off point from main-sequence 主序折向点，主序拐点 | 主序轉折點〔台〕
turn-over radius 转向点半径
turnstile feed 绕杆式馈源
turret dome 角塔圆顶
Turtle beak 觜宿（二十八宿） 又见: *Mouth*
Turyeish 海石二（船底座ι） 又见: *Aspidiske; Scutulum; Turais*
TV camera (=television camera) 电视照相机
TV guider (=television guider) 电视导星镜
TV guiding (=television guiding) 电视导星
twelve Jupiter-stations 十二次 又见: *12 Jupiter-stations; duodenary series*
twenty-eight lunar mansions 二十八宿 又见: *lunar zodiac; 28 lunar mansions*
twenty-four fortnightly periods 二十四节气 又见: *twenty-four solar terms*
twenty-four solar terms 二十四节气 又见: *twenty-four fortnightly periods*
twenty-one centimeter line 21厘米谱线 又见: *21cm line*
twenty-one centimeter radiation 21厘米辐射 | 21公分輻射〔台〕 又见: *21-cm radiation*
twilight 晨昏蒙影 | 曙暮光〔台〕
twilight brightness 曙暮光亮度
twilight correction 曙暮光改正
twilight emission 曙暮光发射
twilight flat 晨昏天光平场
twilight spectrum 曙暮光光谱
twilight zone 曙暮光区
twin-beam model 双射束模型
twin-exhaust model 双排气孔模型
twin galaxy 双重星系 又见: *binary galaxy; double galaxy*
Twin Jet Nebula 双喷流星云（M2-9）
twinkling 闪烁 又见: *scintillation*
twinkling star 闪烁星
twin quasar 双类星体 又见: *binary quasar; double quasar*
twins paradox 双生子佯谬
twisted magnetic field 扭绞磁场
twisting force 扭力 又见: *torque force; torsion; torsional force*
twisting moment 扭矩 又见: *torque; torsional moment*
twist mapping 扭转测绘
TWM (=traveling-wave maser) 行波微波激射器
two-armed spiral pattern 双臂旋涡图像

two-body problem　二体问题　又见：*2-body problem; problem of two bodies*
two-body relaxation　二体弛豫
two-body relaxation time　二体弛豫时间
two-channel noise figure　双通道噪声系数
two-channel receiver　双通道接收机
two-color diagram　两色图 | 兩色圖，色 [指数] 一色 [指數] 圖〔台〕　又见：*color-color diagram; color-color plot*
two-color photometry　两色测光
two-component model　二组元模型（太阳风）
Two-degree-Field Galaxy Redshift Survey　（＝2dFGRS）2度视场星系红移巡天　又见：*2dF Galaxy Redshift Survey*
Two-degree-field redshift survey　（＝2dF Survey）2度视场红移巡天
two-dimensional aperture synthesis　二维综合孔径　又见：*two-dimensional synthesis*
two-dimensional classification　二元光谱分类 | 二維 [光譜] 分類法〔台〕　又见：*two-dimensional spectral classification*
two-dimensional detector　二维探测器
two-dimensional distribution　二维分布
two-dimensional spectral classification　二元光谱分类 | 二維 [光譜] 分類法〔台〕　又见：*two-dimensional classification*
two-dimensional spectroscopy　二维分光 | 二維分光法〔台〕　又见：*bidimensional spectroscopy*
two-dimensional spectrum　二维光谱
two-dimensional synthesis　二维综合孔径　又见：*two-dimensional aperture synthesis*
two-dimension photometry　二维测光
two-element interferometer　双元干涉仪　又见：*double interferometer*
two-fluid model　双流体模型
two-lens objective　双透镜物镜
two-lens ocular　双透镜目镜
Two Micron All Sky Survey　（＝2MASS）2微米全天巡视
Two-Micron Sky Survey　（＝TMSS）2微米巡天　又见：*2-µ Sky Survey*
two-mode turbulence　双模式湍动
two-particle correlation function　两粒子相关函数
two-particle distribution function　两粒子分布函数
two-phase model　两相模型
two-phase structure　双相结构
two-photon emission　双光子发射 | 二光子發射〔台〕

two-point correlation function　两点相关函数
two-reflector antenna　双反射面天线
two-resonance maser　双共振微波激射器
two-ribbon flare　双带耀斑 | 雙帶閃焰〔台〕
two-screen model　双屏模型
two-spectrum binary　双谱 [分光] 双星 | 複綫 [分光] 雙星〔台〕　又见：*double-lined [spectroscopic] binary; double-spectrum binary*
two-stage acceleration　二级加速　又见：*two-step acceleration*
two-step acceleration　二级加速　又见：*two-stage acceleration*
two stream hypothesis　二星流假说
two-stream instability　二星流不稳定性
two-way Doppler　双程多普勒
two-way light time　双程光行时
two-way mode　双路径模式
TWT　（＝traveling-wave tube）行波管
Tycho Catalogue　第谷星表
Tycho crater　第谷环形山（月球）
Tychonic system　第谷体系 | 第谷 [宇宙] 體系〔台〕（宇宙体系）
Tycho's nova　第谷新星（B Cas）　即：*Tycho's supernova*
Tycho's star　第谷星（B Cas）　即：*Tycho's supernova*
Tycho's supernova　第谷超新星（SN 1572）
Tyl　天厨三（天龙座ε）
type A Fermi collision　A型费米碰撞
type B Fermi collision　B型费米碰撞
type Ia supernova　Ia型超新星
type Ib supernova　Ib型超新星
type I Cepheid　I型造父变星
type Ic supernova　Ic型超新星
type II Cepheid　II型造父变星
type III radio burst　III型射电暴
type IIn supernova　IIn型超新星
type II quasar　II型类星体
type II radio burst　II型射电暴
type II radio galaxy　II型射电星系
type II Seyfert　II型赛弗特星系
type II supernova　II型超新星　又见：*SN type II*
type II X-ray burster　II型X射线暴源 | II型X光爆源〔台〕
type I quasar　I型类星体
type I radio burst　I型射电暴
type I radio galaxy　I型射电星系
type I Seyfert　I型赛弗特星系

type I supernova　I 型超新星　又见：*SN type I*
type IV radio burst　IV 型射电暴
type I X-ray burster　I 型 X 射线暴源 | I 型 X 光爆源〔台〕
type V radio burst　V 型射电暴
Typhon　台神星（小行星 42355 号）
typical nova　典型新星
typical spiral galaxy　典型旋涡星系
typical star　典型星
Tzolk'in calendar　卓尔金历（玛雅）

U

U　（=uranium）铀（92 号元素）

Uayeb　瓦耶勃（玛雅历日）　又见：*Wayeb'*

UBC-Laval Telescope　（=University of British Columbia and Laval University Telescope）UBC－拉瓦勒望远镜

U-B color index　U－B 色指数

Ubercal　（=ubercalibration）上定标

ubercalibration　（=Ubercal）上定标

U burst　U 型暴

UBV color system　UBV 颜色系统

UBVGRI photometry　UBVGRI 测光

UBV photometry　UBV 测光

UBVRI color system　UBVRI 颜色系统

UBVRIJKL photometry　UBVRIJKL 测光

UBVRI photometry　UBVRI 测光

UBVRI system　UBVRI 系统

UBV system　UBV 系统

UcBV photometry　UcBV 测光

UCD　①（=Unified Content Descriptors）统一内容描述符　②（=ultra-compact dwarf galaxy）超致密矮星系

U-class asteroid　U 型小行星

UFO　（=unidentified flying object）不明飞行物，未鑑定飛行體，不明飛行物，幽浮〔台〕

UG　①（=U Gem binary）双子 U 型双星 | 雙子[座]U 型雙星〔台〕②（=U Geminorum star）双子 U 型星 | 雙子[座]U 型星〔台〕③（=U Gem variable）双子 U 型变星

UGC　（=Uppsala General Catalogue of Galaxies）乌普萨拉星系总表

U Gem binary　（=UG）双子 U 型双星 | 雙子[座]U 型雙星〔台〕

U Geminorum star　（=UG）双子 U 型星 | 雙子[座]U 型星〔台〕　又见：*U Gem star*

U Gem star　双子 U 型星 | 雙子[座]U 型星〔台〕
又见：*U Geminorum star*

U Gem variable　（=UG）双子 U 型变星
又见：*U Gem variable star*

U Gem variable star　双子 U 型变星　又见：*U Gem variable*

UHECR　（=ultra-high energy cosmic ray）特高能宇宙线

UHF　（=ultra-high frequency）特高频（300-3000 兆赫）

Uhuru　乌呼鲁 X 射线卫星 | 自由號 [X 光天文衛星]〔台〕　又见：*SAS I; SAS-A*

uhuru　乌呼鲁（天体 X 射线流量密度单位，相当于 11μJy）

Uhuru Catalogue of X-ray Sources　乌呼鲁 X 射线源表 | 自由號 X 光源表〔台〕

uinal　乌纳尔（玛雅历月，20 天）

UKIRT　（=United Kingdom Infrared Telescope）英国红外望远镜

UK Schmidt Telescope　（=UKST）英国施密特望远镜

UKST　①（=UK Schmidt Telescope）英国施密特望远镜　②（=United Kingdom Schmidt Telescope）英国施密特望远镜

ULE　（=ultra low expansion glass）超低膨胀玻璃

ULIG　（=ultraluminous infrared galaxy）特高光度红外星系 | 超亮紅外星系〔台〕

U line　U 谱线

ULIRG　（=ultraluminous infrared galaxy）极亮红外星系

ultimate line　驻留谱线

ultrabasic glass　超碱玻璃　又见：*ultramafic glass*

ultra-compact dwarf galaxy　（=UCD）超致密矮星系

ultracompact galaxy nucleus　超密星系核

ultracompact HII region　超致密电离氢区

ultradeep-field observation　特深天区观测

ultradense neutron star　超密中子星

ultradense object　超密天体

ultra-high energy cosmic ray　（=UHECR）特高能宇宙线

ultra-high energy γ-rays　特高能量 γ 射线

ultra-high frequency （＝UHF）特高频（300-3000 兆赫）
ultra low expansion glass （＝ULE）超低膨胀玻璃
ultraluminous galaxy 特高光度星系 | 特亮星系〔台〕
ultraluminous galaxy nucleus 特亮星系核
ultraluminous infrared galaxy ①（＝ULIG）特高光度红外星系 | 超亮红外星系〔台〕②（＝ULIRG）极亮红外星系
ultraluminous X-ray source 极高光度 X 射线源 | 超亮 X 光源〔台〕
ultramafic glass 超碱玻璃　又见：*ultrabasic glass*
ultrametal-poor star 特贫金属星
　　又见：*super-metal-poor star*
ultra-Neptunian planet 海外行星
　　又见：*trans-Neptunian planet*
ultraray 宇宙线　又见：*cosmic ray*
ultra-relativistic electron 极端相对论性电子
ultra-relativistic particle 极端相对论性粒子
ultra-short-period binary 超短周期双星
ultra-short-period B star 超短周期 B 型星
ultra-short-period Cepheid 超短周期造父变星
　　又见：*super-short period Cepheid*
ultrashort-period planet 超短周期行星
ultra-short-period variable 超短周期变星
ultrasonic wave 超声波
ultra-telescopic meteor 微流星　又见：*micrometeor*
ultraviolet astronomy （＝UV astronomy）紫外天文学 | 紫外 [線] 天文學〔台〕
ultraviolet background （＝UV background）紫外背景
ultraviolet band （＝UV band）紫外波段
ultraviolet-bright star （＝UV bright star）紫外亮星
ultraviolet camera （＝UV camera）紫外照相机
ultraviolet color （＝UV color）紫外色
ultraviolet continuum （＝UV continuum）紫外连续谱
ultraviolet counterpart （＝UV counterpart）紫外对应体
ultraviolet cutoff （＝UV cutoff）紫外截断
ultraviolet detector （＝UV detector）紫外探测器
ultraviolet drop-out （＝UV drop-out）紫外漏失
ultraviolet dwarf （＝UV dwarf）紫外矮星
ultraviolet excess （＝UV excess）紫外超 | 紫外 [辐射] 超量〔台〕
ultraviolet-excess object （＝UV-excess object）紫外超天体 | 紫外 [辐射] 超量天體〔台〕
ultraviolet-excess star （＝UV-excess star）紫外超恒星
ultraviolet extinction （＝UV extinction）紫外消光
ultraviolet filter （＝UV filter）紫外滤光片
ultraviolet identification （＝UV identification）紫外证认
ultraviolet image （＝UV image）紫外像
ultraviolet ionization chamber （＝UV ionization chamber）紫外电离室
ultraviolet light （＝UV light）紫外光
ultraviolet line （＝UV line）紫外谱线
ultraviolet luminosity （＝UV luminosity）紫外光度
ultraviolet magnitude （＝UV magnitude）紫外星等
ultraviolet map （＝UV map）紫外天图
ultraviolet object （＝UV object）紫外天体
ultraviolet photography （＝UV photography）紫外照相
ultraviolet photometry （＝UV photometry）紫外测光
ultraviolet radiation （＝UV radiation）紫外辐射
ultraviolet radiation detector （＝UV radiation detector）紫外辐射探测器
ultraviolet satellite （＝UV satellite）紫外卫星
ultraviolet sky （＝UV sky）紫外天空
ultraviolet source （＝UV source）紫外源
ultraviolet spectrograph （＝UV spectrograph）紫外摄谱仪
ultraviolet spectrophotometry （＝UV spectrophotometry）紫外分光光度测量
ultraviolet spectroscopy （＝UV spectroscopy）紫外分光
ultraviolet spectrum （＝UV spectrum）紫外光谱
ultraviolet star （＝UV star）紫外星
ultraviolet surveys （＝UV survey）紫外巡天
ultraviolet telescope （＝UV telescope）紫外望远镜
Ulysses 尤利西斯号太阳探测器 | 尤利西斯號 [太陽探測器]〔台〕
UMa （＝Ursa Major）大熊座
UMa cluster （＝Ursa Major cluster）大熊星团
U-magnitude U 星等
UMa group ①（＝Ursa Major group）大熊星系群 ②（＝Ursa Major group）大熊星群
UMa I （＝Ursa Major I dwarf）大熊矮星系 I
UMa II （＝Ursa Major II dwarf）大熊矮星系 II
umbra 本影 | 本影，暗影〔台〕

umbral dot　本影点
umbral eclipse　本影食
umbral flash　本影闪耀 | 本影閃爍〔台〕
Umbriel　天卫二 | 天 [王] 衛二〔台〕
UMi　（=Ursa Minor）小熊座
UMi system　（=Ursa Minor system）小熊星系
umklapp scattering　反转散射
unannounced satellite　秘密卫星
unbiased estimator　无偏估计量 | 不偏估計量〔台〕
unbound model　非束缚模型
unbound orbit　非束缚轨道
uncertainty　不确定性
uncertainty principle　测不准原理
　　又见：*indeterminate principle*
unclosed orbit　开放轨道
uncorrelated noise　非相关噪声
undercorrection　改正不足
underdensity　欠密度
underexposure　曝光不足
underluminous star　低光度星　又见：*low-luminosity star*
underlying galaxy　基底星系
undermassive star　质量过小星
Undina　波神星（小行星 92 号）
Undina family　波神星族
undisturbed orbit　无摄轨道　又见：*unperturbed orbit*
undisturbed sun　非扰太阳 | 無擾動太陽〔台〕
undriven camera　无跟踪照相机
unevenness　不均匀性　又见：*inhomogeneity; non-uniformity*
unfilled aperture　分立孔径
ungerade　【德】奇态　即：*singular state; odd state*
unidentified flying object　（=UFO）不明飞行物 | 未鑑定飛行體,不明飛行物,幽浮〔台〕
unidentified source　未证认源 | 未識別源〔台〕
unidimension　一维
unidirection　单向
Unified Content Descriptors　（=UCD）统一内容描述符
unified model of active galactic nuclei　活动星系核统一模型
unified standard　统一标准
unified time　统一时间
uniform acceleration　匀加速度
uniform array　等距天线阵
uniform circular motion　匀速圆周运动
uniform-density contraction　匀密度收缩

uniform distribution　均匀分布
uniform induction　均匀感应
uniformization　均匀化,单值化
uniform motion　匀速运动 | 均速運動〔台〕
uniform sidereal time　均匀恒星时 | 平恆星時〔台〕
uniform solar time　均匀太阳时
uniform time　均匀时间
uniform velocity　均匀速度
unilluminated hemisphere　暗半球 | 不照亮半球〔台〕
unipolar group　单极群 | 單極 [黑子] 群〔台〕
unipolar induction　单极感应
unipolar magnetic region　单极磁区
unipolar spot　单极黑子
unipolar sunspot　单极太阳黑子 | 單極黑子〔台〕
uniqueness　唯一性
unique variable　独特变星
unitarity　幺正性 | 么正性〔台〕
unitary spin　幺旋
unitary transformation　幺正变换
unit brightness　单位亮度
unit circle　单位圆
unit distance　单位距离
United Kingdom Infrared Telescope　（=UKIRT）英国红外望远镜
United Kingdom Schmidt Telescope　（=UKST）英国施密特望远镜
United States Naval Observatory　（=USNO）美国海军天文台
unit sphere　单位球
unit vector　单位矢量
univariate distribution　单变量分布
universal background　宇宙背景　又见：*cosmic background*
universal constant　普适常数
universal coupling constant　普适耦合常数
universal covering surface　万有覆盖面
universal dial　通用日晷
universal filter　万用滤波器
universal gravitation　万有引力
universal horizon　宇宙视界　又见：*horizon of the universe*
universal instrument　全能经纬仪　又见：*universal theodolite*
universal theodolite　全能经纬仪　又见：*universal instrument*
universal time　（=UT）世界时
universe　宇宙　又见：*cosmos*

University of British Columbia and Laval University Telescope （=UBC-Laval Telescope）UBC－拉瓦勒望远镜
unlit side 无光照侧边
unmagnetized plasma 非磁化等离子体
unnumbered asteroid 未编号小行星
unobservable quantity 非观测量
unperturbed orbit 无摄轨道 又见: *undisturbed orbit*
unpolarized burst 非偏振暴
unpolarized component 非偏振成分
unpolarized continuum 非偏振连续谱
unpolarized radiation 非偏振辐射
unpolarized wave 非偏振波
unresolved source 不可分辨源
unrestricted orbit 无限制轨道
unrising body 恒隐天体
Unruh radiation 盎鲁辐射
unseen companion ① 未见伴星 ② 未见伴星系
unseen component 未现子星 | 未見子星〔台〕
unseen matter 不可视物质 | 未見物質〔台〕
unsetting body 恒显天体
unsharp masking 模糊遮掩
unstable coordinate 不稳定坐标
unstable equilibrium 不稳定平衡 | 不穩定平衡〔台〕
unstable state 不稳定态
unstationary star 不稳定星 又见: *non-stable star*
un-Sun-like star 非类太阳恒星
Unuk al Hai 天市右垣七，蜀（巨蛇座 α）
  又见: *Cor Serpentis; Unukalhai*
Unukalhai 天市右垣七，蜀（巨蛇座 α）
  又见: *Cor Serpentis; Unuk al Hai*
unwrapping 无卷缠
up-Comptonization 康普顿硬化
up-conversion 上变频
upcrossing of random trajectories 随机轨迹上穿
up-leg light time 上行光行时
upper atmosphere 高层大气
upper chromosphere 色球高层
upper circle 上规，恒显圈 | 恆顯圈〔台〕
  又见: *circle of perpetual apparition*
upper crust 上部地壳
upper culmination 上中天 又见: *superior transit; upper transit*
upper level 高能级 | 高等級，高水準〔台〕
upper limb 上边缘 | 上 [邊] 緣〔台〕
upper limit 上限
upper main sequence 上主星序

upper mantle 上地幔
upper-most crust 最上部地壳
upper tangent arc 上正切晕弧
upper transit 上中天 又见: *superior transit; upper culmination*
Uppsala General Catalogue of Galaxies （=UGC）乌普萨拉星系总表
upward-looking bowl sundial 仰釜日晷 | 仰儀，仰釜日晷〔台〕
upward transition 向上跃迁
Urakhga 天津二（天鹅座 δ） 又见: *Rukh*
Urania 司天星（小行星 30 号）
Uranian ring 天王星环 又见: *ring of Uranus; Uranus' ring*
Uranian ringlet 天王星窄环
Uranian satellite 天卫 | 天 [王] 衛〔台〕
  又见: *satellite of Uranus*
uranicentric coordinate 天心坐标
uranigraphic coordinate 天面坐标
uranium （=U）铀（92 号元素）
uranographer 星图学家
Uranographia 《波德星图》（德国天文学家波德所制星图）
uranographic model 浑天象
uranography 星图学
uranology 天学
Uranometria 《测天图》（德国天文学家约翰·拜尔所制星图）
uranometry 天体测量
Uranus 天王星
Uranus' ring 天王星环 又见: *Uranian ring; ring of Uranus*
Urca process 乌卡过程 | Urca 過程〔台〕（中微子）
ureilite 橄辉无球粒陨石
Ursa Major （=UMa）大熊座
Ursa Major cluster （=UMa cluster）大熊星团
Ursa Major group ① （=UMa group）大熊星系群 ② （=UMa group）大熊星群
Ursa Major I dwarf （=UMa I）大熊矮星系 I
Ursa Major II dwarf （=UMa II）大熊矮星系 II
Ursa Major Moving Cluster 大熊移动星团
Ursa Minor （=UMi）小熊座
Ursa Minorids 小熊流星群 又见: *Ursids*
Ursa Minor system （=UMi system）小熊星系
URSI （=International Union of Radio Science）国际无线电科学联合会
Ursid meteor shower 小熊流星雨

Ursids 小熊流星群　又见：*Ursa Minorids*
Urumqi 乌鲁木齐（小行星 2729 号）
USNO （＝United States Naval Observatory）美国海军天文台
UT （＝universal time）世界时
UTC （＝Coordinated Universal Time）协调世界时
Utopia Planitia 乌托邦平原（火星）
UU Her star 武仙 UU 型星 | 武仙 [座]UU 型星〔台〕
UV astronomy （＝ultraviolet astronomy）紫外天文学 | 紫外 [線] 天文學〔台〕
UV background （＝ultraviolet background）紫外背景
UV band （＝ultraviolet band）紫外波段
UV bright star （＝ultraviolet-bright star）紫外亮星
uvby color system uvby 颜色系统
uvby photometry uvby 测光
uvby system uvby 系统
UV camera （＝ultraviolet camera）紫外照相机
UV Cet star 鲸鱼 UV 型星 | 鯨魚 [座]UV 型星〔台〕
UV color （＝ultraviolet color）紫外色
UV continuum （＝ultraviolet continuum）紫外连续谱
UV counterpart （＝ultraviolet counterpart）紫外对应体
UV cutoff （＝ultraviolet cutoff）紫外截断
UV detector （＝ultraviolet detector）紫外探测器
UV drop-out （＝ultraviolet drop-out）紫外漏失
UV dwarf （＝ultraviolet dwarf）紫外矮星
UV excess （＝ultraviolet excess）紫外超 | 紫外 [辐射] 超量〔台〕
UV-excess object （＝ultraviolet-excess object）紫外超天体 | 紫外 [輻射] 超量天體〔台〕
UV-excess star （＝ultraviolet-excess star）紫外超恒星
UV extinction （＝ultraviolet extinction）紫外消光
UV filter （＝ultraviolet filter）紫外滤光片
UV identification （＝ultraviolet identification）紫外证认
UV image （＝ultraviolet image）紫外像
uviol 紫外玻璃 | 透紫外 [線] 玻璃〔台〕
UV ionization chamber （＝ultraviolet ionization chamber）紫外电离室
UV light （＝ultraviolet light）紫外光
UV line （＝ultraviolet line）紫外谱线
UV luminosity （＝ultraviolet luminosity）紫外光度
UV magnitude （＝ultraviolet magnitude）紫外星等
UV map （＝ultraviolet map）紫外天图
UV object （＝ultraviolet object）紫外天体
UV Per star 英仙 UV 型星 | 英仙 [座]UV 型星〔台〕
UV photography （＝ultraviolet photography）紫外照相
UV photometry （＝ultraviolet photometry）紫外测光
uv plane uv 平面（射电）
UV radiation （＝ultraviolet radiation）紫外辐射
UV radiation detector （＝ultraviolet radiation detector）紫外辐射探测器
UV satellite （＝ultraviolet satellite）紫外卫星
UV sky （＝ultraviolet sky）紫外天空
UV source （＝ultraviolet source）紫外源
UV spectrograph （＝ultraviolet spectrograph）紫外摄谱仪
UV spectrophotometry （＝ultraviolet spectrophotometry）紫外分光光度测量
UV spectroscopy （＝ultraviolet spectroscopy）紫外分光
UV spectrum （＝ultraviolet spectrum）紫外光谱
UV star （＝ultraviolet star）紫外星
UV survey （＝ultraviolet surveys）紫外巡天
UV telescope （＝ultraviolet telescope）紫外望远镜
UX UMa star 大熊 UX 型星

V

V （=vanadium）钒（23号元素）
Vaca Muerta meteorite　巴卡穆埃尔塔陨星
vacant shell　空壳层
vacuum　真空
vacuum energy　真空能量
vacuum expectation value　（=VEV）真空期望值
vacuum solar telescope　真空太阳望远镜
vacuum solar tower　真空太阳塔
vacuum spectrograph　真空摄谱仪
vacuum state　真空态
vacuum telescope　真空望远镜
vacuum tower telescope　真空塔式望远镜
vacuum wavelength　真空波长
Vainu Bappu Observatory　巴普天文台
valence band　价电子带
Valentine Nebula　情人星云（IC 1805）
Valhalla Basin　瓦哈拉盆地（木卫四）
validity check　真实性检验
Valles Marineris　水手谷（火星）
Vallis Alpes　【拉】阿尔卑斯大峡谷（月球）
Vallis Baade　【拉】巴德谷（月球）
Vallis Bohr　【拉】玻尔谷（月球）
Vallis Bouvard　【拉】布瓦尔谷（月球）
Vallis Capella　【拉】卡佩拉谷（月球）
Vallis Christel　【拉】克里斯特尔谷（月球）
Vallis Inghirami　【拉】因吉拉米谷（月球）
Vallis Krishna　【拉】克里希纳谷（月球）
Vallis Palitzsch　【拉】帕利奇谷（月球）
Vallis Planck　【拉】普朗克谷（月球）
Vallis Rheita　【拉】里伊塔月谷（月球）
Vallis Schrödinger　【拉】薛定谔谷（月球）
Vallis Schröteri　【拉】施洛特月谷（月球）
Vallis Snellius　【拉】斯涅尔谷（月球）
vallis（复数：valles）　【拉】谷
value of division　分划值
vanadium　（=V）钒（23号元素）

Van Allen belt　范艾伦带
Van Allen radiation belt　范艾伦辐射带
van Biesbroeck's Star　范比斯布洛克星
van de Graaff crater　范德格拉夫环形山（月球）
van den Bergh classification　范登伯分类法
van den Bergh luminosity class　范登伯光度级
van der Waals crater　范德瓦耳斯环形山（月球）
van der Waals force　范德瓦耳斯力 | 凡得瓦力〔台〕
Vanguard　前锋（美国科学卫星）
van Maanen's star　范玛宁星 | 范瑪倫星〔台〕（HIP 3829）
Vanth　亡卫（小行星Orcus的卫星）
van't Hoff crater　范托夫环形山（月球）
van Vleck relation　范弗莱克关系 | 范扶累克關係〔台〕
VAO　（=Virtual Astronomical Observatory）虚拟天文台
vapour deposition　汽沉积
varactor　可变电抗器
varactor parametric amplifier　变容二极管参量放大器
Variabilis Coronae　北冕座R
variability　变性，变率
variable base-line interferometer　变基线干涉仪 | 變距干涉儀〔台〕
variable base-line interferometry　变基线干涉测量
variable capacitance amplifier　变容放大器
variable component　可变分量
variable infrared object　红外变光天体
variable mass　变质量
variable nebula　变光星云
variable profile antenna　变轮廓天线
variable radio object　（=VRO）射电变化天体 | 變電波體〔台〕
variable radio source　（=VRS）射电变源 | 變電波源〔台〕

variable source 变源
variable-spacing interferometer 变距干涉仪
variable-spacing interferometry 变距干涉测量
variable star （＝VS）变星
variable time scale 可变时标
variable ultraviolet object 紫外变光天体
variable-velocity star 视向速度变星 又见：velocity variable
variable X-ray source X射线变源 | X光變源〔台〕
variable γ-ray source γ射线变源
variance 方差
variational inequality 月行差 又见：lunar inequality
variational method 变分法
variational orbit 变分轨道 又见：variation orbit
variational principle 变分原理
variation of constants 常数变值法
variation of elements 根数变值法
variation of latitude 纬度变化 又见：latitude variation
variation of parameters 参数变值法
variation of the moon 二均差 又见：moon's variation
variation orbit 变分轨道 又见：variational orbit
variometer 可变电感器
varistor 变阻器 又见：rheostat
Varuna 伐楼那（小行星20000号）
Vasco da Gama crater 瓦斯科·达·伽马环形山（月球）
vastitas 【拉】广低平原（行星地貌）
Vastitas Borealis 【拉】北部荒原（火星）
Vatican Advanced Technology Telescope （＝VATT）梵蒂冈高新技术望远镜 | 梵蒂冈先進技術望遠鏡〔台〕
Vatican Observatory 梵蒂冈天文台
VATT （＝Vatican Advanced Technology Telescope）梵蒂冈高新技术望远镜 | 梵蒂冈先進技術望遠鏡〔台〕
Vavilov crater 瓦维洛夫环形山（月球）
VBLUW color system VBLUW颜色系统
VBLUW photometry VBLUW测光
V-class asteroid V型小行星
vector boson 矢量玻色子
vector field 矢量场
vectorial astronomy 矢量天体测量学 | 向量天文測量學〔台〕
vectorial orbital constant 矢量轨道常数
vector meson 矢性介子
vector mode 矢量模
vector perturbation 矢量扰动

vector-point-diagram 矢点图
vector potential 矢量势 | 向[量]势，向[量]位〔台〕
vector product 矢积
vector space 矢量空间
vector spectromagnetograph 矢量分光磁像仪
vector translation 矢量平移
Vega ① 织女一，织女星（天琴座α）② 维加（苏联行星际探测器）
Vega crater 韦加环形山（月球）
Vega-like object 类织女天体
Vega phenomenon 织女星现象
veiling 淡化
Veil Nebula 帷幕星云 | 面紗星雲〔台〕（NGC 6960）
veil variable star 掩食变星 | [掩]食變星〔台〕 又见：occultation variable
Vel （＝Vela）船帆座
Vela （＝Vel）船帆座
Vela pulsar 船帆脉冲星（PSR 0833-45）
Vela supernova remnant 船帆超新星遗迹（NGC 2736）
velocity bias 速度偏袒
velocity component 速度分量
velocity correlation function 速度相关函数
velocity curve 速度曲线
velocity-density reconstruction 速度－密度重构
velocity dispersion 速度弥散度 | 速度彌散[度]〔台〕
velocity-distance relation 速度－距离关系 | 速距關係〔台〕
velocity distribution 速度分布
velocity ellipsoid 速度椭球
velocity field 速度场
velocity gradient 速度梯度
velocity of entry 进入速度 | 進入大氣速度〔台〕
velocity of escape 逃逸速度 | 脫離速度〔台〕 又见：escape velocity; escape speed
velocity of light 光速 又见：speed of light
velocity-of-light cylinder 光速柱面
velocity-of-light radius 光速半径
velocity of recession 退行速度 又见：recession velocity; regression velocity
velocity perturbation 速度扰动
velocity potential 速度势
velocity profile 速度轮廓
velocity space 速度空间
velocity spectrum 速度谱

velocity variable 视向速度变星
又见: *variable-velocity star*
Venera probe 金星号探测器 | 金星號 [行星際探測器] 〔台〕（苏联）
Vening Meinesz crater 芬宁·梅因纳斯环形山（月球）
ventilation 通风
Ventris crater 文特里斯环形山（月球）
Venus 金星，太白
Venus Express 金星快车
Venus globe 金星仪
VERITAS （=Very Energetic Radiation Imaging Telescope Array System）甚高能辐射成像望远镜阵（美国）
Vermilion Bird 朱鸟（四象）又见: *Red Bird*
Vernadskiy crater 韦尔纳茨基环形山（月球）
Vernal Equinox 春分（节气）又见: *Spring Equinox*
vernal equinox 春分点 又见: *vernal point; spring equinox; first point of Aries*
vernal point 春分点 又见: *vernal equinox; spring equinox; first point of Aries*
vernier 游标 | 游標〔台〕
vernier clock 游标钟
vernier-type time signal 游标式时号
versed ascension 负赤经
vertex 奔赴点
vertex angle 顶角
vertical angle 高度角
vertical axis 竖轴 | 縱軸，垂直軸〔台〕
vertical blanking interval 垂直消隐间隔
vertical circle 地平经圈，垂直圈 | 地平經圈〔台〕
又见: *azimuth circle*
vertical distribution 垂直分布
vertical frequency 垂直频率
vertical line 垂线，竖直线 | 垂線，豎 [直] 線〔台〕
vertical parallax screw 竖直视差螺旋
vertical pendulum 竖直摆
vertical plane 竖直面 | 垂直面〔台〕
vertical resonance 竖直共振
vertical revolving circle 立运仪 又见: *vertical revolving instrument*
vertical revolving instrument 立运仪
又见: *vertical revolving circle*
vertical spectrograph 竖直摄谱仪
vertical thread 竖丝 | 縱絲，豎絲〔台〕
very early universe 极早期宇宙

Very Energetic Radiation Imaging Telescope Array System （=VERITAS）甚高能辐射成像望远镜阵（美国）
very hard binary 甚硬双星
very high energy cosmic ray （=VHECR）甚高能宇宙线
very high frequency （=VHF）甚高频 | 特高頻〔台〕（30-300 兆赫）
Very Large Array （=VLA）甚大阵 | 特大天線陣〔台〕
very large-scale phenomenon 甚大尺度现象
Very Large Telescope （=VLT）甚大望远镜 | 特大望遠鏡〔台〕
Very Large Telescope Inteferometer （=VLTI）甚大望远镜干涉仪
very long baseline （=VLB）甚长基线
Very Long Baseline Array （=VLBA）甚长基线 [射电望远镜] 阵 | 特長基線 [電波望遠鏡] 陣〔台〕（美国）
very long baseline interferometer （=VLBI）甚长基线干涉仪 | 特長基線干涉儀〔台〕
Very Long Baseline Interferometry （=VLBI）甚长基线干涉测量 | 特長基線干涉測量〔台〕
very low frequency （=VLF）甚低频 | 特低頻〔台〕
very low-mass star 甚小质量恒星
very massive object （=VMO）甚大质量天体 | 特大質量天體〔台〕
very relativistic electron 极近相对论性电子
very short wave 甚短波
very slow nova 慢变光新星
very small-scale structure 甚小尺度结构
very soft binary 甚软双星
very strong-lined giant （=VSL giant）甚强线巨星
very strong-lined star （=VSL star）甚强线星
Vesalius crater 维萨里环形山（月球）
Vesper 长庚，昏星 | 長庚星〔台〕 又见: *Hesperus; Vesperus*
Vesperus 长庚，昏星 | 長庚星〔台〕
又见: *Hesperus; Vesper*
Vesta 灶神星（小行星 4 号）
Vesta-like asteroid 类灶小行星 又见: *vestoid*
vestoid 类灶小行星 又见: *Vesta-like asteroid*
Vetchinkin crater 韦钦金环形山（月球）
VEV （=vacuum expectation value）真空期望值
VHECR （=very high energy cosmic ray）甚高能宇宙线

VHF （＝very high frequency）甚高频 | 特高频〔台〕（30-300兆赫）
vial 水准器玻璃管
Vibilia 旅神星（小行星144号）
vibration 振动
vibrational constant 振动常数
vibrational energy 振动能　又见：*energy of vibration*
vibrational energy level 振动能级
vibrational instability 振动不稳定性
vibrational line 振动谱线
vibrational quantum number 振动量子数
vibrational spectrum 振动谱　又见：*vibration spectrum*
vibrational structure 振动结构
vibrational transition 振动跃迁
vibration spectrum 振动谱　又见：*vibrational spectrum*
vibration temperature 振动温度
vibrator 振荡器，振子
V-I color index V－I色指数
Victorchang 张任谦（小行星24450号）
Victoria 凯神星（小行星12号）
Victoria crater 维多利亚环形山（火星）
Victor M. Blanco Telescope 布兰科望远镜
video astronomy 视频天文学
video converter 视频变换器
video magnetograph 视频磁像仪
vidicon 视像管
Vienna 维也纳（小行星397号）
Vierter Fundamental Katalog （＝FK4）【德】FK4星表，第四基本星表
Vieta crater 韦达环形山（月球）
viewfinder 寻星镜　又见：*finderscope; finder; star finder*
viewing angle 视角　又见：*apparent angle; visual angle*
viewing distance 视距
viewing field 视场 | 视野〔台〕　又见：*field of view*
vignetting 渐晕
vignetting effect 渐晕效应
vignetting stop 止晕光阑 | 隔晕光欄〔台〕
Viking 海盗号（美国火星探测器）
Vilnius color system 维尔纽斯颜色系统
Vilnius photometry 维尔纽斯测光
Vindemiatrix 东次将，太微左垣四（室女座ε）　又见：*Almuredin*
vinyl cyanide 丙烯腈（C_3H_3N）　又见：*cyanoethylene*

violent galaxy 激变星系
violent relaxation 剧变弛豫
violet layer 紫层
violet shift 紫移
Vir ①（＝Virgo）室女座 ②（＝Virgo）室女宫，鹑尾，巳宫
Virginia 贞女星（小行星50号）
Virginid meteor shower 室女流星雨
Virgo ①（＝Vir）室女座 ②（＝Vir）室女宫，鹑尾，巳宫
virgocentric flow 室女星系团中心流
Virgo cluster 室女星系团　又见：*Virgo galaxy cluster*
Virgo galaxy cluster 室女星系团　又见：*Virgo cluster*
Virgo gravitational wave detector 室女[座]引力波探测器（意大利）
Virgo supercluster 室女超团
virial coefficient 位力系数
virial density 位力密度
virial equation 位力方程
virial equilibrium 位力平衡 | 均功平衡〔台〕
virialization 位力化
virial mass 位力质量 | 均功质量〔台〕
virial parameter 位力参数
virial radius 位力半径 | 均功半径〔台〕
virial temperature 位力温度
virial theorem 位力定理 | 均功定理〔台〕
virial velocity 位力速度
Virtual Astronomical Observatory （＝VAO）虚拟天文台
virtual displacement 虚位移
virtual focus 虚焦点
virtual image 虚像
virtual observatory 虚拟天文台
Virtual Observatory India （＝VO-India）印度虚拟天文台
Virtual Observatory United Kingdom （＝AstroGrid）英国虚拟天文台
virtual particle 虚粒子
virtual phase CCD 虚相CCD
virtual phonon 虚声子
virtual photon 虚光子
virtual wave 虚波
virtual work 虚功
virtuous star 德星
viscidity 黏性，黏度　又见：*viscosity*
viscosity 黏性，黏度　又见：*viscidity*

viscous disk　黏性盘
viscous dissipation　黏性耗散
viscous fluid　黏性流体
viscous force　黏性力
visibility　能见度，可见度
visibility function　能见度函数，可见度函数
visibility of planet　行星可见期
visibility of satellite　卫星可见期，人卫可见期 | 衛星可見期〔台〕
Visible and Infrared Survey Telescope for Astronomy　（＝VISTA）天文可见光及红外巡天望远镜
visible arm　可见臂
visible component　可见子星
visible disk　可见圆面
visible horizon　可见地平
visible light　可见光　又见：*optical light*
visible matter　可见物质
visible nulling coronagraph　可见光消零星冕仪
visible radiation　可见辐射
visible spectrum　可见光谱
VISTA　（＝Visible and Infrared Survey Telescope for Astronomy）天文可见光及红外巡天望远镜
visual acuity　视敏度
visual angle　视角　又见：*apparent angle; viewing angle*
visual astronomy　目视天文学
visual band　目视波段
visual binary　目视双星　又见：*visual double star*
visual catalogue　目视星表
visual classification　目视分类
visual colorimetry　目视色度测量 | 目視色度學〔台〕
visual diameter　视直径 | 視徑〔台〕
　又见：*apparent diameter*
visual double star　目视双星　又见：*visual binary*
visual extinction　目视波段消光
visual light curve　目视光变曲线
visual line　视线　又见：*line of sight; line of vision*
visual luminosity　目视光度
visual magnitude　目视星等
visual meteor　目视流星
visual multiple star　目视聚星
visual object　目视天体
visual objective　目视物镜
visual observation　目视观测
visual photometer　目视光度计

visual photometry　目视测光 | 目视光度测量，目视测光〔台〕
visual polarimeter　目视偏振计
visual polarimetry　目视偏振测量
visual reflector　目视反射望远镜
visual refractor　目视折射望远镜
visual sensitivity curve　目视灵敏度曲线
visual source　目视源
visual spectrophotometer　目视分光光度计
visual spectrophotometry　目视分光光度测量
visual star　目视星
visual telescope　目视望远镜
visual zenith telescope　（＝VZT）目视天顶仪 | 目視天頂筒〔台〕
vis viva equation　活力方程
vis viva integral　活力积分
Vivaldi antenna　维瓦尔第天线
VLA　（＝Very Large Array）甚大阵 | 特大天線陣〔台〕
Vlacq crater　弗拉克环形山（月球）
Vlasov equation　弗拉索夫方程
Vlasov-Maxwell equation　弗拉索夫－麦克斯韦方程
VLB　（＝very long baseline）甚长基线
VLBA　（＝Very Long Baseline Array）甚长基线[射电望远镜]阵 | 特長基線[電波望遠鏡]陣〔台〕（美国）
VLBI　① （＝Very Long Baseline Interferometry）甚长基线干涉测量 | 特長基線干涉測量〔台〕
② （＝very long baseline interferometer）甚长基线干涉仪 | 特長基線干涉儀〔台〕
VLBI Space Observatory Programme　（＝VSOP）空间甚长基线干涉测量天文台计划 | 太空特長基線干涉測量天文台計畫〔台〕
VLF　（＝very low frequency）甚低频 | 特低頻〔台〕
VLT　（＝Very Large Telescope）甚大望远镜 | 特大望遠鏡〔台〕
VLTI　（＝Very Large Telescope Inteferometer）甚大望远镜干涉仪
V-magnitude　V星等
VMO　（＝very massive object）甚大质量天体 | 特大質量天體〔台〕
Vobs.it　（＝Italian Virtual Observatory）意大利虚拟天文台
Vogt-Russell theorem　福格特－罗素定理
　即：*Russell-Vogt theorem*
Void　虚宿（二十八宿）　又见：*Emptiness*
void　巨洞　又见：*cosmic void*

Voigt effect 福格特效应
Voigt line profile 福格特谱线轮廓
Voigt profile 福格特轮廓 | 佛克特線廓〔台〕
VO-India （＝Virtual Observatory India）印度虚拟天文台
Vol （＝Volans）飞鱼座
Volans （＝Vol）飞鱼座
volatile substance 挥发性物质
volcanic activity 火山活动
volcanic ash 火山灰
volcanic glass 火山玻璃
volcanic rock 火山岩
volcanism 火山作用，火山现象　又见：*vulcanism*
volcano 火山
volcano hypothesis 火山假说
volcanology 火山学　又见：*vulcanology*
Volta crater 伏打环形山（月球）
voltage bias 偏压
voltage standing wave ratio （＝VSWR）电压驻波比
Volterra crater 沃尔泰拉环形山（月球）
volume charge 体电荷
volume density 体密度
volume emissivity 体发射率
volume-limited redshift surveys 限定体积红移巡天
volume-limited sample 限定体积样本
volume-phase holographic grating 体相全息光栅
volume-preserving mapping 保体积映射
volvelle 日月升落潮汐仪
Von Baeyer crater 冯・拜耳环形山（月球）
Von Békésy crater 冯・贝凯西环形山（月球）
Von Braun crater 冯・布劳恩环形山（月球）
Von der Pahlen crater 冯・德・帕伦环形山（月球）
Vondrak method 冯德拉克平滑法
Von Kármán crater 冯・卡门环形山（月球）
Von Neumann crater 冯・诺伊曼环形山（月球）
Von Zeipel crater 冯・蔡佩尔环形山（月球）
von Zeipel method 蔡佩尔方法
von Zeipel theorem 蔡佩尔定理
Vorontsov-Velyaminov Catalogue of Interacting Galaxies （＝VV catalogue）沃隆佐夫－威廉明诺夫相互作用星系表
vortex 涡旋
vortex filament 涡丝
vortex flow 涡流 | 旋渦〔台〕　又见：*eddy*; *swirl*

vortex hypothesis 涡流假说
vortex line 涡线
vortex motion 涡动
vortex sheet 涡流片
vortex state 涡旋态
vortex structure 涡旋结构 | 渦流結構〔台〕
vortical perturbations 有旋扰动
vorticity 涡度
vorticity mode 涡度模
Voskhod 上升号（苏联空间飞船）
Vostok 东方号（苏联空间飞船）
Voyager 旅行者号 | 航海家號〔台〕（美国行星际探测器）
V-R color index V－R色指数
Vredefort crater 弗里德堡陨星坑（地球）
VRI color system VRI颜色系统
VRO （＝variable radio object）射电变化天体 | 變電波體〔台〕
VRS （＝variable radio source）射电变源 | 變電波源〔台〕
v/r variable v/r变星
VS （＝variable star）变星
VSL giant （＝very strong-lined giant）甚强线巨星
VSL star （＝very strong-lined star）甚强线星
VSOP （＝VLBI Space Observatory Programme）空间甚长基线干涉测量天文台计划 | 太空特長基線干涉測量天文台計畫〔台〕
VSWR （＝voltage standing wave ratio）电压驻波比
V-type asteroid V型小行星　即：*vestoid*; *Vesta-like asteroid*
Vul （＝Vulpecula）狐狸座
Vulcan 祝融星（假想的一颗位于水星和太阳之间的行星）
vulcanism 火山作用，火山现象　又见：*volcanism*
Vulcan-like asteroid 祝融型小行星 | 祝融型小天體〔台〕　又见：*vulcanoid*
vulcanoid 祝融型小行星 | 祝融型小天體〔台〕　又见：*Vulcan-like asteroid*
vulcanoid zone 祝融区
vulcanology 火山学　又见：*volcanology*
vulgar establishment 朔望平均潮候时差
Vulpecula （＝Vul）狐狸座
VV catalogue （＝Vorontsov-Velyaminov Catalogue of Interacting Galaxies）沃隆佐夫－威廉明诺夫相互作用星系表

VV Cephei star （＝VV Cep star）仙王 VV 型星 | 仙王[座]VV 型星〔台〕

VV Cep star （＝VV Cephei star）仙王 VV 型星 | 仙王[座]VV 型星〔台〕

V/Vm test V/Vm 检验

VZT （＝visual zenith telescope）目视天顶仪 | 目视天頂筒〔台〕

W

W （=tungsten）钨（74号元素）
Wabar crater 瓦巴阴星坑（地球）
Wain 北斗[七星]（星组） 又见：*Triones; Big Dipper; Northern Dipper; Charles' Wain; Plough; Plow*
Waiting-maid 须女 又见：*Lowly concubine of emperor; Serving-maid*
wake 瞬现余迹 | 流星尾〔台〕
Waking of Insects 惊蛰（节气） 又见：*Awakening from Hibernation*
Wall 壁宿（二十八宿）
Wallace crater 华莱士环形山（月球）
wall coating 壁镀膜
wall collision 壁碰撞
walled plain 月面环壁平原 | [月面]圆谷〔台〕
wall shift 壁移
Walraven color system 瓦尔拉文颜色系统
Walraven photometry 瓦尔拉文测光
Walther crater 瓦尔特环形山（月球）
Wanda crater 万达环形山（金星）
wandering of the pole 极移 又见：*motion of earth pole; polar motion; polar displacement; polar shift; polar wandering*
wandering star 游星
Wangchaohao 王超昊（小行星20778号）
Wangdaheng 王大珩（小行星17693号）
Wangganchang 王淦昌（小行星14558号）
Wangshouguan 王绶琯（小行星3171号）
Wangxuan 王选（小行星4913号）
Wangyongzhi 王永志（小行星46669号）
waning crescent 下蛾眉月 | 殘月〔台〕
waning gibbous 亏凸月
waning moon 亏月 又见：*decrescent*
warm dark matter 温暗物质
warm-hot intergalactic medium （=WHIM）星系际温热介质
warm intercloud medium 云际温介质
warm plasma 热等离子体 | 熱電漿〔台〕

warm spot 热斑 又见：*hot spot*
warp 翘曲
warped galaxy 翘曲星系
Wasat 天樽二（双子座δ） 又见：*Wesat*
water 水（H_2O）
water Cherenkov neutrino telescope 水切伦科夫式中微子望远镜
water Cherenkov telescope 水切伦科夫望远镜
water clock 水钟
waterfall field 瀑布场
water ice 水冰
water-jet actinometer 喷水日光能量测定仪
Waterman crater 沃特曼环形山（月球）
water maser 水微波激射，水脉泽 | 水邁射〔台〕
waterspout prominence 龙卷日珥 又见：*tornado prominence*
water-vapour maser 水汽微波激射，水汽脉泽
Watson crater 沃森环形山（月球）
Watt crater 瓦特环形山（月球）
wave aberration 波像差
wave amplitude 波幅
wave band 波段
wave crest 波峰 又见：*crest*
wave energy density 波能密度
wave equation 波动方程
wave form 波形 又见：*wave shape*
wave-form analysis 波形分析
wave front 波前 | 波阵面，波前〔台〕
wave-front sensor 波前传感器
wave-front tilt 波前倾斜
wave function 波函数
waveguide 波导
wave length 波长
wavelet 小波
wave-like behaviour 波动性
wave momentum density 波动量密度

wave motion　波动
wave noise　波动噪声
wave number　波数
wave-number vector　波数矢量
wave packet　波包
wave-particle interaction　波粒相互作用
wave pattern　波图案
wave period　波周期
wave plate analyser　波片检偏器
wave propagation　波传播
wave reflection　波反射
wave refraction　波折射
wave shape　波形　又见: *wave form*
wave spectrum　波谱
wave surface　波面
wave theory　波动说
wave train　波列　又见: *train of waves*
wave trough　波谷
wave vector　波矢
wave velocity　波速
wave-wave interaction　波-波相互作用
wave zone　波动区，远场区（中子星）
waxing and waning of the moon　月相盈亏
waxing crescent　上蛾眉月　又见: *crescent moon*
waxing gibbous　盈凸月
waxing moon　盈月　又见: *increscent moon*
Wayeb'　瓦耶勃（玛雅历日）　又见: *Uayeb*
Wazn　子二（天鸽座 β）　又见: *Wezn*
W. Bond crater　威·邦德环形山（月球）
W boson　W 玻色子
WBVR color system　WBVR 颜色系统
WCN star　CN 型 WR 星
WC star　C 型 WR 星
WD　（＝white dwarf）白矮星
WDC　（＝World Data Center）世界数据中心｜世界资料中心〔台〕
Weak Anthropic Principle　弱人择原理
weak bar　弱棒
weak encounter　弱交会
weak energy condition　弱能量条件
weak equivalence principle　弱等效原理
weak field　弱场
weak-field condition　弱场条件
weak force　弱力
weak G band star　弱 G 谱带星
weak gravitational lensing　弱引力透镜

weak interaction　弱相互作用｜弱[交互]作用〔台〕
weak lensing　（＝WL）弱引力透镜效应
weak-lined star　弱线星
weak-line T Tau star　弱线金牛 T 型星
weakly interacting massive particle　（＝WIMP）弱相互作用大质量粒子
weakly ionized gas　弱电离气体
weak neutral current　弱中性流
weak nuclear force　弱核力
weak nuclear interaction　弱核相互作用
weak source　弱源
weak stability　弱稳定性
weak turbulence　弱湍流
weak turbulence theory　弱湍流理论
weather satellite　气象卫星　又见: *meteorological satellite*
Weber crater　韦伯环形山（月球）
wedge constant　光劈常数｜楔常数〔台〕
wedge photometer　光劈光度计｜楔光度計〔台〕
wedge product　外积　又见: *outer product*
Wegener crater　魏格纳环形山（月球）
Wegener hypothesis　魏格纳假说
Wei　尾宿二（天蝎座 ε）
weighted average　① 加权平均　又见: *weighted averaging* ② 加權平均值｜加權平均值，加權平均數〔台〕　又见: *weighted mean*
weighted averaging　加权平均　又见: *weighted average*
weighted error　加权误差
weighted mean　加权平均值｜加權平均值，加權平均數〔台〕　又见: *weighted average*
weighted observation　加权观测
weight function　权函数
weighting　取权
weighting factor　取权因子
weighting filter　加权滤波器
weighting of observation　观测计权
weightlessness　失重
Weihai　威海（小行星 207931 号）
Weinberg angle　温伯格角
Weinberg-Salam theory　温伯格－萨拉姆理论
Weiss crater　魏斯环形山（月球）
Weizsacker theory　魏扎克理论（太阳系起源）
Well　井宿（二十八宿）
well-tapered illumination　良照明
Weltzeit　（＝W.Z.）【德】世界时　即: *UT*
Wenchuan　汶川（小行星 161715 号）

Wenlingshuguang 温岭曙光（小行星 14147 号）
Wensayling 温世仁（小行星 145545 号）
Wentzel-Kramers-Brillouin approximation （＝WKB approximation）WKB 近似
Wentzel-Kramers-Brillouin-Jeffreys approximation （＝WKBJ approximation）WKBJ 近似
Wentzel-Kramers-Brillouin-Jeffreys method （＝WKBJ method）WKBJ 方法
Wentzel-Kramers-Brillouin method （＝WKB method）WKB 方法
Werner crater 维尔纳环形山（月球）
Werner line 沃纳谱线
Wesat 天樽二（双子座 δ） 又见：*Wasat*
Wesselink analysis 韦塞林克分析
Wesselink mass 韦塞林克质量
Wesselink radius 韦塞林克半径
west 西
Westar 西联星（美国通信卫星）
West comet 韦斯特彗星（C/1975 V1）
Westerbork Radio Observatory 韦斯特博克射电天文台
Westerbork Synthesis Radio Telescope （＝WSRT）韦斯特博克综合孔径射电望远镜 | 韋斯特博克孔徑合成電波遠鏡〔台〕
Westerhout catalogue 韦斯特豪特射电源表 | 韋斯特豪特電波源表〔台〕
western elongation ① 西大距 又见：*greatest western elongation* ② 西角距，西大距 | 西距角〔台〕
western hemisphere 西半球
western quadrature 西方照
West Ford 西福特（美国通信卫星）
west longitude 西经
west point 西点
WET （＝Whole Earth Telescope）全球望远镜
wet merger 湿并合
Wexler crater 韦克斯勒环形山（月球）
Weyl crater 外尔环形山（月球）
Weyl tensor 外尔张量
Weyl unified field theory 外尔统一场论
Weywot 创卫一（小行星 Quaoar 的卫星）
Wezea 弧矢一（大犬座 δ） 又见：*Wezen*
Wezen 弧矢一（大犬座 δ） 又见：*Wezea*
Wezn 子二（天鸽座 β） 又见：*Wazn*
WFI （＝wide field imager）大视场成像器
WFIRST （＝Wide-Field Infrared Survey Telescope）大视场红外巡天望远镜
Whale Galaxy 鲸鱼星系（NGC 4631）
wheel-and-track mount 轮轨式装置

Wheeler-de Witt equation 惠勒－德维特方程
WHIM （＝warm-hot intergalactic medium）星系际温热介质
Whipple 10-m gamma-ray telescope 惠普尔 10 米伽马射线望远镜
Whipple model 惠普尔模型
Whipple Observatory 惠普尔天文台 又见：*Fred Lawrence Whipple Observatory*
Whipple's comet 惠普尔彗星 | 惠普彗星〔台〕
Whirlpool galaxy 涡状星系（NGC 5194）
whistler 啸声 | 電嘯〔台〕
whistler wave 啸声波
white-blue dwarf 蓝白矮星
White Dew 白露（节气）
white dwarf （＝WD）白矮星
white dwarf branch 白矮星支
white frequency noise 白频率噪声
white fringe 白条纹
white Gauss noise 白高斯噪声
white giant 白巨星
white hole 白洞
white-light corona 白光日冕
white-light coronameter 白光日冕光度计 又见：*white-light coronometer*
white-light coronometer 白光日冕光度计 又见：*white-light coronameter*
white light event 白光事件
white-light flare 白光耀斑 | 白光閃焰〔台〕
white night 白夜
white noise 白噪声 | 白雜訊〔台〕
white-noise spectrum 白噪声谱
white phase noise 白相位噪声
white-pupil echelle spectrometer 白瞳阶梯光栅光谱仪
white spot 白斑
white star 白星
White Tiger 白虎（四象）
Whittaker criterion 惠特克判据
Whole Earth Telescope （＝WET）全球望远镜
whole-line width 全线宽
WHT （＝William Herschel Telescope）赫歇尔望远镜 | 赫歇耳望遠鏡〔台〕
wide-angle astrograph 广角天体照相仪
wide-angle camera 广角照相机 | 廣角相機〔台〕
wide-angle eyepiece 广角目镜 又见：*wide-field eyepiece*
wide-angle lens 广角透镜
wide-angle object glass 广角物镜

wide-angle plate　大视场底片 | 廣角底片〔台〕
wide-angle telescope　广角望远镜
wide band filter　宽波段滤光片
wide band photometry　宽波段测光
wide binary　远距双星　又见: *wide pair*
wide binary galaxy　远距双重星系
wide-field eyepiece　广角目镜　又见: *wide-angle eyepiece*
wide field imager　（＝WFI）大视场成像器
wide-field imaging　大视场成像
Wide-Field Infrared Explorer　（＝WIRE）大视场红外探测器
Wide-field Infrared Survey Explorer　（＝WISE）广域红外巡天探测者
Wide-Field Infrared Survey Telescope　（＝WFIRST）大视场红外巡天望远镜
wide-field spectrograph　大视场摄谱仪
wide-field telescope　大视场望远镜
widening　致宽
wide pair　远距双星　又见: *wide binary*
wide visual binary　远距目视双星
Widmanstatten figure　魏德曼花纹　又见: *Widmanstatten pattern*
Widmanstatten pattern　魏德曼花纹　又见: *Widmanstatten figure*
width-shear vibrator　宽度切变振荡器
Wiechert discontinuity　维歇特不连续
Wien displacement law　维恩位移定律
Wiener crater　维纳环形山（月球）
Wiener deconvolution　维纳反卷积法
Wiener filter　维纳滤波器
Wiener-Khinchin theorem　维纳－辛钦定理
Wienphilo　维也纳爱乐（小行星 178263 号）
Wien's displacement law　维恩位移定律
Wien side　维恩侧
WiggleZ survey　WiggleZ 巡天
Wild 2 Comet　怀尔德 2 号彗星
Wild Duck cluster　野鸭星团（NGC 6705）又见: *Wild Duck Nebula*
Wild Duck Nebula　野鸭星团（NGC 6705）又见: *Wild Duck cluster*
Wild's Trio　怀尔德三合星系
Wild's Triplet　怀尔特三合星
Wilhelm crater　威廉环形山（月球）
Wilkins crater　威尔金斯环形山（月球）
Wilkinson Microwave Anisotropy Probe　（＝WMAP）威尔金森微波各向异性探测器

William Herschel Telescope　（＝WHT）赫歇尔望远镜 | 赫歇耳望遠鏡〔台〕
Willman I dwarf galaxy　威尔曼矮星系 I
Willow　柳宿（二十八宿）
Willstrop telescope　威尔斯特罗普望远镜
Wilson-Bappu effect　威尔逊－巴普效应
Wilson crater　威尔逊环形山（月球）
Wilson depression　威尔逊凹陷
Wilson effect　威尔逊效应
Wilson-Harrington Comet　威尔逊－哈林顿彗星
WIMP　（＝weakly interacting massive particle）弱相互作用大质量粒子
WIND　风太阳探测器（美国）
wind-feild interaction　风场互作用
winding dilemma　缠卷疑难
winding mode　缠卷模
winding number　缠卷数
winding problem　缠卷问题
winding rate　缠卷速率
window function　窗函数
Wings　翼宿（二十八宿）
wingspan　翼展
Winnowing-basket　箕宿（二十八宿）又见: *Basket; Dustpan*
wino　W 微子
Winter Solstice　冬至（节气）又见: *December Solstice*
winter solstice　冬至点　又见: *December solstice; first point of Capricornus*
Winter Triangle　冬季大三角
WIRE　（＝Wide-Field Infrared Explorer）大视场红外探测器
wire grid　丝栅 | 線柵〔台〕
wire interval　丝距
wireless time signal　无线电时号 | 無線電時間記錄〔台〕又见: *radio time signal*
Wirtanen Comet　维尔塔宁彗星
Wisconsin, Indiana, Yale and NOAO Observatory　（＝WIYN Observatory）WIYN 天文台
Wisconsin, Indiana, Yale and NOAO Telescope　（＝WIYN Telescope）WIYN 望远镜
WISE　（＝Wide-field Infrared Survey Explorer）广域红外巡天探测者
Witch Head Nebula　女巫头星云（IC 2118）
WIYN Observatory　（＝Wisconsin, Indiana, Yale and NOAO Observatory）WIYN 天文台
WIYN Telescope　（＝Wisconsin, Indiana, Yale and NOAO Telescope）WIYN 望远镜

WKB approximation （＝Wentzel-Kramers-Brillouin approximation）WKB 近似

WKBJ approximation （＝Wentzel-Kramers-Brillouin-Jeffreys approximation）WKBJ 近似

WKBJ method （＝Wentzel-Kramers-Brillouin-Jeffreys method）WKBJ 方法

WKB method （＝Wentzel-Kramers-Brillouin method）WKB 方法

WL （＝weak lensing）弱引力透镜效应

WLM system （＝Wolf-Lundmark-Melotte system）沃尔夫－伦德马克－梅洛特分类法

WMAP （＝Wilkinson Microwave Anisotropy Probe）威尔金森微波各向异性探测器

W. M. Keck Observatory （＝WMKO）凯克天文台

WMKO （＝W. M. Keck Observatory）凯克天文台

WMS （＝World Magnetic Survey）全球地磁普查

WN star WN 型星

wobbling method 摆动法

Wolf Creek crater 沃尔夫－克里克陨星坑（地球）

Wolf diagram 沃尔夫图 | 沃夫圖〔台〕

Wolf-Lundmark-Melotte system （＝WLM system）沃尔夫－伦德马克－梅洛特分类法

Wolf-Lundmark system 沃尔夫－伦德马克系统

Wolf number 沃尔夫数 | 沃夫數〔台〕

Wolf-Rayet galaxy （＝WR galaxy）沃尔夫－拉叶星系，WR 星系 | 沃夫－瑞葉星系，WR 星系〔台〕

Wolf-Rayet nebula （＝WR nebula）沃尔夫－拉叶星云，WR 星云 | 沃夫－瑞葉星雲，WR 星雲〔台〕

Wolf-Rayet star （＝WR star; W star）沃尔夫－拉叶星，WR 型星 | 沃夫－瑞葉星，WR 星〔台〕

Wolf sunspot number 伍尔夫太阳黑子数

Wollaston prism 沃拉斯顿棱镜 | 渥拉斯頓棱鏡〔台〕

Wolter-Schwarzschild X-ray telescope 沃尔特－施瓦西型 X 射线望远镜

Wolter type I X-ray telescope 沃尔特 I 型 X 射线望远镜

Woman 女宿（二十八宿） 又见：Maid; Girl; Damsel

Wongkwancheng 王宽诚（小行星 4651 号）

Wood color system 伍德颜色系统

Wood crater 伍德环形山（月球）

Woodleigh crater 伍德利陨星坑（地球）

Wood photometry 伍德测光

Wood's anomaly 伍德异常

word length 字长

work function 功函数

working area 工作空间

working field 工作像场

working space 工作单元

World Calendar 世界历

World Data Center （＝WDC）世界数据中心 | 世界資料中心〔台〕

world day 世界日

world line 世界线

World Magnetic Survey （＝WMS）全球地磁普查

world model 宇宙模型

world point 世界点

world sheet 世界面

World Space Observatory-Ultraviolet （＝WSO-UV）世界空间紫外天文台 | 世界太空紫外天文台〔台〕

world-wide disturbance 全球扰动

world-wide synchronization 全球同步

WorldWide Telescope （＝WWT）万维望远镜（微软公司软件）

wormhole 虫洞

WO star WO 型星

Wouthuysen-Field process 沃休森－菲尔德过程

wrapping 卷缠

WR galaxy （＝Wolf-Rayet galaxy）沃尔夫－拉叶星系，WR 星系 | 沃夫－瑞葉星系，WR 星系〔台〕

Wright telescope 怀特望远镜

wrinkle ridge 皱脊

WR nebula （＝Wolf-Rayet nebula）沃尔夫－拉叶星云，WR 星云 | 沃夫－瑞葉星雲，WR 星雲〔台〕

Wrongskian 朗斯基行列式

Wrottesley crater 罗茨利环形山（月球）

WR star （＝Wolf-Rayet star）沃尔夫－拉叶星，WR 型星 | 沃夫－瑞葉星，WR 星〔台〕

W Ser star 巨蛇 W 型星 | 巨蛇 [座]W 型星〔台〕

WSO-UV （＝World Space Observatory-Ultraviolet）世界空间紫外天文台 | 世界太空紫外天文台〔台〕

WSRT （＝Westerbork Synthesis Radio Telescope）韦斯特博克综合孔径射电望远镜 | 韋斯特博克孔徑合成電波望遠鏡〔台〕

W star （＝Wolf-Rayet star）沃尔夫－拉叶星，WR 型星 | 沃夫－瑞葉星，WR 星〔台〕

Wu　吴（小行星 2705 号）
Wu Chien-Shiung　吴健雄（小行星 2752 号）
Wuhan　武汉（小行星 3206 号）
Wuheng　武衡（小行星 56088 号）
W UMa binary　大熊 W 型双星
W UMa star　大熊 W 型星
Wuminchun　吴敏骏（小行星 23274 号）
Wunda crater　文达环形山（天卫二）
Wuwenjun　吴文俊（小行星 7683 号）
Wuyeesun　伍宜孙（小行星 3570 号）
W Vir star　室女 W 型星　即：*W Vir type variable; W Vir variable*；又见：*W Vir type star*
W Vir type star　室女 W 型星　即：*W Vir type variable; W Vir variable*；又见：*W Vir star*
W Vir type variable　室女 W 型变星 | 室女 [座]W 型變星〔台〕又见：*W Vir variable*
W Vir variable　室女 W 型变星 | 室女 [座]W 型變星〔台〕又见：*W Vir type variable*
WWT　（＝WorldWide Telescope）万维望远镜（微软公司软件）
Wyld crater　怀尔德环形山（月球）
Wynne corrector　韦恩改正镜组
Wyoming Infrared Observatory　怀俄明红外天文台
Wyoming IR Observatory　怀俄明红外天文台
Wyoming IR Telescope　怀俄明红外望远镜
W.Z.　（＝Weltzeit）【德】世界时　即：*UT*
WZ Sagittae star　（＝WZ Sge star）天箭 WZ 型星 | 天箭 [座]WZ 型星〔台〕
WZ Sge star　（＝WZ Sagittae star）天箭 WZ 型星 | 天箭 [座]WZ 型星〔台〕

X

XANADU （＝X-ray Analysis and Data Ulitilization）X 射线分析和数据应用包
xaser （＝X-ray amplification by stimulated emission of radiation）X 射线激射
X band X 波段
X-band frequency X 带频率
X-boson X 玻色子
Xe （＝xenon）氙（54 号元素）
Xena 齐娜（小行星 Eris 非正式名称）
xenobiology 外空生物学
xenon （＝Xe）氙（54 号元素）
XENON dark matter search experiment 氙暗物质探测实验
Xenophanes crater 色诺芬尼环形山（月球）
XF correlator XF 相关器
Xiajunchao 夏俊超（小行星 20779 号）
Xi'an 西安（小行星 2387 号）
Xianglupeak 香炉峰（小行星 3481 号）
Xingmingzhou 周兴明（小行星 4730 号）
Xinjiang 新疆（小行星 2336 号）
Xiuyanyu 岫岩玉（小行星 21313 号）
Xiwanggongcheng 希望工程（小行星 7494 号）
Xizang 西藏（小行星 2344 号）
Xizezong 席泽宗（小行星 85472 号）
X matter 未名物质
XMM-Newton （＝X-ray Multi-Mirror Mission）多镜面 X 射线空间望远镜，XMM 牛顿望远镜 | 多镜面 X 光衛星〔台〕（欧洲）
X-ray absorption X 射线吸收
X-ray afterglow X 射线余辉
X-ray amplification by stimulated emission of radiation （＝xaser）X 射线激射
X-ray Analysis and Data Ulitilization （＝XANADU）X 射线分析和数据应用包
X-ray astrometry X 射线天体测量
X-ray astronomy X 射线天文学 | X 光天文學〔台〕
X-ray background X 射线背景
X-ray background radiation X 射线背景辐射 | X 光背景輻射〔台〕
X-ray binary X 射线双星 | X 光雙星〔台〕
　又见：binary X-ray source
X-ray bright point X 射线亮点
X-ray burst X 射线暴 | X 光爆發〔台〕
X-ray burster X 射线暴源 | X 光爆發源〔台〕
　又见：X-ray burst source
X-ray burst source X 射线暴源 | X 光爆發源〔台〕
　又见：X-ray burster
X-ray calorimeter X 射线量能器
X-ray catalogue X 射线源表
X-ray cavity X 射线穴
X-ray CCD X 射线 CCD
X-ray cluster X 射线星系团
X-ray continuum X 射线连续谱
X-ray corona X 射线冕 | X 光冕〔台〕
X-ray counterpart X 射线对应体 | X 光對應體〔台〕
X-ray detector X 射线探测器，X 射线检测器
X-ray diffraction X 射线衍射
X-ray eclipse X 射线食 | X 光食〔台〕
X-ray eclipsing star X 射线食变星 | X 光食變星〔台〕 又见：eclipsing X-ray star
X-ray emission X 射线发射
X-ray flare X 射线耀斑 | X 光閃焰〔台〕
X-ray flash （＝XRF）X 闪，X 射线闪变
X-ray fluorescence X 射线荧光
X-ray galaxy X 射线星系 | X 光星系〔台〕
X-ray grating spectrometer X 射线光栅能谱仪
X-ray halo X 射线晕 | X 光暈〔台〕
X-ray identification X 射线证认
X-ray line X 射线谱线
X-ray luminosity X 射线光度
X-ray map X 射线图，X 射线天图

X-ray Multi-Mirror Mission （＝XMM-Newton）多镜面 X 射线空间望远镜，XMM 牛顿望远镜 | 多鏡面 X 光衛星〔台〕（欧洲）
X-ray nebula X 射线星云
X-ray nova X 射线新星 | X 光新星〔台〕
X-ray observation X 射线观测
X-ray observatory X 射线天文台 | X 光天文台〔台〕
X-ray photon X 射线光子
X-ray polarization X 射线偏振
X-ray pulsar X 射线脉冲星 | X 光脈衝星〔台〕
X-ray pulsation X 射线脉动
X-ray pulsator X 射线脉冲体
X-ray pulse X 射线脉冲
X-ray quasar X 射线类星体 | X 光類星體〔台〕
X-ray-rich GRB 富 X 射线 γ 暴
X-ray satellite X 射线卫星
X-ray scattering X 射线散射
X-ray sky X 射线天空
X-ray source X 射线源 | X 光源〔台〕
X-ray spectrograph X 射线摄谱仪
X-ray spectrometer X 射线分光计
X-ray spectroscopy X 射线频谱学
X-ray spectrum X 射线谱
X-ray star X 射线星 | X 光星〔台〕 又见：*extar*
X-ray sun X 射线太阳 | X 光太陽〔台〕
X-ray survey X 射线巡天 | X 光巡天〔台〕
X-ray telescope X 射线望远镜 | X 光望遠鏡〔台〕
X-ray transient X 射线暂现源 | X 光瞬變源〔台〕
XRB （＝binary X-ray source）X 射线双星 | X 光雙星〔台〕
XRF （＝X-ray flash）X 闪，X 射线闪变
X-unit X 单位 | X 光單位〔台〕
XUV （＝extreme-ultraviolet）极紫外
XUV astronomy （＝extreme ultraviolet astronomy）极紫外天文学 | 超紫外天文學〔台〕
XUV light （＝Extreme ultraviolet light）极紫外射线 | 超紫外線〔台〕
Xuyi 盱眙（小行星 4360 号）
Xuzhihong 许智宏（小行星 90826 号）
X wind 未名类太阳风 | 未名類太陽風，X 風〔台〕
X-wind model X 风模式

Y

Y （＝yttrium）钇（39号元素）
Yablochkov crater 雅勃洛奇科夫环形山（月球）
Yagi antenna 八木天线，波道式天线
Yale Bright Star Catalogue 耶鲁天文台亮星星表
Yamamoto crater 山本一清环形山（月球）
Yan'an 延安（小行星2693号）
Yangchenning 杨振宁（小行星3421号）
Yangel' crater 扬格利环形山（月球）
Yangjiachi 杨嘉墀（小行星11637号）
Yangliwei 杨利伟（小行星21064号）
Yang-Mills field 杨—米尔斯场
Yangzhou 扬州（小行星3729号）
Yanhua 华演（小行星11730号）
Yarkovsky effect 雅尔可夫斯基效应
Yarrabubba crater 亚拉布巴陨星坑（地球）
Y array Y形天线阵
Yb ①（＝year book）年历 ②（＝ytterbium）镱（70号元素）
year （＝yr.）年
year book （＝Yb）年历
year cycle 岁周 又见：year revolution
yearly UTC adjustment 每年协调世界时调整
year revolution 岁周 又见：year cycle
year surplus 岁余
Yed Post 楚，天市右垣十（蛇夫座 ε） 又见：Yed Posterior
Yed Posterior 楚，天市右垣十（蛇夫座 ε） 又见：Yed Post
Yed Prior 梁，天市右垣九（蛇夫座 δ） 又见：Jed Prior
Yeduzheng 叶笃正（小行星27895号）
yellow color index 黄色指数
yellow coronal line 日冕黄线
yellow dwarf 黄矮星
yellow giant 黄巨星
yellow glow 黄辉光
yellow star 黄星
yellow straggler 黄离散星
yellow supergiant 黄超巨星
Yenuanchen 严婉祯（小行星20641号）
Yepeiyu 叶佩玉（小行星12881号）
Yerkes classification 叶凯士分类
Yerkes classification system 叶凯士分类系统
Yerkes Observatory 叶凯士天文台
Yeshuhua 叶叔华（小行星3241号）
Yeungchuchiu 杨注潮（小行星19848号）
yielding 沉陷
yield strain 屈服应变
Yildun 勾陈二（小熊座 δ）
Yinhai 银海（小行星3340号）
Yiwu 伊吾（小行星80801号）
Yi Xing 一行（小行星1972号）
ylem 原元素|伊倫〔台〕
Ymir 土卫十九
Yohkoh 阳光号（日本太阳观测卫星） 又见：Solar-A
yoke mounting 轭式装置
Young crater 杨环形山（月球）
young star 年轻星
young star cluster 年轻星团
young stellar object （＝YSO）初期恒星体
yr. （＝year）年
YSO （＝young stellar object）初期恒星体
ytterbium （＝Yb）镱（70号元素）
yttrium （＝Y）钇（39号元素）
Yuanlongping 袁隆平（小行星8117号）
Yuan Tseh Lee Array for Microwave Background Anisotropy （＝AMiBA）李远哲宇宙背景辐射阵|李遠哲宇宙背景輻射陣列〔台〕
Yuchunshun 余纯顺（小行星83600号）
yuga 【梵】瑜伽（5太阳年）
Yukawa theory 汤川理论
Yukawa theory of nuclear forces 汤川核力理论
Yulong 玉龙（小行星31196号）

Yunnan 云南（小行星 2230 号）
Yushan 玉山（小行星 185546 号）
YY Orionis star （＝YY Ori star）猎户 YY 型星 | 獵戶 [座]YY 型星〔台〕

YY Orionis variable （＝YY Ori variable）猎户 YY 型变星
YY Ori star （＝YY Orionis star）猎户 YY 型星 | 獵戶 [座]YY 型星〔台〕
YY Ori variable （＝YY Orionis variable）猎户 YY 型变星

Z

Zach crater 扎赫环形山（月球）
ZAHB （=zero-age horizontal branch）零龄水平支
ZAMS （=zero-age main sequence）零龄主序 | 零龄主星序〔台〕
Zander crater 灿德尔环形山（月球）
 又见：*Tsander crater*
Z And star 仙女Z型星 | 仙女[座]Z型星〔台〕
Zania 左执法，太微左垣一（室女座 η）
 又见：*Zaniah*
Zaniah 左执法，太微左垣一（室女座 η）
 又见：*Zania*
Zanstra's theory 赞斯特拉理论
Zanstra temperature 赞斯特拉温度
Zarya 曙光号（国际空间站功能货舱）
Zaurak 天苑一（波江座 γ）
Zavijava 右执法，太微右垣一（室女座 β）
 又见：*Alaraph*
ZC （=Zodical Catalogue）黄道带星表
Z Cam star 鹿豹Z型星 | 鹿豹[座]Z型星〔台〕
ZD （=zenith distance）天顶距
Z-direction Z向
Zeeman component 塞曼子线
Zeeman crater 塞曼环形山（月球）
Zeeman-Doppler imaging 塞曼—多普勒成像法
Zeeman effect 塞曼效应 | 则曼效應〔台〕
Zeeman spectrogram 塞曼谱图
Zeeman splitting 塞曼分裂
Zeeman state 塞曼态
Zel'dovich approximation 泽尔多维奇近似
Zel'dovich pancake 泽尔多维奇薄饼
Zel'dovich spectrum 泽尔多维奇功率谱
Zelenchukskaya Observatory 泽连丘克斯卡亚天文台（俄罗斯）
Zelentchouk Telescope 俄罗斯6米望远镜
Zengguoshou 曾国寿（小行星21398号）
zenith 天顶

zenith absorption 天顶大气吸收
zenithal equidistant projection 等距天顶投影
zenithal hourly rate （=ZHR）天顶每时出现率
zenithal refraction 天顶大气折射 | 天頂[大氣]折射〔台〕
zenith angle 天顶角
zenith astrograph 天顶照相仪
zenith attraction 天顶引力 | 天頂吸引〔台〕
zenith delay 天顶延迟
zenith discontinuity 天顶间断
zenith distance （=ZD）天顶距
zenith eyepiece 天顶目镜
zenith instrument 天顶仪 又见：*zenith telescope*
zenith magnitude 天顶星等
zenith reading 天顶读数
zenith sector 地平纬仪，象限仪
zenith star 天顶星
zenith sun 天顶太阳
zenith telescope 天顶仪 又见：*zenith instrument*
zenith tube 天顶筒
zenocentric coordinate 木心坐标 | 木星[中心]坐標〔台〕 又见：*jovicentric coordinate*
Zeno crater 芝诺环形山（月球）
zenographic coordinate 木面坐标 | 木[星表]面坐標〔台〕 又见：*jovigraphic coordinate*
zenographic latitude 木面纬度 又见：*jovigraphic latitudem*
zenographic longitude 木面经度 又见：*jovigraphic longitude*
zenography 木面学 | 木[星表]面學〔台〕
Zeno paradox 芝诺佯谬
zero adjustment 零调整
zero-age horizontal branch （=ZAHB）零龄水平支
zero-age main sequence （=ZAMS）零龄主序 | 零龄主星序〔台〕
zero-beat method 零拍法

zero-coefficient	零系数
zero crossing	零交点
zero-crossing discriminator	过零鉴别器
zero date	起算日
zero drift	零点漂移　又见: zero point shift
zerodur	微晶玻璃　又见: cervit; glass-ceramic; sytall
zero elimination	消零
zero energy wave	零能波
zero-expansion glass	零膨胀玻璃
zero field	零势场 | 零勢場，零位场〔台〕
zero geodesic	零短程线
zeroing	调零
zero meridian	零子午线
zero mode	零模
zero order spectrum	零级光谱
zero point	零点　又见: null point
zero point energy	零点能
zero point of photometric system	测光系统零点
zero point shift	零点漂移　又见: zero drift
zero reading	零点读数 | 零[點]讀數，起點讀數〔台〕
zero redundancy array	零重复阵
zero rotation main sequence	（＝ZRMS）无自转主序
zero setting	置零
zero time	时间零点
zero velocity curve	零速度线　又见: curve of zero velocity
zero velocity surface	零速度面　又见: surface of zero velocity
zero year	零黑子年
zero zone	零时区　又见: initial zone
Zeta Aurigae star	御夫 ζ 型星
Zhangdaning	张大宁（小行星 8311 号）
Zhangguoxi	张果喜（小行星 3028 号）
Zhang Heng	张衡（小行星 1802 号）
Zhang Heng crater	张衡环形山（月球）
Zhangjiajie	张家界（小行星 79418 号）
Zhangyi	张翼（小行星 20831 号）
Zhaojiuzhang	赵九章（小行星 7811 号）
Zhejiang	浙江（小行星 2631 号）
Zhongguo	中国（小行星 3789 号）
Zhongkeda	中科大（小行星 19298 号）
Zhongkeyuan	中科院（小行星 7800 号）
Zhongyuechen	钟越尘（小行星 21725 号）
Zhouguangzhao	周光召（小行星 3462 号）
Zhoushan	舟山（小行星 4925 号）
ZHR	（＝zenithal hourly rate）天顶每时出现率
Zhuguangya	朱光亚（小行星 10388 号）
Zhuhai	珠海（小行星 2903 号）
Zhukovskiy crater	茹科夫斯基环形山（月球）
Zhuruochen	朱若辰（小行星 21731 号）
Zhuyuanchen	朱元晨（小行星 20689 号）
Zibal	天苑五（波江座 ζ）
zij	① 历表 ② 积尺
Zīj-i Īlkhānī	《伊尔汗历表》（波斯历书）又见: Al-Zij-Ilkhani
zinc	（＝Zn）锌（30 号元素）
zine-doped germanium	锗掺锌
zino	Z 微子
Zi-qi	紫炁
Zirankexuejijin	自然科学基金（小行星 8425 号）
zirconium	（＝Zr）锆（40 号元素）
zirconium star	锆星
Zn	（＝zinc）锌（30 号元素）
Z-number	原子序数
zodiac	黄道带　又见: zodiacal belt
zodiacal belt	黄道带　又见: zodiac
zodiacal circle	黄道圈
zodiacal cloud	黄道云
zodiacal constellation	黄道星座　又见: zodiac constellation
zodiacal counterglow	对日照　又见: counterglow; counter-twilight
zodiacal dust	黄道尘
zodiacal dust cloud	黄道尘云
zodiacal light	黄道光
zodiacal light bridge	黄道光桥
zodiacal light photometry	黄道光测光
zodiacal signs	黄道十二宫　又见: signs of zodiac
zodiacal star	黄道带恒星 | 黃道[帶恆]星〔台〕
zodiac constellation	黄道星座　又见: zodiacal constellation
Zodical Catalogue	（＝ZC）黄道带星表
Zoe	佐伊（小行星 Logos 的卫星）
zoll	十二分之一食分 | 十二分之一 [食分]〔台〕
zonal harmonic	带谐函数 | 帶諧函數，球帶 [調和] 函數〔台〕
zonal rotation	分带自转
zonal tide	带潮
Zond	探测号（苏联行星际探测器）
zone astrograph	分区天体照相仪
zone catalogue	分区星表
zone meridian	分区子午线

zone noon 分区正午
zone observation 分区观测
zone of annularity 环食带 又见: *path of annular phase; path of annularity; path of annular eclipse*
zone of avoidance 隐带
zone of eclipse 食带 又见: *path of eclipse*
zone of totality 全食带 又见: *path of total eclipse; belt of totality; path of totality*
zone plate 环板
zone star 分区恒星
zone time 区时
zoo hypothesis 动物园假说
zoom lens 变焦透镜
ZPRSN （＝Zurich provisional relative sunspot number）苏黎世临时相对黑子数
Zr （＝zirconium）锆（40 号元素）
ZRMS （＝zero rotation main sequence）无自转主序
Z-shaped nebula Z 形星云
Zsigmondy crater 席格蒙迪环形山（月球）
Z-term Z 项 即: *Kimura term*
Z time （＝Zulu time）格林尼治平时，Z 时
即: GMT
Zubenalgubi 折威七（天秤座 σ，天蝎座 γ）
又见: *Brachium*
Zuben Elakrab 氐宿三（天秤座 γ） 又见: *Zuben el Hakrabi; Zuben Hakraki*
Zuben Elakribi 氐宿增一（天秤座 δ）
Zuben Elgenubi 氐宿一（天秤座 α）
又见: *Elkhiffa Australis*
Zubenelgenubi 氐宿增七（天秤座 α） 又见: *Kiffa Australis*

Zuben el Hakrabi 氐宿三（天秤座 γ）
又见: *Zuben Elakrab; Zuben Hakraki*
Zubeneschamali 氐宿四（天秤座 β） 又见: *Kiffa Borealis*
Zubenhakrabi 西咸四（天秤座 η）
Zuben Hakraki 氐宿三（天秤座 γ） 又见: *Zuben Elakrab; Zuben el Hakrabi*
Zucchius crater 祖基环形山（月球）
Zu Chong-zhi 祖冲之（小行星 1888 号）
Zu Chong-zhi crater 祖冲之环形山（月球）
Zulu time （＝Z time）格林尼治平时，Z 时
即: GMT
Zurich classification 苏黎世分类，苏黎世黑子分类 | 蘇黎世 [黑子] 分類〔台〕
Zurich number 苏黎世数 | 苏黎世黑子相對數〔台〕
Zurich provisional relative sunspot number （＝ZPRSN）苏黎世临时相对黑子数
Zurich relative sunspot number 苏黎世相对黑子数 | 蘇黎世黑子相對數〔台〕
Zurich sunspot number 苏黎世太阳黑子数
Zusatzstern 【德】补充星
Zwicky blue object 兹威基蓝天体
Zwicky catalogue 兹威基星系表
Zwicky classification 兹威基分类
Zwicky compact galaxy 兹威基致密星系
Zwicky crater 兹威基环形山（月球）
ZZ Cet star 鲸鱼 ZZ 型星，鲸鱼 ZZ 型变星 | 鯨魚 [座]ZZ 型星〔台〕 又见: *ZZ Cet variable*
ZZ Cet variable 鲸鱼 ZZ 型星，鲸鱼 ZZ 型变星 | 鯨魚 [座]ZZ 型星〔台〕 又见: *ZZ Cet star*

数字

10-day period　旬　又见：*ten-day period*
10-meter-class telescope　10 米级望远镜
11-year solar cycle　11 年太阳活动周
12 Jupiter-stations　十二次　又见：*duodenary series; twelve Jupiter-stations*
160-minute oscillation　160 分钟振荡 | 160 分 [鐘] 振蕩〔台〕
1/f noise　1/f 噪声
1-mirror telescope　单反光面望远镜
2001 Mars Odyssey　2001 火星奥德赛
21 Centimeter Array　（＝21CMA）21 厘米 [射电望远镜] 阵 | 21 厘米電波望遠鏡陣〔台〕
21CMA　（＝21 Centimeter Array）21 厘米 [射电望远镜] 阵 | 21 厘米電波望遠鏡陣〔台〕
21 cm forest　21 厘米森林，21 厘米线丛
21 cm intensity mapping　21 厘米强度映射
21cm line　21 厘米谱线　又见：*twenty-one centimeter line*
21-cm radiation　21 厘米辐射 | 21 公分輻射〔台〕 又见：*twenty-one centimeter radiation*
21 cm tomography　21 厘米层析
22° halo　22° 日晕
28 lunar mansions　二十八宿　又见：*lunar zodiac; twenty-eight lunar mansions*
2-body problem　二体问题　又见：*problem of two bodies; two-body problem*
2-color photometry　二色测光
2dF Galaxy Redshift Survey　2 度视场星系红移巡天　又见：*Two-degree-Field Galaxy Redshift Survey*
2dFGRS　（＝Two-degree-Field Galaxy Redshift Survey）2 度视场星系红移巡天
2dF Survey　（＝Two-degree-field redshift survey）2 度视场红移巡天
2-dimensional classification　二元分类
2-dimensional photometry　二维测光
2-lens objective　双合物镜
2MASS　（＝Two Micron All Sky Survey）2 微米全天巡视

2-meter-class telescope　2 米级望远镜
2-mirror telescope　二反光面望远镜
2-prism spectrograph　双棱镜摄谱仪　又见：*double-prism spectrograph*
2-μ Sky Survey　2 微米巡天　又见：*Two-Micron Sky Survey*
3-axis mounting　三轴装置　又见：*triaxial mount; triaxial mounting; three-axis mount; three-axis mounting*
3-body problem　三体问题　又见：*three-body problem; problem of three bodies*
3C catalogue　（＝Third Cambridge Catalogue of Radio Sources）第 3 剑桥射电源表，3C 射电源表
3-color photometry　三色测光 | 三色光度測量，三色測光〔台〕 又见：*three-color photometry*
3-dimensional classification　三元分类
3 kpc Arm　三千秒差距臂　又见：*three-kiloparsec arm*
3K radiation　3K 辐射　又见：*three-degree Kelvin radiation*
3-lens objective　三合物镜
3-mirror telescope　三反光面望远镜
3-prism spectrograph　三棱镜摄谱仪
4000 Å break　4000 埃跳变
46° halo　46° 日晕
4-axis mounting　四轴装置　又见：*four-axis mounting; quadraxial mounting*
4-body problem　四体问题　又见：*four-body problem*
4C catalogue　（＝Fourth Cambridge Survey catalogue）第 4 剑桥巡天表，4C 星表
4-color photometry　四色测光　又见：*four-color photometry*
4-dimensional universe　四维宇宙　又见：*four-dimensional universe*
4-meter-class telescope　4 米级望远镜
5-color photometry　五色测光
5-Kilometer Telescope　五千米射电望远镜 | 五千米電波望遠鏡〔台〕 又见：*Five-Kilometer Telescope*
5-minute oscillation　五分钟振荡 | 五分 [鐘] 振蕩〔台〕 又见：*five-minute oscillation*

6-color photometry　六色测光　又见：*six-color photometry*

6C survey　（＝Sixth Cambridge Survey of radio sources）第 6 剑桥射电源巡天

6dF Galaxy Survey　6 度视场星系巡天
　又见：*Six-degree Field Galaxy Survey*

6dFGS　（＝Six-degree Field Galaxy Survey）6 度视场星系巡天

6-meter-class telescope　6 米级望远镜

7-color photometry　七色测光

8-meter-class telescope　8 米级望远镜

希腊字母

α² Canum Venaticorum star　（＝α² CVn star）猎犬 α 型星

α² Canum Venaticorum variable　（＝α² CVn variable）猎犬 α 型变星　即：α² CVn star

α² CVn star　（＝α² Canum Venaticorum star）猎犬 α 型星

α² CVn variable　（＝α² Canum Venaticorum variable）猎犬 α 型变星　即：α² CVn star

α Capricornids　摩羯 α 流星群

α Cygni variable　天鹅 α 型变星

α Persei cluster　英仙 α 星团（Collinder 39）

α-viscosity prescription　α 黏度处理方法

α-β-γ theory　αβγ 理论

β Canis Majoris star　（＝β CMa star）大犬 β 型星

β Cephei star　（＝β Cep star）仙王 β 型星 | 仙王[座]β 型星〔台〕

β Cep star　（＝β Cephei star）仙王 β 型星 | 仙王[座]β 型星〔台〕

β CMa star　（＝β Canis Majoris star）大犬 β 型星

β decay　β 衰变　又见：β disintegration

β-decay electron　β 衰变电子

β disintegration　β 衰变　又见：β decay

β Lyrae star　天琴 β 型星　即：β Lyr-type variable

β Lyr-type variable　天琴 β 型变星 | 天琴[座]β 型變星〔台〕

β Persei star　英仙 β 型星

β transition　β 跃迁

γ Cassiopeiae star　（＝γ Cas star）仙后 γ 型星 | 仙后[座]γ 型星〔台〕

γ Cas star　（＝γ Cassiopeiae star）仙后 γ 型星 | 仙后[座]γ 型星〔台〕

Γ parameter　Γ 参数

Γ-point　Γ 点

γ radiation　γ 辐射

γ-radiation counter　γ 辐射计数器

γ-radiation source　γ 辐射源

γ-ray astronomy　γ 射线天文学

γ-ray background　γ 射线背景

γ-ray burst　（＝GRB）γ 射线暴 | γ 射線爆發〔台〕　又见：Gamma-ray burst

γ-ray burst afterglow　γ 暴余辉

γ-ray Burst Alert Telescope　（＝BAT）γ 暴预警望远镜

γ-ray burst energy　γ 暴能量

γ-ray burster　γ 射线暴源 | γ 射線爆發源〔台〕　又见：γ-ray burst source

γ-ray burst light curve　γ 暴光变曲线

γ-ray burst mass extinction　γ 暴集群灭绝

γ-ray burst optical transient　γ 暴光学暂现源

γ-ray burst progenitor　γ 暴前身天体

γ-ray burst source　γ 射线暴源 | γ 射線爆發源〔台〕　又见：γ-ray burster

γ-ray burst type　γ 暴型

γ-ray counter　γ 射线计数器

γ-ray counterpart　γ 射线对应体

γ-ray detector　γ 射线探测器

γ-ray emission　γ 射线发射

γ-ray identification　γ 射线证认

γ-ray line　γ 射线谱线　又见：γ-ray spectral line

γ-ray line astronomy　γ 射线谱线天文学 | γ[射線]譜線天文學〔台〕

γ-ray line emission　γ 射线谱线辐射 | γ[射線]譜線辐射〔台〕

γ-ray luminosity　γ 射线光度

γ-ray map　γ 射线天图

γ-ray observation　γ 射线观测

γ-ray observatory　γ 射线天文台

γ-ray pulsar　γ 射线脉冲星

γ-ray satellite　γ 射线卫星

γ-ray scattering　γ 射线散射

γ-ray sky　γ 射线天空

γ-ray source　γ 射线源

γ-ray spectral line　γ 射线谱线　又见：γ-ray line

γ-ray spectrometer　γ 射线频谱计

γ-ray spectroscopy γ 射线分频
γ-ray spectrum γ 射线谱
γ-ray star γ 射线星
γ-ray survey γ 射线巡天
γ-ray telescope γ 射线望远镜
γ-ray transient γ 射线暂现
Γ-space Γ 空间
δ Aquarids 宝瓶 δ 流星群
δ Cephei star （＝δ Cep star）仙王 δ 型星
δ Cep star （＝δ Cephei star）仙王 δ 型星
δ Delphini star 海豚 δ 型星
δ Doradus star 剑鱼 δ 型星
δ Sct star （＝δ Scuti star）盾牌 δ 型星 即：dwarf Cepheid
δ Scuti star （＝δ Sct star）盾牌 δ 型星 即：dwarf Cepheid
ζ Aurigae star （＝ζ Aur star）御夫 ζ 型星 | 御夫[座]ζ 型星〔台〕
ζ Aur star （＝ζ Aurigae star）御夫 ζ 型星 | 御夫[座]ζ 型星〔台〕
η Aquarids 宝瓶 η 流星群
η Carina nebula 船底 η 星云（NGC 3372）
η Carina-type object 船底 η 型天体
θ galaxy θ 星系
θ-meson θ 介子
κ mechanism κ 机制
λ Boo star （＝λ Boötis star）牧夫 λ 型星 | 牧夫[座]λ 型星〔台〕
λ Boötis star （＝λ Boo star）牧夫 λ 型星 | 牧夫[座]λ 型星〔台〕
ΛCDM model （＝Λ Cold Dark Matter model）含宇宙学常数的冷暗物质模型
Λ Cold Dark Matter model （＝ΛCDM model）含宇宙学常数的冷暗物质模型
λ doublet λ 双重线 又见：λ doubling line
λ doubling line λ 双重线 又见：λ doublet
λ Eridani star 波江 λ 型星
λ-hyperon λ 超子
μ-meson μ 介子 又见：muon
Ξ-hyperon Ξ 超子
o Ceti star 鲸鱼 o 型星 | 鯨魚[座]o[型]變星〔台〕
  即：Mira Ceti variable; Mira variable; Mira-type variable
π component π 分量
π-meson π 介子 | 派子〔台〕 又见：pion
ρ Cassiopeiae star （＝ρ Cas star）仙后 ρ 型星 | 仙后[座]ρ 型星〔台〕
ρ Cas star （＝ρ Cassiopeiae star）仙后 ρ 型星 | 仙后[座]ρ 型星〔台〕
ρ-meson ρ 介子
ρ Ophiuchi cloud 蛇夫 ρ 星云
σ component σ 分量
τCDM model （＝τ Cold Dark Matter model）τ 冷暗物质模型
τ Cold Dark Matter model （＝τCDM model）τ 冷暗物质模型
τ component τ 分量
τ-meson τ 介子
τ test τ 检验
υ component υ 分量
υ Sagittarii star 人马 υ 型星 | 人馬[座]υ 型星〔台〕
χ Persei cluster 英仙 χ 星团
ω Centauri 半人马 ω 球状星团 | 半人馬 ω[球狀星團]〔台〕（NGC 5139）
ω navigation system ω 导航系统

附 录

表 1 星座 Constellations

拉丁名	所有格	缩写	中文名
Andromeda	Andromedae	And	仙女座
Antlia	Antliae	Ant	唧筒座
Apus	Apodis	Aps	天燕座
Aquarius	Aquarii	Aqr	宝瓶座
Aquila	Aquilae	Aql	天鹰座
Ara	Arae	Ara	天坛座
Aries	Arietis	Ari	白羊座
Auriga	Aurigae	Aur	御夫座
Bootes	Bootis	Boo	牧夫座
Caelum	Caeli	Cae	雕具座
Camelopardalis	Camelopardalis	Cam	鹿豹座
Cancer	Cancri	Cnc	巨蟹座
Canes Venatici	Canum Venaticorum	CVn	猎犬座
Canis Major	Canis Majoris	CMa	大犬座
Canis Minor	Canis Minoris	CMi	小犬座
Capricornus	Capricorni	Cap	摩羯座
Carina	Carinae	Car	船底座
Cassiopeia	Cassiopeiae	Cas	仙后座
Centaurus	Centauri	Cen	半人马座
Cepheus	Cephei	Cep	仙王座
Cetus	Ceti	Cet	鲸鱼座
Chamaeleon	Chamaeleonis	Cha	堰蜓座
Circinus	Circini	Cir	圆规座
Columba	Columbae	Col	天鸽座
Coma Berenices	Comae Berenices	Com	后发座
Corona Austrilis	Coronae Austrilis	CrA	南冕座
Corona Borealis	Coronae Borealis	CrB	北冕座
Corvus	Corvi	Crv	乌鸦座
Crater	Crateris	Crt	巨爵座
Crux	Crucis	Cru	南十字座
Cygnus	Cygni	Cyg	天鹅座
Delphinus	Delphini	Del	海豚座
Dorado	Doradus	Dor	箭鱼座
Draco	Draconis	Dra	天龙座
Equuleus	Equulei	Equ	小马座
Eridanus	Eridani	Eri	波江座
Fornax	Fornacis	For	天炉座
Gemini	Geminorum	Gem	双子座
Grus	Gruis	Gru	天鹤座
Hercules	Herculis	Her	武仙座
Horologium	Horologii	Hor	时钟座
Hydra	Hydrae	Hya	长蛇座
Hydrus	Hudri	Hyi	水蛇座
Indus	Indi	Ind	印地安座
Lacerta	Lacertae	Lac	蝎虎座
Leo	Leonis	Leo	狮子座

拉丁名	所有格	缩写	中文名
Leo Minor	Leonis Minoris	LMi	小狮座
Lepus	Leporis	Lep	天兔座
Libra	Librae	Lib	天秤座
Lupus	Lupi	Lup	豺狼座
Lynx	Lyncis	Lyn	天猫座
Lyra	Lyrae	Lyr	天琴座
Mensa	Mensae	Men	山案座
Microseopium	Microacopii	Mic	显微镜座
Monoceros	Monocerotis	Mon	麒麟座
Musca	Muscae	Mus	苍蝇座
Norma	Normae	Nor	矩尺座
Octans	Octantis	Oct	南极座
Ophiuchus	Ophiuchi	Oph	蛇夫座
Orion	Orionis	Ori	猎户座
Pavo	Pavonis	Pav	孔雀座
Pegasus	Pegasi	Peg	飞马座
Perseus	Persei	Per	英仙座
Phoenix	Phoenicis	Phe	凤凰座
Pictor	Pictoris	Pic	绘架座
Pisces	Piscium	Psc	双鱼座
Piscis Austrinus	Piscis Austrini	PsA	南鱼座
Puppis	Puppis	PuP	船尾座
Pyxis	Pyxidis	Pyx	罗盘座
Reticulum	Reticuli	Ret	网罟座
Sagitta	Sagittae	Sge	天箭座
Sagittarius	Sagittarii	Sgr	人马座
Scorpius	Scorpii	Sco	天蝎座
Sculptor	Sculptoris	Scl	玉夫座
Scutum	Scuti	Sct	盾牌座
Serpens	Serpentis	Ser	巨蛇座
Sextans	Sextantis	Sex	六分仪座
Taurus	Tauri	Tau	金牛座
Telescopium	Telescopii	Tel	望远镜座
Triangulum	Trianguli	Tri	三角座
Triangulum Australe	Trianguli Australis	TrA	南三角座
Tucana	Tucanae	Tuc	杜鹃座
Ursa Major	Ursae Majoris	UMa	大熊座
Ursa Minor	Ursae Minoris	UMi	小熊座
Vela	Velorum	Vel	船帆座
Virgo	Virginis	Vir	室女座
Volans	Volantis	Vol	飞鱼座
Vulpecula	Vulpeculae	Vul	狐狸座

表 2　黄道十二宫 Zodiacal Signs

拉丁名	中文名	明清用名	
Aries	白羊宫	降娄	戌宫
Taurus	金牛宫	大梁	酉宫
Gemini	双子宫	实沈	申宫
Cancer	巨蟹宫	鹑首	未宫
Leo	狮子宫	鹑火	午宫
Virgo	室女宫	鹑尾	巳宫
Libra	天秤宫	寿星	辰宫
Scorpius	天蝎宫	大火	卯宫
Sagittarius	人马宫	析木	寅宫
Capricornus	摩羯宫	星纪	丑宫
Aquarius	宝瓶宫	玄枵	子宫
Pisces	双鱼宫	娵訾	亥宫

表 3　三垣四象七曜

Three Enclosures	三垣	**Seven Luminaries**	七曜
Celestial Market Enclosure	天市垣	Sun	太阳
Supreme Subtlety Enclosure	太微垣	Moon	太阴
Purple Forbidden Enclosure	紫微垣	Mercury	辰星
Four Symbolic Animals	四象	Venus	太白
Azure Dragon	苍龙	Mars	荧惑
White Tiger	白虎	Jupiter	岁星
Vermilion Bird	朱鸟	Saturn	镇星
Black Tortoise	玄武		

表 4　二十八宿 28 Lunar Mansions

Azure Dragon	苍龙	**White Tiger**	白虎
Horn	角宿	Stride; Legs	奎宿
Neck	亢宿	Bond	娄宿
Root	氐宿	Stomach	胃宿
Room	房宿	Ball of wool; Stopping-place	昴宿
Heart	心宿	Net	毕宿
Tail	尾宿	Mouth	觜宿
Winnowing-basket	箕宿	Triads	参宿
Black Tortoise	玄武	**Vermilion Bird**	朱鸟
Dipper	斗宿	Well	井宿
Ox	牛宿	Ghost	鬼宿
Maid	女宿	Willow	柳宿
Void	虚宿	Seven Stars	星宿
Roof	危宿	Extension	张宿
Encampment	室宿	Wings	翼宿
Wall	壁宿	Chariot	轸宿

表 5　节气　Solar Terms

Beginning of Spring	立春	Beginning of Autumn	立秋
Rain Water	雨水	End of Heat	处暑
Awakening from Hibernation	惊蛰	White Dew	白露
Vernal/Spring Equinox	春分	Autumnal Equinox	秋分
Fresh Green	清明	Cold Dew	寒露
Grain Rain	谷雨	First Frost	霜降
Beginning of Summer	立夏	Beginning of Winter	立冬
Lesser Fullness	小满	Light Snow	小雪
Grain in Ear	芒种	Heavy Snow	大雪
Summer Solstice	夏至	Winter Solstice	冬至
Lesser Heat	小暑	Lesser Cold	小寒
Greater Heat	大暑	Greater Cold	大寒

表 6　天然卫星　Natural Satellites

Phobos	火卫一	Hermippe	木卫三十
Deimos	火卫二	Aitne	木卫三十一
		Eurydome	木卫三十二
Io	木卫一	Euanthe	木卫三十三
Europa	木卫二	Euporie	木卫三十四
Ganymede	木卫三	Orthosie	木卫三十五
Callisto	木卫四	Sponde	木卫三十六
Amalthea	木卫五	Kale	木卫三十七
Himalia	木卫六	Pasithee	木卫三十八
Elara	木卫七	Hegemone	木卫三十九
Pasiphae	木卫八	Mneme	木卫四十
Sinope	木卫九	Aoede	木卫四十一
Lysithea	木卫十	Thelxinoe	木卫四十二
Carme	木卫十一	Arche	木卫四十三
Ananke	木卫十二	Kallichore	木卫四十四
Leda	木卫十三	Helike	木卫四十五
Thebe	木卫十四	Carpo	木卫四十六
Adrastea	木卫十五	Eukelade	木卫四十七
Metis	木卫十六	Cyllene	木卫四十八
Callirrhoe	木卫十七	Kore	木卫四十九
Themisto	木卫十八	Herse	木卫五十
Megaclite	木卫十九		
Taygete	木卫二十	Mimas	土卫一
Chaldene	木卫二十一	Enceladus	土卫二
Harpalyke	木卫二十二	Tethys	土卫三
Kalyke	木卫二十三	Dione	土卫四
Iocaste	木卫二十四	Rhea	土卫五
Erinome	木卫二十五	Titan	土卫六
Isonoe	木卫二十六	Hyperion	土卫七
Praxidike	木卫二十七	Iapetus	土卫八
Autonoe	木卫二十八	Phoebe	土卫九
Thyone	木卫二十九	Janus	土卫十

Epimetheus	土卫十一	Desdemona	天卫十
Dione B; Helene	土卫十二	Juliet	天卫十一
Telesto	土卫十三	Portia	天卫十二
Calypso	土卫十四	Rosalind	天卫十三
Atlas	土卫十五	Belinda	天卫十四
Prometheus	土卫十六	Puck	天卫十五
Pandora	土卫十七	Caliban	天卫十六
Pan	土卫十八	Sycorax	天卫十七
Ymir	土卫十九	Prospero	天卫十八
Paaliaq	土卫二十	Setebos	天卫十九
Tarvos	土卫二十一	Stephano	天卫二十
Ijiraq	土卫二十二	Trinculo	天卫二十一
Suttungr	土卫二十三	Francisco	天卫二十二
Kiviuq	土卫二十四	Margaret	天卫二十三
Mundilfari	土卫二十五	Ferdinand	天卫二十四
Albiorix	土卫二十六	Perdita	天卫二十五
Skathi	土卫二十七	Mab	天卫二十六
Erriapus	土卫二十八	Cupid	天卫二十七
Siarnaq	土卫二十九		
Thrymr	土卫三十	Triton	海卫一
Narvi	土卫三十一	Nereid	海卫二
Methone	土卫三十二	Naiad	海卫三
Pallene	土卫三十三	Thalassa	海卫四
Polydeuces	土卫三十四	Despina	海卫五
Daphnis	土卫三十五	Galatea	海卫六
Aegir	土卫三十六	Larissa	海卫七
Bebhionn	土卫三十七	Proteus	海卫八
Bergelmir	土卫三十八	Halimede	海卫九
Bestia	土卫三十九	Psamathe	海卫十
Farbauti	土卫四十	Sao	海卫十一
Fenrir	土卫四十一	Laomedeia	海卫十二
Fornjot	土卫四十二	Neso	海卫十三
Hati	土卫四十三		
Hyrrokkin	土卫四十四	Charon	冥卫一
Kari	土卫四十五	Nix	冥卫二
Loge	土卫四十六	Hydra	冥卫三
Skoll	土卫四十七	Kerberos	冥卫四
Surtur	土卫四十八	Styx	冥卫五
Anthe	土卫四十九		
Jarnsaxa	土卫五十	Pabu	曝卫 (主星：Borasisi)
Greip	土卫五十一	Dysnomia	阋卫 (主星：Eris)
Tarqeq	土卫五十二	Hi'iaka	妊卫一 (主星：Haumea)
Aegaeon	土卫五十三	Namaka	妊卫二 (主星：Haumea)
		Dactyl	艾卫 (主星：Ida)
Ariel	天卫一	Zoe	佐伊 (主星：Logos)
Umbriel	天卫二	Vanth	亡卫 (主星：Orcus)
Titania	天卫三	Weywot	创卫一 (主星：Quaoar)
Oberon	天卫四	Sawiskera	萨维斯克拉 (主星：Teharonhiawako)
Miranda	天卫五	Echidna	台卫 (主星：Typhon)
Cordelia	天卫六	Linus	赋卫一 (主星：Sappho)
Ophelia	天卫七	Remus	林卫二 (主星：Sylvia)
Bianca	天卫八	Romulus	林卫一 (主星：Sylvia)
Cressida	天卫九		

表 7　月面地名　Lunar Features

Catena Abulfeda	艾布·菲达坑链	Lacus Doloris	忧伤湖
Catena Artamonov	阿尔塔莫诺夫坑链	Lacus Excellentiae	秀丽湖
Catena Davy	戴维坑链	Lacus Felicitatis	幸福湖
Catena Dziewulski	杰武尔斯基坑链	Lacus Gaudii	欢乐湖
Catena Humboldt	洪堡坑链	Lacus Hiemalis	冬湖
Catena Krafft	克拉夫特坑链	Lacus Lenitatis	温柔湖
Catena Kurchatov	库尔恰托夫坑链	Lacus Luxuriae	华贵湖
Catena Lucretius	卢克莱修坑链	Lacus Mortis	死湖
Catena Mendeleev	门捷列夫坑链	Lacus Odii	怨恨湖
Catena Sumner	萨姆纳坑链	Lacus Perseverantiae	长存湖
Catena Sylvester	西尔维斯特坑链	Lacus Solitudinis	孤独湖
		Lacus Somniorum	梦湖
Dorsa Aldrovandi	阿尔德罗万迪山脊	Lacus Spei	希望湖
Dorsa Andrusov	安德鲁索夫山脊	Lacus Temporis	时令湖
Dorsa Argand	阿尔甘山脊	Lacus Timoris	恐怖湖
Dorsa Barlow	巴洛山脊	Lacus Veris	春湖
Dorsa Burnet	伯内特山脊		
Dorsa Cato	加图山脊	Mare Anguis	蛇海
Dorsa Dana	达纳山脊	Mare Australe	南海
Dorsa Ewing	尤因山脊	Mare Cognitum	知海
Dorsa Geikie	盖基山脊	Mare Crisium	危海
Dorsa Harker	哈克山脊	Mare Fecunditatis	丰富海
Dorsa Lister	利斯特山脊	Mare Frigoris	冷海
Dorsa Mawson	莫森山脊	Mare Humboldtianum	洪堡海
Dorsa Rubey	鲁比山脊	Mare Humorum	湿海
Dorsa Smirnov	斯米尔诺夫山脊	Mare Imbrium	雨海
Dorsa Sorby	索比山脊	Mare Ingenii	智海
Dorsa Stille	施蒂勒山脊	Mare Insularum	岛海
Dorsa Whiston	惠斯顿山脊	Mare Marginis	界海
Dorsum Arduino	阿尔杜伊诺山脊	Mare Moscoviense	莫斯科海
Dorsum Azara	阿萨拉山脊	Mare Nectaris	酒海
Dorsum Bucher	布赫山脊	Mare Nubium	云海
Dorsum Buckland	巴克兰山脊	Mare Orientale	东海
Dorsum Cayeux	卡耶山脊	Mare Serenitatis	澄海
Dorsum Cloos	克洛斯山脊	Mare Smythii	史密斯海
Dorsum Cushman	库什曼山脊	Mare Spumans	泡沫海
Dorsum Gast	加斯特山脊	Mare Tranquillitatis	静海
Dorsum Grabau	葛利普山脊	Mare Undarum	浪海
Dorsum Heim	海姆山脊	Mare Vaporum	汽海
Dorsum Nicol	尼科尔山脊		
Dorsum Niggli	尼格利山脊	Mons Ampère	安培山
Dorsum Oppel	奥佩尔山脊	Mons Argaeus	阿尔加山
Dorsum Owen	欧文山脊	Mons Huygens	惠更斯山
Dorsum Scilla	希拉山脊	Mons La Hire	拉希尔山
Dorsum Termier	泰尔米埃山脊	Mons Pico	皮科山
Dorsum Von Cotta	冯·科塔山脊	Mons Piton	皮通山
Dorsum Zirkel	齐克尔山脊	Mons Rümker	吕姆克山
Lacus Aestatis	夏湖	Montes Agricola	阿格里科拉山脉
Lacus Autumni	秋湖	Montes Alpes	阿尔卑斯山脉
Lacus Bonitatis	仁慈湖	Montes Apenninus	亚平宁山脉

Montes Archimedes	阿基米德山脉	Rima Oppolzer	奥波尔策溪
Montes Carpatus	喀尔巴阡山脉	Rima Planck	普朗克溪
Montes Caucasus	高加索山脉	Rima Schroedinger	薛定谔溪
Montes Cordillera	科迪勒拉山脉	Rima Sharp	夏普溪
Montes Haemus	海玛斯山脉	Rima Sheepshanks	希普尚克斯溪
Montes Harbinger	前锋山脉	Rima Sirsalis	希萨利斯溪
Montes Jura	侏罗山脉	Rima Suess	休斯溪
Montes Pyrenaeus	比利牛斯山脉	Rima T.Mayer	托·迈耶溪
Montes Recti	直列山脉	Rimae Alphonsus	阿方索溪
Montes Riphaeus	里菲山脉	Rimae Apollonius	阿波罗尼奥斯溪
Montes Rook	鲁克山脉	Rimae Archimedes	阿基米德溪
Montes Secchi	塞奇山脉	Rimae Aristarchus	阿利斯塔克溪
Montes Spitzbergen	施皮茨贝尔根山脉	Rimae Arzachel	阿尔扎赫尔溪
Montes Taurus	金牛山脉	Rimae Atlas	阿特拉斯溪
Montes Teneriffe	特内里费山脉	Rimae Bode	波得溪
		Rimae Bürg	比格溪
Oceanus Procellarum	风暴洋	Rimae Daniell	丹聂耳溪
		Rimae Darwin	达尔文溪
Palus Epidemiarum	疫沼	Rimae Doppelmayer	多佩尔迈尔溪
Palus Putredinis	腐沼	Rimae Fresnel	菲涅尔溪
Palus Somni	梦沼	Rimae Gassendi	伽桑狄溪
		Rimae Grimaldi	格里马尔迪溪
Planitia Descensus	德森萨斯平原	Rimae Gutenberg	谷登堡溪
		Rimae Hase	哈泽溪
Promontorium Agarum	阿格鲁姆海角	Rimae Hevelius	赫维留溪
Promontorium Agassiz	阿加西海角	Rimae Hypatia	希帕蒂娅溪
Promontorium Archerusia	阿切鲁西亚海角	Rimae Janssen	让桑溪
Promontorium Deville	德维尔海角	Rimae Littrow	利特罗夫溪
Promontorium Fresnel	菲涅尔海角	Rimae Maclear	麦克利尔溪
Promontorium Heraclides	赫拉克利德海角	Rimae Maestlin	梅斯特林溪
Promontorium Kelvin	开尔文海角	Rimae Maupertuis	莫佩尔蒂溪
Promontorium Laplace	拉普拉斯岬	Rimae Menelaus	米尼劳斯溪
Promontorium Taenarium	泰纳里厄姆海角	Rimae Mersenius	梅森溪
		Rimae Parry	帕里溪
Rima Agatharchides	阿格瑟奇德斯溪	Rimae Pettit	佩蒂特溪
Rima Agricola	阿格里科拉溪	Rimae Plato	柏拉图溪
Rima Archytas	阿契塔溪	Rimae Plinius	普利纽斯溪
Rima Ariadaeus	阿里亚代乌斯溪	Rimae Prinz	普林茨溪
Rima Artsimovich	阿尔齐莫维奇溪	Rimae Ramsden	拉姆斯登溪
Rima Billy	比伊溪	Rimae Riccioli	里乔利溪
Rima Birt	伯特溪	Rimae Ritter	里特尔溪
Rima Brayley	布雷利溪	Rimae Römer	罗默溪
Rima Cauchy	柯西溪	Rimae Sosigenes	索西琴尼溪
Rima Delisle	德利尔溪	Rimae Sulpicius Gallus	加卢斯溪
Rima Diophantus	丢番图溪	Rimae Theaetetus	特埃特图斯溪
Rima Draper	德雷伯溪	Rimae Vasco da Gama	达·伽马溪
Rima Euler	欧拉溪	Rimae Zupus	祖皮溪
Rima Flammarion	弗拉马里翁溪		
Rima G.Bond	乔·邦德溪	Rupes Altai	阿尔泰峭壁
Rima Galilaei	伽利略溪	Rupes Boris	鲍里斯峭壁
Rima Hadley	哈德利溪	Rupes Cauchy	柯西峭壁
Rima Hesiodus	赫西奥德溪	Rupes Kelvin	开尔文峭壁
Rima Hyginus	希吉努斯溪	Rupes Liebig	李比希峭壁
Rima Marius	马里乌斯溪	Rupes Mercator	墨卡托峭壁
Rima Messier	梅西叶溪	Rupes Recta	直壁

Rupes Toscanelli	托斯卡内利峭壁	Vallis Alpes	阿尔卑斯大峡谷
		Vallis Baade	巴德谷
Sinus Aestuum	浪湾	Vallis Bohr	玻尔谷
Sinus Amoris	爱湾	Vallis Bouvard	布瓦尔谷
Sinus Asperitatis	狂暴湾	Vallis Capella	卡佩拉谷
Sinus Concordiae	和谐湾	Vallis Christel	克里斯特尔谷
Sinus Fidei	信赖湾	Vallis Inghirami	因吉拉米谷
Sinus Honoris	荣誉湾	Vallis Krishna	克里希纳谷
Sinus Iridum	虹湾	Vallis Palitzsch	帕利奇谷
Sinus Lunicus	眉月湾	Vallis Planck	普朗克谷
Sinus Medii	中央湾	Vallis Rheita	里伊塔月谷
Sinus Roris	露湾	Vallis Schrödinger	薛定谔谷
Sinus Successus	成功湾	Vallis Schröteri	施洛特月谷
		Vallis Snellius	斯涅尔谷

表 8　小行星　Astroids

编号	拉丁名	中文名	编号	拉丁名	中文名
1	Ceres	谷神星	35	Leukothea	沉神星
2	Pallas	智神星	36	Atalante	驰神星
3	Juno	婚神星	37	Fides	忠神星
4	Vesta	灶神星	38	Leda	卵神星
5	Astraea	义神星	39	Laetitia	喜神星
6	Hebe	韶神星	40	Harmonia	谐神星
7	Iris	虹神星	41	Daphne	桂神星
8	Flora	花神星	42	Isis	育神星
9	Metis	颖神星	43	Ariadne	爱女星
10	Hygiea	健神星	44	Nysa	侍神星
11	Parthenope	海妖星	45	Eugenia	香女星
12	Victoria	凯神星	46	Hestia	司祭星
13	Egeria	芙女星	47	Aglaja	仁神星
14	Irene	司宁星	48	Doris	昏神星
15	Eunomia	司法星	49	Pales	牧神星
16	Psyche	灵神星	50	Virginia	贞女星
17	Thetis	海女星	51	Nemausa	禽神星
18	Melpomene	司曲星	52	Europa	欧罗巴
19	Fortuna	命神星	53	Kalypso	岛神星
20	Massalia	王后星	54	Alexandra	哲女星
21	Lutetia	司琴星	55	Pandora	祸神星
22	Kalliope	司赋星	56	Melete	中神星
23	Thalia	司剧星	57	Mnemosyne	龙女星
24	Themis	司理星	58	Concordia	协神星
25	Phocaea	福后星	59	Elpis	乾神星
26	Proserpina	冥后星	60	Echo	司音星
27	Euterpe	司箫星	61	Danaë	囚神星
28	Bellona	战神星	62	Erato	效神星
29	Amphitrite	海后星	63	Ausonia	奥索尼亚
30	Urania	司天星	64	Angelina	安杰利纳
31	Euphrosyne	丽神星	65	Cybele	原神星
32	Pomona	果神星	66	Maja	光神星
33	Polyhymnia	司瑟星	67	Asia	亚细亚
34	Circe	巫神星	68	Leto	明神星

表 8　小行星

编号	拉丁名	中文名	编号	拉丁名	中文名
69	Hesperia	夕神星	150	Nuwa	女娲星
70	Panopaea	蟹神星	153	Hilda	希尔达
71	Niobe	司石星	171	Ophelia	奥菲丽亚
72	Feronia	期女星	197	Arete	阿雷特
73	Klytia	芥神星	202	Chryseis	赫露斯
74	Galatea	巫女星	221	Eos	曙神星
75	Eurydike	狱神星	232	Russia	俄罗斯
76	Freia	舒女星	241	Germania	德意志
77	Frigga	寒神星	243	Ida	艾达
78	Diana	月神星	250	Bettina	贝蒂
79	Eurynome	配女星	279	Thule	图勒
80	Sappho	赋神星	301	Bavaria	巴伐利亚
81	Terpsichore	司舞星	323	Brucia	布鲁斯
82	Alkmene	怨女星	324	Bamberga	班贝格
83	Beatrix	欣女星	397	Vienna	维也纳
84	Klio	史神星	433	Eros	爱神星
85	Io	犊神星	434	Hungaria	匈牙利
86	Semele	化女星	451	Patientia	忍神星
87	Sylvia	林神星	511	Davida	戴维
88	Thisbe	尽女星	518	Halawe	芝麻片糖
89	Julia	姝女星	532	Herculina	大力神星
90	Antiope	休神星	588	Achilles	阿基里斯
91	Aegina	河神星	617	Patroclus	帕特洛克鲁斯
92	Undina	波神星	624	Hektor	赫克托
93	Minerva	慧神星	659	Nestor	内斯托
94	Aurora	彩神星	719	Albert	阿尔伯特
95	Arethusa	源神星	849	Ara	阿拉
96	Aegle	辉神星	862	Franzia	法兰西
97	Klotho	纺神星	883	Matterania	马特尔
98	Ianthe	佳女星	884	Priamus	普丽阿姆斯
99	Dike	泰神星	911	Agamemnon	阿伽梅农
100	Hekate	权神星	944	Hidalgo	希达尔戈
101	Helena	拐神星	951	Gaspra	加斯普拉
102	Miriam	圣女星	1000	Piazzia	皮亚齐
103	Hera	后神星	1001	Gaussia	高斯
104	Klymene	伴女星	1002	Olbersia	奥伯斯
105	Artermis	群女星	1086	Nata	娜塔
106	Dione	坤神星	1125	China	中华
107	Camilla	驶神星	1134	Kepler	开普勒
108	Hecuba	犬后星	1143	Odysseus	奥德赛
109	Felicitas	祥神星	1172	Aneas	阿涅阿斯
110	Lydia	吕女星	1173	Anchises	安希塞斯
111	Ate	苟神星	1208	Troilus	特洛伊鲁斯
112	Iphigenia	祭神星	1221	Amor	阿莫尔
113	Amalthea	羊神星	1437	Diomedes	迪奥梅黛斯
114	Kassandra	见神星	1486	Marilyn	玛莉琳
115	Thyra	赛娜	1566	Icarus	伊卡洛斯
116	Sirona	细女星	1583	Antilochus	安蒂洛库斯
117	Lomia	罗女星	1620	Geographos	地理星
127	Johanna	约翰娜	1665	Gabi	加比
135	Hertha	沃神星	1802	Zhang Heng	张衡
139	Juewa	九华星	1862	Apollo	阿波罗
142	Polana	波兰	1881	Shao	邵正元
144	Vibilia	旅神星	1888	Zu Chong-zhi	祖冲之

表 8　小行星

编号	拉丁名	中文名	编号	拉丁名	中文名
1972	Yi Xing	一行	2899	Runrun Shaw	邵逸夫
2012	Guo Shou-Jing	郭守敬	2903	Zhuhai	珠海
2027	Shen Guo	沈括	2963	Chen Jiageng	陈嘉庚
2045	Peking	北京	3011	Chongqing	重庆
2051	Chang	张钰哲	3014	Huangsushu	黄授书
2060	Chiron	喀戎	3024	Hainan	海南
2062	Aten	阿登	3028	Zhangguoxi	张果喜
2077	Kiangsu	江苏	3048	Guangzhou	广州
2078	Nanking	南京	3051	Nantong	南通
2085	Henan	河南	3088	Jinxiuzhonghua	锦绣中华
2101	Adonis	阿多尼斯	3089	Oujianquan	区鉴泉
2162	Anhui	安徽	3136	Anshan	鞍山
2169	Taiwan	台湾	3139	Shantou	汕头
2184	Fujian	福建	3171	Wangshouguan	王绶琯
2185	Guangdong	广东	3187	Dalian	大连
2197	Shanghai	上海	3200	Phaethon	法厄松
2201	Oljato	奥加托	3206	Wuhan	武汉
2209	Tianjin	天津	3221	Changshi	常熟
2212	Hephaistos	冶神星	3239	Meizhou	梅州
2215	Sichuan	四川	3241	Yeshuhua	叶叔华
2230	Yunnan	云南	3297	Hong Kong	香港
2240	Tsai	蔡章献	3317	Paris	帕里斯
2255	Qinghai	青海	3335	Quanzhou	泉州
2263	Shaanxi	陕西	3340	Yinhai	银海
2336	Xinjiang	新疆	3388	Tsanghinchi	曾宪梓
2344	Xizang	西藏	3405	Daiwensai	戴文赛
2355	Nei Monggol	内蒙古	3421	Yangchenning	杨振宁
2380	Heilongjiang	黑龙江	3443	Leetsungdao	李政道
2387	Xi'an	西安	3462	Zhouguangzhao	周光召
2398	Jilin	吉林	3463	Kaokuen	高锟
2425	Shenzhen	深圳	3476	Dongguan	东莞
2503	Liaoning	辽宁	3481	Xianglupeak	香炉峰
2505	Hebei	河北	3494	Purple Mountain	紫金山
2510	Shandong	山东	3502	Huangpu	黄浦
2514	Taiyuan	太原	3509	Sanshui	三水
2515	Gansu	甘肃	3513	Quqinyue	曲钦岳
2539	Ningxia	宁夏	3542	Tanjiazhen	谈家桢
2547	Hubei	湖北	3543	Ningbo	宁波
2592	Hunan	湖南	3556	Lixiaohua	李晓华
2617	Jiangxi	江西	3560	Chenqian	陈骞
2631	Zhejiang	浙江	3570	Wuyeesun	伍宜孙
2632	Guizhou	贵州	3609	Liloketai	李陆大
2655	Guangxi	广西	3611	Dabu	大埔
2693	Yan'an	延安	3613	Kunlun	昆仑
2705	Wu	吴	3643	Tienchanglin	田长霖
2719	Suzhou	苏州	3650	Kunming	昆明
2729	Urumqi	乌鲁木齐	3678	Mongmanwai	蒙民伟
2743	Chengdu	成都	3704	Gaoshiqi	高士其
2752	Wu Chien-Shiung	吴健雄	3729	Yangzhou	扬州
2778	Tangshan	唐山	3746	Heyuan	河源
2789	Foshan	佛山	3751	Kiang	江涛
2790	Needham	李约瑟	3763	Qianxuesen	钱学森
2851	Harbin	哈尔滨	3789	Zhongguo	中国
2886	Tinkaping	田家炳	3797	Ching-Sung Yu	余青松

表 9　陨星坑　Meteorite Craters

英文名	中文名	所在国家	半径（千米）	年龄 (亿年)
Acraman	阿克拉曼	澳大利亚	90	5.9
Amelia Creek	阿米利亚溪	澳大利亚	20	6～16.4
Araguainha	阿拉瓜伊尼亚	巴西	40	2.54
Beaverhead	比弗黑德	美国	60	6
Boltysh	波泰士	乌克兰	24	0.65
Carswell	卡斯韦尔	加拿大	39	1.15
Charlevoix	夏洛瓦	加拿大	54	3.42
Chesapeake Bay	切萨皮克湾	美国	40	0.355
Chicxulub	奇克苏鲁布	墨西哥	180	0.65
Clearwater East	清水湾东	加拿大	26	2.9
Clearwater West	清水湾西	加拿大	36	2.9
Gosses Bluff	戈斯崖	澳大利亚	22	1.425
Haughton	霍顿	加拿大	23	0.39
Kamensk	卡缅斯克	俄罗斯	25	0.49
Kara	卡拉河	俄罗斯	120	0.703
Karakul	卡拉库尔湖	塔吉克斯坦	52	0.25
Keurusselkä	凯乌鲁塞尔凯湖	芬兰	30	14～15
Lappajärvi	拉帕湖	芬兰	23	0.733
Logancha	洛甘察	俄罗斯	20	0.40
Manicouagan	马尼夸根	加拿大	85	2.15
Manson	曼森	美国	35	0.74
Mistastin	米斯塔斯汀湖	加拿大	28	0.364
Mjølnir	米约涅尔	挪威	40	1.42
Montagnais	蒙塔格尼	加拿大	45	0.505
Morokweng	莫罗昆	南非	70	1.45
Nördlinger Ries	诺德林格里斯	德国	24	0.14
Obolon'	奥博隆	乌克兰	20	1.69
Popigai	波皮盖	俄罗斯	90	0.357
Presqu'île	半岛	加拿大	24	<5
Puchezh-Katunki	普切日-卡通基	俄罗斯	40	1.67
Rochechouart	罗什舒阿尔	法国	23	2.01
Saint Martin	圣马丁	加拿大	40	2.20
Shoemaker	舒梅克	澳大利亚	30	16.30
Siljan Ring	锡利扬湖	瑞典	52	3.77
Slate Islands	斯莱特群岛	加拿大	30	4.50
Steen River	斯廷河	加拿大	25	0.91
Strangways	斯特兰韦斯	澳大利亚	25	6.46
Sudbury	萨德伯里	加拿大	250	18.49
Tookoonooka	图库努卡	澳大利亚	55	1.12～1.33
Tunuunik	图努尼克	加拿大	25	1.3～4.5
Vredefort	弗里德堡	南非	300	20.23
Woodleigh	伍德利	澳大利亚	40	3.64
Yarrabubba	亚拉布巴	澳大利亚	30	11.30～26.0

表 10　流星群　Meteoric Streams

Andromedids	仙女流星群	Lyrids	天琴流星群
Aquarids	宝瓶流星群	Monocerids	麒麟流星群
Arietids	白羊流星群	Ophiuchids	蛇夫流星群
Bielids	比拉流星群	Orionids	猎户流星群
Bootids	牧夫流星群	Perseids	英仙流星群
Capricornids	摩羯流星群	Phoenicids	凤凰流星群
Cassiopeids	仙后流星群	Piscis Australids	南鱼流星群
Cepheids	仙王流星群	Quadrantids	象限仪流星群
Cetids	鲸鱼流星群	Sculptorids	玉夫流星群
Cygnids	天鹅流星群	Taurids	金牛流星群
Draconids	天龙流星群	Ursa Minorids	小熊流星群
Geminids	双子流星群	α Capricornids	摩羯 α 流星群
Leonids	狮子流星群	δ Aquarids	宝瓶 δ 流星群
Librids	天秤流星群	η Aquarids	宝瓶 η 流星群

表 11　希腊字母　Greek Alphabet

大写	小写	英文名	中文名	大写	小写	英文名	中文名
A	α	alpha	阿尔法	N	ν	nu	纽
B	β	beta	贝塔	Ξ	ξ	xi	克西
Γ	γ	gamma	伽马	O	o	omicron	奥密克戎
Δ	δ	delta	德尔塔	Π	π	pi	派
E	ϵ,ε	epsilon	伊普西龙	P	ρ,ϱ	rho	肉
Z	ζ	zeta	截塔	Σ	σ,ς	sigma	西格马
H	η	eta	艾塔	T	τ	tau	套
Θ	θ,ϑ	theta	西塔	Υ	υ	upsilon	宇普西龙
I	ι	iota	约塔	Φ	ϕ,φ	phi	佛爱
K	κ	kappa	卡帕	X	χ	chi	西
Λ	λ	lambda	兰布达	Ψ	ψ	psi	普赛
M	μ	mu	缪	Ω	ω	omega	欧米伽

表 12　罗马数字　Roman Numerals

I	II	III	IV	V	VI	VII	VIII	IX	X	XI	XX	XL	L	C	D	M
1	2	3	4	5	6	7	8	9	10	11	20	40	50	100	500	1000

中国科协三峡科技出版资助计划
2012 年第一期资助著作名单

1. 包皮环切与艾滋病预防
2. 东北区域服务业内部结构优化研究
3. 肺孢子菌肺炎诊断与治疗
4. 分数阶微分方程边值问题理论及应用
5. 广东省气象干旱图集
6. 混沌蚁群算法及应用
7. 混凝土侵彻力学
8. 金佛山野生药用植物资源
9. 科普产业发展研究
10. 老年人心理健康研究报告
11. 农民工医疗保障水平及精算评价
12. 强震应急与次生灾害防范
13. "软件人"构件与系统演化计算
14. 西北区域气候变化评估报告
15. 显微神经血管吻合技术训练
16. 语言动力系统与二型模糊逻辑
17. 自然灾害与发展风险

中国科协三峡科技出版资助计划
2012 年第二期资助著作名单

1. BitTorrent 类型对等网络的位置知晓性
2. 城市生态用地核算与管理
3. 创新过程绩效测度——模型构建、实证研究与政策选择
4. 商业银行核心竞争力影响因素与提升机制研究
5. 品牌丑闻溢出效应研究——机理分析与策略选择
6. 护航科技创新——高等学校科研经费使用与管理务实
7. 资源开发视角下新疆民生科技需求与发展
8. 唤醒土地——宁夏生态、人口、经济纵论
9. 三峡水轮机转轮材料与焊接
10. 大型梯级水电站运行调度的优化算法
11. 节能砌块隐形密框结构
12. 水坝工程发展的若干问题思辨
13. 新型纤维素系止血材料
14. 商周数算四题
15. 城市气候研究在中德城市规划中的整合途径比较
16. 心脏标志物实验室检测应用指南
17. 现代灾害急救
18. 长江流域的枝角类

中国科协三峡科技出版资助计划
2013 年第三期资助著作名单

1. 蛋白质技术在病毒学研究中的应用
2. 当代中医糖尿病学
3. 滴灌——随水施肥技术理论与实践
4. 地质遗产保护与利用的理论及实证
5. 分布式大科学项目的组织与管理：人类基因组计划
6. 港口混凝土结构性能退化及耐久性设计
7. 国立北平研究院史稿
8. 海岛开发成陆工程技术
9. 环境资源交易理论与实践研究——以浙江为例
10. 荒漠植物蒙古扁桃生理生态学
11. 基础研究与国家目标——以北京正负电子对撞机为例的分析
12. 激光火工品技术
13. 抗辐射设计与辐射效应
14. 科普产业概论
15. 科学与人文
16. 空气净化原理、设计与应用
17. 煤炭物流——基于供应链管理的大型煤炭企业分销物流模式及其风险预警研究
18. 农产品微波组合干燥技术
19. 配电网规划
20. 腔静脉外科学
21. 清洁能源技术政策与管理研究——以碳捕集与封存为例
22. 三峡水库生态渔业
23. 深冷混合工质节流制冷原理及应用
24. 生物数学思想研究
25. 实用人体表面解剖学
26. 水力发电的综合价值及其评价
27. 唐代工部尚书研究
28. 糖尿病基础研究与临床诊治
29. 物理治疗技术创新与研发
30. 西双版纳傣族传统灌溉制度的现代变迁
31. 新疆经济跨越式发展研究
32. 沿海与内陆就地城市化典型地区的比较
33. 疑难杂病医案
34. 制造改变设计——3D 打印直接制造技术
35. 自然灾害会影响经济增长吗——基于国内外自然灾害数据的实证研究
36. 综合客运枢纽功能空间组合设计——理论与实践
37. TRIZ——推动创新的技术（译著）
38. 从流代数到量子色动力学：结构实在论的一个案例研究（译著）
39. 风暴守望者——天气预报风云史（译著）
40. 观测天体物理学（译著）
41. 可操作的地震预报（译著）
42. 绿色经济学（译著）
43. 谁在操纵碳市场（译著）
44. 医疗器械使用与安全（译著）
45. 宇宙天梯 14 步（译著）
46. 致命的引力——宇宙中的黑洞（译著）

中国科协三峡科技出版资助计划
2014 年第四期资助著作名单

1. 科学的学派（译著）
2. 河流健康的法制化管理
3. 水资源系统决策分析方法及应用
4. 中国特色现代农业建设路径研究
5. 碳排放规律与经济发展路径研究
6. 武夷岩茶（大红袍）研究
7. 生态型地面停车场绿化
8. 英汉天文学名词
9. SAR 与光学影像融合的变化信息提取
10. 云设计——工业设计新模式
11. 光学分子影像外科学
12. 定向木塑复合刨花板热压成型机理

发行部

地址：北京市海淀区中关村南大街 16 号
邮编：100081
电话：010 – 62103130

办公室

电话：010 – 62103166
邮箱：kxsxcb@ cast. org. cn
网址：http：//www. cspbooks. com. cn